Introduction to Environmental Health: A Global Perspective

Edited by Anne Marie Zimeri
University of Georgia

Bassim Hamadeh, CEO and Publisher
Christopher Foster, General Vice President
Michael Simpson, Vice President of Acquisitions
Jessica Knott, Managing Editor
Kevin Fahey, Cognella Marketing Manager
Jess Busch, Senior Graphic Designer
Zina Craft, Acquisitions Editor
Jamie Giganti, Project Editor
Brian Fahey, Licensing Associate

First published in the United States of America in 2013 by Cognella, Inc.

Printed in the United States of America

ISBN: 978-1-60927-592-1 (pbk) / 978-1-60927-591-4 (br)

www.cognella.com 800.200.3908

Contents

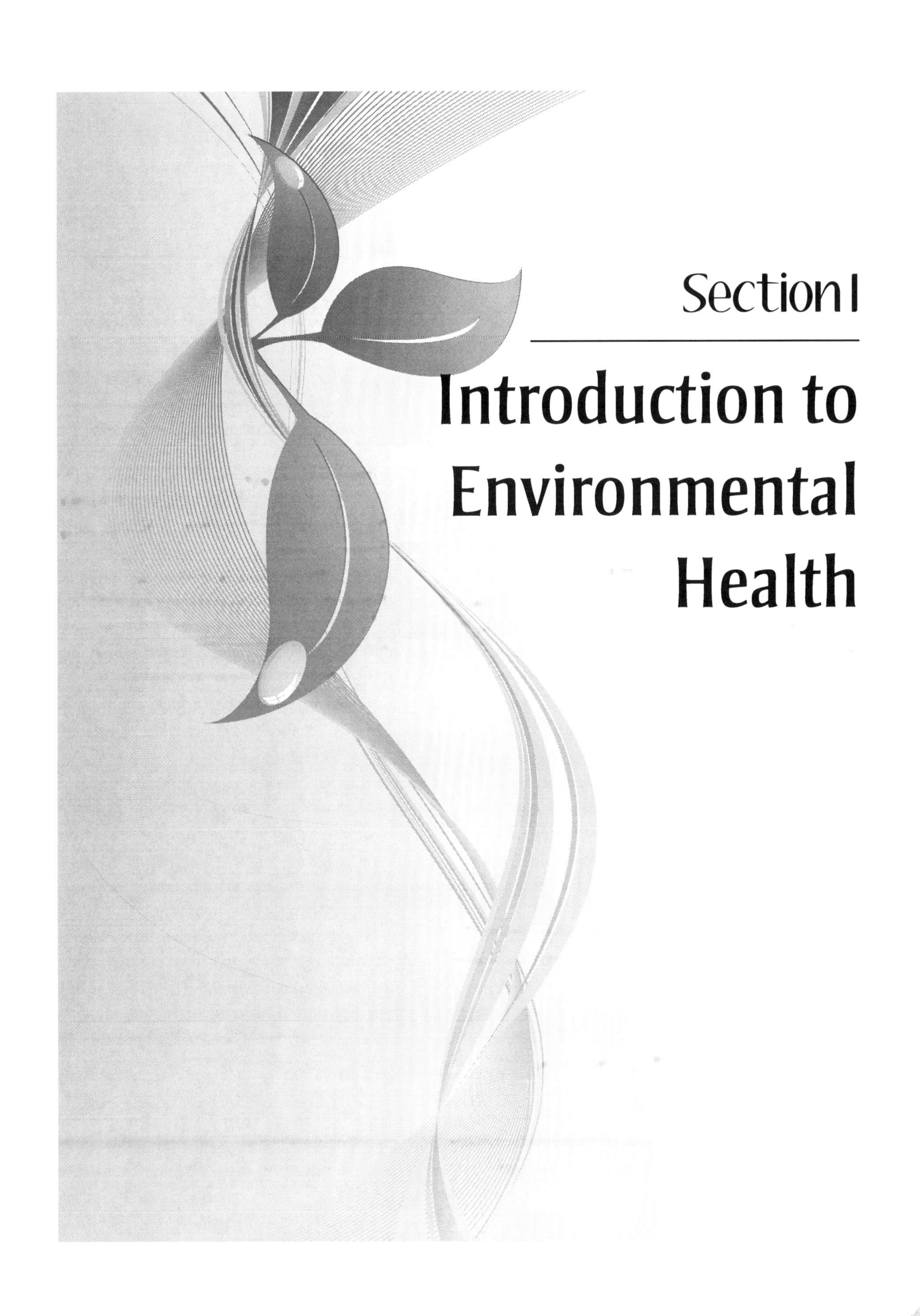

Section I

Introduction to Environmental Health

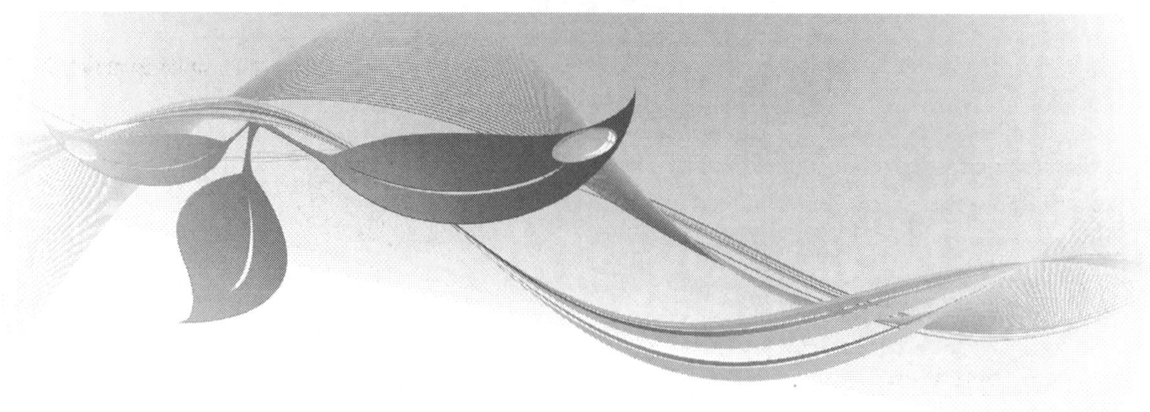

By Anne Marie Zimeri

DEFINING ENVIRONMENTAL HEALTH

According to the World Health Organization (WHO), environmental health addresses all the physical, chemical, and biological factors external to a person, and all the related factors impacting behaviors. It encompasses the assessment and control of those environmental factors that can potentially affect health. It is targeted toward preventing disease and creating health-supportive environments. Studying the environment and its relationship to humans can lead to better overall public health. There are environmental health researchers and practitioners all over the world whose goal is to better public health by addressing environmental issues.

Did you know...?

- Worldwide, 13 million deaths could be prevented every year by making our environments healthier.
- In children under the age of five, one-third of all disease is caused by environmental factors such as unsafe water and air pollution.
- Every year, the lives of 4 million children under five years—mostly in developing countries—could be saved by preventing environmental risks such as unsafe water and polluted air.
- In developing countries, the main environmentally caused diseases are diarrheal disease, lower respiratory infections, unintentional injuries, and malaria.
- Better environmental management could prevent 40% of deaths from malaria, 41% of deaths from lower respiratory infections, and 94% of deaths from diarrheal disease—three of the world's biggest childhood killers.
- In the least developed countries, one-third of death and disease is a direct result of environmental causes.
- In developed countries, healthier environments could significantly reduce the incidence of cancers, cardiovascular diseases, asthma, lower respiratory infections, musculoskeletal diseases, road-traffic injuries, and poisonings.
- Environmental factors influence 85 out of the 102 categories of diseases and injuries listed in *The World Health Report*.
- Much of this death, illness, and disability could be prevented through well-targeted interventions such as promoting safe household water storage, better hygiene measures, and the use of cleaner and safer fuels.
- Other interventions that can make environments healthier include increasing the safety of buildings; promoting safe, careful use and

management of toxic substances at home and in the workplace; and better water resource management.

ENVIRONMENTAL HEALTH SCIENCE IS A SCIENCE

Environmental health scientists systematically study the environment and its relationship to human health using the scientific method. Truths and theories are derived from experimental data that have been critically reviewed by other scientists prior to being published. This data is often derived in a laboratory setting using a model system. Model systems are created to mimic nature as much as possible in a controlled setting where single variables can be easily manipulated. Scientists set up these experiments to be as objective as possible, often employing blind or double-blind studies. In a blind experiment, the scientist does not know whether s/he is observing the experimental treatment or the control until after data is collected. This way, potential bias is removed as much as possible. Double-blind studies are often employed when human subjects are being tested in clinical trials. In this case, neither the test subject nor the researchers know who has been given the experimental treatment and who is in the control group.

Environmental health scientists strive to design experiments that derive quantitative data, i.e., data that can be quantified or verified on a numerical scale as opposed to qualitative data, which is descriptive, yet not on a numerical scale. For example, if a scientist wants to observe the toxicity of a compound that may be applied to an agricultural plant, s/he may observe that the plant leaves begin to look yellow and develop some lesions. To make that qualitative observation quantitative, the yellowing should be measured by using a spectrophotometer that will yield a wavelength for that color. Lesion size should

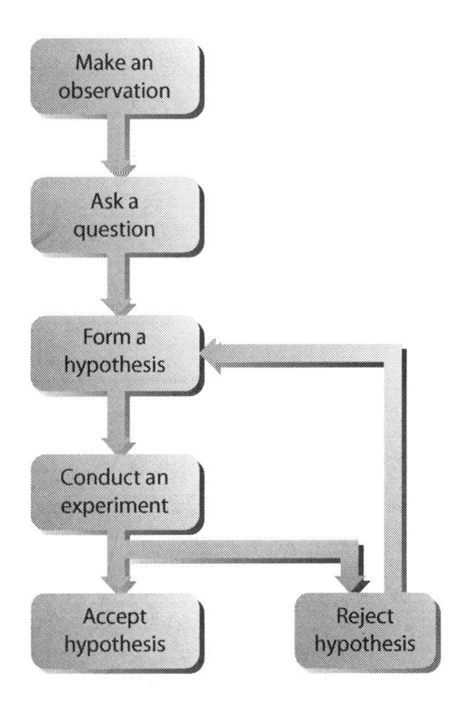

Figure 1.1. The scientific method is a technique used to develop, modify, and test hypotheses.

be measured and the number of lesions per leaf should be counted. This quantitative data will serve as a baseline for replicates of the original experiment, but will also be able to serve as a baseline for scientists all over the globe.

There are circumstances when model systems cannot be developed to answer a scientific question. Such circumstances can prompt data collection from a natural experiment. An example of natural data in environmental health science is the collection of ice cores from Antarctica that date back hundreds of thousands of years. From these cores, atmospheric CO_2 and temperature can be calculated. No laboratory experiment will provide historical data, so researchers must develop this observational data. Natural data may have less strength when it comes to establishing cause and effect because, unlike model systems, variables cannot be manipulated one at a time.

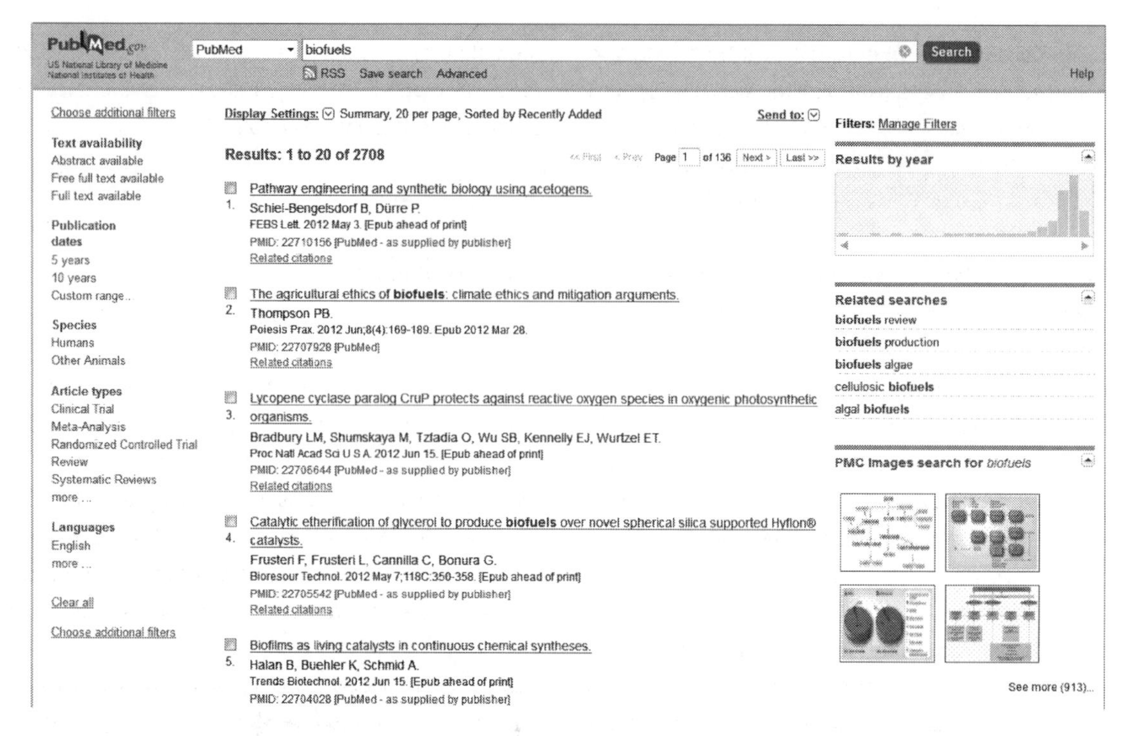

Figure 1.2. Searching peer-reviewed papers can be easy using a search engine such as the NCBI Pubmed database.

Communicating findings is one of the most important aspects of being a scientist. Once scientists replicate their experiments and have applied statistical algorithms to show that their data is significant, they proceed to the next step, that of publishing their results. Scientists publish their work in peer-reviewed scientific journals based on subject. There are journals specific to most any field of science, and a few journals such as *Science* and *Nature* that take the most groundbreaking discoveries from all fields. Once a manuscript is submitted to a journal, the editor of that journal will send it out to several experts in the field for peer review. These reviewers, sometimes called referees, will look over the paper with a critical eye to determine whether it should be published. Oftentimes they will suggest additional experiments prior to publication. The work is published only after satisfying the reviewers.

Would you like to look at some of the peer-reviewed scientific papers in the field? Go to the National Center for Biotechnology Information website at http://www.ncbi.nlm.nih.gov/. Search the Pubmed database by keyword or author, and it will give you a list of publications. If you click on the title of a publication, an expanded abstract will appear to let you know more details about the work. Many of the abstract pages will have a direct link to the article on the right-hand side of the page on a publisher button.

GLOBAL CITIZENSHIP AND ENVIRONMENTAL HEALTH

As the world population continues to expand in the new millennium, we see the need arise for each person to participate in the sustainability of the earth. Otherwise, we will likely see environmental degradation to the point where it greatly affects the quality of our lives. Environmental health scientists play a role in research that can help each person develop a sense of personal responsibility when it comes to sustaining our environment. Certainly, in the United States, citizens have the luxury of making choices that can serve us in the long term by being environmentally responsible; we are in this fortunate position because we do not have to spend all of our energy and resources to feed our families on a daily basis, as is the case in many areas of the world. In lesser developed countries (LDCs), citizens spend the better part of their day and up to 90% of their income simply to house and feed themselves. When a population is faced with such circumstances, sustainable choices are only made when they are congruent with immediate benefits to the individual. Environmental health scientists want to address resource disparities so that we can have an environmentally responsible population. This ideal population would contribute to a sustainable world.

TWO DEMOGRAPHIC WORLDS

A consistent theme in each of the chapters of this book is the discrepancy between the two demographic worlds. More developed (richer) countries and lesser developed (poorer) countries (LDCs) do not share access to the same resources, including food, fossil fuels, strategic minerals, education, health care, and general wealth. The gap between the two worlds will likely continue to widen as the population rapidly expands in LDCs in the coming years (figure 1.3).

Addressing environmental issues requires different approaches in each of these types of countries. The chapters that follow will look at current topics in environmental health from a global perspective that includes issues and data from both developed nations and LDCs.

SUSTAINABILITY

A sustainable future is the goal of environmental health scientists. Research and practice in the field seeks to create and maintain the conditions under which humans and nature can exist in productive harmony, such that we can fulfill the social, economic, and environmental requirements of present and future generations. This book will touch on

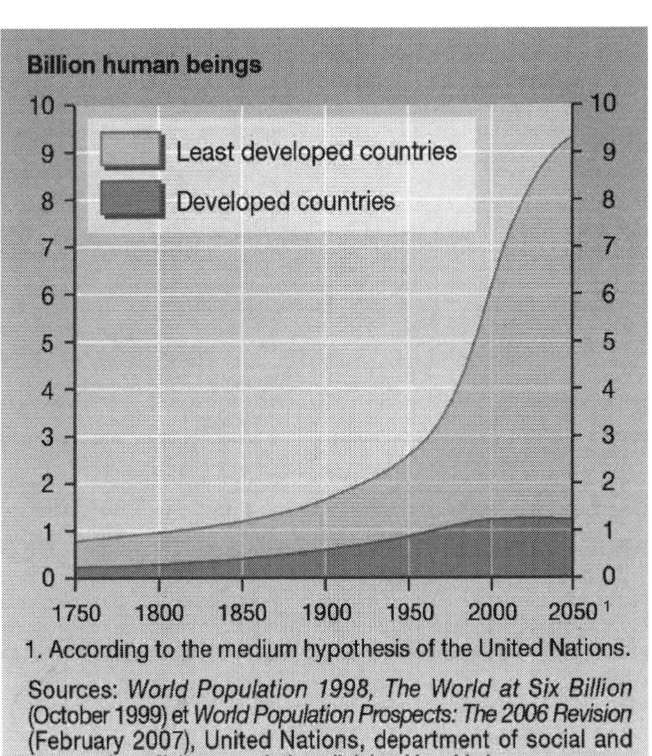

Figure 1.3 shows not only the proportion of the population in LDCs today, but also the projected proportion in the future. Copyright © 2008 by United Nations Environment Programme. Reprinted with permission.

the social and economic aspects of sustainability, but it will mainly focus on the environment. Each chapter will relate to practices that may currently be unsustainable for generations to come, whether that is land availability to grow our food, resource extraction so that we can heat our homes, water usage, or other environmental practices that cannot last far into the future.

QUESTIONS FOR THOUGHT

1. How does your life differ from the life of a person who may be living in a lesser developed country?
2. What activities/practices in our life are unsustainable? What differences do you foresee in your lifestyle in 20 years and 50 years, based on those activities/practices?

Examples of Scientific Method

By Bertrand Russell

I. GALILEO

Scientific method, although in its more refined forms it may seem complicated, is in essence remarkably simple. It consists in observing such facts as will enable the observer to discover general laws governing facts of the kind in question. The two stages, first of observation, and second of inference to a law, are both essential, and each is susceptible to almost indefinite refinement; but in essence the first man who said "fire burns" was employing scientific method, at any rate if he had allowed himself to be burnt several times. This man had already passed through the two stages of observation and generalization. He had not, however, what scientific technique demands—a careful choice of significant facts on the one hand, and, on the other hand, various means of arriving at laws otherwise than by mere generalization. The man who says "unsupported bodies in air fall" has merely generalized, and is liable to be refuted by balloons, butterflies, and aeroplanes; whereas the man who understands the theory of falling bodies knows also why certain exceptional bodies do not fall.

Scientific method, simple as it is in essence, has been acquired only with great difficulty, and is still employed only by a minority, who themselves confine its employment to a minority of the questions upon which they have opinions. If you number among your acquaintances some eminent man of science, accustomed to the minutest quantitative precision in his experiments and the most abstruse skill in his inference from them, you will be able to make him the subject of a little experiment which is likely to be by no means unilluminating. If you tackle him on party politics, theology, income tax, house-agents, the bumptiousness of the working-classes and other topics of a like nature, you are pretty sure, before long, to provoke an explosion, and to hear him expressing wholly untested opinions with a dogmatism which he would never display in regard to the well-founded results of his laboratory experiments.

As this illustration shows, the scientific attitude is in some degree unnatural to man; the majority of our opinions are wish-fulfilments, like dreams in the Freudian theory. The mind of the most rational among us may be compared to a stormy ocean of passionate convictions based upon desire, upon which float perilously a few tiny boats carrying a cargo of scientifically tested beliefs. Nor is this to be altogether deplored: life has to be lived, and there is no time to test rationally all the beliefs by which our conduct is regulated. Without a certain wholesome rashness, no one could long survive. Scientific method, therefore, must, in its very nature, be confined to the more solemn and official of our opinions. A medical man who gives advice on diet should give it after full consideration of all that science has to say on the matter, but the man who follows his advice cannot stop to verify it, and is obliged to rely, therefore, not upon science, but upon his belief that his medical adviser is scientific. A community impregnated with science is one in which the recognized experts have arrived at their opinions by scientific methods, but it is impossible for the ordinary citizen to repeat the work of the experts for himself. There is, in the modern world, a great body of well-attested knowledge on all kinds of subjects, which the ordinary man accepts on authority without any need for hesitation; but as soon as any strong passion intervenes to warp the expert's judgment he becomes unreliable, whatever scientific equipment he may possess. The views of medical men on pregnancy, child-birth, and lactation were until fairly recently impregnated with sadism. It required, for example, more evidence to persuade them that anesthetics may be used in childbirth than it would have required to persuade them of the opposite. Anyone who desires an hour's amusement may be advised to look up the tergiversations of eminent craniologists in their attempts to prove from brain measurements that women are stupider than men.

It is not, however, the lapses of scientific men that concern us when we are trying to describe scientific method. A scientific opinion is one which there is some reason to believe true; an unscientific opinion is one which is held for some reason other than its probable truth. Our age is distinguished from all ages before the seventeenth century by the fact that some of our opinions are scientific in the above sense. I except bare matters of fact, since generality in a greater or less degree is an essential characteristic of science, and since men (with the exception of a few mystics) have never been able wholly to deny the obvious facts of their everyday existence.

The Greeks, eminent as they were in almost every department of human activity, did surprisingly little for the creation of science. The great intellectual achievement of the Greeks was geometry, which they believed to be an a priori study proceeding from self-evident premises, and not requiring experimental verification. The Greek genius was deductive rather than inductive, and was therefore at home in mathematics. In the ages that followed, Greek mathematics were nearly forgotten, while other products of the Greek passion for deduction survived and flourished, notably theology and law. The Greeks observed the world as poets rather than as men of science, partly, I think, because all manual activity was ungentlemanly, so that any study which required experiment seemed a little vulgar. Perhaps it would be fanciful to connect with this prejudice the fact that the department in which the Greeks were most scientific was astronomy, which deals with bodies that only can be seen and not touched.

However that may be, it is certainly remarkable how much the Greeks discovered in astronomy. They early decided that the earth is round, and some of them arrived at the Copernican theory that it is the earth's rotation, and not the revolution of the heavens, that causes the apparent diurnal motion of the sun and stars. Archimedes, writing to King Gelon of Syracuse, says: "Aristarchus of Samos brought out a book consisting of some hypotheses of which the premises lead to the conclusion that the universe is many times greater than that now so called. His hypotheses are that the fixed stars and the sun remain unmoved, that the earth revolves about the sun in the circumference of a circle, the sun lying in the centre of the orbit." Thus the Greeks discovered not only the diurnal rotation of the earth, but also its annual revolution about the sun. It was the discovery that a Greek had held this opinion which gave Copernicus courage to revive it. In the days of the Renaissance, when Copernicus lived, it was held that any opinion which had been entertained by an ancient might be true, but an opinion which no ancient had entertained could not deserve respect. I doubt whether Copernicus would ever have become a Copernican but for Aristarchus, whose opinion had been forgotten until the revival of classical learning.

The Greeks also discovered perfectly valid methods of measuring the circumference of the earth. Eratosthenes the Geographer estimated it at 250,000 stadia (about 24,662 miles), which is by no means far from the truth.

The most scientific of the Greeks was Archimedes (257–212 B.C.). Like Leonardo da Vinci in a later period, he recommended himself to a prince on the ground of his skill in the arts of war, and like Leonardo he was granted permission to add to human knowledge on condition that he subtracted from human life. His activities in this respect were, however, more distinguished than those of Leonardo, since he invented the most amazing mechanical contrivances for defending the city of Syracuse against the Romans, and was finally killed by a Roman soldier when that city was captured. He is said to have been so absorbed in a mathematical problem that he did not notice the Romans coming. Plutarch is very apologetic on the subject of the mechanical inventions of Archimedes, which he feels to have been hardly worthy of a gentleman; but he considers him excusable on the ground that he was helping his cousin the king at a time of dire peril.

Archimedes showed great genius in mathematics and extraordinary skill in the invention of mechanical contrivances, but his contributions to science, remarkable as they are, still display the deductive attitude of the Greeks, which made the experimental method scarcely possible for them. His work on Statics is famous, and justly so, but it proceeds from axioms like Euclid's geometry, and the axioms are supposed to be self-evident, not the result of experiment. His book *On Floating Bodies* is the one which according to tradition resulted from the problem of King Hiero's crown, which was suspected of being not made of pure gold. This problem, as everyone knows, Archimedes is supposed to have solved while in his bath. At any rate, the method which he proposes in his book for such cases is a perfectly valid one, and although the book proceeds from postulates by a method of deduction, one cannot but suppose that he arrived at the postulates experimentally. This is, perhaps, the most nearly scientific (in the modern sense) of the works of Archimedes. Soon after his time, however, such feeling as the Greeks had had for the scientific investigation of natural phenomena decayed, and though pure mathematics continued to flourish down to the capture of Alexandria by the Mohammedans, there were hardly any further advances in natural science, and the best that had been done, such as the theory of Aristarchus, was forgotten.

The Arabs were more experimental than the Greeks, especially in chemistry. They hoped to transmute base metals into gold, to discover the philosopher's stone, and to concoct the elixir of life. Partly on this account chemical investigations were viewed with favour. Throughout the Dark Ages it was mainly by the Arabs that the tradition of civilization was carried on, and it was largely from them that Christians such as Roger Bacon acquired whatever scientific knowledge the later Middle Ages possessed. The Arabs, however, had a defect which was the opposite of that of the Greeks: they sought detached facts rather than general principles, and had not the power of inferring general laws from the facts which they discovered.

In Europe, when the scholastic system first began to give way before the Renaissance, there came to be, for a time, a dislike of all generalizations and all systems. Montaigne illustrates this tendency. He likes queer facts, particularly if they disprove something. He has no desire to make his opinions systematic and coherent. Rabelais also, with his motto: "Fais ce que voudras," is as averse from intellectual as from other fetters. The Renaissance rejoiced in the recovered liberty of speculation, and was not anxious to lose this liberty even in the interests of truth. Of the typical figures of the Renaissance far the most scientific was Leonardo, whose notebooks are fascinating and contain many brilliant anticipations of later discoveries, but he brought almost nothing to fruition, and remained without effect upon his scientific successors.

Scientific method, as we understand it, comes into the world full-fledged with Galileo (1564–1642), and, to a somewhat lesser degree, in his contemporary, Kepler (1571–1630). Kepler is known to fame through his three laws: he first discovered that the planets move round the sun in ellipses, not in circles. To the modern mind there is nothing astonishing in the fact that the earth's orbit is an ellipse, but to minds trained on antiquity anything except a circle, or some complication of circles, seemed almost incredible for a heavenly body. To the Greeks the planets were divine, and must therefore move in perfect curves. Circles and epicycles did not offend their esthetic susceptibilities, but a crooked, skew orbit such as the earth's actually is would have shocked them deeply. Unprejudiced observation without regard to esthetic prejudices required therefore, at that time, a rare intensity of scientific ardour. It was Kepler and Galileo who established the fact that the earth and the other planets go round the sun. This had been asserted by Copernicus, and, as we have seen, by certain Greeks, but they had not succeeded in giving proofs. Copernicus, indeed, had no serious arguments to advance in favour of his view. It would be doing Kepler more than justice to suggest that in adopting the Copernican hypothesis he was acting on purely scientific motives. It appears that, at any rate in youth, he was addicted to sun-worship, and thought the centre of the universe the only place worthy of so great a deity. None but scientific motives, however, could have led him to the discovery that the planetary orbits are ellipses and not circles.

He, and still more Galileo, possessed the scientific method in its completeness. While much more is known than was known in their day, nothing essential has been added to method. They proceeded from observation of particular facts to the establishment of exact quantitative laws, by means of which future particular facts could be predicted. They shocked their contemporaries profoundly, partly because their conclusions were inherently shocking to the beliefs of that age, but partly also because the belief in authority had enabled learned men to confine their researches to libraries, and the professors were pained at the suggestion that it might be necessary to look at the world in order to know what it is like.

Galileo, it must be confessed, was something of a gamin. When still very young he became Professor of Mathematics at Pisa, but as the salary was only 7½d. a day, he does not seem to have thought that a very dignified bearing could be expected of him. He began by writing a treatise against the wearing of cap and gown in the University, which may perhaps have been popular with undergraduates, but was viewed with grave disfavour by his fellow-professors. He would amuse himself by arranging occasions which would make his colleagues look silly. They asserted, for example, on the basis of Aristotle's Physics, that a body weighing ten pounds would fall through a given distance in one-tenth of the time that would be taken by a body weighing one pound. So he went up to the top of the Leaning Tower of Pisa one morning with a ten-pound shot and a one-pound shot, and just as the professors were proceeding with leisurely dignity to their respective lecture-rooms in the presence of their pupils, he attracted their attention and dropped the two weights from the top of the tower to their feet. The two weights arrived practically simultaneously. The professors, however, maintained that their eyes must have deceived them, since it was impossible that Aristotle could be in error.

On another occasion he was even more rash. Giovanni dei Medici, who was the Governor of Leghorn, invented a dredging machine of which he was very proud. Galileo pointed out that whatever else it might do it would not dredge, which proved to be a fact. This caused Giovanni to become an ardent Aristotelian.

Galileo became unpopular and was hissed at his lectures—a fate which also befell Einstein in Berlin. Then he made a telescope and invited the professors to look through it at Jupiter's moons. They refused on the ground that Aristotle had not mentioned these satellites, and therefore anybody who thought he saw them must be mistaken.

The experiment from the Leaning Tower of Pisa illustrated Galileo's first important piece of work, namely, the establishment of the Law of Falling Bodies, according to which all bodies fall at the same rate in a vacuum and at the end of a given time have a velocity proportional to the time in which they have been falling, and have traversed a distance proportional to the square of that time. Aristotle had maintained otherwise, but neither he nor any of his successors throughout nearly two thousand years had taken the trouble to find out whether what he said was true. The idea of doing so was a novelty, and Galileo's disrespect for authority was considered abominable. He had, of course, many friends, men to whom the spectacle of intelligence was delightful in itself. Few such men, however, held academic posts, and university opinion was bitterly hostile to his discoveries.

As everyone knows, he came in conflict with the Inquisition at the end of his life for maintaining that the earth goes round the sun. He had had a previous minor encounter from which he had emerged without great damage, but in the year 1632 he published a book of dialogues on the Copernican and Ptolemaic systems, in which he had the temerity to place some remarks that had been made by the Pope into the mouth of a character named Simplicius. The Pope had hitherto been friendly to him, but at this point became furious. Galileo was living at Florence on terms of friendship with the Grand Duke, but the Inquisition sent for him to come to Rome to be tried, and threatened the Grand Duke with pains and penalties if he continued to shelter Galileo. Galileo was at this time seventy years old, very ill, and going blind; he sent a medical certificate to the effect that he was not fit to travel, so the Inquisition sent a doctor of their own with orders that as soon as he was well enough he should be brought in chains. Upon hearing that this order was on its way, he set out voluntarily. By means of threats he was induced to make submission.

The sentence of the Inquisition is an interesting document:

> ... Whereas you, Galileo, son of the late Vincenzio Galilei, of Florence, aged 70 years, were denounced in 1615, to this Holy Office, for holding as true a false doctrine taught by many, namely, that the sun is immovable in the centre of the world, and that the earth moves, and also with a diurnal motion; also, for having pupils whom you instructed in the same opinions; also, for maintaining a correspondence on the same with some German mathematicians; also for publishing certain letters on the sunspots, in which you developed the same doctrine as true; also for answering the objections which were continually produced from the Holy Scriptures, by glozing the said Scriptures according to your own meaning; and whereas thereupon was produced the copy of a writing, in form of a letter, professedly written by you to

a person formerly your pupil, in which, following the hypothesis of Copernicus, you include several propositions contrary to the true sense and authority of the Holy Scriptures; therefore (this Holy Tribunal being desirous of providing against the disorder and mischief which were thence proceeding and increasing to the detriment of the Holy Faith) by the desire of his Holiness and of the Most Eminent Lords Cardinals of this supreme and universal Inquisition, the two propositions of the stability of the sun, and the motion of the earth, were qualified by the Theological Qualifiers as follows:

1. The proposition that the sun is in the centre of the world and immovable from its place is absurd, philosophically false, and formally heretical; because it is expressly contrary to the Holy Scriptures.

2. The proposition that the earth is not the centre of the world, nor immovable, but that it moves, and also with a diurnal action, is also absurd, philosophically false, and, theologically considered, at least erroneous in faith.

But whereas, being pleased at that time to deal mildly with you, it was decreed in the Holy Congregation, held before his Holiness on the twenty-fifth day of February, 1616, that his Eminence the Lord Cardinal Bellarmine should enjoin you to give up altogether the said false doctrine; and if you should refuse, that you should be ordered by the Commissary of the Holy Office to relinquish it, not to teach it to others, nor to defend it; and in default of acquiescence, that you should be imprisoned; and whereas in execution of this decree, on the following day, at the Palace, in the presence of his Eminence the said Lord Cardinal Bellarmine, after you had been mildly admonished by the said Lord Cardinal, you were commanded by the Commissary of the Holy Office, before a notary and witnesses, to relinquish altogether the said false opinion, and, in future, neither to defend nor teach it in any manner, neither verbally nor in writing, and upon your promising obedience you were dismissed.

And, in order that so pernicious a doctrine might be altogether rooted out, not insinuate itself further to the heavy detriment of the Catholic truth, a decree emanated from the Holy Congregation of the Index prohibiting the books which treat of this doctrine, declaring it false, and altogether contrary to the Holy and Divine Scripture.

And whereas a book has since appeared published at Florence last year, the title of which showed that you were the author, which title is *The Dialogue of Galileo Galilei, on the two principal Systems of the World—the Ptolemaic and Copernican;* and whereas the Holy Congregation has heard that, in consequence of printing the said book, the false opinion of the earth's motion and stability of the sun is daily gaining ground, the said book has been taken into careful consideration, and in it has been detected a glaring violation of the said order, which had been intimated to you; inasmuch as in this book you have defended the said opinion, already, and in your presence, condemned; although, in the same book, you labour with many circumlocutions to induce the belief that it is left undecided and merely probable; which is equally a very grave error, since an opinion can in no way be probable which has been already declared and finally determined contrary to the Divine Scripture. Therefore, by Our order, you have been cited to this Holy Office, where, on your examination upon oath, you have acknowledged the said book as written and printed by you. You also confessed that you began to write the said book ten or twelve years ago, after the order aforesaid had been given. Also, that you had demanded licence to publish it, without signifying to those who granted you this permission that you had been commanded not to hold, defend, or teach, the said doctrine in any manner. You also confessed that the reader might think the arguments adduced on the false side to be so worded as more effectually to compel conviction than to be easily refutable, alleging, in excuse, that you had thus run into an error, foreign (as you say) to your intention, from writing in the form of a dialogue, and in consequence of the natural complacency which everyone feels with regard to his own subtleties, and in showing himself more skilful than the generality of mankind in contriving, even in favour of false propositions, ingenious and plausible arguments.

And, upon a convenient time being given you for making your defence, you produced a certificate in the handwriting of his Eminence the Lord Cardinal Bellarmine, procured, as you said, by yourself, that you might defend yourself against the calumnies of your enemies, who reported that you had abjured your opinions, and had been punished by the Holy Office; in which certificate it is declared that you had not abjured nor had been punished, but merely that the declaration made by his Holiness, and promulgated by the Holy Congregation of the Index, had been announced to you, which declares that the opinion of the motion of the earth and stability of the sun is contrary to the Holy Scriptures, and, therefore, cannot be held or defended. Wherefore, since no mention is there made of two articles of the order, to wit, the order "not to teach" and "in any manner," you argued that we ought to believe that, in the lapse of fourteen or sixteen years, they had escaped your memory, and that this was also the reason why you were silent as to the order when you sought permission to publish your book, and that this is said by you, not to excuse your error, but that it may be attributed to vainglorious ambition rather than to malice. But this very certificate, produced on your behalf, has greatly aggravated your offence, since it is therein declared that the said opinion is contrary to the Holy Scriptures, and yet you have dared to treat of it, and to argue that it is probable. Nor is there any extenuation in the licence artfully and cunningly extorted by you, since you did not intimate the command imposed upon you. But whereas it appeared to Us that you had not disclosed the whole truth with regard to your intention, We thought it necessary to proceed to the rigorous examination of you, in which (without any prejudice to what you had confessed, and which is above detailed against you, with regard to your said intention) you answered like a good Catholic.

Therefore, having seen and maturely considered the merits of your cause, with your said confessions and excuses, and everything else which ought to be seen

> and considered, We have come to the underwritten final sentence against you:
>
> Invoking, therefore, the most holy name of our Lord Jesus Christ, and of His Most Glorious Virgin Mother, Mary, We pronounce this Our final sentence, which, sitting in council and judgment with the Reverend Masters of Sacred Theology and Doctors of both Laws, Our Assessors, We put forth in this writing in regard to the matters and controversies between the Magnificent Carlo Sincereo, Doctor of both Laws, Fiscal Proctor of the Holy Office, of the one part, and you, Galileo Galilei, defendant, tried and confessed as above, of the other part, We pronounce, judge, and declare, that you, the said Galileo, by reason of these things which have been detailed in the course of this writing, and which, as above, you have confessed, have rendered yourself vehemently suspected by this Holy Office of heresy, that is of having believed and held the doctrine (which is false and contrary to the Holy and Divine Scriptures), that the sun is the centre of the world, and that it does not move from east to west, and that the earth does move, and is not the centre of the world; also, that an opinion can be held and supported and probable, after it has been declared and finally decreed contrary to the Holy Scripture, and, consequently, that you have incurred all the censures and penalties enjoined and promulgated in the sacred canons and other general and particular constitutions against delinquents of this description. From which it is Our pleasure that you be absolved, provided that with a sincere heart and unfeigned faith, in Our presence, you abjure, curse, and detest, the said errors and heresies, and every other error and heresy, contrary to the Catholic and Apostolic Church of Rome, in the form now shown to you.
>
> But that your grievous and pernicious error and transgression may not go altogether unpunished, and that you may be made more cautious in future, and may be a warning to others to abstain from delinquencies of this sort. We decree that the book *Dialogues of Galileo Galilei* be prohibited by a public edict, and We condemn you to the formal prison of this Holy Office for a period determinable at Our pleasure; and by way of salutary penance, We order you during the next three years to recite, once a week, the seven penitential psalms, reserving to Ourselves the power of moderating, commuting, or taking off, the whole or part of the said punishment or penance.

The formula of abjuration, which, as a consequence of this sentence, Galileo was compelled to pronounce, was as follows:

> I, Galileo Galilei, son of the late Vincenzio Galilei of Florence, aged seventy years, being brought personally to judgment, and kneeling before you, Most Eminent and Most Reverend Lords Cardinals, General Inquisitors of the Universal Christian Republic against heretical depravity, having before my eyes the Holy Gospels which I touch with my own hands, swear that I have always believed, and, with the help of God, will in future believe, every article which the Holy Catholic and Apostolic Church of Rome holds, teaches, and preaches. But because I have been enjoined, by this Holy Office, altogether to abandon the false opinion which maintains that the sun is the centre

and immovable, and forbidden to hold, defend, or teach, the said false doctrine in any manner; and because, after it had been signified to me that the said doctrine is repugnant to the Holy Scripture, I have written and printed a book, in which I treat of the same condemned doctrine, and adduce reasons with great force in support of the same, without giving any solution, and therefore have been judged grievously suspected of heresy; that is to say, that I held and believed that the sun is the centre of the world and immovable, and that the earth is not the centre and movable, I am willing to remove from the minds of your Eminences, and of every Catholic Christian, this vehement suspicion rightly entertained towards me, therefore, with a sincere heart and unfeigned faith, I abjure, curse, and detest the said errors and heresies, and generally every other error and sect contrary to the said Holy Church; and I swear that I will never more in future say, or assert anything, verbally or in writing, which may give rise to a similar suspicion of me; but that if I shall know any heretic, or anyone suspected of heresy, I will denounce him to this Holy Office, or to the Inquisitor and Ordinary of the place in which I may be. I swear, moreover, and promise that I will fulfil and observe fully all the penances which have been or shall be laid on me by this Holy Office. But if it shall happen that I violate any of my said promises, oaths, and protestations (which God avert!), I subject myself to all the pains and punishments which have been decreed and promulgated by the sacred canons and other general and particular constitutions against delinquents of this description. So, may God help me, and His Holy Gospels, which I touch with my own hands, I, the above-named Galileo Galilei, have abjured, sworn, promised, and bound myself as above; and, in witness thereof, with my own hand have subscribed this present writing of my abjuration, which I have recited word for word.

At Rome, in the Convent of Monerva, June 22, 1633, I, Galileo Galilei, have abjured as above with my own hand.

It is not true that, after reciting this abjuration, he muttered: "*Epper si muove*." It was the world that said this—not Galileo.

The Inquisition stated that Galileo's fate should be "a warning to others to abstain from delinquencies of this sort." In this they were successful, so far, at least, as Italy was concerned. Galileo was the last of the great Italians. No Italian since his day has been capable of delinquencies of his sort. It cannot be said that the Church has altered greatly since the time of Galileo. Wherever it has power, as in Ireland and Boston, it still forbids all literature containing new ideas.

The conflict between Galileo and the Inquisition is not merely the conflict between free thought and bigotry or between science and religion; it is a conflict between the spirit of induction and the spirit of deduction. Those who believe in deduction as the method of arriving at knowledge are compelled to find their premises somewhere, usually in a sacred book. Deduction from inspired books is the method of arriving at truth employed by jurists, Christians, Mohammedans, and Communists. Since deduction as a means of obtaining knowledge collapses when doubt is thrown upon its premises, those who believe in deduction must necessarily be bitter against men who question the authority of the sacred books. Galileo questioned both Aristotle and the Scriptures, and thereby destroyed the whole edifice of medieval knowledge. His predecessors had known how the world was created, what was man's destiny, the deepest mysteries of metaphysics, and the hidden principles governing the behaviour of bodies. Throughout the moral and material universe nothing was mysterious to them, nothing hidden, nothing incapable of exposition in orderly syllogisms. Compared with all this wealth, what was left to the followers of Galileo?—a law of falling bodies, the theory of the pendulum, and Kepler's ellipses. Can it be wondered at that the learned cried out at such a destruction of their hard-won wealth? As the rising sun scatters the multitude of stars, so Galileo's few proved truths banished the scintillating firmament of medieval certainties.

Socrates had said that he was wiser than his contemporaries because he alone knew that he knew nothing. This was a rhetorical device. Galileo could have said with truth that he knew something, but knew he knew little, while his Aristotelian contemporaries knew nothing, but thought they knew much. Knowledge, as opposed to fantasies of wish-fulfilment, is difficult to come by. A little contact with real knowledge makes fantasies less acceptable. As a matter of fact, knowledge is even harder to come by than Galileo supposed, and much that he believed was only approximate; but in the process of acquiring knowledge at once secure and general, Galileo took the first great step. He is, therefore, the father of modern times. Whatever we may like or dislike about the age in which we live, its increase of population, its improvement in health, its trains, motor-cars, radio, politics, and advertisements of soap—all emanate from Galileo. If the Inquisition could have caught him young, we might not now be enjoying the blessings of air-warfare and atomic bombs, nor, on the other hand, the diminution of poverty and disease which is characteristic of our age.

It is customary amongst a certain school of sociologists to minimize the importance of intelligence, and to attribute all great events to large impersonal causes. I believe this to be an entire delusion. I believe that if a hundred of the men of the seventeenth century had been killed in infancy, the modern world would not exist. And of these hundred, Galileo is the chief.

II. NEWTON

Sir Isaac Newton was born in the year in which Galileo died (1642). Like Galileo he lived to be a very old man, as he died in the year 1727.

In the short period between these two men's activities, the position of science in the world was completely changed. Galileo, all his life, had to fight against the recognized men of learning, and in his last years had to suffer persecution and condemnation of his work. Newton, on the other hand, from the moment when, at the age of eighteen, he became an undergraduate at Trinity College, Cambridge, received universal applause. Less than two years after he had taken his M.A. degree the Master of his College was describing him as a man of incredible genius. He was acclaimed by the whole learned world; he was honoured by monarchs; and, in the true English spirit, was rewarded for his work by a Government post in which it could not be continued. So important was he, that when George I ascended the throne, the great Leibniz had to be left behind in Hanover because he and Newton had quarrelled.

It is fortunate for succeeding ages that Newton's circumstances were so placid. He was a timorous, nervous man, at once quarrelsome and afraid of controversy. He hated publication because it exposed him to criticism, and had to be bullied into publishing by kind friends. *A* propos of his Opticks he wrote to Leibniz: "I was so persecuted with discussions arising from the publication of my theory of light, that I blamed my own imprudence for parting with so substantial a blessing as my quiet to run after a shadow." If he had encountered the sort of opposition with which Galileo had to contend, it is probable that he would never have published a line.

Newton's triumph was the most spectacular in the history of science. Astronomy, since the time of the Greeks, had been at once the most advanced and the most respected of the sciences. Kepler's laws were still fairly recent, and the third of them was by no means universally accepted. Moreover, they appeared strange and unaccountable to those who had been accustomed to circles and epicycles. Galileo's theory of the tides was not right, the motions of the moon were not properly understood, and astronomers could not but feel the loss of that epic unity that the heavens possessed in the Ptolemaic system. Newton, at one stroke, by his law of gravitation brought order and unity into this confusion. Not only the major aspects of the motions of the planets and satellites were accounted for, but also all the niceties at that time known; even the comets, which, not so long ago, had "blazed forth the death of princes," were found to proceed according to the law of gravitation. Halley's comet was one of the most obliging among them, and Halley was Newton's best friend.

Newton's Principia proceeds in the grand Greek manner: from the three laws of motion and the law of gravitation, by purely mathematical deduction, the whole solar system is explained. Newton's work is statuesque and Hellenic, unlike the best work of our own time. The nearest approach to the same classical perfection among moderns is the theory of relativity, but even that does not aim at the same finality, since the rate of progress nowadays is too great. Everyone knows the story of the fall of the apple. Unlike most such stories, it is not certainly known to be false. At any rate, it was in the year 1665 that Newton first thought of the law of gravitation, and in that year, on account of the Great Plague, he spent his time in the country, possibly in an orchard. He did not publish his Principia until the year 1687: for twenty-one years he was content to think over his theory and gradually perfect it. No modern would dare to do such a thing, since twenty-one years is enough to change completely the scientific landscape. Even Einstein's work has always contained ragged edges, unresolved doubts, and unfinished speculations. I do not say this as a criticism; I say it only to illustrate the difference between our age and that of Newton. We aim no longer at perfection, because of the army of successors whom we can scarcely outstrip, and who are at every moment ready to obliterate our traces.

The universal respect accorded to Newton, as contrasted with the treatment meted out to Galileo, was due in part to Galileo's own work and to that of the other men of science who filled the intervening years, but it was due also, and quite as much, to the course of politics. In Germany, the Thirty Years' War, which was raging when Galileo died, halved the population without achieving the slightest change in the balance of power between Protestants and Catholics. This caused even the least reflective to think that perhaps wars of religion were a mistake. France, though a Catholic power, had supported the German Protestants, and Henry IV, although he became a Catholic in order to win Paris, was not led by this motive into any great bigotry with regard to his new faith. In England the Civil War, which began in the year of Newton's birth, led to the rule of the saints, which turned everybody except the saints against religious zeal. Newton entered the University in the year after that in which Charles II returned from exile, and Charles II, who founded the Royal Society, did all in his power to encourage science, partly, no doubt, as an antidote to bigotry. Protestant bigotry had kept him an exile, while Catholic bigotry caused his brother to lose the throne. Charles II, who was an intelligent monarch, made it a rule of government to avoid having to set out on his travels again. The period from his accession to the death of Queen Anne was the most brilliant, intellectually, in English history.

In France, meanwhile, Descartes had inaugurated modern philosophy, but his theory of vortices proved an obstacle to the acceptance of Newton's ideas. It was only after Newton's death, and largely as a result of Voltaire's Lettres Philosophiques, that Newton gained vogue, but when he did his vogue was terrific; in fact, throughout the following century down to the fall of Napoleon, it was chiefly the French who carried on Newton's work. The English were misled by patriotism into adhering to his methods where they were inferior to those of Leibniz, with the result that after his death English mathematics were negligible for a hundred years. The harm that in Italy was done by bigotry was done in England by nationalism. It would be hard to say which of the two proved the more pernicious.

Though Newton's Principia retains the deductive form which was inaugurated by the Greeks, its spirit is quite different from that of Greek work, since the law of gravitation, which is one of its premises, is not supposed to be self-evident, but is arrived at inductively from Kepler's laws. The book, therefore, illustrates scientific method in the form which is its ideal. From observation of particular facts, it arrives by induction at a general law, and by deduction from the general law other particular facts are inferred. This is still the ideal of physics, which is the science from which, in theory, all others ought to be deduced; but the realization of the ideal is somewhat more difficult than it seemed in Newton's day, and premature systemization has been found to be a danger.

Newton's law of gravitation has had a peculiar history. While it continued for over two hundred years to explain almost every fact that was known in regard to the motions of the heavenly bodies, it remained itself isolated and mysterious among natural laws. New branches of physics grew to vast proportions; the theories of sound, heat, light, and electricity were successfully explored; but no property of matter was discovered which could be in any way connected with gravitation. It was only through Einstein's general theory of relativity (1915) that gravitation was fitted into the general scheme of physics, and then it was found to belong rather to geometry than to physics in the old-fashioned sense. From a practical point of view, Einstein's theory involves only very minute corrections of Newtonian results. These very minute corrections, so far as they are measurable, have been empirically verified; but while the practical change is small, the intellectual change is enormous, since our whole conception of space and time has had to be revolutionized. The work of Einstein has emphasized the difficulty of permanent achievement in science. Newton's law of gravitation had reigned so long, and explained so much, that it seemed scarcely credible that it should stand in need of correction. Nevertheless, such correction has at last proved necessary, and no one doubts that the correction will, in its turn, have to be corrected.

IV. PAVLOV

Each fresh advance of science into a new domain has produced a resistance analogous in kind to that encountered by Galileo, but growing gradually less in vehemence. Traditionalists have always hoped that somewhere a region would be found to which scientific method would prove inapplicable. After Newton, they abandoned the heavenly bodies in despair; after Darwin, most of them admitted the broad fact of evolution, though they continue, to this day, to suggest that the course of evolution has not been guided by mechanistic forces, but has been directed by a forward-looking purpose. The tapeworm, we are to suppose, has become what it is, not because it could not otherwise have survived in human intestines, but because it realizes an idea laid up in Heaven, which is part of the Divine Mind. As the Bishop of Birmingham says: "The loathsome parasite is a result of the integration of mutations; it is both an exquisite example of adaptation to environment and ethically revolting." This controversy is not yet wholly concluded, though there can be little doubt that mechanistic theories of evolution will prevail completely before long.

One effect of the doctrine of evolution has been to compel men to concede to animals some portion, at least, of the merits that they claim for homo sapiens. Descartes maintained that animals are mere automata, while human beings have free will. Views of this kind have lost their plausibility, though the doctrine of "emergent evolution," which we shall consider at a later stage, is designed to rehabilitate the view that men differ qualitatively from other animals. Physiology has been the battleground between those who regard all phenomena as subject to scientific method, and those who still hope that, among vital phenomena, there are some, at least, which demand mystical treatment. Is the human body a mere machine, governed wholly by the principles of physics and chemistry? Wherever it is understood, it is found to be so, but there are still processes which are not completely understood: perhaps in them a vital principle will be found to be lurking? In this way, the champions of vitalism become the friends of ignorance. Let us not, they feel, know too much about the human body, lest we should discover to our dismay that we can understand it. Every fresh discovery makes this view less plausible, and restricts the territory still open to the obscurantists. There are some, however, who are willing to surrender the body to the tender mercies of the scientist, provided they can save the soul. The soul, we know, is immortal, and has cognizance of right and wrong. The soul, if it belongs to the right person, is aware of God. It reaches out after higher things, and is informed by a divine spark. This being the case, it surely cannot be governed by the laws of physics and chemistry, or, indeed, by any laws at all? Psychology, therefore, has been more obstinately defended by the enemies of scientific method than any other department of human knowledge. Nevertheless, even psychology is becoming scientific; many men have contributed to this result, but none more than the Russian physiologist, Pavlov.

Pavlov was born in the year 1849, and devoted the bulk of his working life to the investigation of the behaviour of dogs. This, however, is too wide a statement—the bulk of his work consisted merely of observing when dogs' mouths water, and how much. This illustrates one of the most important characteristics of scientific method, as opposed to the methods of metaphysicians and theologians. The man of science looks for facts that are significant, in the sense of leading to general laws; and such facts are frequently quite devoid of intrinsic interest. The first impression of any non-scientific person, when he learns what is being done in some famous laboratory, is that all the investigators are wasting their time on trivialities; but the facts that are intellectually illuminating are often such as they are, in themselves, trivial and uninteresting. This applies in particular to Pavlov's speciality, namely, the flow of saliva in dogs. By studying this, he arrived at general laws governing a great deal of animal behaviour, and of the behaviour of human beings likewise.

The procedure is as follows. Everyone knows that the sight of a juicy morsel will make a dog's mouth water. Pavlov puts a tube into the dog's mouth, so that the amount of saliva to which the juicy morsel gives rise can be measured. The flow of saliva, when there is food in the mouth, is what is called a reflex; that is to say, it is one of those things that the body does spontaneously, and without the influence of experience. There are many reflexes, some very specific, some less so. Some of these can be studied in new-born infants, but some only arise at later stages of growth. The infant sneezes, and yawns, and stretches, and sucks, and turns its eyes towards a bright light, and performs various other bodily movements at the appropriate occasions, without the need of any previous learning. All such actions are called reflexes, or, in Pavlov's language, unconditioned reflexes. They cover the ground that was formerly covered by the somewhat vague appellation of instinct. Complicated instincts, such as nest-building in birds, appear to consist of a series of reflexes. In the lower animals, reflexes are very little modified by experience: the moth continues to fly into the flame, even after it has singed its wings. But in higher animals, experience has a great effect upon reflexes, and this is most of all the case with man. Pavlov studied the effect of experience upon the salivary reflexes of dogs. The fundamental law in this subject is the law of conditioned reflexes: when the stimulus to an unconditioned reflex has been repeatedly accompanied, or immediately preceded, by some other stimulus, this other stimulus alone will, in time, equally produce the response which was originally called forth by the stimulus to the unconditioned reflex. The flow of saliva is originally called forth only by the actual food in the mouth; later on, it comes to be called forth by the sight and smell of the food, or by any signal which habitually precedes the giving of food. In this case, we have what is called a conditioned reflex; the response is the same as in the unconditioned reflex, but the stimulus is a new one, which has become associated with the original stimulus through experience. This law of the conditioned reflex is the basis of learning, of what the older psychologists called the "association of ideas," of the understanding of language, of habit, and of practically everything in behaviour that is due to experience.

On the basis of the fundamental law, Pavlov has built up, experimentally, all kinds of complications. He uses not only the stimulus of agreeable food, but also of disagreeable acids, so that he can build up in the dog responses of avoidance as well as responses of approach. Having formed a conditioned reflex by one set of experiments, he can proceed to inhibit it by another. If a given signal is followed sometimes by pleasant results, and sometimes by unpleasant ones, the dog is apt to suffer in the end a nervous breakdown; he becomes hysterical or neurasthenic, and, indeed, a typical mental patient. Pavlov does not cure him by making him reflect upon his infancy, or confess to a guilty passion for his mother, but by rest and bromide. He relates a story which should be studied by all educationists. He had a dog to whom he always showed a circular patch of bright light before giving him food, and an elliptical patch before giving him an electric shock. The dog learned to distinguish clearly between circles and ellipses, rejoicing in the former, and avoiding the latter with dismay. Pavlov then gradually diminished the eccentricity of the ellipse, making it more and more nearly resemble a circle. For a long time the dog continued to distinguish clearly:

> As the form of the ellipse was brought closer and closer to that of the circle, we obtained more or less quickly an increasingly delicate differentiation. But when we used an ellipse whose two axes were as 9:8, i.e. an ellipse which was nearly circular, all this was changed. We obtained a new delicate differentiation, which always remained imperfect, lasted two or three weeks, and afterwards not only disappeared spontaneously, but caused the loss of all earlier differentiations, including even the less delicate ones. The dog which formerly stood quietly on his bench, now was constantly struggling and howling. It was necessary to elaborate anew all the differentiations and the most unrefined now demanded much more time than at first. On attempting to obtain the final differentiation the old story was repeated, i.e. all the differentiations disappeared and the dog fell again into a state of excitation.

I am afraid a similar procedure is habitual in schools, and accounts for the apparent stupidity of many of the scholars.

Pavlov is of the opinion that sleep is essentially the same thing as inhibition, being, in fact, a general, instead of a specific, inhibition. On the basis of his study of dogs, he accepts the view of Hippocrates that there are four temperaments, namely, choleric, melancholic, sanguine, and phlegmatic. The phlegmatic and sanguine he regards as the saner types, while the melancholic and choleric are liable to nervous disorders. He finds his dogs divisible into these four types, and believes the same to be true of human beings.

The organ through which learning takes place is the cortex, and Pavlov considers himself as being engaged upon the study of the cortex. He is a physiologist, not a psychologist, but he is of the opinion that, where animals are concerned, there cannot be any psychology such as we derive from introspection when we study human beings. With human beings, it would seem that he does not go so far as Dr. John B. Watson. "Psychology," he says, "in so far as it concerns the subjective state of man, has a natural right to existence; for our subjective world is the first reality with which we are confronted. But though the right of existence of human psychology be granted, there is no reason why we should not question the necessity of an animal psychology." Where animals are concerned, he is a pure Behaviourist, on the ground that one cannot know whether an animal has consciousness, or, if it has, of what nature this consciousness may be. In regard to human beings also, in spite of his theoretical concession to introspective psychology, all that he has to say is based upon his study of conditioned reflexes, and it is clear that, in regard to bodily behaviour, his position is entirely mechanistic.

> One can hardly deny that only a study of the physico-chemical processes taking place in nerve tissue will give us a real theory of all nervous phenomena, and that the phases of this process will provide us with a full explanation of all the external manifestations of nervous activity, their consecutiveness and their interrelations.

The following quotation is interesting, not only as illustrating his position on this point, but as showing the idealistic hopes for the human race which he bases upon the progress of science:

> ... At the beginning of our work and for a long time afterwards we felt the compulsion of habit in explaining our subject by psychological interpretations. Every time the objective investigation met an obstacle, or when it was halted by the complexity of the problem, there arose quite naturally misgivings as to the correctness of our new method. Gradually with the progress of our research these doubts appeared more rarely, and now I am deeply and irrevocably convinced that along this path will be found the final triumph of the human mind over its uttermost and supreme problem—the knowledge of the mechanism and laws of human nature. Only thus may come a full, true and permanent happiness. Let the mind rise from victory to victory over surrounding nature, let it conquer for human life and activity not only the surface of the earth, but all that lies between the depth of the seas and the outer limits of the atmosphere, let it command for its service prodigious energy to flow from one part of the universe to the other, let it annihilate space for the transference of its thoughts—yet the same human creature, led by dark powers to wars and revolutions and their horrors, produces for itself incalculable material losses and inexpressible pain, and reverts to bestial conditions. Only science, exact science about human nature itself, and the most sincere approach to it by the aid of the omnipotent scientific method, will deliver man from his present gloom, and will purge him from his contemporary shame in the sphere of interhuman relations.

In metaphysics, he is neither a materialist nor a mentalist. He holds the view that I firmly believe to be the right one, that the habit of distinguishing between mind and matter is a mistake, and that the reality may be considered as both or neither with equal justice. "We are now coming," he says, "to think of the mind, the soul, and matter as all one, and with this view there will be no necessity for a choice between them."

As a human being, Pavlov has the simplicity and regularity of learned men of an earlier time, such as Immanuel Kant. He lived a quiet home life, and was invariably punctual at his laboratory. Once, during the Revolution, his assistant was ten minutes late, and adduced the Revolution as an excuse, but Pavlov replied: "What difference does a Revolution make when you have work in the laboratory to do?" The only allusion to the troubles of Russia to be found in his writings is in connection with the difficulty of feeding his animals during the years of food- shortage. Although his work has been such as might be held to give support to the official metaphysics of the Communist Party, he thought very ill of the Soviet government, and denounced it vehemently both publicly and privately. In spite of this, the Government treated him with every consideration, and supplied his laboratory generously with everything that he needed.

It is typical of the modern attitude in science, as compared with that of Newton, or even Darwin, that Pavlov has not attempted a statuesque perfection in the presentation of his theories. "The reason that I have not given a systematic exposition of our results during the last twenty years is the following. The field is an entirely new one, and the work has constantly advanced. How could I halt for any comprehensive conception, to systematize the results, when each day new experiments and observations brought us additional facts!" The rate of progress in science nowadays is much too great for such works as Newton's Principia, or Darwin's Origin of Species. Before such a book could be completed, it would be out of date. In many ways this is regrettable, for the great books of the past possessed a certain beauty and magnificence which is absent from the fugitive papers of our time, but it is an inevitable consequence of the rapid increase of knowledge, and must therefore be accepted philosophically.

Whether Pavlov's methods can be made to cover the whole of human behaviour is open to question, but at any rate they cover a very large field, and within this field they have shown how to apply scientific methods with quantitative exactitude. He has conquered a new sphere for exact science, and must therefore be regarded as one of the great men of our time. The problem which Pavlov successfully tackled is that of subjecting to scientific law what has hitherto been called voluntary behaviour. Two animals of the same species, or one animal on two different occasions, may respond differently to the same stimulus. This gave rise to the idea that there is something called a will, which enables us to respond to situations capriciously and without scientific regularity. Pavlov's study of the conditioned reflex has shown how behaviour which is not determined by the congenital constitution of an animal may nevertheless have its own rules, and be as capable of scientific treatment as is the behaviour governed by unconditioned reflexes. As Professor Hogben says:

> In our generation, the work of Pavlov's school has successfully tackled, for the first time in history, the problem of what Dr. Haldane calls "conscious behaviour" in non-teleological terms. It has reduced it to the investigation of the conditions under which new reflex systems are brought into being.

The more this achievement is studied, the more important it is seen to be, and it is on this account that Pavlov must be placed among the most eminent men of our time.

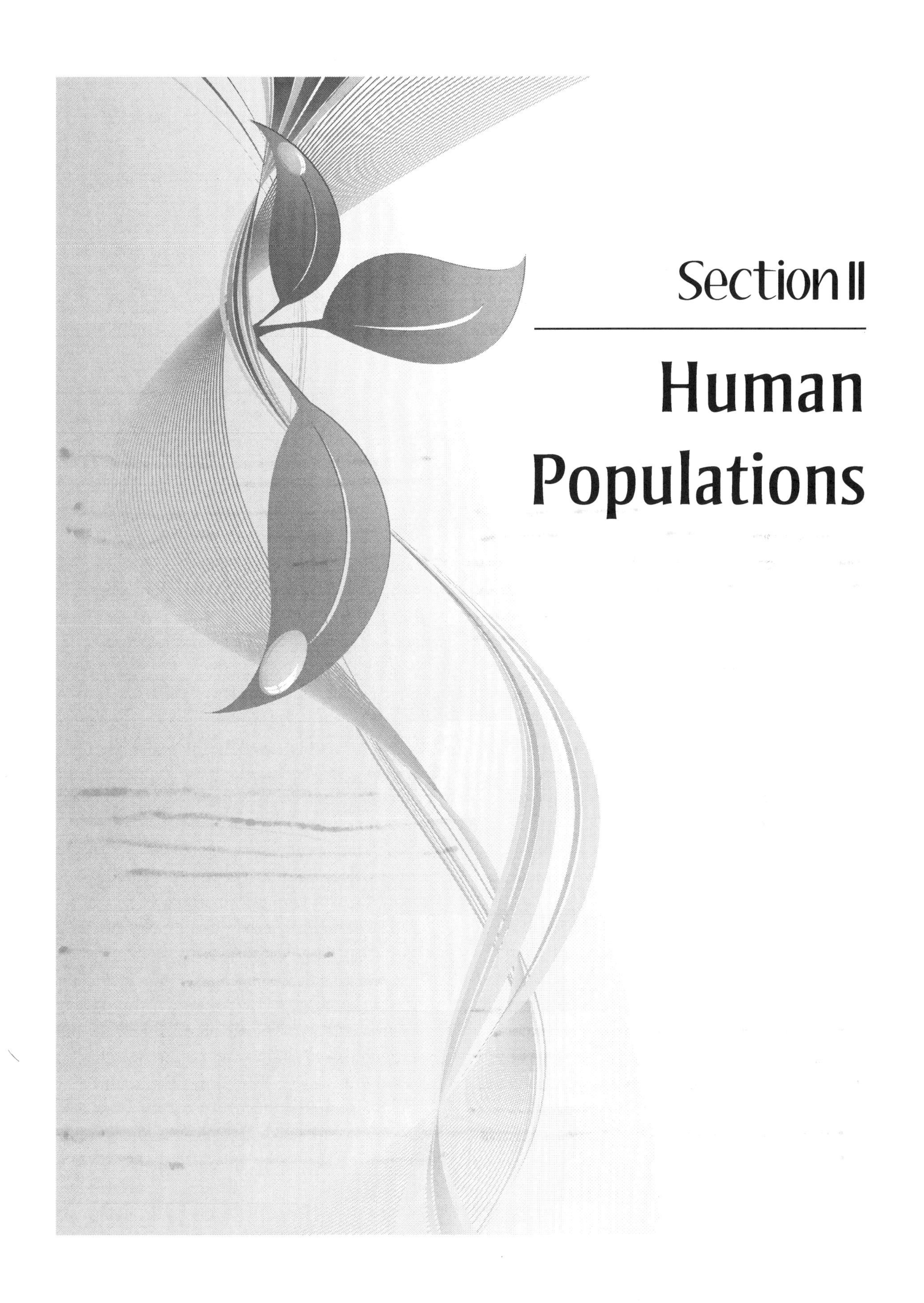

Section II

Human Populations

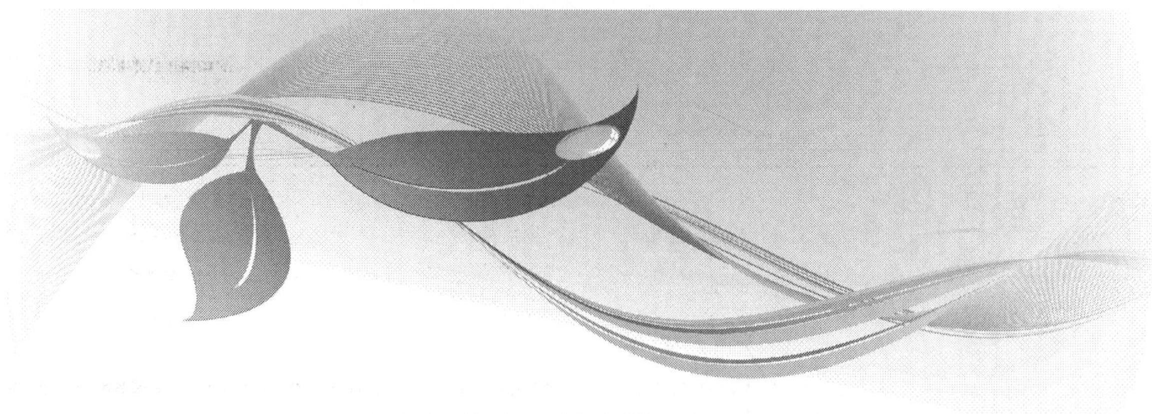

By Anne Marie Zimeri

Central to the field of environmental health is the issue of the expanding human population. There are currently more than 7 billion people on the planet, 307 million of whom live in the United States. Environmental degradation, uneven distribution of resources, and food security are just a few of the problems that may arise in the near future as the human population is expected to reach 9 billion within the next two decades. In this chapter, we will discuss how the population is growing, the ways in which we characterize populations, predictions for population growth based on demographics, and approaches to population control.

CHARACTERISTICS OF POPULATIONS

Typically, population growth is tracked by characteristics such as birth rates, death rates, doubling time, and fertility rates among females. When demographers use the term "total fertility rate," they are referring to the number of children a woman will have in her reproductive lifetime. It is expressed as the ratio of live births in an area to the population of that area; expressed per 1,000 people of that population per year. Fertility rates are almost always related to females because females almost always know how many children they have had, whereas men may not be certain. New molecular techniques that can test for paternity make it possible to verify male fertility rates for some studies, but unless otherwise stated, it can be safely assumed that fertility rates are referring to the number of children born to women. Table 2.1 shows a range of fertility rates in several areas of the world.

Birth rates and death rates can be used to track the speed at which a population is growing by natural increase. These rates are also expressed per 1,000 people per year. The typical range of these rates goes from a low birth or death rate of 10 per thousand per year to a high rate of 40 per thousand per year. These rates can be quite different among countries in the developed world compared to LDCs. More developed countries have lower birth and death rates; LDCs have higher birth and death rates. To calculate the rate of natural increase, simply subtract the death rate from the birth rate. If the number is positive, the population is growing. If the number is negative, the population is in decline. From this data, total growth rates can be calculated. Table 2.2 shows some current birth and death rates, as well as growth rates around the world. In addition, knowing a population's growth rate can be useful. Growth rate is calculated in three steps: 1) take the present

COUNTRY	FERTILITY RATE	COUNTRY	FERTILITY RATE
Korea	1.23	Guatemala	3.27
China	1.54	Nigeria	4.73
United States	2.06	Afghanistan	5.39
India	2.62	Ethiopia	6.02
Egypt	2.97	Niger	7.60

Table 2.1. Fertility rates for 2011 by country. Notice that more developed countries tend to have lower fertility rates than lesser developed countries. The total fertility rate for the world in 2011 was 2.52.

COUNTRY	BIRTH RATE	DEATH RATE	GROWTH RATE	DOUBLING TIME
China	12.29	7.03	0.493	142
United States	13.69	8.38	0.963	73
India	20.97	7.48	1.344	52
Afghanistan	39.53	14.84	2.375	29
Uganda	47.49	11.71	3.576	20

Table 2.2. Population expansion by country can be characterized by birth and death rates, growth rate, and doubling time (in years). Tables 2.1. and 2.2. adapted from the CIA World Factbook.

population, then subtract the past population, 2) divide that number by the past population, 3) multiply that by 100. Though the range of growth rates can be narrow, most often between 3.5% and 0.1%, the difference in time it can take that population to double can be dramatic, from 20 years to 700 years. A simple formula, based on "the rule of 70," can be used to give an estimate of doubling times for populations. This calculation divides the number 70 by the growth rate of the population.

For example, the growth rate of Japan is 0.1%; therefore its doubling time is 700 years. The growth rate of Niger is 3.5%; therefore its doubling time in years is 20. The lower the doubling time, the faster a population is growing. Countries with low doubling times are challenged with providing food, energy, and resources for a population that is quickly outgrowing community housing, sewer, sanitation, and hospital systems. When such challenges exist, and when populations are faced with shortages, it is difficult to be good stewards of the environment.

POPULATION DYNAMICS

Populations that are growing unrestrictedly, without environmental resistance, tend to grow exponentially. Environmental resistance factors include limited space and resources, disease, and other factors that tend to limit growth. Until a population's growth is limited by these factors, it is in what is called a J-curve. Currently, the world's population is in a J-curve (Figure 2.1) because we are heading toward biotic potential or the maximum reproductive capacity of a population if resources are unlimited.

The great question for human demographers is whether we will have a massive dieback, or whether we will move into an S-curve (Figure 2.2) and slowly approach our carrying capacity. Carrying capacity is the maximum population size of the species that the environment can sustain indefinitely.

Carrying capacity is not a number set in stone. It can oscillate depending on resource availability.

For example, in drought years, when fewer edible crops are produced, the carrying capacity of a population will decrease. If a species, including humans, passes carrying capacity, there is a risk of dieback (Figure 2.3).

Many times overshooting carrying capacity is related to doubling time. When a population is reproducing quickly, it is more likely to overshoot its carrying capacity when compared to a population that is growing slowly. Ideally, a population will approach carrying capacity and level out into an S-curve, thus avoiding a massive dieback. Another factor that influences the speed at which populations approach carrying capacity is the age at which reproduction begins. When the average reproductive age begins early (in teenage years) versus in a woman's twenties or thirties, the population tends to grow quickly. Delaying reproductive years is one way that the United States is trying to approach the issue of population control. The teen birth rate in the United States ranges from 30 to 165 births per thousand girls from age 15 to 19 (Figure 2.4). Though population control is a highly politicized subject, both major political parties in the United States agree that they would like to reduce the teen birth rate.

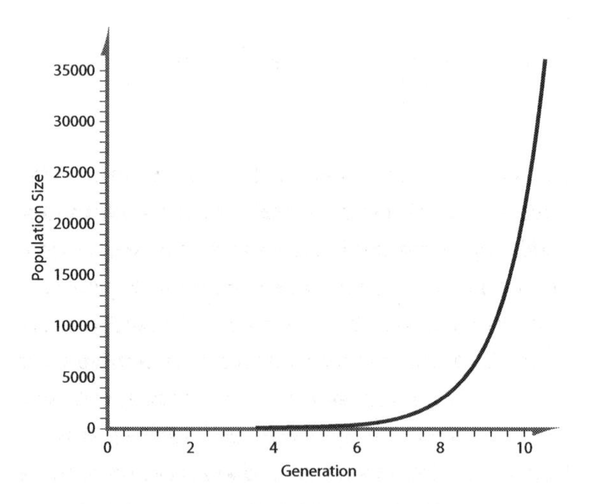

Figure 2.1. J-curves are curves that are J-shaped because of exponential growth.

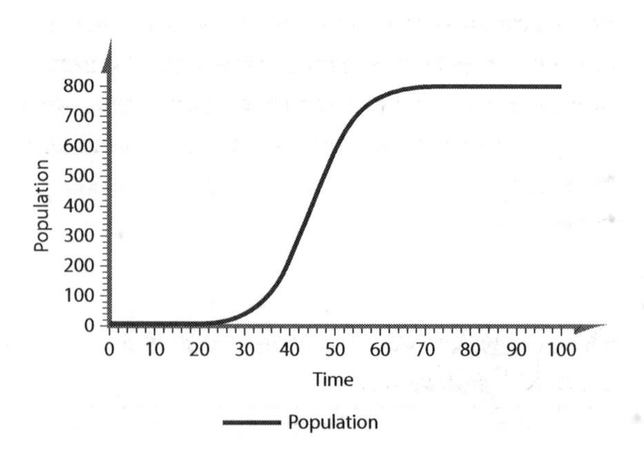

Figure 2.1. S-curves are curves that are S-shaped because after a period of time in a J-curve, growth slows and stabilizes.

PREDICTING POPULATION GROWTH BASED ON DEMOGRAPHICS

Based on the average reproductive age of a population, and what percent of the population is in that age range versus the percent of the population about to enter reproductive age, predictions can be made as to whether the population will expand,

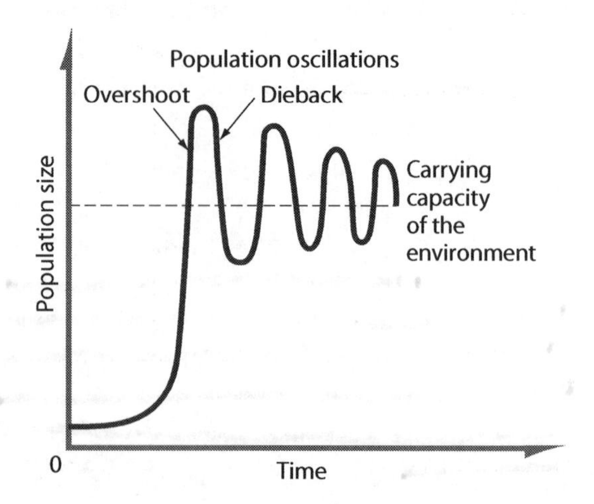

Figure 2.3. Populations can oscillate around carrying capacity in a series of overshoots and diebacks.

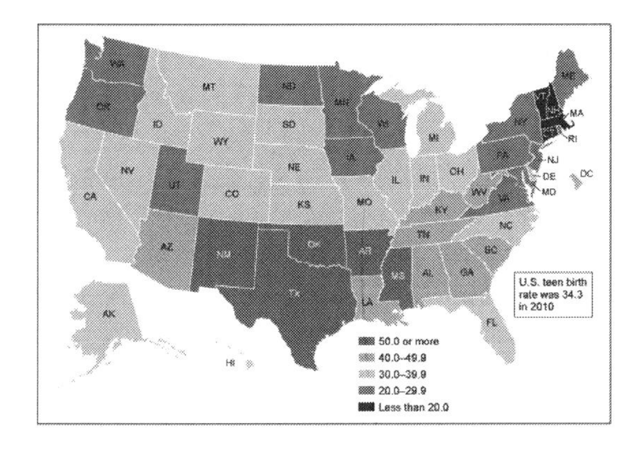

Figure 2.4. Teen birth rate in the United States tends to be higher in southern states.

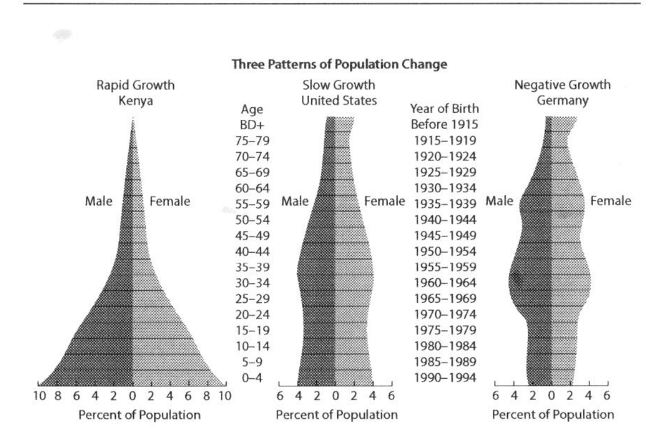

Figure 2.5. The demographics of a population can help to predict whether a population is going to rapidly expand, stabilize, or decline. Source: United Nations.

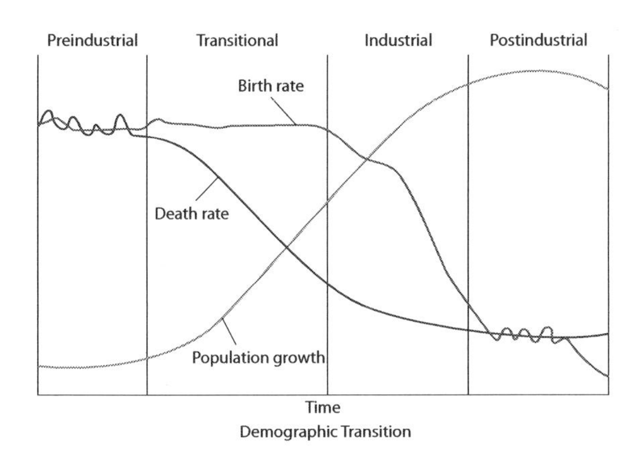

Figure 2.6. Demographic transition often accompanies the industrialization of a nation. During this transition time, birth and death rates are lowered, though the population continues to grow.

stay the same, or decline. When more people are approaching reproductive age than are already there, a population is facing expansion. The inertia of this system is difficult to alter because to change it, a dramatic lowering of the fertility rate must occur. When there are about the same number of people coming into reproductive age as the number already there, a population tends to be stable. If there are fewer people in the lower age range, then a population may be facing a decline (Figure 2.5).

DEMOGRAPHIC TRANSITION

Demographic transition, or the transition of high birth and death rates to low birth and death rates, usually accompanies industrialization as well as general economic and social development (Figure 2.6). The history of U.S. birth rates over the last century shows that the country went through demographic transition during its industrial age from about 1910 to 1930. During this time, birth and death rates went from a high to a low rate. Following that time, there have been other fluctuations in the birth rate such as the increase during the baby boom after World War II and the decrease in the late 1960s and 1970s when more women entered the workforce, and when the birth control pill became readily available. It is important to remember, however, that though these rates declined over the last century, the total population increased.

POPULATION INCREASES DUE TO LIFE EXPECTANCY

Another factor that has contributed to the number of people on the planet is the increase in life expectancy.

When people live longer, there are more people to be counted. Advances in medicine, the green revolution, and demographic transitions have all contributed to this increase. Table 2.3 shows some life expectancies by country.

Notice that more developed countries have higher life expectancies. Many of these dramatic differences can be directly predicted, up to a point, using annual per capita income (Figure 2.7). Increased life expectancy can, in some cases, contribute to an increased "dependency ratio," which is the number of nonworking individuals compared to the number of working individuals.

A person born in a rich country can expect to live longer than a person born in a poor country up to a point where the link flattens out. This means that at low levels of per capita income, increases in income are associated with large gains in life expectancy, but at high levels of income, increased income will show a smaller change in life expectancy.

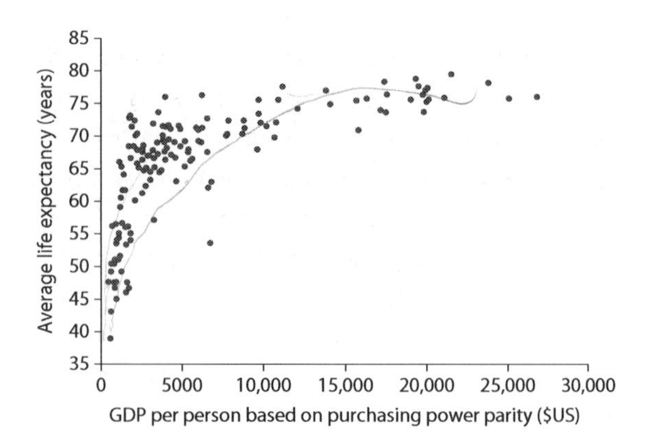

Figure 2.7. Life expectancy correlates with income.

PRONATALIST PRESSURES

There are many reasons why humans have children. Those that tend to increase a person's desire to have more children or those that exalt the role of parenthood are considered pronatalist pressures. Cultural and societal factors exerted on a person can be extremely deep rooted and serve as strong pronatalist pressures. Social status is often elevated when a person has children, especially male children. One could argue that even in the United States, status is related to the number of children one has. Consider your preconceived notions about couples or people who have children or who do not have children, especially those who have passed reproductive age. Why do you assume they did not have children? Do you first think that they could not, or did not have the opportunity, or do you think that they chose not to have children? Do you expect all young couples to have children one day? Do couples get asked whether they are going to have children or when are they going to have children?

Religion can play a role in applying pronatalist pressure because several mainstream religions have edicts that require followers to have as many children as possible, or to reject the idea of modern

COUNTRY	LIFE EXP. FEMALE	LIFE EXP. MALE	COUNTRY	LIFE EXP. FEMALE	LIFE EXP. MALE
Japan	86.1	79	Bolivia	67.7	63.4
South Korea	82.2	75	Ethiopia	54.3	51.7
United States	80.8	75.6	Nigeria	47.3	46.4
Guatemala	73.8	66.7	Afghanistan	43.8	43.9
Egypt	73.1	69.1	Sierra Leone	44.1	41

Table 2.3. Life expectancy (by country): the average number of years to be lived by a group of people born in the same year.

birth control methods. This issue can make the idea of any government role in population control quite controversial because it may impede religious freedom. Politics can also influence pronatalist pressure, especially by legislating pronatalism. In some countries, tax laws, laws limiting nonreproductive sex, restrictive abortion, and laws limiting the availability of contraceptives are common and have been considered pronatalist pressures. Less common is political pressure to increase the population so that a larger military force can amass.

Relevant everywhere in the world is the need for old-age security. Once a person is elderly and no longer able to work, many times he will rely on his children for financial support. Having more children to provide for an individual once he is elderly makes sense and is a forceful pronatalist pressure. Generally, a population that is growing has more young people than elderly. However, in some countries with low growth rates, the number of older, nonworking people compared to younger, working people is declining. This ratio of nonworking compared to working individuals is called the dependency ratio. In the United States, it can be costly to raise children, and the financial benefits of having a large family may not be seen until one is in retirement or when it is time to cash in on old-age security. In developing nations, however, immediate family finances can be a pronatalist pressure because children can contribute to the family's wealth at a much younger age whether by working in countries without child labor laws, by working in the family business or farm, or simply by begging.

BIRTH REDUCTION PRESSURES

Birth reduction pressures are factors that tend to reduce fertility. As more people are added to the planet in the next few decades, more will live in cities. Even in developing nations, a growing share of the population will live in urban areas. Therefore, a common birth-reduction pressure in the future will be space limitation in cities.

IMMIGRATION AND EMIGRATION

Populations in the future will not only expand and decline due to the rate of natural increase, but people may also move into an area (immigrate) or move away from an area (emigrate). The bulk of U.S. growth in the next 30 years is predicted to arise from immigrants and the descendants of those immigrants. Nearly one in five Americans (19%) will be an immigrant in 2050, compared with one in eight (12%) in 2005. The largest group of immigrants to the United States is Hispanic. Demographic projections suggest that the Latino population, already the nation's largest minority group, will triple in size and will account for most of the nation's population growth from 2005 through 2050. Hispanics will make up 29% of the U.S. population in 2050, compared with 16% in 2012. Much of this immigration is due to economic opportunity. However, in other parts of the world, immigration and emigration can be driven by other factors including food and water shortages, war, and political oppression. For example, Sudanese emigration driven by violence and war is increasing immigration in Kenya, Egypt, Israel and other nations.

VARYING PERSPECTIVE ON POPULATION CONTROL TECHNIQUES: COERCIVE VS. EMPOWERMENT

There have been many effective approaches used to address population growth in several countries.

Success or effectiveness was determined by a dramatic reduction in growth rate. Granted, the population in many of these areas is still growing, but with a marked reduction that has increased the quality of life in the area. One set of statistics that has been used to cultivate programs to reduce birth rates is the relationship between the level of education a woman has and her fertility rate. Time and time again, in hundreds of studies, data has shown that the more education a woman has, the fewer children she will have. She will also have those children later in life. Therefore, one strategy to reduce growth rate is to empower women with education.

Government mandates and incentivized sterilization programs that reduce fertility are generally considered coercive because women have a directive of how many children they may have, or they are so economically pressured into being sterilized that they may have no children at all. Yet these mandates are quite effective. China's family planning policy (also known as the one-child policy) was instituted in 1978. Though not uniformly enforced outside of urban settings, it would fine and stigmatize families with more than one child. The goal was to alleviate economic and environmental problems. It is estimated that this policy has reduced China's population since its induction by 400 million, more than the number of people living in the United States.

QUESTIONS FOR THOUGHT

1. Have you thought about an ideal family size for you in the future? Would you be willing to limit your family size based on what you have learned in this chapter?

ACTIVITY

1. Take a minute to look back in your family history. Characterize each generation based on the demographics discussed in this chapter. What were the fertility rates of each generation?

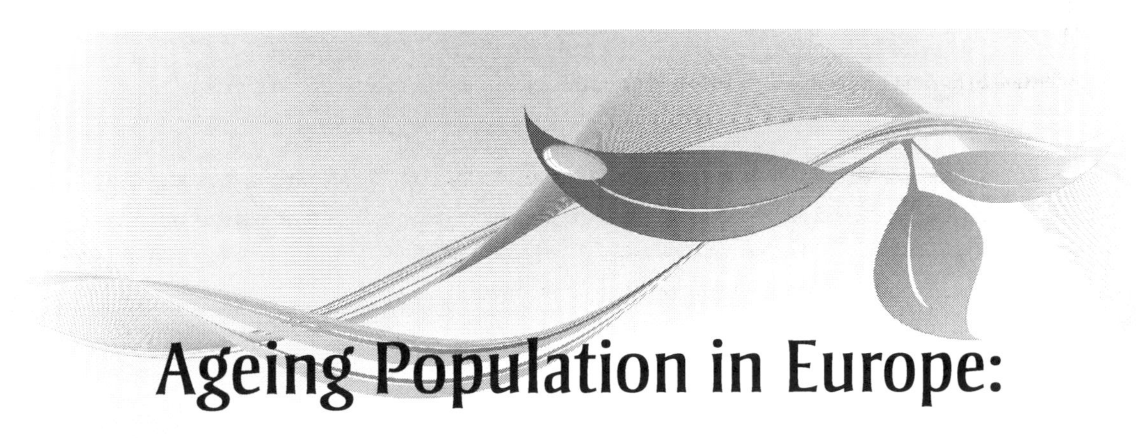

Ageing Population in Europe:

The Economic Challenges

By Souvik Dhar

"The EU will undergo significant demographic changes in coming decades, which will lead to fewer people of working age and significant pressures for increased public spending, especially on pensions and healthcare. There is a narrow, but rapidly closing window of opportunity available to prepare for the inevitably serious economic and budgetary repercussions of these developments.'[1]

—The European Commission

Europe would be faced with problems because of increase in the average age of its population. Decrease in population growth rate, as a result of lower birth rate and rise in life expectancy of its population, were cited as the causes.

Ageing population would have a negative impact on the European economy because of increased expenditure towards healthcare and pension payments, and thus would result in lower economic growth. The corporate sector would also be affected because of slowdown in growth.

To minimise the effects of the ageing population several steps were taken by the European governments. Further measures were also suggested for solving the problems. But, not all analysts were unanimous in their opinion that the proposed measures would be an answer to the problems of Europe's ageing population.

1 Brooksbank, Daniel. "Population ageing could hit euro—Commission," www.ipe.com, July 20, 2004.

CAUSES OF AN AGEING POPULATION

One of the causes of an ageing population was decline in population growth rate. Data from European Union's (EU) statistical office showed that although the rate of population growth from 1975 to 1995 was 6%, from 1995 to 2025 it was expected to decrease to 3.7% [Exhibit 1].[2] According to United Nation's (UN) projections, by 2025 Italy's population would be reduced from 57.5 million people to 45 million, Hungary from 10 million to 7.5 million, Poland from 39 million to 33 million, while Russia's population would decrease from 145 million to 100 million.[3] Population growth rate depends on the birth rate, which was decreasing in Europe.

In 2004, the average birth rate[4] for Europe was 1.4.[5] In Germany, Spain and Italy it was even lower at 1.3.[6] According to the UN, birth rate of 2.1 is considered to be healthy to maintain a stable population growth rate in a country. This is called "replacement-fertility." However, most of the European countries had a lower replacement fertility rate than required.[7]

Another cause for Europe's ageing population was an increase in life expectancy. By 2050, .one-third of the European population is expected to be 60 years plus[8] and life expectancy is expected to be greater than 80.5 years as compared to 73.2 years in 2002.[9] Increasing life expectancy, leading to a surge in old age population, would impose many kinds of problems on the European economy.

2 Geddes, Andrew. "Europe's ageing workforce," http://news.bbc.co.uk, June 20, 2002.

3 Colebatch, Tim. "Europe paying for ageing population," www.theage.com.au, April 10, 2004.

4 Number of children born per 1000 persons in a year.

5 Fennel, Tom. "The Numbers Game," www.walrusmagazine.com, October 5, 2004.

6 "Old Europe," www.economist.com, September 30, 2004.

7 "European Pension Systems Set to Collapse/Low Fertility Blamed," www.overpopulation.org, May 4, 2000.

8 "Healthy ageing," www.euractiv.com, October 22, 2004..

9 "Healthy ageing," op.cit.

EFFECTS OF EUROPE'S AGEING POPULATION

Ageing population in Europe would result in a long-term negative effect on the European economy. Ageing population demands an increase in healthcare expenditures apart from investment for pension payments and thus put significant pressure on the European state budgets. Above all, according to projections by experts, there would be a huge decline in the number of workers and children. Thus, there could be a great decrease in the "dependency ratio"[10/11] [Exhibit 2]. It was estimated that by 2050 there would be about 75 pensioners to 100 workers.[12] The UN Population Division forecasted that by 2050 the ratio of working population to those aged over 65 would reduce to 2:1 from 4:1 (in 2002) in France, 1:8 from 4:2 in Germany and 1:5 from 3:7 in Italy. To maintain the same ratio as of 2002, people would have to work till they are 74 in France, 76 in Germany and 76.5 in Italy.[13]

European economists warned that in the absence of reforms in the pension and healthcare sector, extra costs had to be incurred, which would amount to 4%–8% of GDP. Further, to meet such increased expenditure, taxes would also have to be increased ranging between 10%–20%.[14] Increase in tax rates

10 Dependency Ratio is the ratio between population aged between 15 and 64 and population above age 65.

11 "Commission says immigration cannot correct the effects of Europe's ageing population," http://europa.eu.int, May 29, 2002.

12 "Europe's population implosion," www.economist.com, July 17, 2003.

13 Equeter, Jonathan. "Can migrant workers save an ageing Europe?" www.ilo.org.

14 "Europe paying for ageing population," op.cit.

would trigger a strong public opposition, as many European countries were already burdened with high taxes. In France, the rate of pension expenditure was expected to rise faster than the growth rate of GDP from year 2005 and difference in growth rate is expected to widen from 2010.[15] In 2003, the British Government spent 9.3% of the country's GDP to maintain the living standards of the pensioners on par with the rest of the population. According to the government's actuary department, this would increase to 13.5% of GDP within the next 25 to 30 years.[16] From 2003 to 2030, the cost of maintaining the state funded pensions system was expected to increase from 11% to 16.5% of GDP in Germany, from 10% to 13.5% in France and from 13% to more than 20% in Italy.[17]

Increased expenditure by the governments towards pension payments would lead to reduced public expenditures in other sectors and therefore could slow down GDP growth rate. According to the EU, Europe's GDP growth rate could slow down from an average of 2.3% a year to 1.8% a year as a result of its ageing population.[18] Also to maintain a standard of living on par with the rest of the population, the employees would have to contribute more from their earnings towards provident fund contributions [Exhibit 3]. This would mean lesser disposable incomes and thus could hurt businesses and have a negative impact on the entire economy. Experts predicted that Europe's share of the world economy would be reduced from 22% in 2003 to 12% by 2050.[19]

With increased longevity, more spending has to be made for healthcare. Public expenditure on healthcare was expected to increase between 0.7% and 2.3% of GDP from 2000 to 2050 in the European Union (EU).[20] Another effect of Europe's ageing population could be that it could be detrimental to the Euro, EU's single currency. According to the EU, ageing of the population could lead to real-exchange rate adjustments and push up interest rates of the European countries undermining the Euro.[21] The corporate sector could also be adversely affected as a result of the ageing population.

EFFECT ON THE CORPORATE SECTOR

The biggest problem that European companies could face was the reduction of Europe's share of the world economic output.[22] Faced with a shortage in the working population, companies across Europe would have to make better use of older workers. For this purpose they would have to increase the retirement age, provide training facilities and flexible working environment. The companies would also have to invest a considerable amount to retain their skilled staff as a considerable shortfall in skilled manpower availability is forecasted.

The companies' finances would also be affected as a result of governments across countries shifting the burden of paying pension on to the companies. For example, in the UK, companies had to initiate defined-benefit pension schemes for managing their employees' pensions. However, because of the prolonged downturn in the capital markets, huge deficits had arisen in such pension schemes in 2003. According to the Confederation of British Industry, the total shortfall in British corporate pension funds was $265 billion.[23] Pension obligations had become a

15 "Ageing populations," www.economics.strath.ac.uk.

16 Elliot, Larry. "Baby boom or pensions bust," www.guardian.co.uk, December 1,2003.

17 Smith, David. "Europe—Getting Older and Slower," www.economicsuk.com.

18 Ibid.

19 Parker, Randall. "Europe Set For Prolonged Population Decline," www.futurepandit.com, May 21, 2003.

20 "Healthy ageing," op.cit.

21 "Population ageing could hit euro-Commission," op.cit.

22 "Europe Set For Prolonged Population Decline," op.cit.

23 "State pensions in Europe—The crumbling pillars of old age," www.economist.com, September 25, 2003.

heavy pull on companies' earnings. As a result many companies were switching over from a defined-benefit to defined-contribution scheme. Some of these companies were Sainsbury's, Marks and Spencer, HSBC, BT and GlaxoSmithKline.

To counter the negative impact on the corporate sector and the economy, several corrective measures were taken by the various European countries.

CORRECTIVE MEASURES ALREADY TAKEN

Since 1980, UK carried out reforms in its pension sector through systematic cuts in pension provision made by the state. The job of providing pension was transferred to the private sector. The Netherlands and the Scandinavian countries shifted much of their pension burden from the state to the employers as well as to the future individual pensioners. Countries like Bulgaria, Estonia, Hungary, Latvia, and Poland scaled down public pension schemes and started commercially managed pension schemes. In 1993, Finland introduced an additional scheme of pension contribution by the employees besides the existing one.

According to experts, Sweden was the most successful European country in reforming its pension sector.[24] In 1994 the Swedish Parliament passed a resolution for the reform of its pension sector, whereby all working Swedish citizens were allowed to invest up to 2.5% of their contributions to state related pension contributions, to funds of their choice.[25] In 1997, the Netherlands increased the pensions levy on wages from 15.4% to 18.25%.[26] Some European governments started special funds, like Fonds de Reserve pour les Retraites (FRR) by the French Government, to meet their future pension commitments.

Germany came out with a reforms package to revive its economy in 2003. The reform package was called Agenda 2010 and reform of the pension system was also one of its objectives. Three steps were suggested in Agenda 2010 for reform of the pension sector. First was to keep people employed, it suggested to stop levying taxes on contributions to pension funds. Instead, to levy taxes on the payouts that people would receive after retirement. The second was to provide incentives to those who retire late by rewarding them with an extra five percent of their pension for each year beyond the official retirement age of 65. A disincentive for those retiring early was suggested. Early retirees would lose 3.6% of their entitlement for each year before the official retirement age. The reform also suggested that no fixed entitlement would be paid to retirees. Rather each year calculations would be made to determine how much could be paid for that particular year as retirement benefits. The German Government also introduced a tax on energy use for funding pension payments and announced tax breaks for encouraging private superannuation.[27]

The German Government also took steps to encourage immigration to improve the proportion of working age population to the total population. Special work permits for immigrants called "Green Cards" were issued.

The French Government also carried out reforms in the pension sector. It increased the number of years before a worker could become entitled to receive his pension benefits. The government also over the years improved birth rate in that country by offering incentives like child allowance and progressive

24 Traynor, Ian and Henley, Jon. "Europe's ageing population revolts at longer work and lower pensions," www.ireport.net, June 11,2003.

25 "CSC Wins Contract To Support Sweden State Pension Reform," www.csc.com, October 14, 1998.

26 "State pensions in Europe—The crumbling pillars of old age," op.cit.

27 "Europe paying for ageing population," op.cit.

reduction in tax burden according to family size and composition. The government also provided extensive support to women employees.

Italy's proposed reforms included increasing the mandatory retirement age from 60 to 65 from the year 2008. Other steps, like increasing the number of years a person was required to work to be eligible for pension and also scrapping the seniority pensions whereby Italians were allowed to retire at the age of 57 if they had worked for 35 years, were suggested.[28] Italy also announced in its annual budget in 2003 a cash bonus for the birth of a second child.

In addition to the measures already taken by the various European Governments, several others were suggested by experts for solving the problem of Europe's ageing population like encouraging immigration, further reforms in the pension sector, labor market reforms, increasing the birth rate and healthcare sector reforms and outsourcing.

FURTHER MEASURES SUGGESTED FOR SOLVING THE PROBLEM

A 2002 UN report on replacement migration suggested that immigration could help solve the ageing population problem.[29] A report published by the European Commission stated that Germany would need 324000 immigrants per year till 2050 to maintain the population size as of 2002. The report also said that half a million immigrations ever year would be required to maintain the average age of its working age population as of 2002.[30]

However, not all experts are convinced that immigration alone would be able to solve the problem of ageing population. *The Social Situation Report 2002,* published by the European Commission, stated even if the rate of immigration to Europe is doubled, that alone would not be able to make any significant contribution towards solving the problem. Researchers from the Rand–Europe, said, "The sheer number of immigrants required to offset population ageing ... would be unacceptable in Europe's socio-political climate."[31] The 2002 UN report on "Replacement Migration" suggested that to maintain a balanced ratio of working to non-working population 700 million immigrant workers would be needed by 2050. In such a scenario immigrants would represent three quarters of the European population.[32] There were also negative sentiments against immigration among the Europeans. So, whether encouraging more immigration into the European countries could be a possible solution or not is being widely debated.

It was also suggested that the pension sector requires further reforms. Privatization of the pension schemes, where they would gradually shift from a state backed "pay-as-you-go" scheme to a privately run "advance funds" scheme, had been suggested as the important measure that could be adopted by all the countries across Europe. However, a study by the International Labour Organization (ILO) stated that both the schemes required the present working population to support the pensions of those who were retiring by making contributions from their incomes. The study also predicted that East European and Mediterranean countries would have to face excess fiscal burden of 0.5% to 2.5% of GDP a year for several decades as a result of such a shift.[33]

Another suggested solution was that retirees could be asked to take lesser retirement benefits

28 "State pensions in Europe—The crumbling pillars of old age," op.cit.

29 "Replacement Migration: Is it a Solution to Declining and Ageing Populations?" www.un.org.

30 Ibid.

31 "Europe's ageing workforce," op.cit.

32 "Replacement Migration: Is it a Solution to Declining and Ageing Populations?" op.cit.

33 "Privatising pensions not effective answer to ageing populations: ILO study," www.ebusiness.com, June 10, 2003.

in the future. But, besides tremendous public opposition to this it was also feared that lowering retirement benefits in the future would increase income inequality and poverty among the elderly.[34] Reforming the pension sector was also difficult from the political point of view. Proposals for pension reforms in Austria and France had led to violent protests and demonstrations against the governments.

Therefore, according to some experts, reforming the pension sector would not be a possible solution, as it does not address the problem of decrease in dependency ratio, which would make pension payments in the future very difficult to the elderly. Also, because of the huge opposition to radical reforms in the pension sector, bringing about dramatic changes would not be feasible.

Several labor market reforms were suggested such as pushing more working age population into employment. For this the overall employment rate had to be increased. The EU in the Lisbon Summit of 2000 had set itself the target of raising the overall employment rate to 70%, and 50% for 55–64 years age group till 2010. However till 2003, the rates were 65% and 41 % respectively.[35] Another solution could be to increase the retirement age to increase the size of the working population. However, experts predict that even modest increase in retirement age would be difficult for countries where there are no proper legislation to protect the elderly against discrimination in employment and low levels of employment.[36]

Increasing active participation by women in the labour force, particularly in Southern Europe, was also suggested.[37] However some experts feared that employing women would adversely affect fertility rate as many women may choose career over starting a family.[38]

One other solution suggested was increasing the birth rate. This could be achieved by encouraging marriage or cohabitation among younger couples.[39] Monetary incentives could also be provided for increasing birth rates and for having a larger family.[40] Providing improved child care facilities like creches and kindergartens, flexible working hours for mothers and generous maternal leaves would also go a long way in increasing birth rates. Biotech advances for increasing fertility rates for women in their 30s, 40s, 50s and later, was also suggested as options to increase birth rate.[41]

To enable the elderly to have more and easy access to healthcare it was suggested to provide cost-efficient ways to supply healthcare provisions. To achieve this, financial viability of the healthcare system had to be ensured.[42] To shoulder increased burden for healthcare of the elderly, compulsory health insurance for the employees and the pensioners was also proposed. Contributions to such insurance would be in proportion to gross income and hence would become smaller after retirement.[43] Allowing entry of trained immigrants, like doctors and nurses, would also arm the healthcare sector to meet increased future needs of an ageing population.[44]

To counter the problem of gradual ageing of population and the resultant shortfall in the supply of labor, outsourcing jobs to cheaper locations had also

34 "Unequal Welfare States—Distributive Consequences of Population Ageing in Six European Countries," www.ibpeurope.com, 2004.

35 "Old Europe," www.economist.com, September 30, 2004.

36 Rix, Sara E. "Raising Retirement Age: How Much of a Solution to Rising Support Burdens in the Developed World?," www.research.aarp.org.

37 "Pensions and Savings," www.pwcglobal.com.

38 "Low Fertility and Population Ageing—Causes, Consequences and Policy Options," www.rand.org, 2004.

39 Ibid.

40 Ford, Peter. "Can a graying Europe still support itself ?"www.overpopulation.org, January 21, 2000.

41 "Europe Set For Prolonged Population Decline," op.cit.

42 "Healthy ageing," op.cit.

43 Hoehn, Charlotte. "Policy Responses To Population Ageing And Population Decline In Germany," www.un.org, October 2000.

44 "After the flood," www.economist.com, September 7, 2000.

been suggested.[45] But outsourcing of jobs to other countries would face stiff opposition from locals.

It is clear that the experts are not unanimous in their opinion regarding policy option to be adopted to solve the problem of Europe's ageing population. Though various measures had been suggested, many experts are of the view that these would not be a possible solution to meet the challenge. Some experts are again of the view that no single measure could alone solve the problem. According to them, combination of different suggested measures is required. According to Anna Diamantopoulou, former Member of the European Commission in charge of Employment and Social Affairs," Immigration will help fill some gaps in our labour market but it has no impact on our basic employment policy message: we still need radical reform, with a focus on increased participation rates for women and older workers, if we are to achieve sustainable labour markets and pensions systems."[46]

45 Merrell, Caroline. "Britain warned of huge skills shortage," http://business.timesonline.co.uk, January 28, 2004.

46 "Commission says immigration cannot correct the effects of Europe's ageing population,", op.cit.

Fatherhood and Fertility

By Trude Lappegård, Marit Rønsen and Kari Skrede

This article is about men's childbearing behaviour, which is a relatively unexplored area in fertility research. Traditionally, this research has been highly gendered with a strong focus on women's childbearing. Consequently, shifting fertility trends have usually been ascribed to changes in female behaviour, while male fertility behaviour have been regarded as more or less constant (Goldscheider & Kaufman, 1996). An obvious reason why fertility research has remained highly gendered is that entry into parenthood continues to have more consequences for women than for men, as the mother is still the main caregiver. However, changing gender roles have brought more attention to fatherhood and men's role in fertility decisions and over the years more studies of female fertility have incorporated men in a couple perspective (e.g., Liefbroer & Corijn, 1999; Sorensen, 1989; Thomson and Hoem 1998; Winkler-Dworak & Toulemon, 2007). Still, analyses of male fertility behaviour *per se* are relatively uncommon, except for some recent contributions mainly from the U.S. (e.g., Guzzo & Furstenberg, 2007; Hynes, Joyner, Peters, & Delone, 2008; Manlove, Logan, Ikramullah, & Holcombe, 2008), and Europe (Martfn-Garcia, 2009; Puur, Olah, Tazi-Preve, & Dorbritz, 2008).

Another reason why analyses of male fertility are few and far between is a lack of appropriate data. So far, most analyses of fertility behaviour have been based on survey data, but some authors have questioned the quality of such data for studies of male fertility (Rendal, Clarke, Peters, Ranjit, & Verropoulou, 1999). There seems to be a tendency of underreported men's biological children, especially if the father no longer co-resides with the child (Juby & Le Bourdais, 1999). We are in a better position in this respect, as we have access to high-quality, administrative, register data on the whole population of Norway where the underreporting of men's children is very modest. Only about 1–1.5 percent of the total number of children has no registered father in our data.

Trude Lappegård, Marit Rønsen, and Kari Skrede, "Fatherhood and Fertility," *Fathering*, vol. 9, no. 1, pp. 103-120.

The focus of our study is on childlessness and multi-partner fertility—two phenomena that have been on the increase, especially among men (Lappegard, 2007; Skrede, 2005). Our point of departure is the various roles men and women play within families, and how these roles create different selection processes *into* fatherhood as well as different self-selection processes *away from* fatherhood. In countries like Norway, where the prominent provider model is the dual-earner/dual-carer family, we argue that two aspects in particular deserve attention, namely men's potentials as economic providers on the one hand (economic parenting) and their preferences and opportunities for child-caring on the other hand (practical parenting). Using register data, we do not know anything about personal preferences, nor do we have access to register information on income or occupation, but we do have detailed longitudinal information on an individual's level and field of education. Together these variables reflect both human capital resources and income potential, and the kind of job a man is likely to have in the labour market. The analysis is mainly descriptive and exploratory, that is we do not develop and test specific hypothesis, but when interpreted within a broader context as elaborated upon below, the results contribute to a broader understanding of education-related differentials in men's childbearing behaviour.

BACKGROUND

Changes in family structure in the industrialised and post-industrialised world involve unstable marriages, higher union dissolution rates, postponement of childbearing, rising trends of childlessness, and declining fertility. For instance, mean age at becoming first time fathers in Norway were 31 years in 2009, while it was 26 years in the beginning of the 1970s. Also, even if Norway is characterised as a country with highest-low fertility (Billari & Kohler, 2004), the total fertility rate has decreased from 2.25 children per women in the beginning of the 1970s to 1.98 in 2009. Decreasing fertility rates can be related to societal changes such as shifts in birth technology, female emancipation, and changes in norms and values with a greater emphasis on individualisation. One of the most important technological innovations for women in the last century was the introduction of new contraceptives which meant that women now got a genuine choice about whether and when to bear a child. At the same time, increased educational attainment and labour market participation led to greater female autonomy and more alternatives to marriage and parenthood than before. In tandem with changing gender roles and family structures new expectations towards parenthood emerged.

In much of the public debate in Nordic societies, a family model where both fathers and mothers combine income generating work and unpaid family work has been an implicit ideal (Kitterod & Kjeldstad, 2003). New norms of motherhood and fatherhood challenged the old breadwinner model with the father as the main income provider and the mother as the main caregiver, bringing on an influx of mothers into the labour market and in the public sphere, and more involvement from fathers in household tasks, in particular in childcare. New expectations to the fatherhood role have also promoted the introduction of novel legislative rights that have strengthened men's positions as fathers. The right of fathers to share most of the parental leave was, for example, introduced as early as in the mid- to late-1970s in most Nordic countries. In the 1990s, Norway and Sweden furthermore reserved four weeks of the common parental leave for the sole use of the father (the so-called "daddy-quota"), and if not taken by the father, the family will forfeit this part of the leave. These changes has led to more father involvement and strengthened fathers' position.

On the other hand, increasing divorce and union dissolution rates have led to more single living and more lone-parent households. In Norway in 2008,21 percent of marriages that have lasted for 10 years were estimated to end in divorce, given today's divorce pattern. In 1981 the same proportion was 13 percent and in 1960 5 percent (Statistics Norway, 2009a). Also, in 2008, 25 percent of all children aged 0–17 years were not living with both biological parents. Among these, 64 percent were living primarily with the mother, while 16 percent were living primarily with the father (Statistics Norway, 2009b). Since children often end up living mainly with their mother after union dissolutions, increasing divorce and union dissolutions have led to a more distant father role, which contrasts sharply with the political goal of more father involvement as described above.

In this connection, a couple of trends in male fertility arouse further interest. First, more men than women remain childless, and this gender gap has increased in younger cohorts (Lappegard, 2007; Kravdal & Rindfuss, 2008). This indicates that the threshold to become a father has become higher. Second, there is evidence of an increased propensity to have children with more than one partner, so-called multi-partner fertility, a phenomenon that has also been observed in the U.S. (see Carlson & Furstenberg, 2006; Guzzo & Furstenberg, 2007). In this article we shall look closer at these trends and their manifold associations with education, distinguishing between both level and field of education. Level of education or educational attainment is commonly used as a predictor of fertility behaviour, but recent analyses show that field of education may be at least as powerful a predictor as educational level (Hoem, Neyer, & Andersson, 2006a, 2006b; Lappegard, 2002; Lappegard & Ronsen, 2005; Martfn-Garcfa, 2009; Martfn-Garci'a & Baizan, 2006; van Bavel, 2010). Educational attainment reflects primarily human capital resources and income prospects, while field of education has been shown to be related to several aspects of men's lives such as political orientation, lifestyle and labour market outcomes (van de Werfhorst, 2004). Being closely correlated with future occupation, a man's field of education will also say something about the type of job he is likely to have, indicating for example whether it will be secure, well paid, and family friendly and flexible. Besides, the choice of educational field reflects personal preferences, which may also be related to men's attitudes to childbearing (Hoem, et al., 2006b). All these features might have important bearings on men's child-bearing behaviour.

CONCEPTUAL FRAMEWORK

An underlying assumption of our analysis is that male fertility is closely linked to men's preferences for partnership and fatherhood on the one side and their attractiveness to women as partners and potential fathers to future common children on the other. In societies with a growing dual-earner/dual-carer family structure, the opportunities for both economic parenting (breadwinning) and practical parenting (childcare) become crucial. A man's resources for economic and practical parenting will amongst others be reflected in his position in the labour market and in his work-environment. We take into account that there might be aspects that may affect the relationship between educational attainment and childbearing behaviour through the link between education and the labour market. Different features such as income prospects, job security, job flexibility and the gender composition of the job may be more or less important for a man's capacity for breadwinning and childcare, and thereby more or less important for his attractiveness as potential marriage partner and father of future common children. These characteristics are not easily observable, however, but all of them are closely associated with occupation and sector of work. As mentioned, we do not have access to information on occupation or sector of work, but use level and field of education as proxies. Below we elaborate on these associations and discuss how a man's capacity for economic and practical parenting can be related to the complex interrelationships existing between childbearing behaviour and educational attainment (level) and educational orientation (field).

The Economic Parenting Argument

Traditionally, men have a strong identity as main breadwinners and their role as fathers is embedded in their availability to support a family (Nolan, 2005). One feature that is obviously important for a man's ability to support a family is his income prospects, and previous research from the U.S. corroborates that income is an important determinant factor of multi-partner fertility among men (Carlson & Furstenberg, 2006; Guzzo & Furstenberg, 2007; Manlove, et al., 2008). In a traditional family with gender-specific division of labour, men specialize in market work and women in housework and childrearing. According to economic theory, the spouses specialize in the fields in which they have a comparative advantage and by doing so they maximize the joint utility of the household (Becker, 1981). Other theories contend that the gender division of work is determined by the gender system, constituted by common beliefs, norms and practices that define the meaning of being men and women (Mason, 2001). Within the "doing gender" theory the basic argument is that both men and women continuously construct and reconstruct their gender identity (West & Zimmermann, 1987). In couples this means that men and women "do gender" as part of dialectic process, interpreting and interacting with their partners. For men this entails undertaking activities that are seen as typical masculine tasks such as economic breadwinning, and avoiding activities with feminine connotations such as childcare.

Women's work role has changed during the last few decades, and as a result, the work-family dynamic of both men and women have changed. However, even though women increasingly contribute to the family-income, men are still the main providers in most couples and are also expected to be so. Although dual breadwinning has become an ideal and a more common family type in Norway, part-time adjustments in the labour market are still very common among mothers, and mothers continue to do most of the household work (Kitterod & Pettersen, 2006). The Norwegian family model is therefore far from gender-neutral, and the division of labour between women and men has been characterised as "gender-equality light" (Ronsen & Skrede, 2006). In order to achieve such a family model, the income prospect of the male partner still plays a crucial role. Generally, level of education is a good predictor of a man's income potential, but field of education gives additional information, as some jobs at a given level are paid better than others, for example engineering, business, finance and law.

Job security is another important feature that influences a man's prospects of supporting a family. In the Scandinavian countries, the public sector in general offers better job security than the private sector and examples of fields of educations that leads to job with high job security are teaching, health-care and protection (police and firemen). Most Norwegian men work in the private sector, while the public sector is dominated by women. In the private sector, job security will vary with the business cycle, and some sectors may be more exposed than others. If the downturn is global, jobs within the export industry will e.g., be particularly hard hit, whereas a more national-specific decline in demand also will affect other types of private sector jobs, e.g., within engineering and construction, and business and finance. Other fields of education commonly lead to jobs that score low on several dimensions, e.g. educations in arts and music that generally are associated with very low job security and also relatively poor income prospects.

The Practical Parenting Argument

Care-giving is part of the new father's role, and a man's sector of employment may also influence his opportunities to be engaged in childcare. The public sector in Norway offers better parental leave arrangements than the private sector, thereby increasing fathers' opportunities to take (longer) parental leave. Another feature that is obviously important for a man's prospects of being an active care-giver is job flexibility. Generally, the public sector is characterised as more flexible than the private sector, in the sense that there are more opportunities for part-time work. However, sometimes the public sector can be described as less flexible than the private sector as more occupations have very fixed working hours (e.g., teachers and hospital workers).

Jobs with flexible working hours give more opportunities for practical parenting than jobs with fixed working hours, for example, by enabling employees to take mornings off or staying home from work when the child is sick. However, jobs with a high degree of flexibility also entrust employees with much responsibility and encourage their active involvement in the formulation of strategies and plans for the future of the organization. This may result in work-places that have been referred to as "greedy" organisations, making high demands on their employees (Brandth & Kvande, 2002). If this implies longer hours at work, it contrasts sharply with a more compatible work/family-life balance.

As discussed above, the gender division of work is determined by the gender system, which means that the gender practises in the work-place may influence a man's desire for fatherhood and availability for care-giving. The Norwegian labour market is very sex-segregated, partly as a result of traditional choices in fields of education. The high degree of sex-segregation and a high proportion of female part-time workers have been used to explain the high share of mothers continuing in the labour market after and between childbirths (Ellingsseter & Ronsen, 1996). But as discussed above, there is no obvious coherence between a female-dominated job (in the public sector) and a work-family adaptive job. Nevertheless, female-dominated jobs tend to create workplace environments that are beneficial for both mothers and fathers of young children. If social norms of becoming a father are closely linked to his identity as a man, such norms may also be maintained in a "masculine" work environment with a large share of male workers. A "masculine" work culture may therefore also be associated with strong preferences for fatherhood.

DATA, METHODS AND CLASSIFICATIONS

Data

Our analyses are based on individual-level data extracted from the Norwegian Central Population Register, and the Norwegian Educational Database. The population-register system has a long history of full and reliable coverage of the resident populations and their vital events. Each resident has a unique identifying code, which makes it possible to link information from different data sources to each other. The population database originates from the census held in 1960 and contains longitudinal information on each date of recorded childbirth of every person who has ever lived in the country since then, including the personal identification number of the mother and the father of the respective child. For each childbirth, we are therefore able to link the father to the mother to determine whether the respective birth is with the same or with a new partner. Individuals who died or emigrated (without a subsequent re-entry) prior to 1960 do not appear in our calculations. This means that the fertility rates for the oldest cohorts have been computed conditional on survival and non-migration until the census year. Earlier investigations have shown that this effect is negligible (Andersson & Sobelev, 2001; Brunborg & Kravdal, 1986). We have access to fertility histories up to 2007. Individual data on childbearing histories have been linked to individual data on educational histories. These data originate from the Population Census held in 1970, and have thereafter been updated annually from 1974. The information we have access to include education up to 2005.

Methods

Our study is based on original male birth cohorts, i.e., we observe the birth histories of men born in the country and calculate cohort fertility measures from age-specific parity-progression rates cumulated over their life course (ages 15–59). The present analysis is based on native male cohorts born in 1935 or later. Age is defined as age by the end of a calendar year (calendar year minus birth year). Men who die or emigrate before age 59 are censored at the time of death or emigration.

In the analyses we condition on the educational level attained at age 30, when most men have finished their educational activity, and study the cumulated fertility outcome beyond that age. In this way we avoid most of the common problem of seeking to explain fertility behaviour at a certain age by the educational level reported and possibly attained at a later stage, which is a form of anticipatory analysis that can produce misleading results on the interrelationship between education and fertility (Hoem & Kreyenfeld, 2006a, 2006b). Since there is no information on educational attainment before 1970, cohorts born before 1940 are excluded from the analyses when studying the association with education. People with missing information on educational attainment have been excluded from our analyses, but they constitute a very small group (less than 1%).

Classification

To illuminate important contrasts in men's fertility behaviour, we have constructed several groups based on level and field of education that are meant to capture the various dimensions of a man's capacity for economic and practical parenting. Field and level of education are classified using the Norwegian standard classification of education (Statistics Norway, 2001). We use a recent version of the standard where the levels of education have been revised to be more compatible with international standards (see http://www.ssb.no/utniv_en/) and distinguish between the following four levels: (i) primary and lower secondary (10 years of compulsory schooling, labelled "Primary" in figures), (ii) upper secondary and post-secondary, non-tertiary (11–13 years of schooling, labelled "Secondary" in figures), (iii) lower tertiary (some college or university, up to and including a bachelor's degree, 14–17 years of schooling, labelled "University I"), and higher tertiary university II (all college or university education taking 5 years or more, i.e., 18 years or more of schooling, labelled "University II"). When fields of education are concerned, we have constructed groups that are meant to reflect differences in labour market prospects and work-place environments as discussed above. Since primary and lower and upper secondary education mainly are general programmes without specific vocational directions, we do not subdivide these levels further into fields of education, and to avoid too small groups, we collapse all post-secondary and tertiary level education before splitting into fields of education. The resulting cross-tabulation of level and field of education are as follows:

The group *Humanities and Arts* captures both degrees that lead to no obvious set of occupations, e.g. general language skills, and degrees where there is a clearer link between the education and set of occupational outcomes, e.g. theology and musicians.

In general the group can be characterized as fields of education that lead to occupations with low job security and low income prospects, i.e., fields of education with no clear job prospects or occupations that are more loosely connected to the labour market than others (maybe with the exception of theology). The group *Teaching, Health and Welfare* captures fields of education that in general lead to occupations within the public sector with good opportunities for both economic- and practical parenting. The group *Social science and Journalism* captures fields of education with both employment possibilities in the public sector, e.g., bureaucracy, and the private sector, e.g., media. The groups *Business, Finance and Law, Science and Computing*, and *Engineering and Construction* capture fields of education that lead to occupations with high income prospects and thereby high provider ability. In general they can also be described as high-flexibility jobs, in the sense of flexible hours, but they vary in job-security as some occupations are more exposed to business cycle fluctuations than others. The *Agriculture* group captures fields of education that lead to occupations within farming, fishing and forestry. For many of these occupations the income prospects may vary due to changing crops and harvests, but for many men within these occupations, the choice of life-style is probably more important than positions and income in the labour market. The agricultural population is also characterised by more traditional family forms and a closer attachment to their place of origin than people in general (Jervell, 2002). The last group *Sports, Transports and Protections* captures fields of education that generally lead to male-dominated occupations in a "masculine" work environment. Occupations within the police and the military are further in the public sector with good job security and ample opportunities for economic and practical parenting.

There has been an increase in the proportion of men that complete higher levels of education among the cohorts included in our analysis (Figure 1). From the cohorts 1940–44 to the cohorts 1960–62 the proportion with higher education increased from 21 to 27 percent. This is mostly due to a higher proportion with lower tertiary education. The higher proportion with primary education in the youngest group is mainly an artefact of the new standard of education, however. Because of several changes in the school system since the early 1970s, the new standard assign courses to different categories depending on the calendar period in which they were completed, and if completed after the mid-1970s, short courses at the secondary level have been assigned to the primary level.

The composition of fields of education at the post-secondary and tertiary level has also changed (Figure 2). In particular, there has been an increase in the proportion of men within business, finance and law, while there has been a decrease in the proportion of men within teaching, health and welfare.

LEVEL OF EDUCATION	FIELD OF EDUCATION	DETAILS OF FIELD OF EDUCATION
Primary and lower secondary		
Upper secondary		Language skills, theology, musicians, actors
Post-secondary and tertiary	*Humanities, Arts Teaching, Health,*	Teaching, medicine, dentists, social work
	Welfare Social science, Journalism	Social science, journalism & information
	Business. Finance, Law	Business & administration, finance, *banking, management, law*
	Science, Computing Engineering, Construction Agriculture	*Biology, physics, computing Mechanics, electricity, construction Farming, fishing, forestry*
	Sports, Transport, Protection	*Sports, post, military, police, firemen*

RESULTS

The Overall Picture

As has been reported elsewhere for selected Norwegian male cohorts (Skrede, 2005) or groups of cohorts (Kravdal & Rindfuss, 2008), the proportion childless was lower for men born in the mid and late 1940s than for men born in the beginning of that decade. Our calculations of completed fertility at ages 40, 45 and 50 years for single-year male cohorts born 1935 to 1967 show that childlessness was at its lowest among men born in 1943 with a fairly rapid decrease over cohorts born a few years before and a renewed increase over cohorts born later (Figure 3). Evidently, some men wait a long time before they become fathers, as is reflected in the reduction in the proportion childless from age 40 to age 50, and this pattern seems to be getting more pronounced in the younger cohorts. However, to be able to include those born in the early 1960s in our analyses, we shall mainly focus on completed fertility at age 45 in the following.

It is worth noticing that childlessness among men has accelerated in the younger cohorts. From a fairly low level of 13.3 percent in the 1943 cohort, the proportion with no children rose to s15.9 percent among those born 10 years later and to 19 percent in the 1962 cohort. Judged by the observed childlessness in the 1967-cohort at age 40, this trends seems to continue, as 22.3 percent of them had no biological children, while the corresponding proportion among men born just five years before was 21.3 percent.

Turning to multi-partner fertility (Figure 4), we notice that this phenomenon has increased continuously across our cohorts. At age 45, the proportion of men who had children with more than one partner had risen from less than 4 percent in cohorts born before the Second World War to about 11 percent in cohorts born in the early 1960s. Calculated as percentage of those who had become fathers, multi-partner fertility rose from about 5 percent in the oldest cohorts to about 13 percent in the youngest cohorts.

Contrasts by Educational Level

A well-established finding from studies of female cohort fertility in most countries is that women with short education have lower childlessness and more children than women with longer education. For men, we see the opposite pattern: In all cohorts from the early 1940s to the early 1960s the highest proportion with no children is found among men with compulsory schooling only (primary and lower secondary level) and the lowest proportion among men with a postgraduate university or college degree (higher tertiary level) (Figure 5). At age 45, 22.1 percent of men with compulsory schooling and 13.2 percent of men with a postgraduate degree were childless in the youngest cohort (1960–62). Traditionally, high levels of education have been linked to high income prospects and good provider abilities. The persistent differences in childlessness by educational level therefore suggest that provider ability is still an important determinant of men's reproductive behaviour.

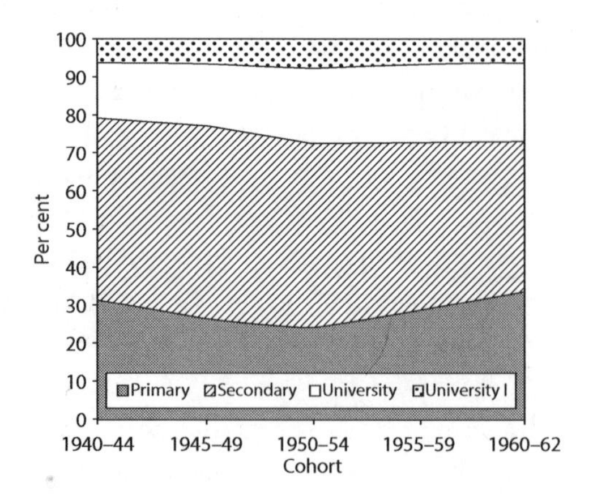

Figure 1. Trends in level of education by cohort, men.

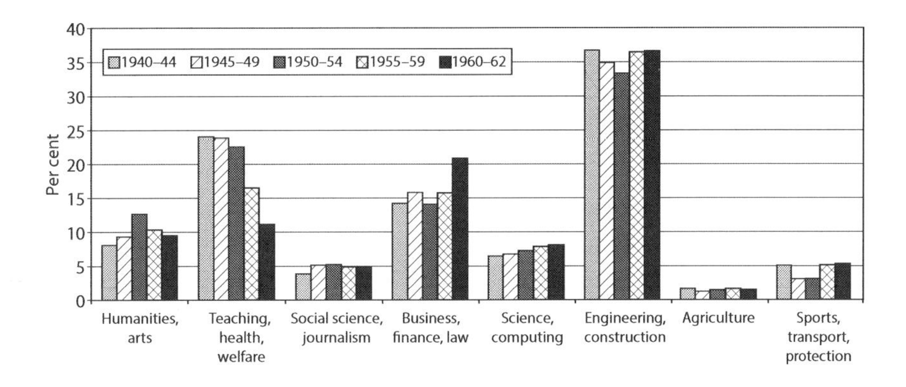

Figure 2. Trends in field of education for men with post-secondary and tertiary education by cohort.

There has been an increase in childlessness in all educational groups, also among those with high education. In fact, childlessness has increased most among men with a lower tertiary education, and least in the group with compulsory education. In the former group the proportion with no children rose from 9.8 percent in the 1940–44 cohort to 16.9 percent in the 1960–62 cohort, while the proportion in the latter group increased from 18.9 to 22.1 percent. However, it is important to underline that educations at the same level may lead to a variety of jobs in different segments of the labour market with different opportunities for economic and practical parenting. In order to get better insight into the reproductive behaviour of men we therefore also need to study variations due to field of education, which we return to shortly.

In spite of the fact that men with low education are the most likely to remain childlessness, multi-partner fertility is more widespread in this group than in the other educational groups. At age 45, about 15 percent of all men in the 1960–62 cohort with a compulsory education had had children with more than one woman, compared to about 5 percent among men with a tertiary degree. If looking at fathers only (Figure 6), the pattern becomes even more pronounced. At the lowest educational level, 19.3 percent of those who had become fathers, had children with more than one woman, compared to 6.1 percent of those at the highest educational level. In the following we shall stick to fathers only, since this does not change the main pattern, but merely enhances the contrasts.

Like childlessness, multi-partner fertility has increased across cohorts, but unlike childlessness it has increased more among men with lower education than among those with higher education. From the 1940–44 cohort to the 1960–62 cohort the proportion of fathers who had children with more than one woman more than doubled (from 8.9% to 19.3%) in the compulsory schooling group, while it only rose by about 30% in the highest tertiary group, from 4.7 to

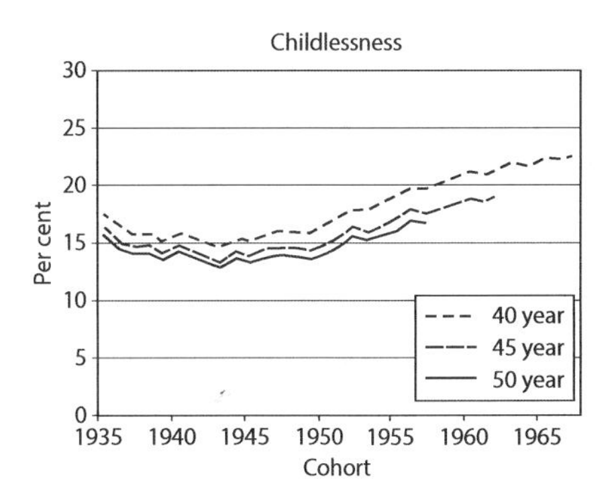

Figure 3. Childlessness at age 40, 45 and 50 years by cohort, men.

6.1 percent. There has also been almost a doubling of multi-partner fertility in the upper secondary groups (from 6.5% to 12-13%) and about a 70 percent increase in the lowest tertiary group, from 4.7 to 8 percent.

Multi-partner fertility is obviously closely linked to marital and non-marital union dissolution. A common finding from the Nordic countries is that there is an inverse relationship between educational attainment and union dissolution: the lower the education of either partner, the higher the break-up rates (Hoem, 1997; Jalovaara, 2003; Lyng- stad, 2004). This gradient is clearly reflected in the multi-partner fertility pattern reported above. However, when considering both childlessness and multi-partner fertility together, we would like to stress the more bifurcated pattern of the lower educated group: While more than 20 percent never become fathers, those who do so are much more likely than higher educated men to have children with more than one woman. As mentioned earlier, the majority of children end up living without their father in the household after union dissolution, and therefore it has been argued that it is important to better understand the factors associated with multi-partner fertility among men in particular (Manlove et al., 2008). The growing trend towards increasing multi-partner fertility in the lowest education group is an indication that the family formation and dissolution processes among men have become more selective, and the low-educated group may be more heterogeneous than the other educational groups. This raises important questions about men's capacity for economic parenting and the implication for children's outcomes. In a study of multi-partner fertility in the U.S. it is argued that to the extent that childrearing across households diminishes parental resources, multi-partner fertility can have important negative consequences for children's well-being (Carlson & Furstenberg, 2006). Our results indicate that the consequences may be particularly grave if the fathers have low education. One potentially confounding factor that we have not been able to control for so far is income differentials, and this is an obvious task for future research. However, from an earlier analysis based on Norwegian Tax Register data we know that the income differences between men living with and providing for children (own or stepchildren) and men not living with children were larger among men with only compulsory education than for among men with longer education (Skrede, 2002). This indicates that there is a stronger selection by income into co-resident fatherhood among men with only compulsory education, but we also suspect that there is a lot of remaining unobserved heterogeneity in this group.

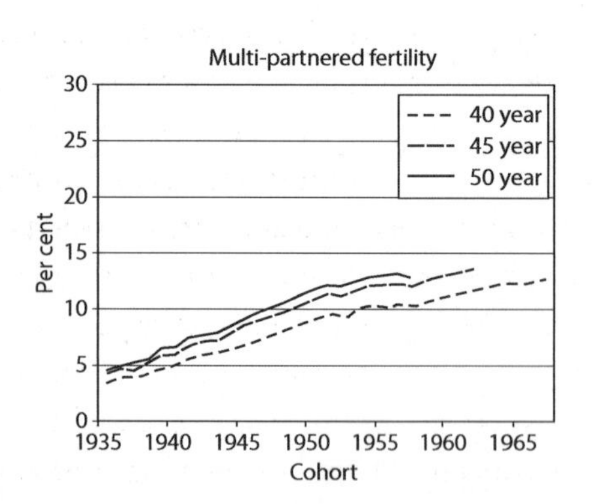

Figure 4. Multi-partner fertility at age 40, 45 and 50 years by cohort, men

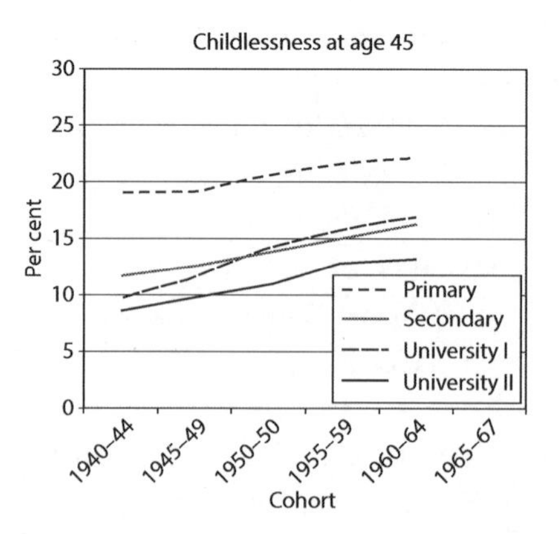

Figure 5. Childlessness at age 45 by level of education and cohort, men.

Contrasts by Field of Education

The childlessness pattern of men within different levels and fields of education is displayed in Figure 7. Using register data with such a vast number of observations, most of these differences are significant both within and between cohorts. The highest childlessness of all groups is found among men with an education in humanities and arts (e.g., language skills, music and performing arts, theology etc.). In cohorts born since the mid-1950s the proportion with no children at age 45 is approaching 25 percent, which is even higher than in the group with compulsory schooling only. Other fields of education with relatively high male childlessness are social science and journalism, and science and computing. On the other end of the scale, we have fields of education like agriculture, sports, transport and protection and partly also teaching, health and welfare, which have childlessness proportions ranging from 10.5 to 13 percent in the youngest cohorts. This is even lower than among men with high tertiary education in general (ref. last section). Most of the differences within the cohorts are significant, especially within the youngest cohorts. There are however no significant differences between men within social science and journalism and men within science and computing, and between men within business and finances and men within engineering and agriculture in the youngest cohort. In line with the results for level of education, we observe a rising trend of childlessness across cohorts, and the increase has been particularly large for humanities and arts, and science and computing. This trend is significant across the cohorts for more or less all groups. However, among men within business and finances, and within social science there are no significant differences between the 1950–54 cohort and the two youngest cohorts, indicating a stable level of childlessness among men within these two fields of education in the younger cohorts. It is interesting to observe that the rising trend across cohorts seems to have been broken for two groups, namely for men within sports, transport and protection and for men within agriculture, although the difference between the 1955–59 cohort and 1960–62 cohort is not significant for the agricultural group. A changing composition of the groups may also have contributed to the observed pattern, as closer investigations reveal that there are fewer men within transport and more men within protection in the younger than in older cohorts, and in the agricultural field there has been a switch from farming to fishing, mainly because of growing job opportunities in the expanding fish faming industry.

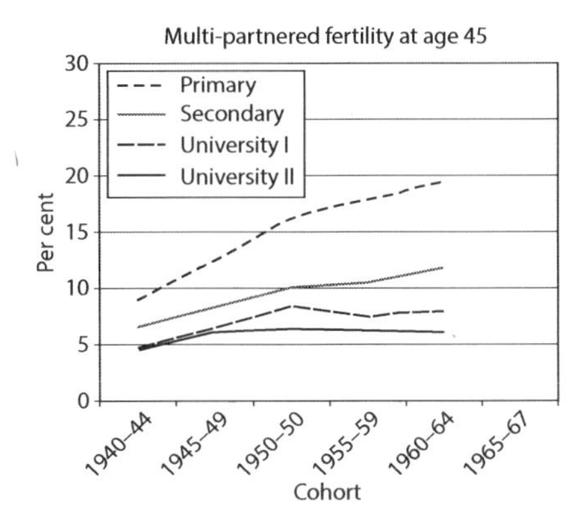

Figure 6. Multi-partner fertility at age 45 by level of education and cohort, men.

When multi-partner fertility is concerned we saw that the general trend was increasing (Figure 4). However, from the 1950–54 cohort and onwards there are few significant differences across cohorts within fields of education (Figure 8). The differences between fields of education within cohorts hold for all cohorts though. One point worth noticing is that the behaviour of the groups with lowest childlessness, the sports, protection and transport field, and the agricultural field, is quite opposite. Whereas the former group has the very highest proportion of fathers who have children with more than one woman, the latter group has the very lowest proportion. In fact, none of the educational level groups discussed above have a multi-partner fertility that is as low as within the agricultural field, and the sports, transport and protection fields have a proportion that is on par with fathers with an upper-secondary education. Obviously, we here have an example of two groups with very different family formation and family dissolution patterns and practices. As previously discussed, multi-partner fertility is closely linked to marital and non-marital dissolution. Unfortunately we are not aware of any studies relating union dissolution to different fields of education, so we have no evidence of lower break-up rates in the agriculture group than in the sports, transport and protection group, but it is likely that this is a confounding factor. In the latter group, multi-partner fertility has increased substantially in the youngest cohort. As already mentioned, there has been a switch in the composition of this group with a larger proportion belonging to fields within protection, e.g., police, firemen. The majority of the men in this group are in male-dominated jobs with a "masculine" work environment. As argued initially, social norms of fatherhood may be strong in such environments, and closely linked to men's identity as men. Besides these occupations also mainly belong to the public sector, and are characterised by good job security and fairly family friendly work schedules. Thus there are good opportunities for both economic and practical parenting.

Our analysis of childlessness and multi-partner fertility among men in Norway demonstrates that education influences men's childbearing behaviour in multiple ways.

In contrast to the well-documented positive relationship between educational level and childlessness among women, childlessness among men is most pronounced among those with low education and least pronounced among those with high education. This is in line with economic theories suggesting that a man with higher earnings power (education) is potentially more able to support a family and therefore more attractive as a partner and father to a future child. But at a given educational level, we also observe contrasting behaviour between men within different fields of education. These contrasts have become more pronounced over time, and may be related to at least three factors.

First, provider availability of the male partner still seems to be crucial among couples, and this is reflected in his labour market position and workplace environment. Different positions in the labour market give different opportunities for economic parenting. Since job security and income prospects are important ingredients in provider availability, we expected two groups in particular to be more likely to become fathers than others. Due to better job security, the first group would be men with an education leading to work in the public sector, and due to higher income potential, the second group would be men with an education within engineering, business, finance and law. Both groups turn out to be at the very low end of the childlessness scale, which indicates that provider availability is still a determining element in men's reproductive behaviour. During the last decades, the labour market has become more competitive, and this might explain why the fertility behaviour of men at the same educational level, but with different fields of education, has become more divergent.

Second, during the last decades, more women participate in the labour market, also when they have small children, and the compatibility between family and work has become crucial, for women as well as for men. Different positions in the labour market also give different opportunities for practical parenting. Generally, the public sector offers better arrangements for childcare, e.g., better parental leave benefits, and therefore we expected to find lower childlessness among men in the public sector. Furthermore, men's gender role attitudes can be reflected in the gender composition of the job and influence their desire for economic parenting and childcare. Female-dominated work-places may create environments that are beneficial for parents of young children, whereas masculine work-places may

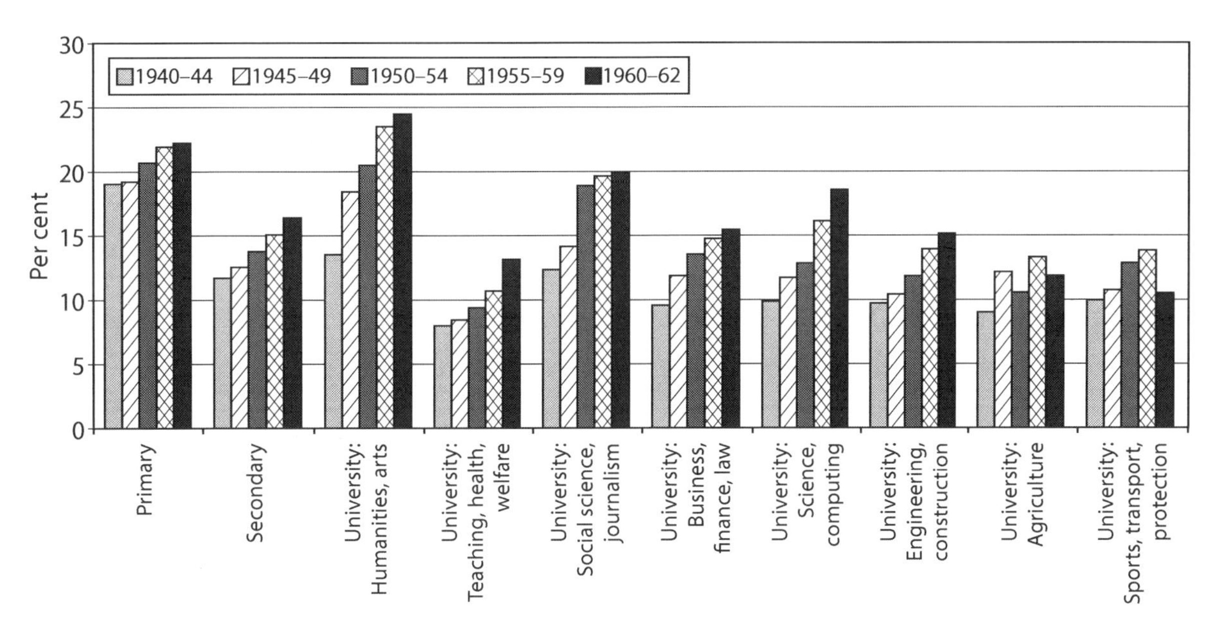

Figure 7. Childlessness at age 45 by level and field of education and cohort, men.

create environments where fatherhood is a strong social norm. The fact that the lowest childlessness proportions were found among men with educations for respectively the agricultural- and the transport- and protection-sectors confirms that social norms play a part, as the former sector is characterised by strong traditions and family-orientation, and the latter by a distinct masculine work environment.

The educational pattern of multi-partner fertility is different from childlessness, as the propensity to have children with more than one woman is most pronounced among those with low education. Becoming a father is thus more of a selective process for men with low education than for men with higher education, but having become fathers, low-educated men are much more likely to have another child with a new partner. Obviously, multi-partner fertility is closely linked to union dissolution, but we should underline that some of these men have never been in a stable relationship with the mother (Skrede, 2005). This has grave implication both for the children and the fathers themselves. Similar to childlessness, there is much variation across fields of education in multi-partner fertility. Interestingly, one of the groups with the lowest proportion of childlessness, men within transport and protection, have the highest proportion of multi-partner fertility. These fields of education mainly lead to public sector jobs with good opportunities for both economic and practical parenting. Furthermore, these jobs are usually in "masculine" work environments where fatherhood is a strong social norm and closely linked to their identity as men.

The contrasting outcomes across fields of education suggest that the underlying processes behind both childlessness and multi-partner fertility are similar, depending on the one side on men's preferences for partnership and fatherhood and on the other side on their attractiveness to women as partners and potential fathers to future common children. Conditional on their work- and family-life strategies, some women may have stronger preferences for a main provider, while others may have stronger preferences for a co-childcarer. In order to get a better understanding of educational differentials in men's childbearing behaviour (as well as women's)

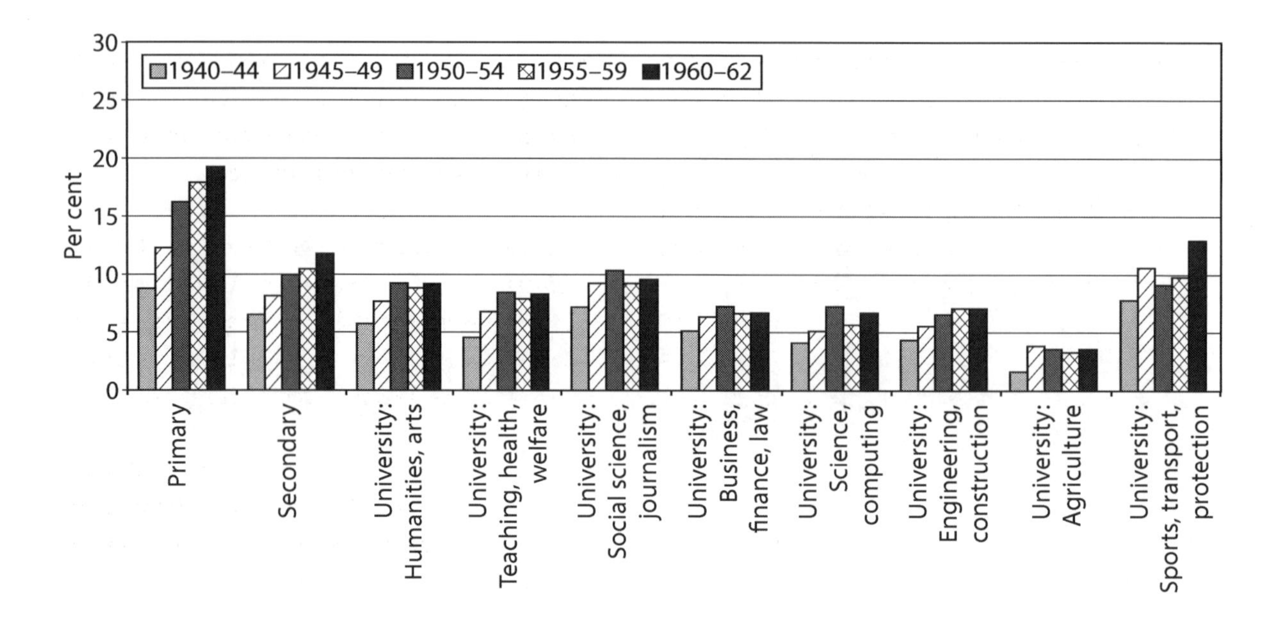

Figure 8. Multi-partner fertility at age 45 by level and field of education and cohort, men.

we would need data on couples and explore fertility outcomes among couples with different combinations of educational level and field.

This analysis has two main limitations. First, using register data we only have access to observable behaviour and no information about the men's attitudes and preferences towards fatherhood and economic and practical parenting. In order to get better insights into how these mechanisms are influencing the processes into as well as away from fatherhood we need data that illuminate more of the factors that determine men's fertility behaviour. The second limitation is linked to our use of field of education as a proxy of the type of job a man is likely to hold in the labour market. Even if there is likely to be a large correspondence between field of education and occupation for the majority of our population extract, some will have ended up in other jobs than they are educated for. Moreover, field of education tells us little about the occupations of men with primary or secondary education, as these are mainly general programmes with no job-specificity. For a more comprehensive analysis of the direct relationship between men's position in the labour market and their capacity for economic parenting and practical parenting we would therefore need data on occupation, as well as information about income.

REFERENCES

Andersson, G. and B. Sobelev. (2001). Small effects of selective migration and selective survival in retrospective studies of fertility. In MPIDR Working Paper WP 2001-031. Rostock: Max Planck Institute for Demographic Research. Becker, G. (1981). *A treatise on the family.* Cambridge: Harvard University Press. Billari, F.C., & Kohler, H-P. (2004). Patterns of low and lowest-low fertility in Europe. *Population Studies* 58(2), 161–176.

Brandth, B. and E. Kvande. (2002). Reflexive fathers: Negotiating parental leave and Working life. *Gender, Work and Organization* 9(2), 186–203.

Brunborg, H. and O. Kravdal. (1986). *Fertility by birth order in Norway.* A register based analysis. In Report 86/27. Oslo: Statistics Norway.

Carlson, M.J. and F.F. Furstenberg. (2006). The prevalence and correlates of multipartnered fertility among urban U.S. parents. *Journal of Marriage and the Family* 68,18–732.

Ellingsffiter, A.L. and M. Ronsen. (1996). The dual strategy: Motherhood and the work contract in Scandinavia. *European Journal of Population* 12, 239–260.

Goldscheider, F. and G. Kaufman. (1996). Fertility and commitment: Bringing men back in. *Population and Development Review,* 22 (Supplement), 87–99.

Guzzo, K.B. and F.F. Furstenberg. (2007). Multipartnered fertility among American men. *Demography* 44(3), 583–601.

Hoem, J.M. (1997). Educational gradients in divorce risks in Sweden in recent decades. *Population Studies* 51, 19–27.

Hoem, J.M. and M. Kreyenfeld. (2006a). Anticipatory analysis and its alternatives in life-course research. Part 1: The role of education in the study of first childbearing. *Demographic Research 15(16),* 461–484.

Hoem, J.M. and M. Kreyenfeld. (2006b). Anticipatory analysis and its alternatives in life-course research. Part 2: Two interacting processes. *Demographic Research* 15(11), 485–498.

Hoem, J.M., G. Neyer, and G. Andersson. (2006a). Educational attainment and ultimate fertility among Swedish women born in 1955–59. *Demographic Research* 14, 381-403.

Hoem, J.M., G. Neyer, and G. Andersson. (2006b). Education and childlessness: The relationship between educational field, educational level, and childlessness among Swedish women born in 1955-59. *Demographic Research* 14, 331–380.

Hynes, K., K. Joyner, H.E. Peters, and F.Y. Delone. (2008). The transition to early fatherhood: National estimates based on multiple surveys. *Demographic Research* 18(12), 337–376.

Jalovaara, M. (2003). The joint effects of marriage partners' socioeconomic positions on the risk of divorce. *Demography* 40, 67–81.

Jervell, A.M. (2002). Tradisjon og forandring - generasjonsskifte som rekruttering til landsbruk [Tradition and change—generational change as recruitment to agriculture]. In Rodseth, T. (ed.), *Landbruket ved en korsvei [Farming at a crossroads]* (pp. 91–106). Bergen: Fagbokforlaget.

Juby, H. and C. Le Bourdais. (1999). Where have all the children gone?—Comparing mothers' and fathers' declarations in retrospective surveys. *Canadian Studies in Population* 26, 1–20.

Kitterod, R.H. and R. Kjeldstad. (2003). A new father's role? Employment pattern among Norwegian fathers 1991–2001. *Economic Survey* 1, 39–51.

Kitterod, R.H. and S.V. Pettersen. (2006). Making up for mothers' employed working hours: housework and childcare among Norwegian fathers. *Work Employment and Society* 20, 473492.

Kravdal, O. and R.R. Rindfuss. (2008). Changing relationship between education and fertility–a study of women and men born 1960–64. *American Sociological Review* 73, 854–873.

Lappegård, T. (2002). Educational attainment and fertility patterns among Norwegian mothers. Document 2002/18. *Statistics Norway*, Oslo.

Lappegåard, T. and M. Rønsen. (2005). The Multifaceted impact of education on entry into motherhood. *European Journal of Population* 21, 31–49.

Lappegård, T. (2007). Sosiologiske forklaringer pa fruktbarhetsendring i Norge i nyere tid [Sociological explanations on fertility changes in contemporary Norway]. *Sosiologisk tidsskrift* 15, 55–71.

Liefbroer, A.C. and M. Corijn. (1999). Who, what, where and when? Specifying the impact of educational attainment and labour force participation on family formation. *European Journal of Population* 15, 45–75.

Lyngstad, T.H. (2004). The impact of parents' and spouses' education on divorce rates in Norway. *Demographic Research*, 10, 122–142.

Manlove, J., C. Logan, E. Ikramullah, and E. Holcombe. (2008). Factors associated with multiple-partner fertility among fathers. *Journal of Marriage and the Family* 70, 536–548.

Martfn-Garcfa, T. (2009). Bring men back in: A re-examination of the impact of type of education and educational enrolment on first births in Spain. *European Sociological Review* 2, 199213.

Martfn-Garcfa, T.and P. Baizan. (2006). The impact of type of education and of educational enrolment on first births. *European Sociological Review* 22(3), 259–275.

Mason, K.O. (2001). Gender and family system in the fertility transition. In R.A. Bulatao & J.B. Casterline (eds.), *Global fertility transition.* Supplement to Population and Development Review (pp. 160–176). New York: Population Council.

Nolan, J. (2005). Job insecurity, gender and work orientation: An exploratory study of breadwinning and care-giving identity. *GeNet Working Paper no. 6.* University of Cambridge.

Puur, A., L.S. Olah, M.I. Tazi-Preve, and J. Dorbritz. (2008). Men's childbearing desires and views of the male role in Europe at the dawn of the 21st century. *Demographic Research* 79(56), 1883–1912.

Rendal, M., L. Clarke, H.E. Peters, N. Ranjit, and G. Verropoulou. (1999). Incomplete reporting of men's fertility in the United States and Britain: A research note. *Demography* 36(1), 135-–44.

Rønsen, M. and K. Skrede. (2006). Nordic fertility patterns: compatible with gender equality? In A.-L Ellingsæter and A. Leira (eds.). *Politicising parenthood: Gender relations in Scandinavian welfare state restructuring* (pp. 53–76). Bristol: Policy Press.

Skrede, K. (2002). Towards gender equality in Norway's young generations? *Scandinavian Population Studies* 13, 191–218.

Skrede, K. (2005). Foreldreskap i forandring—fasrre menn blir fedre [Change is parenthood—fewer men becomes fathers], *Tidsskrift for kjfinnsforskning* 2,6–22.

Sorensen, A.M. (1989). Husbands' and wives' characteristics and fertility decisions: A diagonal mobility model. *Demography* 26(1), 125–135.

Statistics Norway. (2001). Norwegian standard classification of education. Official Statistics of Norway.

"Statistics Norway," (2009a,). www.ssb.no/ekteskap/arkiv/tab-2009-08-27-ll.html.

"Statistics Norway," (2009b), www.ssb.no/barn/arkiv/tab-2009-04-30-01 .html.

Thomson, E. and J.M. Hoem. (1998). Couple childbearing plans and births in Sweden. *Demography* 35(3), 315–322.

Van Bavel, J. (2010). Choice of study discipline and the postponement of motherhood in Europe: The impact of expected earnings, gender composition, and family attitudes. *Demography* 47(2), 439–458.

van de Werfhorst, H.G. (2004). Systems of eudcational specialization and labor market outcomes in Norway, Australia, and the Netherlands. *International Journal of Comparative Sociology* 45, 315–235.

West, C. and D.H. Zimmermann. (1987). Doing gender. *Gender & Society* 1(2), 125–151.

Winkler-Dworak, M. and L. Toulemon. (2007). Gender differences in the transition to adulthood in France: Is there convergence over recent period?. *European Journal of Population* 23,273–314.

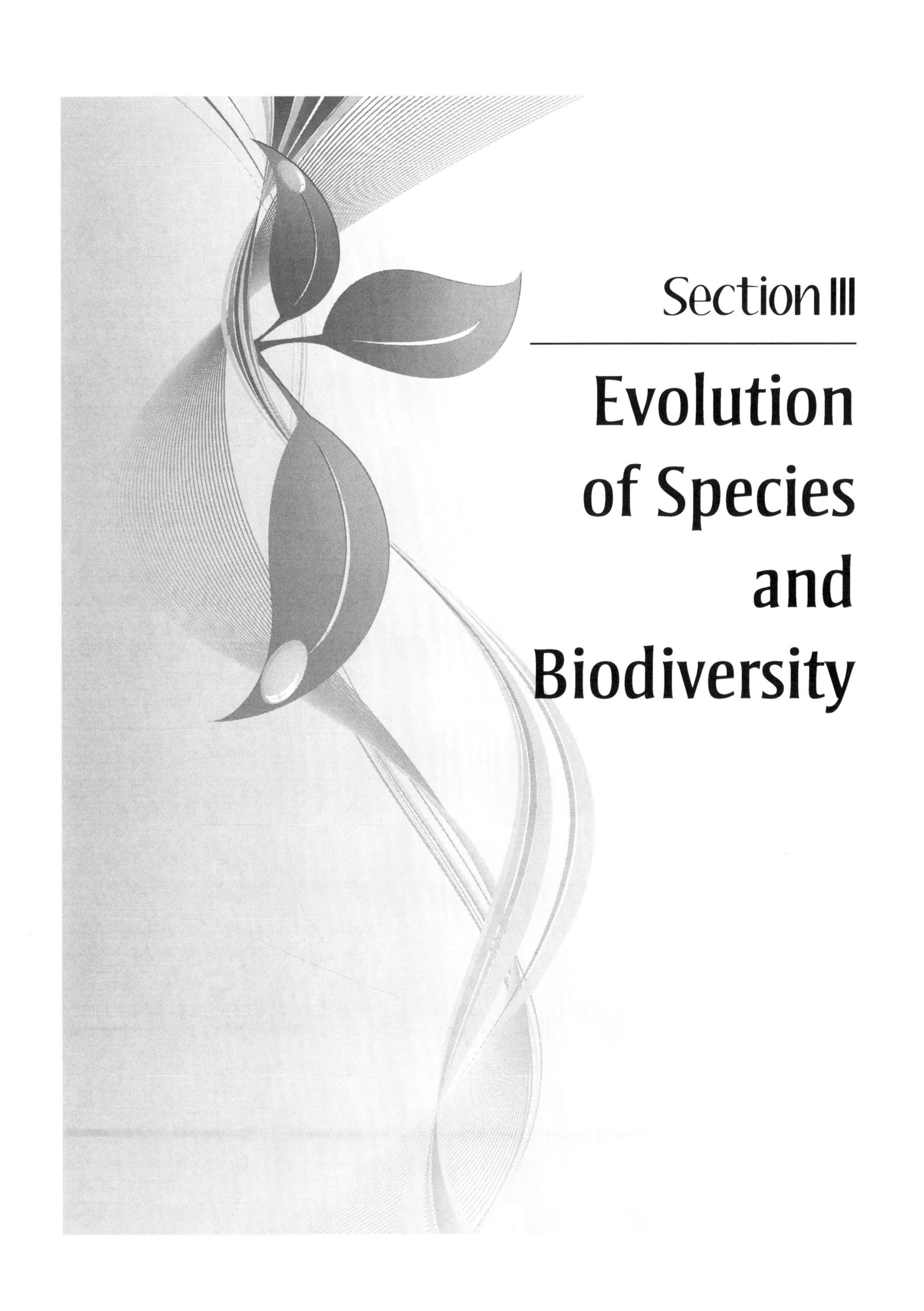

Section III

Evolution of Species and Biodiversity

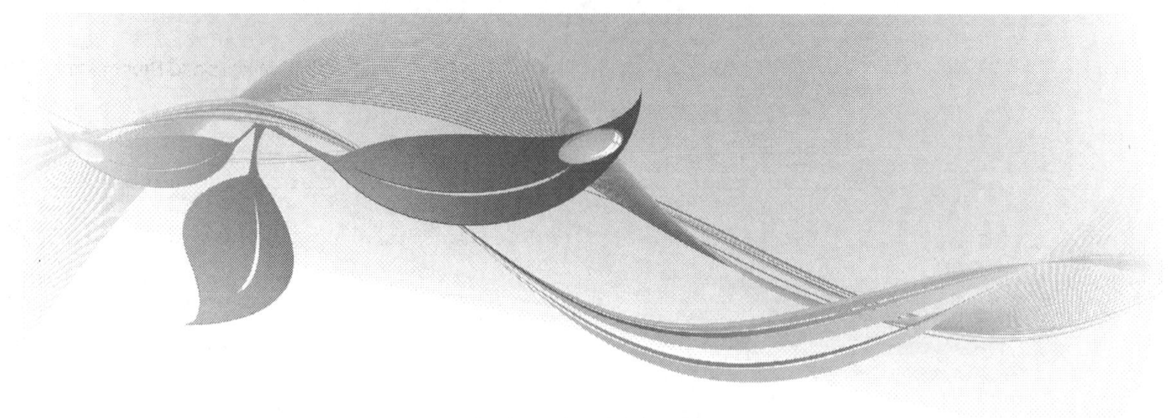

By Anne Marie Zimeri

Understanding the conditions that are best for each species can help us understand the balance among members of an ecosystem or a food web. But how did species adapt to those ideals? In this chapter we will discuss the concept of evolution and how species interactions affect evolution.

For most species on the planet, there is a set of ideal conditions in which a particular species can live, and straying from that ideal can physiologically stress members of the population until a point is reached where no member of that species can survive. When discussing the range of an environmental factor that affects a species' ability to survive, it is common to do so in terms of "tolerance limits" (Figure 3.1). For example, pocket gophers in Colorado can be found in abundance in deep, light soils. Using soil condition as an environmental factor, as we stray from the optimum soil condition, we will see fewer members of the population. At some point, the soil condition will be so far from the optimum that no pocket gophers will be able to survive. This is the zone of intolerance. Other examples of environmental factors that affect tolerance limits of species include temperature, pH, moisture, and oxygen availability. While it is important to understand tolerance limits to understand species distribution, tolerance limits can also be exploited in

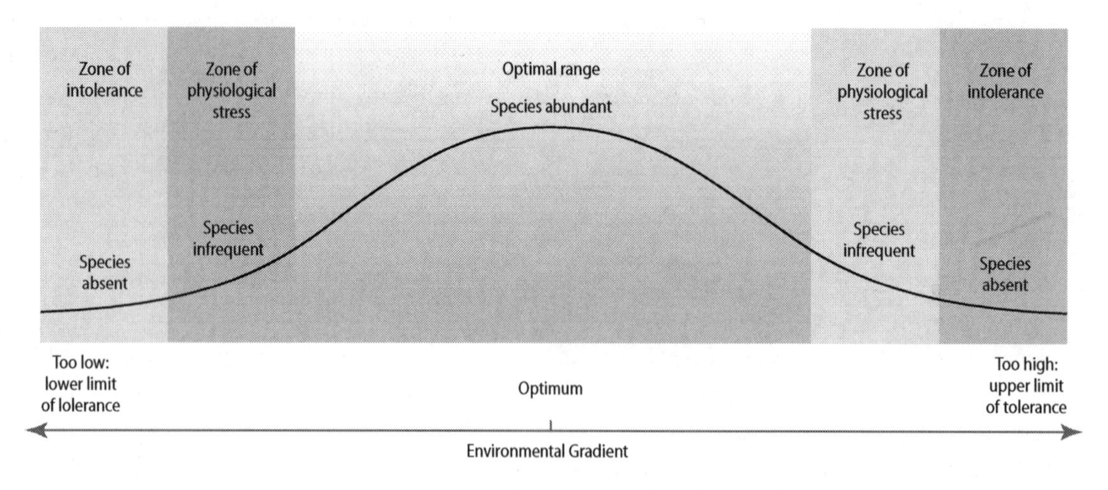

Figure 3.1. Tolerance limits to environmental factors can determine species abundance.

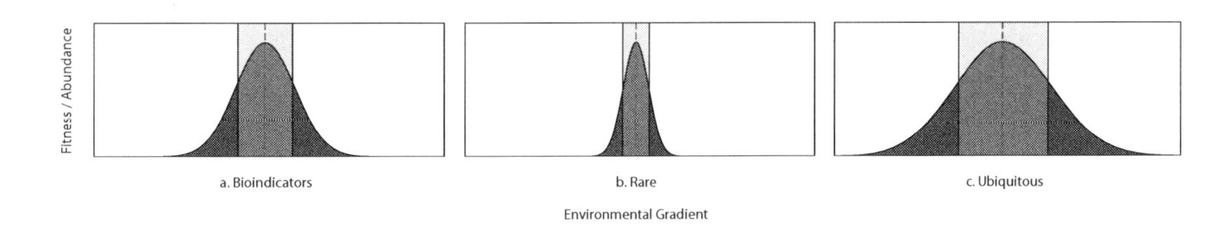

Figure 3.2. Understanding of tolerance limits can be exploited in order to use species abundance as a bioindicator for environmental factors including many environmental pollutants.

order to use species as bioindicators (Figure 3.2). For example, many species of lichen, a symbiotic organism between a fungus and a photosynthetic partner, are so sensitive to air pollution that they cannot grow when it is above a specific threshold. This is especially true for sulfur dioxide. Therefore, if lichens are present, one can surmise that air quality is such that pollution is below that threshold. Another example of a bioindicator that takes advantage of tolerance limits is the presence or absence of freshwater clams that are only able to grow when the concentration of metals is below a specific threshold. The presence of these clams indicates metal concentrations below a specific value.

How is it that some species can have some members of the population living in a zone of physiological stress while others can live only under the optimum conditions for that environmental factor? Each species has some level of genetic diversity in the population that allows for some members to adapt better to conditions of change or stress. This variation occurs through random and spontaneous mutations in the genome. Certainly some mutations can be detrimental, but others are beneficial. When conditions are such that those members of the population with the genetic background to adapt to a circumstance are better able to survive and pass their genes on to the next generation, we begin to see evolution. This evolution, or change in the hereditary material over time, usually happens over an extensive period of time, and thousands upon thousands of generations, but there are certainly many examples of fast, punctuated evolutionary events as well. When these genetic combinations help an organism adapt to its environmental conditions, it is called "natural selection." In the three panels below, we see an abbreviated version of evolution. Panel one has a dinosaur that eats leaves from a tree. Both the dinosaur and the tree in the picture represent the average-sized member of the population. Because of genetic diversity, there will certainly be taller and shorter members as well. In this example, as the dinosaur eats the leaves, the trees will die and not be able to pass their genes on to the next generation. As a result, most of the trees of that height and those that are shorter will not be present to feed that average-height dinosaur. These average dinosaurs, when faced with a food shortage, will no longer be able to pass their genes on to the next generation. However,

Figure 3.3. Species interaction can drive evolution.

the few dinosaurs that are taller than average will be able to survive because they will be able to reach the new taller average population of trees. This example also illustrates how species interactions often help direct evolution (Figure 3.3).

There are three types of selection that we will discuss to further illustrate how species can evolve in relationship around an original optimum of any given trait. The first example is common in predator-prey relationships. Imagine a lion that must chase down its prey in order to survive. The lion will typically catch the slowest member of a herd or group for its meal. As these slower members are killed, the genes that encode for speed are "selected for." That is to say, the fast members of the herd, with the genetic background that allows them to go faster, are able to pass on their genes to the next generation. Over time, the prey species evolves to be faster because only those fast enough to capture prey are able to eat, survive, and pass on their genes. This increase of speed will affect the evolution of the lion because only the fastest lions will be able to catch the prey, survive, and pass their genes on to subsequent generations. This is an example of "directional selection," which is seen when a trait moves in one direction from its original optimum (Figure 3.4).

In directional selection, there is selective pressure forcing the evolution of a trait in one direction, but there are many circumstances where selective pressure acting on the trait comes from two directions. Suppose we have a bird species whose beak is the perfect size to obtain and eat the average-sized seeds in its environment. These average birds must compete with one another for their food. What if in the average bird population there were a few birds with smaller, narrower beaks that allowed them to crack and eat a smaller seed, and there were some birds in the population that had beaks large enough and strong enough to handle larger seeds? Those birds would have less competition and might be better able to survive and pass their genes on to the next generation than the birds with the original optimum beak. Thus, two beak types would evolve away from the optimum in what is called disruptive selection (Figure 3.5).

In disruptive selection, the two forces of the on the optimum are moving away from the original, but what if the two forces acting on the optimum were in the opposite direction (Figure 3.5)? This would brace or maintain the original optimum in what is called "stabilizing selection." The classic example of stabilizing selection is birth weight. Low-weight

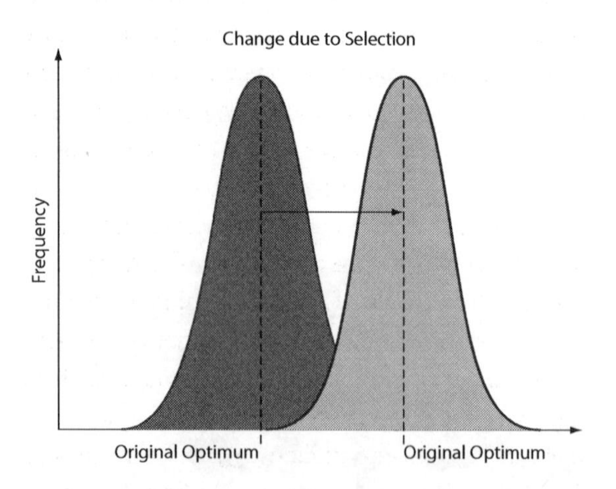

Figure 3.4. Directional selection occurs when hereditary changes move in one direction away from an original optimum.

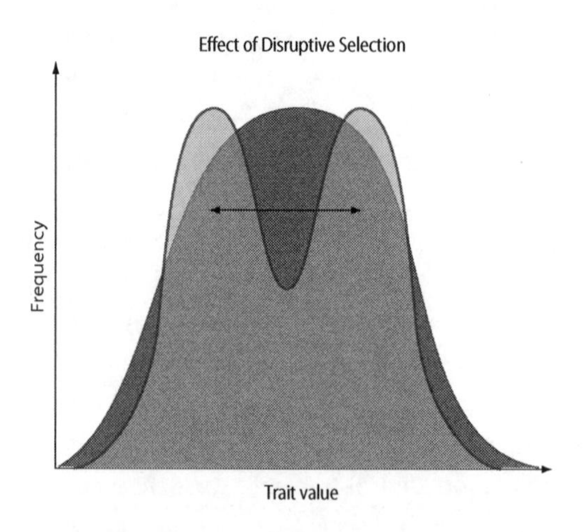

Figure 3.5. Disruptive selection describes changes in which extreme values for a trait are favored over the original, intermediate value.

Adapted from: www.oregonstate.edu/instruct/css/330/one/index2.htm

babies and high-weight babies have a higher infant mortality rate, so we see pressure to maintain the optimum.

There are changes in species that occur during their lifetime that are not heritable and do not influence evolution. For example, if a man or woman were to have rhinoplasty surgery to reduce the size of his or her nose, the nose of the man's or woman's offspring would not be affected by the surgery. Certainly there are activities that can cause heritable changes in gene expression for several generations, but this falls into the category of epigenetics and does not drive evolution in the classical way. Selective pressures that drive heritable changes can be physiological, due to predation or competition, and in some cases can be the result of bad luck.

Species Interactions Drive Evolution

There are many types of interactions among species that drive evolution. Predator-prey interactions and competition have already been discussed, but what about relationships of a more beneficial nature? There are hundreds of examples of symbiotic relationships that are mutually beneficial to both participants (mutualism). This reciprocal relationship enhances participants' survival, and fitness (ability to pass their genes on to the next generation). One such example is the mutualistic interaction between flowering plants and their animal pollinators. Pollinators can receive food from the flower in the form of nectar, but some bees also use waxes and resins from flowers to build their hives; other animals use the pollen itself as food. The plant, of course, is benefited by the dispersal of its pollen (Figure 3.6).

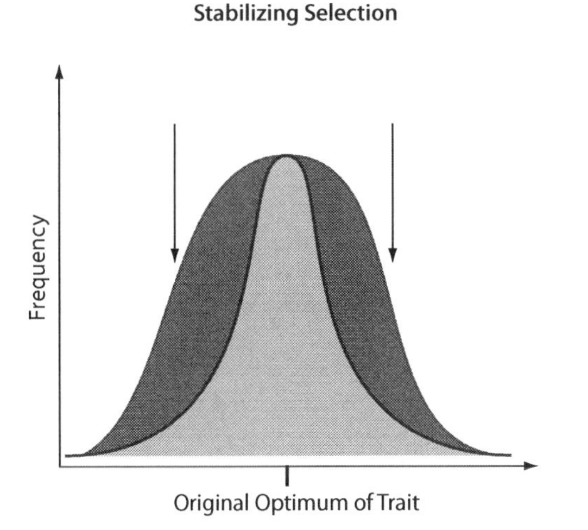

Figure 3.6. Stabilizing selection occurs when diversity in a population stabilizes around an original optimum.

Other symbiotic relationships, called commensalism, occur where one member benefits and the other member is neither benefited nor harmed. Epiphytic plants (some orchids and ferns) are plants that are benefited by living on larger plants high in the canopy where they are protected and have access to sunlight. The host trees, however, are not affected by this relationship. In all of these relationships, the evolution, or change, in one species can drive the evolution of another.

Batesian Mimicry

One of the most fascinating types of evolution involves mimicry, i.e., when a harmless species mimics a harmful model species such that it reaps the benefits of the model's defense mechanisms. Most often Batesian mimicry involves visual cues, though there are examples of evolving scents and sounds in mimicking species that are similar to those in model species. For example, warning colors are present on many species of poisonous reptiles that warn away predator species (Figure 3.7). In nature, the predator species will stay clear of anything that looks like the model species, which displays its aposematic signal. Species that have random, spontaneous mutations that make them resemble a model species more so than other members of their own population are less likely to be eaten, and more likely to be able to pass their genes on to the next generation.

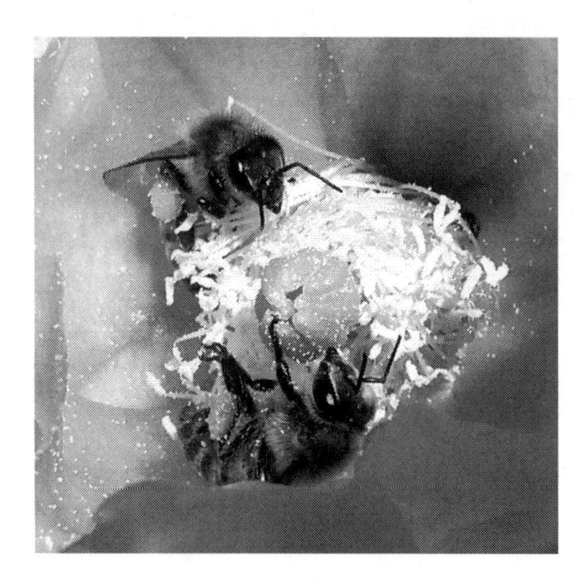

Figure 3.7. Mutualistic pollinators play a significant role in the production of more than 150 food crops in the United States, which is beneficial to the pollinators and the crop plants, and ultimately, to humans.

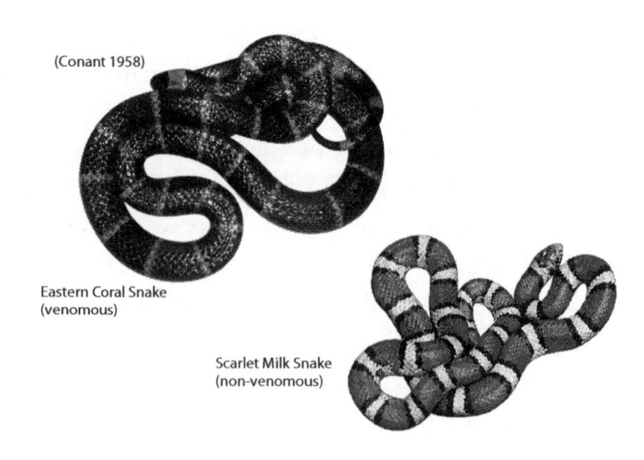

Figure 3.8. Batesian mimicry has occurred often in the reptile world such that a nonvenemous snake may evolve to look like a venomous snake.

QUESTIONS FOR THOUGHT

1. Given today's environmental conditions, select a trait that you think will continue to be present in a species of your choice due to natural selection.
2. As an extreme example, how could you modify a relay race such that it would drive evolution? What trait would you be selecting for? How will this trait be preferentially passed on to future generations?

Examples Of Scientific Method

By Bertrand Russell

III. DARWIN

The earliest triumphs of scientific method were in astronomy. Its most noteworthy triumphs in quite recent times have been in atomic physics. Both these are matters requiring much mathematics for their treatment. Perhaps in its ultimate perfection all science will be mathematical, but in the meantime there are vast fields to which mathematics is scarcely applicable, and among these are to be found some of the most important achievements of modern science.

We may take Darwin's work as illustrative of the nonmathematical sciences. Darwin, like Newton, dominated the intellectual outlook of an epoch, not only among men of science, but among the general educated public; and, like Galileo, he came into conflict with theology, though with results less disastrous to himself. Darwin's importance in the history of culture is very great, but the value of his work from a strictly scientific point of view is difficult to appraise. He did not invent the hypothesis of evolution, which had occurred to many of his predecessors. He brought a mass of evidence in its favour, and he invented a certain mechanism which he called "natural selection" to account for it. Much of his evidence remains valid, but "natural selection" is less in favour amongst biologists than it used to be.

He was a man who travelled widely, observed intelligently, and reflected patiently. Few men of his eminence have had less of the quality called brilliance; no one thought much of him in his youth. At Cambridge he was content to do no work and take a pass degree. Not being able, at that time, to study biology in the University, he preferred to spend his time walking round the country collecting beetles, which was officially a form of idleness. His real education he owed to the voyage of the *Beagle*, which gave him the opportunity of studying the flora and fauna of many regions, and of observing the habitats of allied, but geographically separated, species. Some of his best work was concerned with what is now called ecology, i.e. the geographical distribution of species and genera. He observed, for example, that the vegetation of the High Alps resembles that of the Polar regions, from which he inferred a common ancestry at the time of the glacial epoch.

Apart from scientific details, Darwin's importance lies in the fact that he caused biologists, and through them, the general public, to abandon the former belief in the immutability of species, and to accept the view that all different kinds of animals have been developed by variation out of a common ancestry. Like every other innovator of modern times, he had to combat the authority of Aristotle. Aristotle, it should be said, has been one of the great misfortunes of the human race. To this day the teaching of logic in most universities is full of nonsense for which he is responsible.

The theory of biologists before Darwin was that there is laid up in Heaven an ideal cat and an ideal dog, and so on; and that actual cats and dogs are more or less imperfect copies of these celestial types. Each species corresponds to a different idea in the Divine Mind, and therefore there could be no transition from one species to another, since each species resulted from a separate act of creation. Geological evidence made this view increasingly difficult to maintain, since the ancestors of existing widely separated types were found to resemble each other much more closely than do the species of the present day. The horse, for example, once had his proper complement of toes; early birds were scarcely distinguishable from reptiles, and so on. While the particular mechanism of "natural selection" is no longer regarded by biologists as adequate, the general fact of evolution is now universally admitted among educated people.

In regard to animals other than man, the theory of evolution might have been admitted by some people without too great a struggle, but in the popular mind Darwinism became identified with the hypothesis that men are descended from monkeys. This was painful to our human conceit, almost as painful as the Copernican doctrine that the earth is not the centre of the universe. Traditional theology, as is natural, has always been flattering to the human species; if it had been invented by monkeys or inhabitants of Venus, it would, no doubt, not have had this quality. As it is, people have always been able to defend their self-esteem, under the impression that they were defending religion. Moreover, we know that men have souls, whereas monkeys have none. If men developed gradually out of monkeys, at what moment did they acquire souls? This problem is not really any worse than the problem as to the particular stage at which the fœtus acquires a soul, but new difficulties always seem worse than old ones, since the old ones lose their sting by familiarity. If, to escape from the difficulty, we decide that monkeys have souls, we shall be driven, step by step, to the view that protozoa have souls, and if we are going to deny souls to protozoa, we shall, if we are evolutionists, be almost compelled to deny them to men. All these difficulties were at once apparent to the opponents of Darwin, and it is surprising that the opposition to him was not even more fierce than it was.

Darwin's work, even though it may require correction on many points, nevertheless affords an example of what is essential in scientific method, namely, the substitution of general laws based on evidence for fairy-tales embodying a fantasy of wish-fulfilment. Human beings find it difficult in all spheres to base their opinions upon evidence rather than upon their hopes. When their neighbours are accused of lapses from virtue, people find it almost impossible to wait for the accusation to be verified before believing it. When they embark upon a war, both sides believe that they are sure of victory. When a man puts his money on a horse, he feels sure that it will win. When he contemplates himself, he is convinced that he is a fine fellow who has an immortal soul. The objective evidence for each and all of these propositions may be of the slightest, but our wishes produce an almost irresistible tendency to believe. Scientific method sweeps aside our wishes and endeavours to arrive at opinions in which wishes play no part. There are, of course, practical advantages in the scientific method; if this were not so, it would never have been able to make its way against the world of fantasy. The bookmaker is scientific and grows rich, whereas the ordinary better is unscientific and grows poor. And so in regard to human excellence, the belief that men have souls has produced a certain technique for the purpose of improving mankind, which, in spite of prolonged and expensive effort, has hitherto had no visible good result. The scientific study of life and of the human body and mind, on the contrary, is likely, before very long, to give us the power of producing improvements beyond our previous dreams, in the health, intelligence, and virtue of average human beings.

Darwin was mistaken as to the laws of heredity, which have been completely transformed by the Mendelian theory. He had also no theory as to the origin of variations, and he believed them to be much smaller and more gradual than they have been found to be in certain circumstances. On these points modern biologists have advanced far beyond him, but they would not have reached the point at which they are but for the impetus given by his work; and the massiveness of his research was necessary in order to impress men with the importance and inevitability of the theory of evolution.

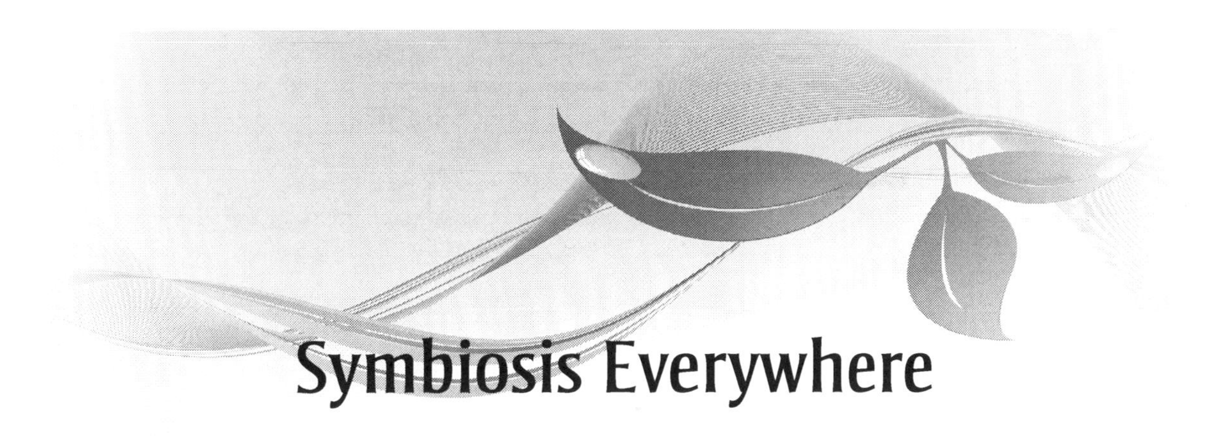

Symbiosis Everywhere

By Lynn Margulis

A Bee his burnished Carriage
Drove boldly to a Rose—
Combinedly alighting—
Himself—(1339)

Symbiosis, the system in which members of different species live in physical contact, strikes us as an arcane concept and a specialized biological term. This is because of our lack of awareness of its prevalence. Not only are our guts and eyelashes festooned with bacterial and animal symbionts, but if you look at your backyard or community park, symbionts are not obvious but they are omnipresent. Clover and vetch, common weeds, have little balls on their roots. These are the nitrogen-fixing bacteria that are essential for healthy growth in nitrogen-poor soil. Then take the trees, the maple, oak, and hickory. As many as three hundred different fungal symbionts, the mycorrhizae we notice as mushrooms, are entwined in their roots. Or look at a dog, who usually fails to notice the symbiotic worms in his gut. We are symbionts on a symbiotic planet, and if we care to, we can find symbiosis everywhere. Physical contact is a nonnegotiable requisite for many differing kinds of life.

Practically everything I work on now was anticipated by unknown scholars or naturalists. One of my most important scientific predecessors thoroughly understood and explained the role of symbiosis in evolution. The University of Colorado anatomist Ivan E. Wallin (1883–1969) wrote a fine book arguing that new species originate through symbiosis. *Symbiogenesis*, an evolutionary term, refers to the origin of new tissues, organs, organisms—even species—by establishment of long-term or permanent symbiosis. Wallin never used the word *symbiogenesis*, but he entirely understood the idea. He especially emphasized animal symbiosis with bacteria, a process he called "the establishment of microsymbiotic complexes" or "symbionticism." This is important,.

Although Darwin entitled his magnum opus *On the Origin of Species,* the appearance of new species is scarcely even discussed in his book.

Symbiosis, and here I fully agree with Wallin, is crucial to an understanding of evolutionary novelty and the origin of species. Indeed, I believe the idea of species itself requires symbiosis. Bacteria do not have species. No species existed before bacteria merged to from larger cells including ancestors to both plants and animals, In this book I will explain how long-standing symbiosis led first to the evolu-Lion of complex cells with nuclei and from there to other organisms such as fungi, plants, and animals.

That animal and plant, cells originated through symbiosis is no longer controversial. Molecular biology, including gene sequencing, has vindicated this aspcct of my theory of cell symbiosis. The permanent incorporation of bacteria inside plant and animal cells as plastids and mitochondria is the part of my serial endosymbiosis theory that now appears even in high school textbooks. But the full impact of the symbiotic view of evolution has yet to be felt. And the idea that new species arise from symbiotic mergers among members of old ones is still not even discussed in polite scientific society.

Here is an example. I once asked the eloquent and personable paleontologist Niles Eldredge whether he knew of any case in which the formation of a new species had been documented. I told him I'd be satisfied if his example were drawn from the laboratory, from the field, or from observations from the fossil record. He could muster only one good example: Theodosius Dobzliansky's experiments with *Drosophila,* the fruit fly. In this fascinating experiment, populations of fruit flies, bred at progressively hotter temperatures, became genetically separated. After two years or so the hot-bred ones could no longer produce fertile offspring with their cold-breeding brethren. "But," Eldredge quickly added, "that turned out to have something to do with a parasite!" Indeed, it was later discovered that the hot-breeding flies lacked an intracellidar symbiotic bacterium found in the cold breeders, Eldredge dismissed this case as an observation of speciation because it entailed a microbial symbiosis! He had been taught, as we all have, that microbes are germs, and when you have germs, you have a disease, not a new species. And he had been taught that evolution through natural selection occurs by the gradual accumulation, over eons, of single gene mutations.

Ironically, Niles Eldredge is author with Stephen Jay Gould of the theory of "punctuated equilibrium." Eldredge and Gould argue that the fossil record shows evolution to be static most of the time and to

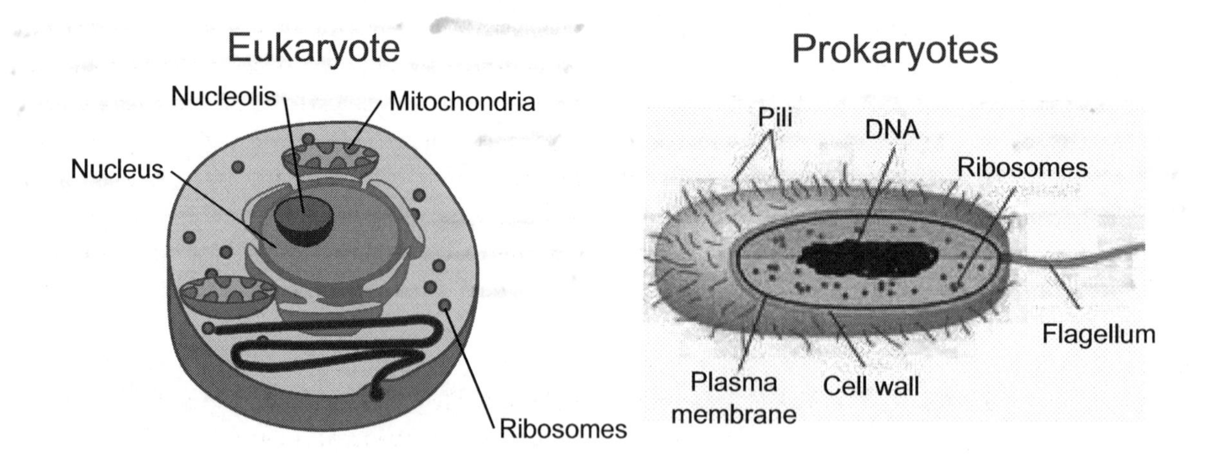

Figure 1. Prokaryotic and eukaryotic cells compared.

proceed suddenly: rapid change in fossil populations occurs over brief time spans; stasis then prevails for extended periods. From the long view of geological time, symbioses are like flashes of evolutionary lightning. To me symbiosis as a source of evolutionary novelty helps explain the observation of "punctuated equilibrium," of discontinuities in the fossil record.

Among the only other organisms besides fruit flies in which species have been seen to originate in the laboratory are members of the genus *Amoeba,* and symbiosis was involved. Symbiosis is a kind, but not the notorious kind, of Lamarckianism. "Lamarckianism," named for Jean Baptiste Lamarck, who the French claim was the first evolutionist, is often dismissed as "inheritance of acquired characteristics." In simple Lamarckianism, organisms inherit traits induced in their parents by environmental conditions, whereas through symbiogenesis, organisms acquire not traits but entire other organisms, and of course, their entire sets of genes! I could say, as my French colleagues often have, that symbiogenesis is a form of neo-Larmarckianism. Symbiogenesis is evolutionary change by the inheritance of acquired gene sets.

Living beings defy neat definition. They fight, they feed, they dance, they mate, they die. At the base of the creativity of all large familiar forms of life, symbiosis generates novelty. It brings together different life-forms, always for a reason. Often, hunger unites the predator with the prey or the mouth with the photosynthetic bacterium or algal victim. Symbiogenesis brings together unlike individuals to make large, more complex entities. Symbiogenetic life-forms are even more unlike than their unlikely "parents." "Individuals" permanently merge and regulate their reproduction. They generate new populations that become multiunit symbiotic new individuals. These become "new individuals" at larger, more inclusive levels of integration. Symbiosis is not a marginal or rare phenomenon. It. is natural and common. We abide in a symbiotic world.

In Brittany, on the northwest coast of France, and along beaches bordering the English Channel is found a strange sort of "seaweed" that is not seaweed at all. From a distance it is a bright green patch on the sand. The patches slosh around, shimmering in shallow puddles. When you pick up ihe green water and let it slip through your fingers you notice gooey ribbons much like seaweed. A small hand lens or low-power microscope reveals that what looked like seaweed are really green worms. These masses of sunbathing green worms, unlike any seaweed, burrow into the sand and effectively disappear.They were first described in the 1920s by an Englishman, J. Keeble, who spenL his summers at Roscoff. Keeble called them "plant-animals" and diagrammed them splendidly in the color frontispiece of his book, *Plant-Animals.* The flatworms of the species *Convoluta roscoffensis* are all green because their tissues are packed with *Platymonas* cells; as the worms are translucent, the green color of *Platymonas*, photosynthesizing algae, shows through. Although lovely, the green algae are not merely decorative: they live and grow, die and reproduce, inside the bodies of the worms. Indeed they produce the food that the worms "eat." The mouths of the worms become superfluous and do noy function after the worm larvae hatch. Sunlight reaches the algae inside their mobile greenhouses and allows them to grow and feed themselves as they leak photosynthetic products and feed their hosts from the inside. The symbiotic algae even do the worm a waste management favor: they recycle the worm's uric acid waste into nutrients for themselves. Algae and worm make a miniature ecosystem swimming in the sun. Indeed, these two beings are so intimate that it is difficult, without very high-power microscopy, to say where the animal ends and the algae begin.

Such partnerships abound. Bodies of *Plachobranchus,* snails, harbor green symbionts growing in such even rows they appear to have been planted. Giant clams act as living gardens, in which their bodies hold algae toward the light. *Mastigias* is a man-of-war type of medusoid that swims in the Pacific Ocean. Like myriad small green umbrellas, *Mastigias* modusoids float through the light beams near the water's surface by the thousands.

Similarly, freshwater tentacled hydras may be white or green, depending on whether or not their bodies are packed with green photosynthetic partners. Are hydras animals or plants? When a green hydra is permanently inhabited by its food producing partners (called *Chlorella),* it is hard to tell. Hydras, if green, are symbiotic. They are capable of photosynthesizing, of swimming, of moving, and of staying put. They have remained in the game of life because they become individuals by incorporation.

We animals, all thirty million species of us, emanate from the microcosm. The microbial world, the source and well-spring of soil and air, informs our own survival. A major theme of the microbial drama is the emergence of individuality from the community interactions of once-independent actors.

I love to gaze on the daily life struggles of our nonhuman planetmates. For many years Lorraine Olendzenski, my former student, now at the University of Connecticut, and I have videographed life in the microcosm. More recently we have worked with Lois Byrnes, the vivacious former associate director of the New England Science Center in Worcester, Massachusetts. Together we and a fine group of U MASS students make films and videos that introduce people to our microbial acquaintances.

Ophrydium, a pond water scum that, upon close inspection, seems to be countable green "jelly ball" bodies is an example of emergent individuality that we recently discovered in Massachusetts and redescribed. Our films show these water balls with exquisite clarity. The larger "individual" green jelly ball is composed of smaller cone-shaped actively contractile "individuals." These in turn are composite: green *Chlorella* dwell inside ciliates, all packed into rows. Inside each upside-down cone are hundreds of spherical symbionts, cells of *Chlorella. Chlorella* is a common green alga; the algae of *Ophrydium* are trapped into service for the jelly ball community. Each "individual organism" in this "species" is really a group, a membrane-bounded packet of microbes that looks like and acts as a single individual.

A nutritious drink called kefir consumed in the Caucasus Mountains is also a symbiotic complex, Kefir contains grainy curds the Georgians call "Mohammed pellets." The curd is an integrated packet of more than twenty-five different kinds of yeast and bacteria. Millions of individuals make up each curd. From such interactive bodies of fused organisms new beings sometimes emerge. The tendency of "independent" life is to bind together and reemerge in a new wholeness at a higher, larger level of organization. I suspect that the near future of *Homo sapiens* as a species requires our reorientation toward the fusions and mergers of the planetmates that have preceded us in the microcosm. One of my ambitions is to coax some great director into producing evolutionary history as the microcosmic image in the 72mm format film (IMAX or OMNIMAX), showing spectacular living relationships as they form and dissolve.

Now as throughout Earth's history, living associations form and dissolve. Symbioses, both stable and ephemeral, prevail. Such evolutionary tales deserve broadcasting.

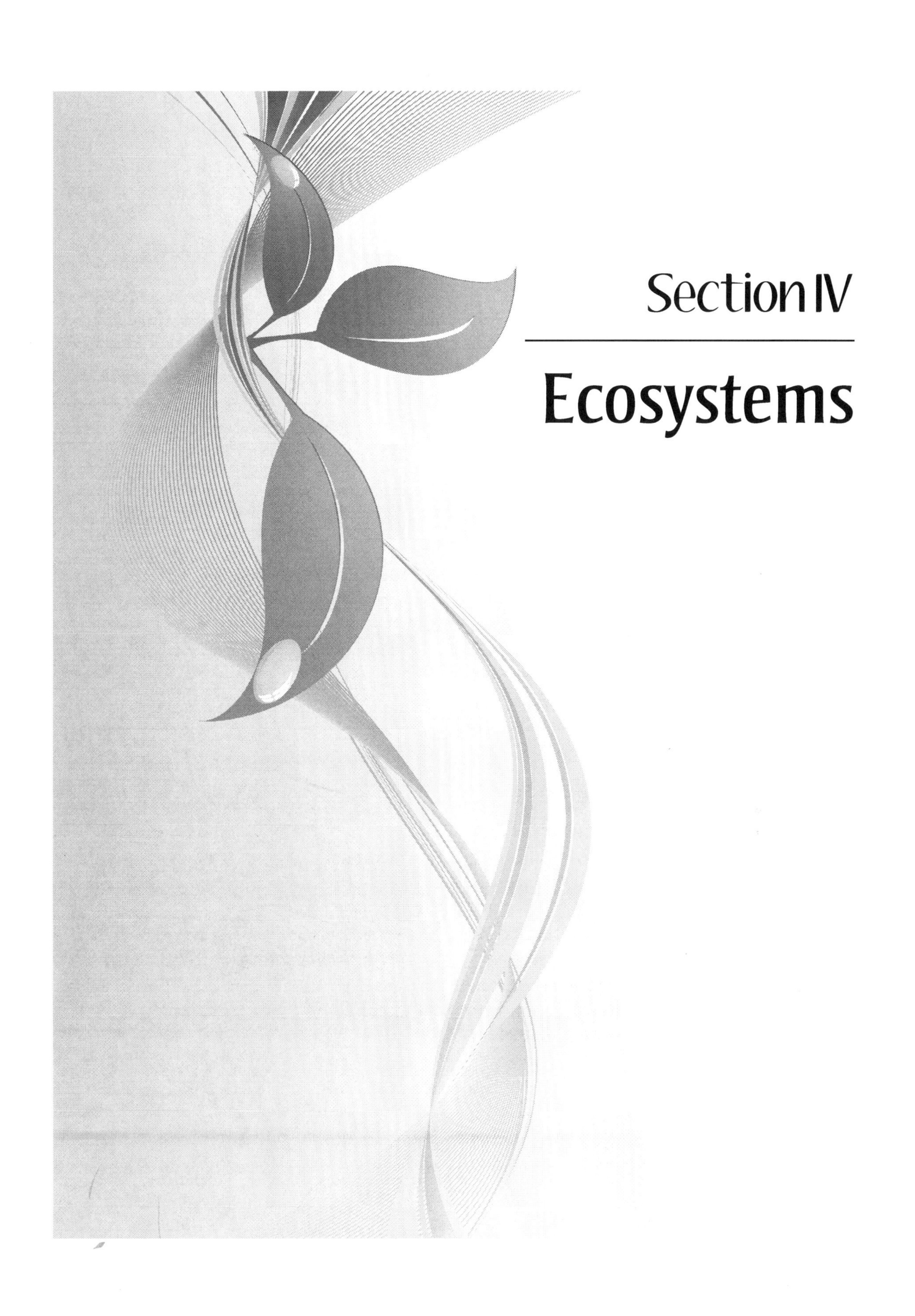

Section IV

Ecosystems

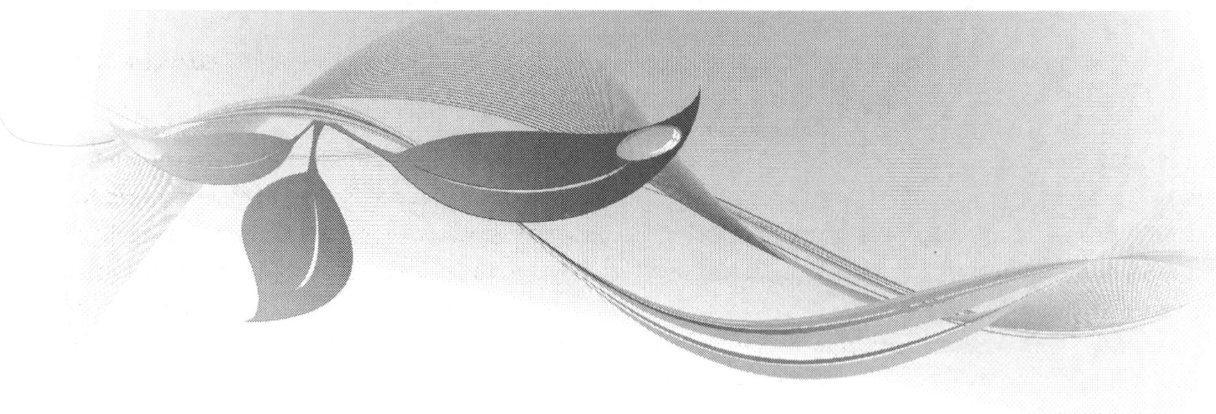

By Anne Marie Zimeri

Ecosystems are sets of complex relationships among living organisms, their habitats, and resources. When an ecosystem is healthy, it is sustainable. To sustain an ecosystem, resources, especially energy, must be able to be transferred among trophic levels. Almost all energy for ecosystems originates from the sun and is stored when carbon from the atmosphere is "fixed" during photosynthesis. A few exceptions occur in deep-sea floor vents and in other extreme circumstances, but for the most part all life on earth is dependent on energy that originates from the sun.

Photosynthesis

Photosynthesis is the process of converting light energy to chemical energy and storing it in the chemical bonds of sugars. This process occurs in plants and some algae that have chloroplasts. Plants need only light energy, CO_2, and H_2O to make sugar (Figure 4.1). Photosynthesizing organisms are the "producers" in ecosystems.

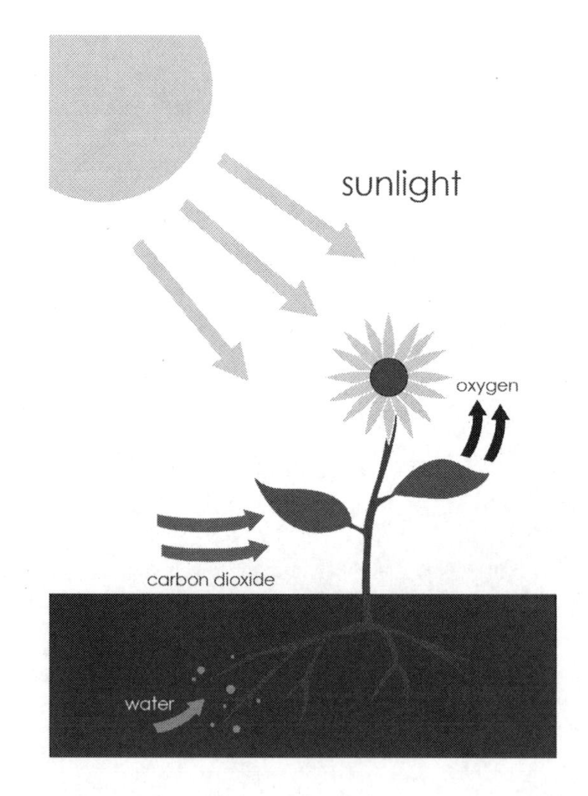

Figure 4.1. Photosynthesis converts light energy into energy-rich sugars.

Once energy has been stored in the chemical bonds in plants, it can begin its movement through the food web, i.e., a complex and interrelated food chain. Plants (or producers) are consumed by herbivores, which make up the second trophic level in ecosystems. Some of the energy (calories) from plants goes toward biomass, while the bulk of the energy (~90%) is lost during respiration and locomotion. As we move a step farther away from producers, we begin to see carnivores or secondary consumers

(Figure 4.2). Secondary consumers are considered part of the tertiary trophic level. These consumers also retain some of the energy stored in their food as biomass and again lose the bulk of it to metabolic processes and movement.

Continuing with the concept of how energy flows up the trophic levels in an ecosystem, we see that the higher the trophic level, the lower the percentage of energy derived from the original plant. For example, a lion that consumes a zebra would be a secondary consumer from the tertiary trophic level. Because 90% of an organism's energy is lost during its movement, etc., each trophic level would retain only 10% of the plant's energy. Therefore, the lion would receive 1/10 of the energy from the zebra that retained 1/10 of the energy from the plant, or 1/100 of the original energy from the plant. This is the basis for increasing plant production over animal production in the world in order to more efficiently feed the growing population. For example, it may take 3,000 lbs. of corn to feed one steer in a feeding operation, while that amount of grain would feed 20 people over the same period of time.

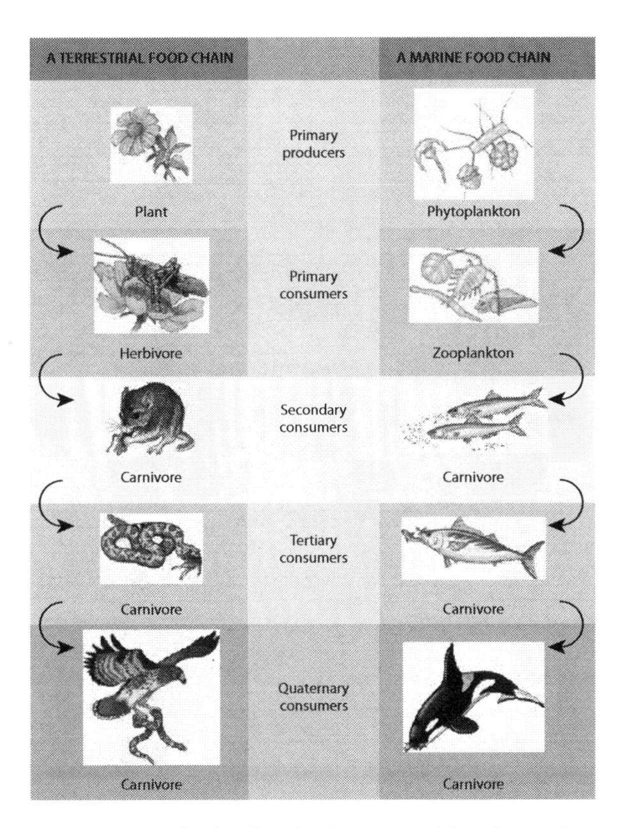

Figure 4.2. Trophic levels in both a terrestrial and a marine ecosystem. Energy is transferred from trophic level to trophic level, beginning with primary producers.

Biogeochemical Cycling

Besides energy's flow through ecosystems, other nutrients must make their way through the trophic levels. To be available for movement in the food web, these nutrients must be cycled/recycled through a process called biogeochemical cycling. This cycling involves biological, geological, and chemical processes. Photosynthesis is one stage of the cycling of carbon throughout the planet. It takes atmospheric carbon (CO_2) and fixes it such that plants accumulate that molecule. We know that plant molecules make their way through the trophic levels, but how does that carbon cycle back to the atmosphere so that it can be used again? Atmospheric carbon comes from respiration, decomposition of organic matter/deceased organisms, and from industry (mainly the burning of fossil fuels). There is also a substantial amount of carbon present in the oceans, and this dissolved CO_2 can evaporate and make its way back into the atmosphere.

Other nutrients that are cycled/recycled through ecosystems and around the planet in solid, liquid or dissolved, and gaseous form are nitrogen (Figure 4.3) and sulfur. Phosphorus is yet another important nutrient that is cycled, but it is not typically found in gaseous form during biogeochemical cycling. Each of these nutrients must be incorporated into molecules in primary producers (photosynthesizing organisms) prior to entering the higher trophic levels in the food web.

It is important to remember that all of these nutrients are cycled. They are neither created nor destroyed; they are just cycled. Human activities can alter the concentrations of these nutrients in certain areas of the cycle, and this can have negative effects on ecosystems. For example, burning excessive fossil fuels has increased the percentage of CO_2 in the atmosphere, contributing to climate change and some plant processes. Mining for phosphorus and nitrate so that agricultural fields can be fertilized has contributed to nutrient runoff that can cause eutrophication of our rivers, lakes, and streams.

QUESTIONS FOR THOUGHT

1. Consider the constituents of your own body. Where might they have resided before they were a part of you?
2. In Figure 4.2, an eagle is shown catching a snake. If that eagle ate the snake, what percent energy would it have derived from plants/producers?

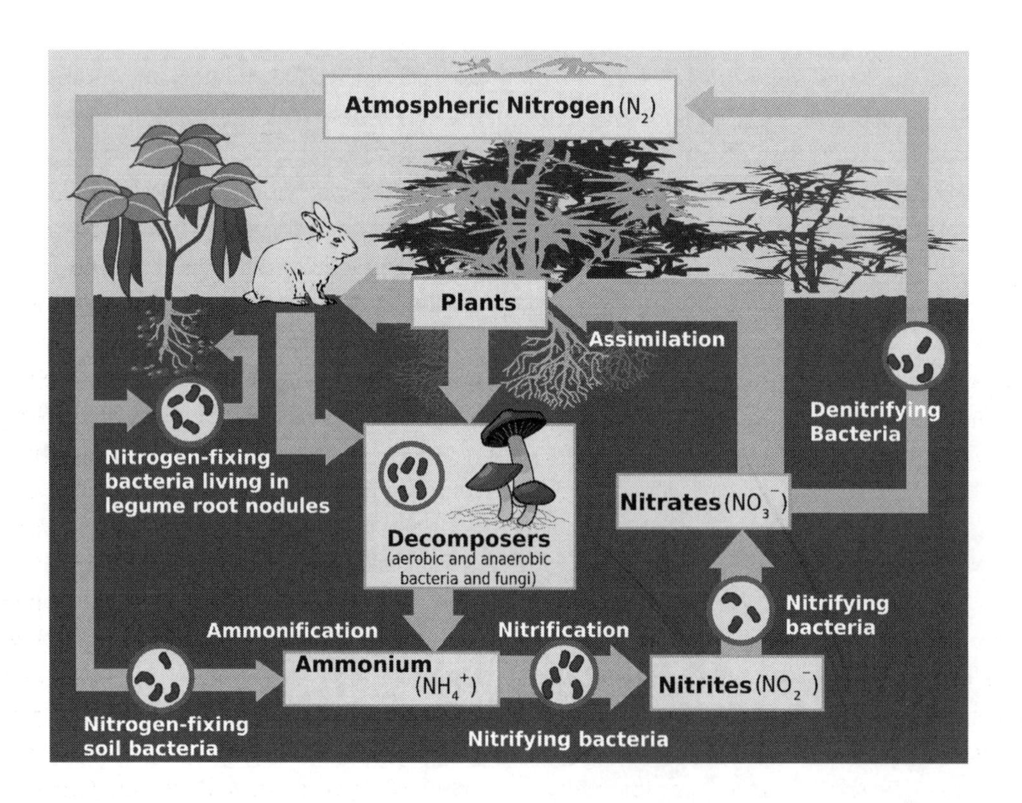

Figure 4.3. Nitrogen makes up 78% of the atmosphere, but it can be absorbed by plants only in the form of nitrate (NO_3^-) or ammonium (NH_4^+).

Ecosystems:

Sustaining Life on Earth

By Edward A. Keller and Daniel Botkin

THE ECOSYSTEM: SUSTAINING LIFE ON EARTH

So far we have talked about species interacting in pairs, or, through the community effect, in larger groups, and we have discussed the dependence of species on characteristics of their environments. But there is a much more important connection between species and their environment: the ecosystem, which makes it possible for life to persist.

The oldest fossils are more than 3.5 billion years old, so life has persisted on Earth for an incredibly long time. We struggle to maintain threatened species for a decade or two and consider ourselves fortunate if we succeed. But for at least 4.5 billion years, life has been sustainable! What accounts for that?

Ecosystems are crucial to sustaining life. To understand how life persists on Earth, we have to understand ecosystems. We tend to think about life in terms of individuals, because it is individuals that are alive. Btit sustaining life on Earth requires more than individuals or even single populations or species. As we learned in Chapter 3, living things require 24 chemical elements, and these must cycle from the environment into organisms and back to the environment. Life also requires a flow of energy, as we will learn in this chapter. Although alive, an individual cannot by itself maintain all the necessary chemical cycling or energy flow. Those processes are maintained by a group of individuals of various species and their nonliving environment. We call that group and its local environment an ecosystem. Sustained life on Earth, then, is a characteristic of ecosystems, not of individual organisms or populations.

BASIC CHARACTERISTICS OF ECOSYSTEMS

Ecosystems have three fundamental characteristics: structure, processes, and change.

Structure. An ecosystem has two major parts: nonliving and living. The nonliving part is the physical-chemical environment, including the local atmosphere, water, and mineral soil (on land). As we have already seen, the living part, called the ecological community, is the set of species interacting within the ecosystem.

Trees in a forest, grasses in a prairie, and kelp in an ocean create a biological structure that provides habitats for many species. The ecological community is a living part of an ecosystem, made up of individuals of a number of interacting species. The individuals interact by feeding on one another (*predation/parasitism)*, by competing for resources (competition), and by helping one another (symbiosis). A diagram

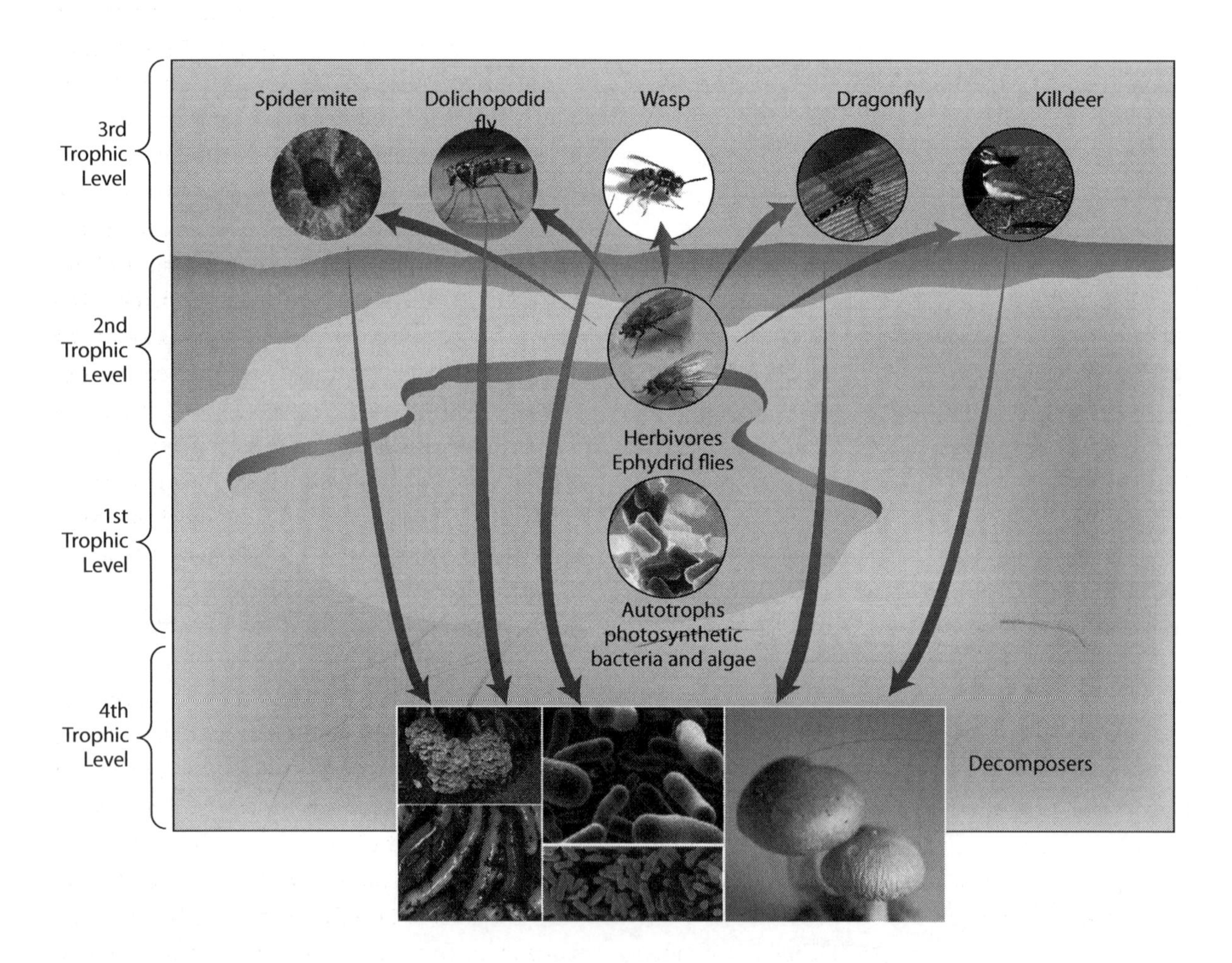

Figure 4.7. Structure and function of an ecosystem. Diagram of the food web and the trophic levels in a Yellowstone National Park hot spring.

of who feeds on whom is called a food web (Figure 4.7). It shows us how chemicals cycle and energy flows within an ecological community. A food web is divided into trophic levels. A trophic level is all the organisms that are the same number of feeding levels away from the original source of energy.

Processes. Two basic kinds of processes must occur in an ecosystem: a cycling of chemical elements and a flow of energy. Related to this is the concept of *ecosystem function* which is the rates of chemical cycling and flow of energy.

Change. An ecosystem changes and develops through a process called *succession,* which is discussed in Chapter 6. How can an ecosystem be sustainable under such variable conditions?

To understand the idea of an ecosystem, it is helpful to consider one of the simplest ecosystems. Perhaps the simplest of all is a Yellowstone National Park hot spring that has among the fewest species (Figure 4.8).

FOOD WEBS

Food webs are even more complicated than they look. The introductory history of the acorn connection described one food web: acorns eaten by mice and deer, which in turn are food for a variety of carnivores, including people who hunt deer.

A diagram of a food web and its trophic levels seems simple and neat, as in the Yellowstone hot springs (Figure 4.8) or as usually shown for a forest ecosystem (Figure 4.9), but in reality food webs are complex, because most creatures feed on several trophic levels. For example, consider the food web of the harp seal (Figure 4.10). The harp seal is at the fifth trophic level. It feeds on flatfish (fourth trophic level), which feed on sand lances (third level), which feed on euphausiids (second level), which feed on phytoplankton (first level). But the harp seal actually feeds at several trophic levels, from the second through the fourth, so it feeds on predators of some of its own prey and thus is a competitor with some of its own food. (Note that a species that feeds on several trophic levels typically is classified as belonging to the trophic level above the highest level from which it feeds. Thus, we place the harp seal on the Fifth level.)

ECOSYSTEM ENERGY FLOW

In a food web, energy and chemical elements are transferred up through trophic levels. All life requires energy, and the role of energy in life brings us to one of the most philosophical topics in ecology: life and the laws of thermodynamics.

Energy is the ability to do work, to move matter. Ecosystem energy flow is the movement of energy through an ecosystem from the external environment, through a series of organisms, and back to the external environment (Figure 4.11). It is one of the fundamental processes common to all ecosystems. Energy enters an ecosystem when it is "fixed" by organisms—meaning that it is *put into and stored in organic compounds.* This fixation of energy is called biological production, which we will explain shortly.

Life and the Laws of Thermodynamics

Energy can only flow one way through an ecosystem—it cannot be reused. If it could be reused, then it might be possible to have the kind of ecosystem shown in Figure 4.12, which would never require an input of energy but could keep running forever on its own. In that diagram, two facts are illustrated: Frogs eat insects, including mosquitoes, and mosquitoes bite frogs. Why then couldn't there be an ecosystem that was just frogs eating mosquitoes and mosquitoes eating frogs? This would be an

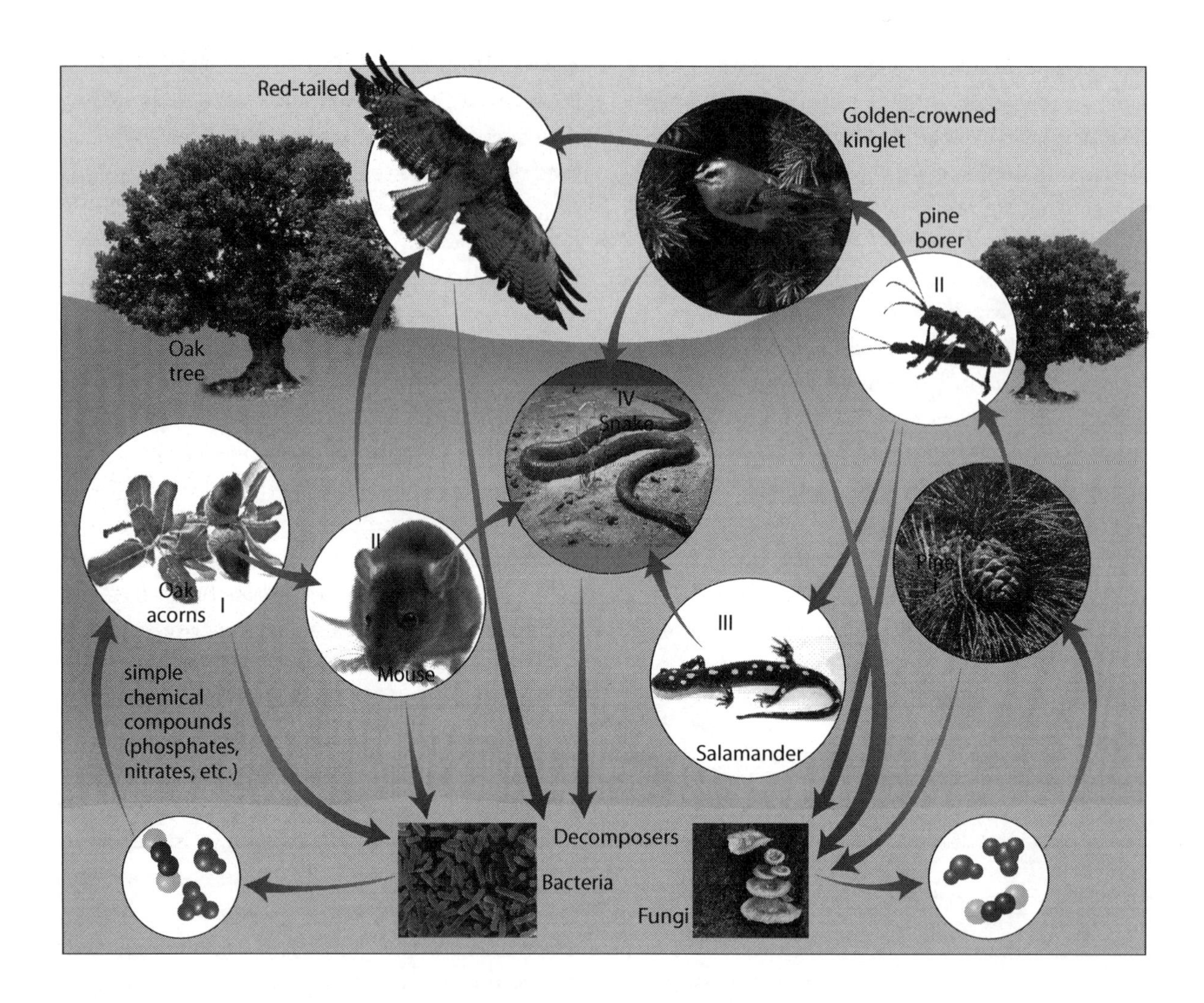

Figure 4.9. A food web in a forest ecosystem where mice eat acorns and there may be Lyme disease. The Roman numeral near the picture of an organism is its trophic level.

ecological perpetual motion machine, which the laws of thermodynamics tell us is impossible, as we will explain here. (There are also other reasons that this ecosystem couldn't work, including that only female mosquitoes bite vertebrates to get proteins required for reproduction, and this biting, although a problem for us, is not the major food source for mosquitoes during their lifetimes.)

All life is governed by the laws of thermodynamics, which are fundamental physical laws about energy. There are three thermodynamic laws: (1) the conservation of energy, (2) the increase in entropy, and (3) what is usually called the "zeroth" law—the law of absolute zero temperature. We will discuss only the first two. These laws are important not only for the chapters in this book about ecosystems but also for the chapter on energy sources for human uses.

The law of conservation of energy states that in any physical or chemical change, energy is neither created nor destroyed but merely changed from one form to another. Here is a basic question that

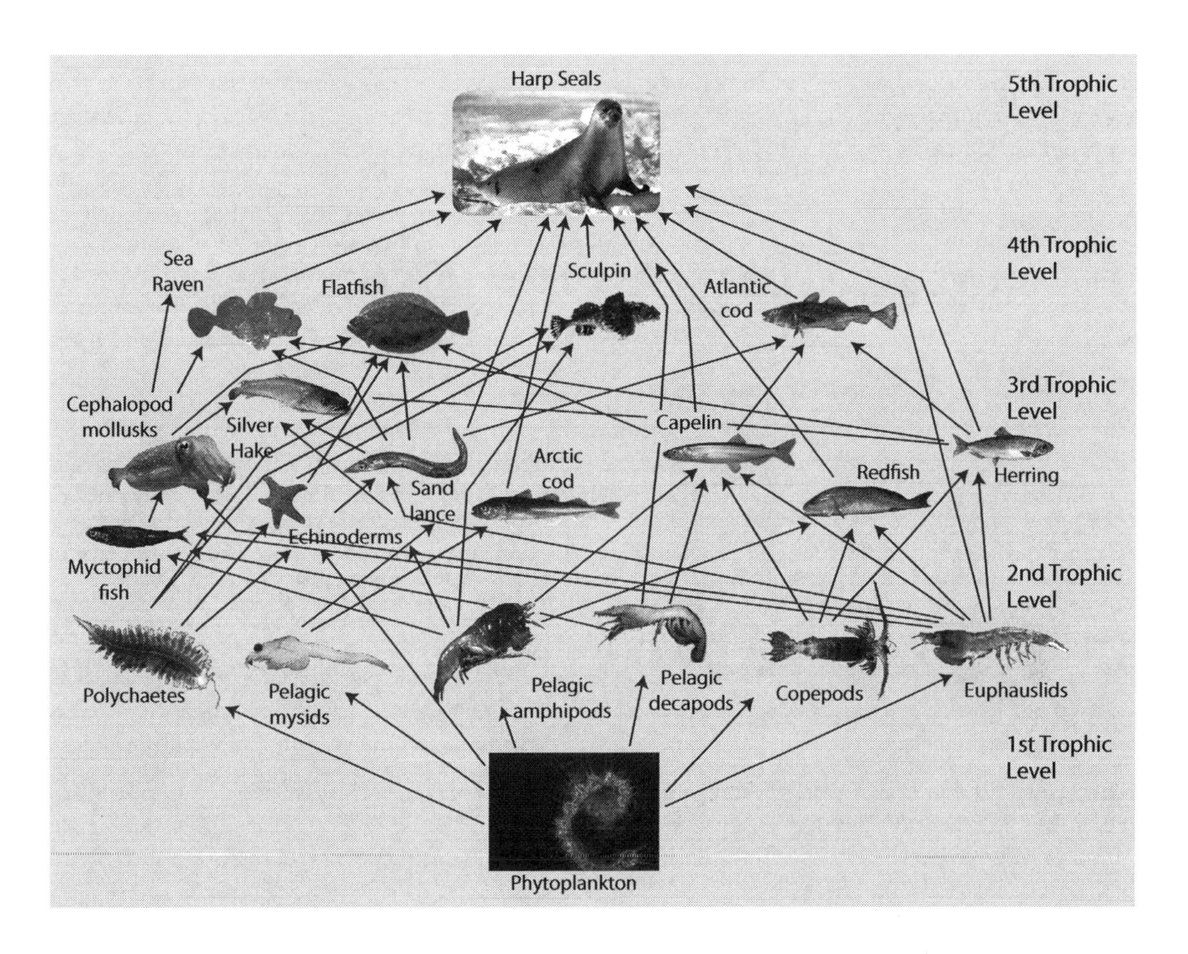

Figure 4.10. Food web of the harp seal.

arises from the law of the conservation of energy: If the total amount of energy is always conserved—if it remains constant—then why can't we just recycle energy inside our bodies and why can't energy be recycled in ecosystems? Let us return to Figure 4.12 and consider that imaginary ecosystem consisting only of frogs, mosquitoes, water, air, and a rock for the frogs to sit on. Frogs eat insects, including mosquitoes. Mosquitoes suck blood from vertebrates, including frogs. In our imaginary ecosystem, the frogs get their energy from eating the mosquitoes, and the mosquitoes get their energy from biting the frogs (Figure 4.12). Such a closed system would be a biological perpetual-motion machine: It could continue indefinitely without an input of any new material or energy.

The law of entropy tells us that this is impossible. This second law of thermodynamics addresses how energy changes in form. It is a sad reality of our universe that energy always changes from a more useful, more highly organized form to a less useful, disorganized form. This means that energy cannot be completely recycled to its original state of organized, high-quality usefulness. Whenever useful work is done, heat is released to the environment, and the energy in that it can never be completely recycled. The amount of able energy gets less and less. For this reason, the mosquito-frog system will eventually stop working when not enough useful energy is left.

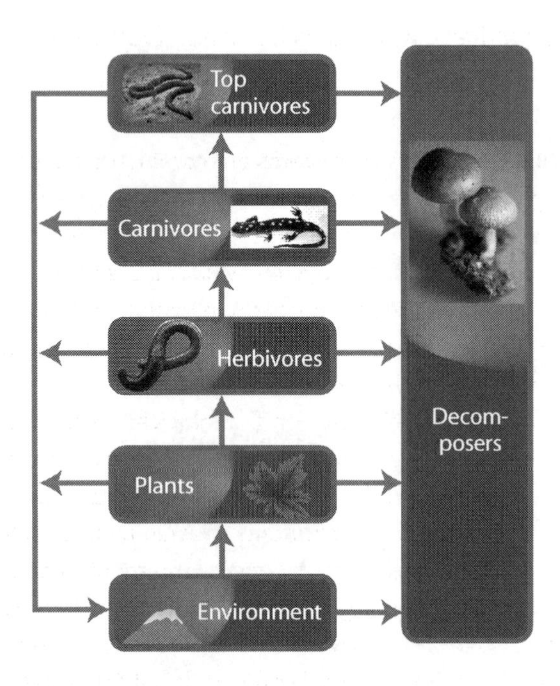

Figure 4.11. Energy pathways through an ecosystem. Usable energy flows from the external environment (the sun) to the plants, then to the herbivores, carnivores, and top carnivores. Death at each level transfers energy to decomposers. Energy lost as heat is returned to the external environment.

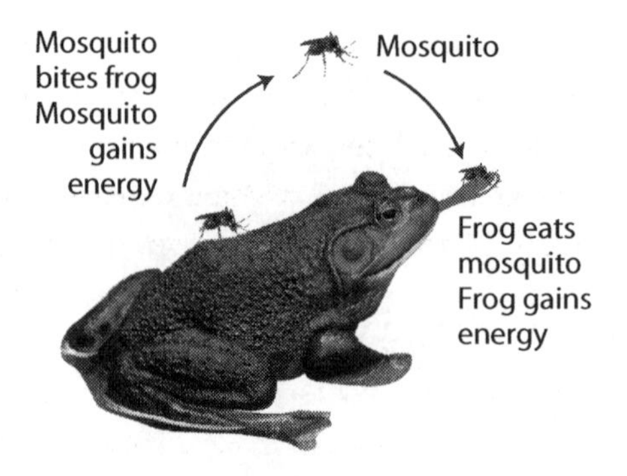

Figure 4.12. An impossible ecosystem. Frogs eat mosquitoes, and mosquitoes bite frogs. Why then couldn't an ecosystem just consist of frogs and mosquitoes, each feeding on the other? The laws of thermodynamics tell us this is impossible, as the text explains.

The net flow of energy through an ecosystem, then, is a one-way flow, from a source of usable energy to a place where heat can be released (Figure 4.11). *An ecosystem must lie between a source of usable energy and a sink for degraded energy (heat).* You can view the ecosystem as an intermediate system between the energy source and the energy sink. The energy source, ecosystem, and energy sink together form a thermodynamic system.

Producing New Organic Matter

Producing organic matter requires energy; organic matter stores energy. The total amount of organic matter in any ecosystem or area is called its *biomass.* Biomass increases through biological production (growth). Change in biomass over a given period is called *net production.* There are two kinds of biological production: primary and secondary.

Primary production. Some organisms make their own organic matter from a source of energy and inorganic compounds. These organisms are called *autotrophs* (meaning self-nourishing). The autotrophs include (1) green plants (plants containing chlorophyll), such as herbs, shrubs, and trees; (2) algae, which are usually found in water but occasionally grow on land; and (3) certain kinds of bacteria. The production carried out by autotrophs is called primary production. Most autotrophs make sugar and oxygen from sunlight, carbon dioxide, and water in a process called *photosynthesis.*

Secondary production. Other kinds of life cannot make their own organic compounds from inorganic ones and must feed on other living things. These are called *heterotrophs.* All animals, including human beings, are heterotrophs, as are fungi, many kinds of bacteria, and many other forms of life. Production by heterotrophs is called secondary production because it depends on production by autotrophic organisms (Figure 4.11).

Living things use energy from organic matter through respiration. Once an organism has obtained new organic matter, it can use the energy in that organic matter to do things—to move, to make new kinds of compounds, to grow, to reproduce—or store it for future uses. In *respiration*, an organic compound combines with oxygen to release energy and produce carbon dioxide and water. The process is similar to the burning of fuels, like a fire in a fireplace, or gasoline burned in an automobile engine, but it takes place within cells at much lower temperatures with the help of organic chemicals called *enzymes.* Respiration is the use of biomass to release energy that can be used to do work. Complete respiration releases energy, carbon dioxide, and water into the environment. Incomplete respiration also releases a variety of organic compounds into the environment.

Gross and net production. The production of biomass and its use as a source of energy by autotrophs includes three steps:

1. An organism produces organic matter within its body.
2. It uses some of this new organic matter as a fuel in respiration.
3. It stores some of the newly produced organic matter for future use.

The first step, production of organic matter before use, is called *gross production.* The amount left over is called *net production.*

The difference between gross and net production is like the difference between a person's gross and net income. Your gross income is the amount you are paid. Your net income is what you have left after taxes and other fixed costs. Respiration is like the necessary expenses that are required in order for you to do your work.

Most primary production takes place through photosynthesis, which, as we have said, is the process by which sunlight, carbon dioxide, and water are combined to produce sugar and oxygen. Green plants, algae, and certain bacteria use photosynthesis.

Practical Implication I: Human Domination of Ecosystems

Many of Earth's ecosystems are dominated directly by human beings, and essentially no ecosystem in the oceans or on land is free of human influence. A recent study reached the following conclusions:

Human domination of ecosystems is not yet a global catastrophe, although serious environmental degradation has resulted.

Earth's ecological and biological resources have been greatly modified by human use of the environment, and this modification and its impact are growing. An important human-induced alteration of Earth's ecosystems is land modification. Approximately 12% of the land surface of Earth is now occupied by agriculture (row crops such as corn, beans, or cotton) as well as urban-industrial uses. An additional 7% has been converted to pastureland. Although at first glance this may seem a small percentage of Earth's total land area, its impacts are large because much of Earth's land is not suitable for agriculture, pasture, or other urban uses.

Is there anything we can do to cause less damage? Having recognized that our activities can have significant global consequences for ecosystems, what can we do? First, we can reduce the rate at which we are altering Earth's ecosystems. This includes reducing the human population, finding ways of using fewer resources per person more efficiently, and better managing our waste. Second, we can try to better understand ecosystems and how they are linked to human-induced global change.

Practical Implication II: Ecosystem Management

Ecosystems can be natural or artificial or a combination of both. An artificial pond that is a part of a waste-treatment plant is an example of an artificial ecosystem. Ecosystems can also be managed, and management can include a large range of actions. Agriculture can be viewed as partial management of certain kinds of ecosystems (see Chapter 7), as can forests managed for timber production. Wildlife preserves are examples of partially managed ecosystems.

Sometimes, when we manage or domesticate individuals or populations, we separate them from their ecosystems. We also do this to ourselves (see Chapter 2). When we do this, we must replace the ecosystem functions of energy flow and chemical cycling with our own actions. This is what happens in a zoo, where we must provide food and remove the wastes for individuals separated from their natural environments.

The ecosystem concept is central to management of natural resources. When we try to conserve species or manage natural resources so that they are sustainable, we must focus on the ecosystem and make sure that it continues to function. If it doesn't, we must replace or supplement ecosystem functions ourselves.

Ecosystem management, however, involves more than just compensating for changes we make in ecosystems. It means managing and conserving life on Earth by considering chemical cycling, energy flow, community-level interactions, and the natural changes that take place within ecosystems.

RETURN TO THE BIG QUESTION

What is necessary to sustain life on Earth?

In this chapter we have discussed some of the basic features of ecosystems that make it possible for life to persist. We learned that certain kinds of interactions among species are necessary, in particular those across trophic levels—predation and parasitism that move energy and chemical elements up food webs. We learned that the ecosystem is the basic unit for the persistence of life, because life requires a cycling of chemical elements and a flow of energy, and this must take place among species and between species and their environment. The ecosystem is a set of species (an ecological community) and the local, nonliving environment.

One of the questions that will concern us most as we explore environmental issues about life on Earth is how a population's size changes over time. It is clear that populations do not remain constant, but change in abundance almost all the time. We want to be able to forecast changes in abundance, but to do that we need to know what factors cause populations to swell or shrink. We learned that populations respond to environmental change, and to changes in the abundance of other populations, especially changes in the abundance of species that one competes with or interacts with through predation/parasitism or symbiosis. We also learned that populations can change as a result of indirect interactions—community effects—the way sea otters affect kelp.

So the answer to the question of what is needed to sustain life on Earth is, in sum, an ecosystem with its chemical cycling and energy flow.

SUMMARY

- Populations, once believed to be constant in abundance if people did not affect them, are now known to change continually.
- Populations of different species affect each other directly through competition, predation/parasitism, and symbiosis.
- Populations also affect each other indirectly through the community effect.
- An ecosystem is the simplest entity that can sustain life. At its most basic, an ecosystem consists of several species and a fluid medium

(air, water, or both). The ecosystem must sustain two processes—the cycling of chemical elements and the flow of energy. Biological production is the production of new organic-matter, which we measure as a change in biomass.

- In every ecosystem, energy flow provides a foundation for life.
- The living part of an ecosystem is the ecological community, a set of species connected by food webs and trophic levels. A food web or chain shows who feeds on whom. A trophic level consists of all the organisms that are the same number of feeding steps from the initial source of energy.
- Community-level effects result from indirect interactions among species, such as those that occur when sea otters reduce the abundance of sea urchins.
- Ecosystems are real and important, but it is often difficult to define the limits of a system or to pinpoint all the interactions that take place.

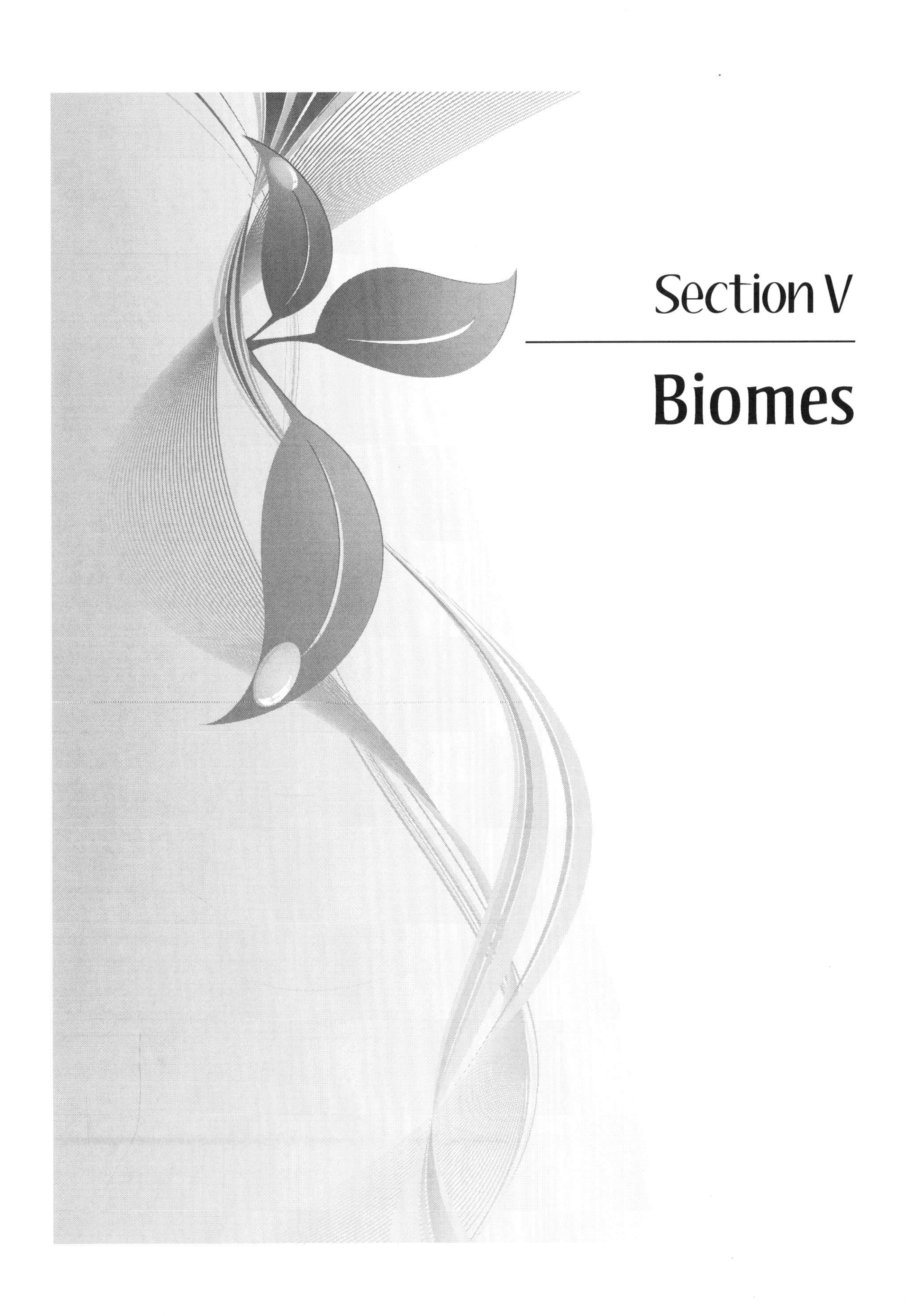

Section V

Biomes

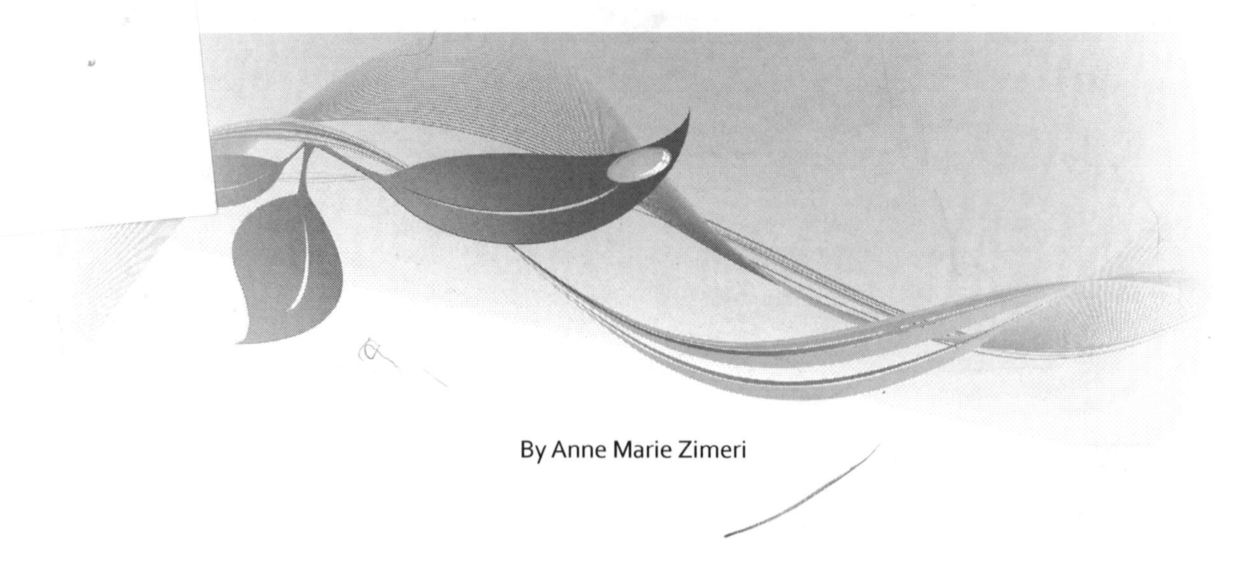

By Anne Marie Zimeri

Preserving biodiversity is an important part of maintaining environmental and public health. Differences in temperature or precipitation as well as topography, soil conditions, and abiotic features determine the types of organisms that grow in a given area. These features help define a broadly defined life zone called a "biome." Understanding biomes and biome preservation is key to maintaining biodiversity around the world. In this section, we will discuss several important biomes around the world as well as biodiversity. Each biome will be characterized in this chapter by precipitation and temperature as well as by other notable features specific to each. Typically, the more moisture present and the less seasonality we see with temperature and rainfall, the more biodiversity is present. Most biodiversity, therefore, is in a band around the equator and begins to diminish as we move farther north or south. The following chart (Figure 5.1) can be used to reference the general rainfall and temperature profiles of the major biomes discussed in this chapter.

Terrestrial Biomes

Tropical rain forests are characterized by consistent year-round temperatures and excessive rainfall. The typical annual rainfall in the rain forest can exceed 2,600 mm. Because of this excess moisture and these stable temperatures, trees are able to grow quite large. There is an enormous amount of biodiversity in the rain forests. Estimates suggest that more than two-thirds of the world's species reside in this biome. Less known about the rain forest, however, is that most of the nutrients and the biodiversity are in the canopy. The topsoil tends to be acidic and quite thin. This causes repercussions that can be detrimental when the rain forest is cut down to convert rain forest into agricultural land. Whether the land will be used for growing crops such as soy or corn, or used for grazing livestock, the nutrients can provide for only a couple of seasons. Beyond that, the soil requires extensive fertilizers and amendments in order to remain fruitful. This is often cost prohibitive, and the land is abandoned. Without plantings or the nutrients to support them, the area can become dry and barren. In fact, the dry hot air emitted from the

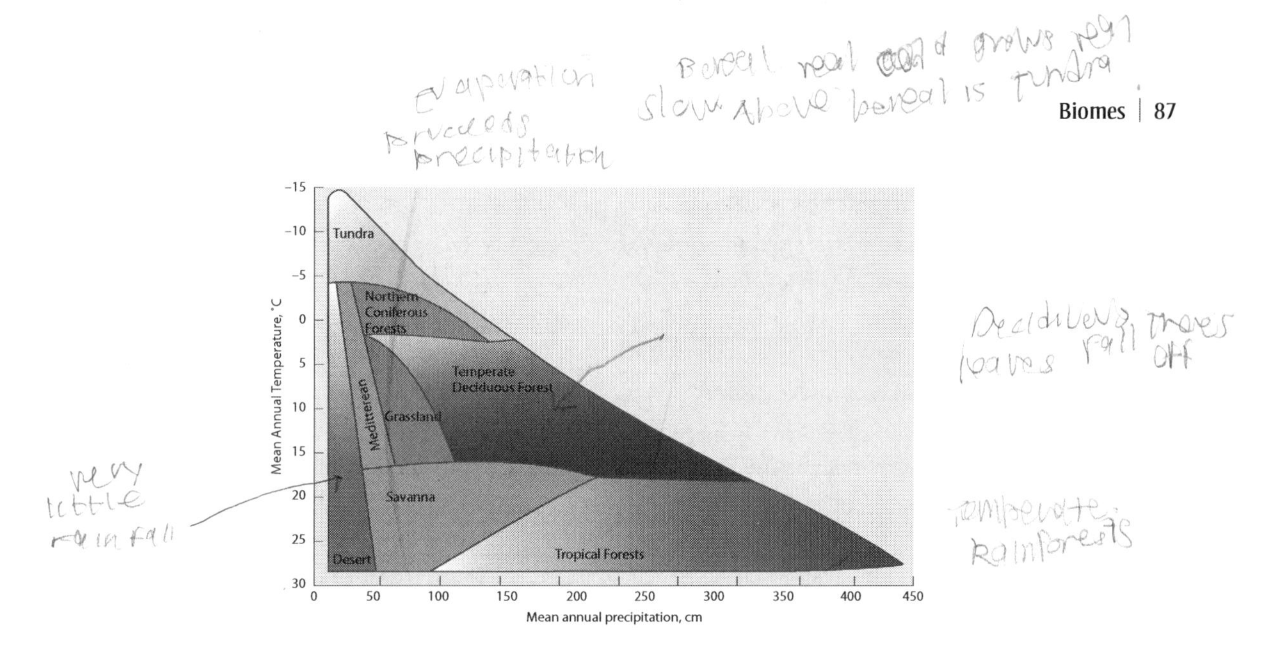

Biome Chart Figure 5.1. Biome distribution based on mean annual temperature and precipitation.

area can drive away moisture-laden clouds, thereby perpetuating the land conversion to something desertlike. This phenomenon is call **desertification**.

Tropical savannahs are close to the equator and have a relatively warm, stable temperature year round, but they differ from the rain forests in that they have a lengthy dry season in the winter months. The decrease in moisture prevents the growth of the same type of foliage in the savannahs. Deep-rooted grasses and smaller shrubs that can regenerate after a brush fire are common in the tropical savannahs. Migratory grazers such as buffalo are common in this biome because they are well adapted to the available grasses for food.

The desert is a biome mainly characterized by constantly dry weather, i.e., evaporation exceeds precipitation throughout the year. Because of this extreme lack of moisture, we see some interesting adaptations in the life there. Most organisms successful in this biome have adaptations that allow them to conserve water. Many animals do not have sweat glands and excrete crystals instead of liquid urine. They also tend to be nocturnal. As is the case with many cacti and succulents, many plants have adapted to store water.

Temperate deciduous forests have seasonal temperatures and rainfall. In the United States, these forests can be seen in the Virginias. Though there are coniferous trees in these forests, the predominant species are deciduous, i.e., they drop their leaves during the winter months.

Temperate grasslands differ from tropical grasslands in that they have seasonal temperatures and no dry season. Moisture is such that it can support abundant grasses and small shrubs, but not forests. In the United States, Kansas is a state belonging to this biome. Temperate grasslands often have a thick, rich organic layer of topsoil well suited for growing agricultural crops such as corn, wheat, and soy.

There is very little precipitation in the **tundra**, the coldest biome we will discuss. Though the precipitation exceeds evaporation, it does so because temperatures are so low year round. These low temperatures allow the perpetuation of a frozen layer of topsoil called "permafrost."

Aquatic Biomes

The previously discussed biomes were all terrestrial, but aquatic biomes also house a wealth of biodiversity. Aquatic communities also have some key characteristics that include temperature, nutrient availability (similar to soil richness in terrestrial biomes), current, and depth. As one goes from shallow depths to deeper areas in aquatic ecosystems, the body of water becomes stratified by oxygen and light as well. The area near the surface of a body of water, the epilimnion layer, has the most light penetration, and less light can penetrate as you move closer toward the bottom or benthos layer. Typically, temperature changes as you move deeper as well. As you progress from the epilimnion toward the benthos layer, you will pass through an abrupt temperature gradient, the thermocline, below which temperatures are much colder. The colder layer is referred to as the hypolimnion. In addition to light and temperature changes, oxygen is stratified based on depth. The epilimnion is the most oxygen-rich layer in part because it is near the surface and is able to mix with air. Progression toward the hypolimnion will show a decline in oxygen concentration.

Three examples of freshwater biomes/ecosystems are swamps, marshes, and bogs. Swamps are low-lying, often-saturated wetlands. Swamps are forested wetlands. Should a low-lying region that is seasonally or always saturated not have woody species present, it is referred to as a marsh. A bog is an area that is saturated and too soft to support building and structures or other heavy bodies, such as a swamp or a marsh, but also is characterized by an acidic substrate composed chiefly of sphagnum moss and peat (undecayed vegetation), in which characteristic shrubs, herbs, and trees can grow.

Biodiversity

There are many levels of diversity that, taken together, define the biodiversity of an area. At the population level, there are several versions of the same genes and sometimes absence of some genes at the genetic level. This can be seen in the different morphology of some members of a population as well as in differences in disease susceptibility. Species diversity refers to the number of different types of organisms living in an ecosystem. The more diversity an ecosystem has, the more easily it can resist perturbations. Ecological diversity takes into account the complexity of a system by looking for the number of different trophic levels and niches present. The more complex an ecosystem, the more inertia it has and the better able it is to persist in the face of challenges. Biodiversity quantification can be a daunting task and often involves taxonomists. To date, approximately 1.8 million species have been identified, but that is not nearly the total number of species existing on the planet. Taxonomists estimate that the total number of different kinds of species on the planet may range from 8 to 50 million. These unidentified species may be important to ecosystem and human health. They may be useful as a food source or may hold important chemicals that can be used as pharmaceuticals. The task of many environmental health scientists is to preserve this biodiversity so that we may study these organisms and perhaps derive some use for them. Certainly, though, the most biodiversity is found in the tropical rain forest and equatorial biomes. Unfortunately, these are the biomes threatened with the most land loss due to conversion to agriculture and urban areas (Figure 5.2).

The food that we eat on the planet may seem as if it comes from a diverse set of plants and animals, but Americans typically eat little variety. Certainly we manage to package the 12 to 18 items we do eat into a myriad of products, but we are beholden to relatively few crop plants and livestock for our nutrients. In fact, just three crop plants, (rice, wheat, and corn) account for more than half of the world's food supply. Should something such as a disease, blight, or even extreme drought threaten one of our major food sources, we will have to look to other sources for something to eat. Therefore, preserving biodiversity is important in order for us to feed the growing population on the planet. Not only is biodiversity important with respect to what we eat, but also with respect to how we produce our food.

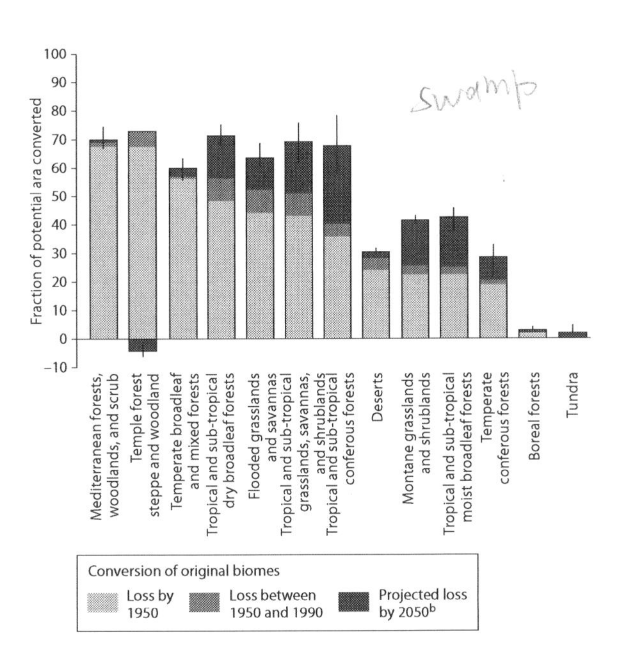

Figure 5.2. Past land loss and projected future land loss by biome.

Agricultural systems rely on the diversity of ecosystems to produce the food. For example, a diverse set of pollinators from insect to mammals is required for plants to come to fruition and produce the portions that we deem edible. A diverse set of microbes must be present in the soil to make nutrients available for the plants we eat directly and for the plants that we feed to livestock.

Biodiversity has been critical in the discovery of most all of the drugs and medicines used by man. Until recent advances in chemistry allowed for the synthesis of novel compounds, scientists relied almost completely on natural products for our pharmaceuticals. Antibiotics have been derived from bacteria and fungi, digitalis was derived from the foxglove plant, and the anticancer drug Taxol was extracted from the bark of the Pacific yew. In fact, approximately 119 pure-chemical substances extracted from higher plants are used in medicine throughout the world.

Ecological and Economic Benefits

The ecological benefits to preserving biodiversity are many. Areas with high species diversity have higher quality soil, more remediation of chemical pollutants, and increased resistance to disease in crop plants and livestock.

The economic benefit of retaining our biodiversity in the United States totals more than $300 billion each year. This includes everything from the savings wetlands allow by cleaning up our water, biological pest control, and, as previously mentioned, pharmaceuticals. In addition, tourism in our national parks and the industry related to outdoor activities in these parks contribute to the billions of dollars in economic benefits.

Predator and Pest Control

It has been estimated that there are more than 70,000 pest species worldwide that can attack agricultural crops. Ninety-nine percent of these pests can be controlled by natural predators and natural plant resistance. Under natural conditions, when pesticides are not heavily sprayed, each of these pests can have up to 15 natural predators. Without natural predators of pests and host-plant resistance, crop losses would reach up to 67% per year.

Why Study Biodiversity?

In order to preserve biodiversity, we must understand the circumstances that threaten it. Many of these threats, such as weather conditions and geological disturbances, are natural and are mostly out of our control. However, it is thought that most threats to biodiversity are currently caused by humans.

QUESTIONS FOR THOUGHT

1. Which biomes have you visited? What plant and animal species did you notice there? What were the temperature and precipitation profiles?
2. What human activities do you think are the most threatening to each of the biomes described in this chapter?

Biodiversity

By Antje Brown

The term "biodiversity," though used widely and liberally by researchers and practitioners, refers to a complex and under-researched environmental policy area. Deforestation, habitat destruction, wildlife conservation, overfishing, species extinction, and the introduction of genetically modified organisms (GMOs) have all necessitated the adoption of a biodiversity regime at the UN level in the form of the Biodiversity Convention of 1992 and subsequent Biosafety Protocol of 2000, as well as various ad hoc working groups and thematic programmes.[1] This chapter explores the different aspects involved in the biodiversity debate and identifies key actors and their interests involved. It also provides an overview of the UN policy to date and highlights unresolved issues that are likely to occupy stakeholders in the near future. Ultimately, biodiversity is positioned at the policy periphery and is not integrated properly into political and economic paradigms of societies. This neglect is somewhat surprising, considering that biodiversity involves—in true "think global act local" fashion—both local communities and intergovernmental organizations such as the UN, the EU, and the WTO. Biodiversity has caused much controversy over the years and continues to throw up a number of economic, political, and ethical questions.

In policy terms, biodiversity refers to the notion that the access and use of shared natural resources should be regulated carefully and in a sustainable manner following centuries of unsustainable resource exploitation and degradation. More importantly, worldwide biodiversity loss suggests that the problem needs to be solved at the international level through regime-building and a system of global governance.

The objective of the UN policy is threefold: conservation, the sustainable use of natural resources, and benefit-sharing. These three objectives should ensure that natural resources benefit both current and future generations. In order to be effective, however, the policy has to be applied evenly in every part of the world, and this, in turn, suggests that the UN policy affects (or interferes with) the sovereignty and decision-making powers of national and local governments. Member states are meant to ratify an international agreement through a vote in their parliaments and arrive at a democratic policy; however, complex negotiations at the international level involve many trade-offs. It is therefore inevitable that policies raise questions of equity and justice. Furthermore, in light of modern GM technology and intellectual property rights, biodiversity increasingly affects relationships between transnational actors representing environmental and economic interests. In a climate where economic actors are allowed to patent natural processes, the lines between nature and business become ever more blurred. This obfuscates the supposed open-access, common character of "natural processes" and transforms them into private property, denying universal access. In addition, policies influence relations between environmentally rich and economically poor countries in the South and economically rich and environmentally poor countries in the North. It therefore touches upon issues such as multilevel governance, economic relations, development, and environmental justice.

The chapter explores the different aspects of the biodiversity debate and identifies key actors and their interests involved in regime formation. Biodiversity is an area that makes it particularly clear that the traditional state-centric way of organizing and regulating an environmental problem is facing severe limits. It also assesses to what extent international relations and environmental theory can explain policy developments while also considering the lessons that can be learned from the biodiversity case study for the discipline of international environmental politics.

DEFINING BIODIVERSITY

According to the UN Convention on Biological Diversity of 1992, biodiversity can be defined as: "variability among all living organisms from all sources ... ; this includes diversity within species, between species and of ecosystems." Put simply, biodiversity is about diversity in species, ecosystems, landscapes, and genetic resources. This means it is not just about endangered animals or plants but also about their genetic material and about the relations between species, plants, and humans in particular areas. Scientists estimate the total number of species as somewhere between 5 and 15 million, 1.75 million of which have been formally identified worldwide, and that there remain a greater number than this that are unknown.[2] Ecosystems vary from forests, deserts, rivers, mountains, oceans, swamps, and more—all consisting of carefully balanced and interdependent components. Ecosystems and species have come under increasing pressure in recent decades. Pressure can occur in many forms, predominantly man-made, and can be caused by hunting, overfishing, logging, industrial pollution, climate change, intensive farming, and the commercial exploitation of particular regions. The current number of species officially recognized as threatened is large. One just needs to look at the list provided by the Convention on International Trade in Endangered Species of Wild Fauna and Flora (CITES) to realize that the rate at which ecosystems and species are destroyed is alarming.

CITES is an international treaty on endangered species which was signed by eighty parties in March 1973 in Washington, DC. It entered into force in July 1975 and now has 175 parties. The convention is a voluntary agreement to "ensure that international trade in specimens of wild fauna and flora does not threaten their survival." This is achieved by identifying different types of endangered species and regulating (and, if necessary, banning) their trade worldwide. Endangered species—to date, CITES has formally identified 5,000 animal species and 28,000 plant species—are listed in three appendices according to the level of the threat against them. It has to be noted that these lists include whole groups of species, such as primates, therefore increasing the actual number of species covered by the convention. Endangered species are the most visible form of biodiversity loss and were recognized as an environmental problem earlier than other biodiversity issues.

Biodiversity loss as a whole is projected to accelerate tenfold by 2050. A recent UN report, *Global Environmental Outlook GEO-4* (UNEP 2007), pointed out that many animal and plant species have seen a dramatic decline both in terms of numbers *and* in terms of geographical spread. For example, in regions with a dense population of diverse species, such as areas threatened by deforestation, there is an acute danger of losing substantial numbers of species and genetic material for good. One-quarter of mammal species are currently threatened by extinction. Loss of biodiversity is no longer a problem limited to local communities or regions; it has become a global problem which requires collective action at the international level on account of the nature of the global political economy and also the interconnectedness of ecosystems.

The problem of biodiversity loss is therefore recognized not only as a scientific fact but also as a "human-made" problem that calls for collective action. It is at this point that biodiversity moves beyond pure *science* and becomes a matter of *politics* and *international relations.* Collective action starts with the premise that biodiversity requires a regime that seeks to regulate the conservation, access, and use of shared natural resources. This should be conducted in a sustainable and fair manner after centuries of resource exploitation and degradation. Indeed, UN documents confirm and emphasize this train of thought. According to the UN, current biodiversity loss is not only unprecedented but it takes an entirely new form: "all available evidence points to a sixth major extinction event currently under way. Unlike the previous five events, which were due to natural disasters and planetary change, the current loss of biodiversity is mainly due to human activities" (UNEP 2007). These human impacts are either the result of direct intervention, for example through deforestation by large timber companies, or the indirect result of wider developments such as climate change and its effects on flora and fauna.

FACING POLITICAL CONUNDRUMS

One might assume that the objective of preserving the diversity of our species and ecosystems is simple and straightforward. However, in reality, regime-building at the UN level and compliance on the ground has produced a number of conundrums in a similar yet unique way compared with other attempts to curb and regulate environmental problems.

First, the UN convention seeks to ensure the *sensible* and sustainable use of resources and thereby provide for biodiversity for generations to come. In other words, it establishes an intergenerational responsibility. However, in order to be successful in the long term, the policy has to be applied *evenly* and effectively in every country and at every level of government. It is therefore inevitable that the UN policy touches upon the sovereignty and decision-making powers of national and local governments. In this case, biodiversity is an excellent example of how the term "think global, act local" should be understood. Indeed, the policy is dependent on the commitment of national and local actors. Or, as Le Prestre (2002) puts it, "the policy succeeds or fails at the national level." In other words, it takes cooperation and commitment at the international level, but the action then has to follow at the national and subnational level to implement what was agreed. However, this commitment toward biodiversity in every part of the world is difficult to achieve. Many researchers have documented variations in environmental policy commitment and have discussed the ever apparent "tragedy of the commons," which suggests that individuals will always seek to maximize their short-term (economic) interests even if this is to the long-term detriment of the environment and society at large.[4] Biodiversity is no exception to these observations; indeed, in many respects the biodiversity regime illustrates the "tragedy" in following short-term profits rather than long-term goals.

So part of the reason why the international community has failed to establish an effective biodiversity regime is because of actors' interests and their relationships. At first glance it may not be immediately obvious, but biodiversity touches upon vital economic interests, particularly in the areas of modern GM technology and intellectual property rights. Here the UN policy seeks to mediate between economic and preservation interests—that is, between actors who aim to control and use animal and plant species for economic profit (e.g. pharmaceutical companies) and actors who seek to protect species and ecosystems for their own sake and with them wider environmental values (e.g. non-governmental organizations). While there are instances where economic interests and environmental protection can be combined in line with sustainable development—ecotourism is just one example—there are many other instances where economic and environmental interests clash, as will be discussed further below.

Another complexity concerns *intra*-generational relations and responsibilities: the UN policy raises questions regarding equity and justice, as it challenges the already tense relationship between developed and developing countries. Paradoxically, developing countries tend to be rich in terms of natural resources but economically poor while developed (industrial) countries tend to be comparatively poor in terms of natural resources but economically rich.[5] In other words, potentially profitable natural or genetic resources tend to be located in developing countries, while the large economic players able to take advantage of them tend to be based in developed countries. Over the years, actors from the developed world have proven to be better equipped for UN negotiations than their counterparts from developing countries. While actors from developing countries (e.g. representatives from indigenous communities) have started to improve their participation and representation skills in biodiversity regime-building, they are still lagging behind actors from the developed North.[6]

The above complexities and diverse interests have contributed toward a rather cumbersome regime-building process and a UN policy that is far from being effective.

REGIME-BUILDING AMONG A DIVERSITY IN INTERESTS

There is no doubt among the international community that biodiversity loss is real and man-made, and that its prevention requires collective action. The international community also acknowledges the responsibility of the current generation to ensure biodiversity for future generations. However, in light of the above conundrums, there are major questions regarding sovereignty, equity, and interest representation that prevent actors from agreeing on common or shared policy details which in turn would contribute toward an effective regime. In essence, differences exist in terms of:

- the prioritization of the problem;
- a commitment to regulate the use of natural resources and, if necessary, to make (economic) adjustments;
- accepting concessions with regard to national sovereignty for the sake of the wider international good.

To date these differences have contributed toward a somewhat ineffective international regime,[7] and it is unlikely that, under current systems of political and economic governance, the international community will adopt a system that will halt the trend of biodiversity loss in the long term. In order to understand the shortfalls, it is necessary to look at the regime-building process so far and the interests involved in shaping the policy.

Despite biodiversity loss being an ongoing process spanning several centuries, it was not until the early 1980s that the international community saw a consensus among scientists that it was under threat. After a number of initiatives—one of which, interestingly, came from the US[8]—the UN Environment Programme (UNEP) set up a working group to formulate an international convention on biodiversity. The working group at this stage had already encountered a North-South divide over genetic resources and intellectual property rights. This would resurface during the 1992 Rio Summit negotiations.

Basically, the main obstacle in agreeing on a collective plan of action concerned the issue of *ownership of genetic resources:* governments and actors in the developing world wanted to protect their (sovereign) rights over the access and use of their natural resources, while governments and actors in the developed North argued that these resources formed part of a "common heritage of mankind" or were simply "owned by no one." In other words, the developed countries' perspective considered natural resources as common or shared resources that should be freely accessible to all. Previously, representatives from developed countries had won the argument and had benefited economically from this arrangement. This time, however, the South sought to regain control over their natural resources and thereby tackle what they perceived to be "biopiracy" and "genetic robbery."[9] Decision-makers trying to develop good, coherent, consistent policy on genetic resources continue to be faced with a multitude of interconnected issues. Even discerning which are relevant to conserving and managing these resources, and integrating them into policy, is extremely difficult.

A fundamental area of dispute is who should share the benefits derived from the exploitation of genetic resources and biotechnology. The lines are usually drawn between developed and developing countries and, within countries, between local communities—usually rural and indigenous peoples—and their better-off, more powerful compatriots.

The concerns are numerous. Some people, for example, consider that the expanded scope of intellectual property rights and their extension to biological materials enable institutions and researchers to appropriate and limit access to the resources and knowledge of farmers and indigenous communities without compensation or consent, especially in the developing world. Others point to the practices of multinational seed companies that are developing, implementing, and promoting a variety of technological tools to restrict the rights of farmers to save and reuse seeds from their harvests. This is taking place against a backdrop of accelerating loss of biological diversity, as forests are felled for timber or to make way for agriculture, fisheries collapse, and the number of plant and animal species facing extinction grows.

Not far from this ownership debate was the debate over the distinction between "old" and "new" species. Here, representatives from the North insisted on a clear distinction between "old" and "new" in their favor. They argued that any species identified recently were effectively the private property of those who had discovered them (often large transnational corporations), while species that had been long established and formed part of traditional knowledge were common and therefore shared (or free) property. In other words, while those from developed countries (e.g. large pharmaceutical companies) wanted free access to long-established genetic resources, they sought control over (and to benefit from) newly identified resources. This approach led to some astonishing decisions, the most publicized one being the US decision of 1995 to grant a patent for turmeric (a herb) to two US doctors even though turmeric's anti-inflammatory properties had been known for centuries as part of the Indian Ayurvedic tradition.[10] Understandably, actors from the South tried to rebalance this approach at the time of negotiation. And, equally understandably, those from the North tried to resist the pressure and focus instead on less controversial but media-friendly issues such as the protection of mammals and the conservation of rainforests.

What actors, including UN officials, national governments, and transnational actors (transnational corporations and environmental NGOs), had to thrash out was a UN convention that would ensure not only the *sustainable* use of natural resources but also the *fair and equitable* use of these resources. However, influenced by their respective "rational" self-interests, the majority of governments stayed well clear of quantifying specific and binding targets of biodiversity conservation and regulation, as had been suggested by some governments and NGOs. Setting watertight targets and thereby committing all sides to clear objectives proved to be a step too far for key actors in both North *and* South.

The actual convention document reflects the attitudes and interests of the main negotiators. It is essentially about good intentions regarding biodiversity but also about maintaining the existing economic paradigm. It advocates economic incentives to encourage biodiversity conservation rather than imposing tough regulation. Signed in 1992 by 150 governments at the Rio Earth Summit, the convention consists of a preamble, forty-two short articles, and three annexes covering identification and monitoring, arbitration, and conciliation. The protocol is a compromise between divergent interests. It also highlights the paradox between national sovereignty and international responsibility and cannot hide its undercurrent economic agenda:

> The objectives of this Convention are the conservation of biological diversity, the sustainable use of its components and the fair and equitable sharing of the benefits arising out of the utilisation of genetic resources, including by appropriate access to genetic resources and by appropriate transfer of relevant technologies, taking into account all rights over those resources and to technologies, and by appropriate funding.
>
> (Convention on Biological Diversity, 1992, Article 1)

Furthermore, Article 3 states the following:

> States have ... the sovereign right to exploit their own resources pursuant to their own environmental policies, and the responsibility to ensure that activities within their jurisdiction or control do not cause damage to the environment of other states or of areas beyond the limits of national jurisdiction.

To put it bluntly, the convention offers something for everybody. The document goes on to outline the "intrinsic ecological, genetic, social, economic, scientific, educational, cultural, recreational and aesthetic values" inherent in biodiversity and argues that biodiversity strengthens friendly relations among states and contributes toward peace for humankind. These objectives would be met through (loosely formulated) policy tools such as national strategies, plans, and programs as well as biodiversity identification, monitoring, conservation, and impact assessment.

Beyond good intentions, and in more practical terms, the convention appeals to governments and their economic actors to consider the environment first, *before* exploiting natural resources. Further, they should take into account what their local or national actions might do to the *international* environment. And, as far as access to genetic resources is concerned, the convention manages to find a compromise between North and South by leaving decision-making to the parties involved: they should regulate the access and use of resources on "mutually agreed terms." In effect, this compromise does not limit resource exploitation; it simply calls upon interested parties—often governments in the South and transnational corporations from the North—to act *sensibly* (whatever this may entail) and work out a deal between themselves.

Apart from the questions of ownership and access to natural resources, two other issues had to be resolved: technology transfer and funding. First, negotiators from the United States expressed concern that the inclusion of *technology transfer* from North to South would jeopardize intellectual property rights and therefore economic interests in the North.

India, on behalf of the South, represented the other side of the argument and settled in the end for a compromise: the convention acknowledges intellectual property rights but also insists that these should not contradict the aims and objectives of biodiversity.

The second issue surrounded the question of whether or not the convention deserved its own *funding mechanism*. The North insisted that funding should be provided through the already existing Global Environment Facility (GEF). From a Northern point of view, a new funding mechanism might have involved more financial commitment and less control over the policy itself, and this was an option the developed countries could not accept. It was the then executive director of UNEP, Mostafa Tolba, who put pressure on negotiators to accept a "take it or leave it" compromise. He proposed the GEF as funding mechanism in the first instance but also a concession for the South in the form of more transparency. To date the GEF is still involved in biodiversity funding; in fact, biodiversity now constitutes its largest portfolio (Le Prestre 2002).

While Tolba's proposals were duly adopted by 150 governments, interpretations of convention details continued to differ after the event depending on actors' respective sets of interests. Notably, the document was not signed by the then US president, George H. W. Bush, because the points of funding and intellectual property rights were still deemed unacceptable. While his successor, Bill Clinton, signed at a later stage, the US has not to date ratified the convention and is therefore not party to it. Considering that, by 2005, 188 countries had ratified, it can be argued that the US stands very much alone and isolated on the issue of biodiversity.

Of course, the convention signalled a starting point. Over the years, the various parties have had to fill in policy gaps with details in the form of protocols and follow-up work programs. To date we have seen only one protocol, the Cartagena Protocol of 2000, which focuses on living modified organisms and their regulation vis-a-vis biodiversity. The first work program was adopted in 1998, and so far such programs have proven to be rhetoric at best.

One major obstacle in filling the convention's gaps with effective policy details constituted the formation of so-called veto coalitions that pursue their own specific economic interests. For instance, on the issue of deforestation, the parties saw a veto coalition made up of Brazil, Canada, and Malaysia, whose governments made sure that any follow-up policies would not interfere unnecessarily with their large timber and logging industries (Chasek, et al. 2006: 163).

A welcome distraction from the above controversies of ownership and access to natural/genetic resources was biosafety, more specifically the regulation of genetically modified organisms. The resulting Cartagena Biosafety Protocol of 2000, however, was not free of controversy either. The protocol consists of a preamble, forty articles, and three annexes, which cover information required in notifications concerning *living modified organisms* (LMOs) as well as risk assessments. Initially intended to regulate the trade in GMOs and their impacts on biodiversity, the document is, yet again, a watered-down compromise between transnational actors. Much to the disappointment of environmental NGOs, it effectively legitimizes the use and trade of GMOs for transnational corporations. Its key policy tool is the Advanced Informed Agreement (AIA), which comprises a prior notification and consent procedure. This procedure is applied mainly to LMOs such as GM seeds. During the formulation stage of the protocol, controversy flared up regarding the inclusion of other GMOs and the question of whether or not to integrate the precautionary principle in addition to the issue of clarifying the relationship between this UN protocol and relevant WTO rules. Not surprisingly, these issues would be resolved with a watered-down compromise in the final version of the protocol (outlined below).

Apart from typical divisions between environmental and economic transnational actors, two large groups of states were formed in the course of negotiations. To a certain extent, these two groups adopted the roles of environmental and economic advocates. The so-called Miami Group, consisting of the US, Canada, Australia, Chile, Argentina, and Uruguay, pursued a lenient policy approach by keeping the number of GMOs to be included under the protocol to an absolute minimum while at the same time opposing the precautionary principle. At the opposite end of the negotiating table was the Like-Minded Group, consisting of developing countries and EU member states; this group pursued a more regulated approach, with more GMOs to be included, and insisted on the precautionary principle.

It is striking how the protocol was shaped not only by government representatives and environmental NGOs (which operated very much in the background) but also by transnational corporations, such as Monsanto, DuPont, and Sungenta (formerly Novartis and Zeneca), with vital interests in GM technology. These transnational corporations went on to create a Global Industry Coalition which would achieve a more coordinated voice and a strong force during negotiations. The coalition, chaired by BioteCanada, involved some 2,200 companies, thereby forming an unprecedented, united, and proactive economic front. This observation is similar to Jennifer Clapp's "enhanced lobby power," described in Chapter 3.

Considering the different forces in biodiversity regime-building, it is not surprising that the final version of the protocol was a compromise solution. It does cover a number of GMOs; however, these are put in different categories and governed by different criteria and, more importantly, exemptions. LMOs come under the formal AIA procedure for first international boundary movement. Other GMOs exempted from AIA are processed through a separate Biosafety Clearing House, which is a simplified notification mechanism through an internet database. As a concession to the Like-Minded Group, the precautionary principle is included in the document; however, the principle is considered only if scientific evidence can be presented to justify its application and only if the matter of cost-effectiveness is taken into consideration. The protocol also leaves existing WTO rules intact. It could be argued that it has, in effect, given transnational corporations more certainty and stability rather than providing a "green obstacle" in the trade of GMOs.

Elsewhere, discussions still continue on the issue of access to genetic resources, which begs the question whether negotiators are truly committed to biodiversity protection or whether their real interest lies in the maximization of economic interests. In 2002, the parties to the convention agreed to the Bonn Guidelines on the Access to Genetic Resources and the Fair and Equitable Sharing of the Benefits Arising out of their Utilization. Meanwhile, the great expectations on either side of the North-South divide to exploit and benefit from the "green gold" have not materialized.

In recent years, the UN has attempted to establish a link between climate change and biodiversity loss. UNEP set up an ad hoc Technical Expert

Group on Biological Diversity and Climate Change, which produced a document in October 2003 titled "Interlinkages between biological diversity and climate change: advice on the integration of biodiversity considerations into the implementation of the UN FCCC and its Kyoto Protocol." The document identifies specific links between climate change and biodiversity loss—for instance, climate change impacts on species migration as well as species extinction and ecosystem changes. Furthermore, the document considers indirect impacts through human population adjustments and adaptations. Among proposals for change are policy integration, sustainable management, including *afforestation* (to counteract deforestation), and—last but not least—an effective climate change policy.

May 2008 saw the ninth official meeting of the parties in Bonn, where, again, the main themes were the interlinkage between climate change and biodiversity loss and attaching a monetary value to natural resources, thereby internalizing natural resources as part of the dominant economic paradigm. Attaching a value to biodiversity, an exercise increasingly popular since the 1990s, is currently gaining in momentum.[11] Among recent estimates are the following gains or values:

- the annual world fish catch—$58 billion;
- anti-cancer agents from marine organisms—up to $1 billion per year;
- global herbal medicine market—circa $43 billion in 2001;
- honeybees as pollinators for agriculture crops—$2-8 billion per year;
- coral reefs for fisheries and tourism—$30 billion per year.

In similar fashion, the projected costs include some very specific estimates, such as:

- mangrove degradation in Pakistan—$20 billion in fishing losses, $500,000 in timber losses, $1.5 million in feed and pasture losses;
- Newfoundland cod fishery collapse—$2 billion and tens of thousands of job losses.

(UNEP 2007)

By attaching a monetary or economic value to natural resources, it is hoped that actors use these resources, or "capital," sensibly and not as "free-for-all" commodities/common goods.[12] In its efforts to attach a monetary value to biodiversity, the UN is also looking into turning so-called carbon sinks into profitable protected areas that can be bought and traded, just like carbon permits. Already, some investors and businesses have shown an interest, as carbon-sink trading may become a convenient way of offsetting greenhouse gas emissions, which is in line with the Kyoto Protocol and its climate change targets. Yet, one all-too-familiar question is already being raised: once they have been identified as market commodities, who will own these carbon sinks—private investors, such as transnational corporations from the North, or local communities/countries from the South? There are already pilot schemes, such as the Forest Carbon Partnership Facility in the Republic of Congo, which offer interesting testing grounds. Another example is the ADB carbon sequestration project in Indonesia, which is funded by the Asian Development Bank and is intended to earn Indonesia emission reduction credits. The Global Forest Coalition, a coalition of NGOs and indigenous peoples' organizations (founded in 2000), is currently looking into these carbon-sink projects, and their initial assessment appears to be skeptical.[13] The projects have not entirely addressed the question of ownership, nor do they help remove all the obstacles that have prevented actors from establishing a proper and collective mediating and policing system.

It is, nevertheless, noticeable how developments in the climate change policy area have fed into biodiversity. There are, indeed, obvious scientific as well as political interconnections. A chapter on biodiversity in *GEO-4* (UNEP 2007) offers examples of climate change-biodiversity interlinkages. *GEO-4* reports, for instance, on extinction cases in the amphibians category as well as alterations in species distribution (e.g. Arctic foxes, mountain plants, Northern temperate butterflies, and British birds) and variations in tree distribution in Europe resulting from climate change. Furthermore, it highlights modifications in species behaviour; these include earlier flight times in insects and differences in breeding patterns of amphibians and flowering of trees. And, finally, *GEO-4* points out changes in population demography—for instance, transformation in population sex ratios in reptiles. From a slightly different angle, but conveying a similar message, the *Carbon and Biodiversity* atlas (UNEP 2008) highlights so-called biodiversity hotspots, or areas threatened by biodiversity loss, whose protection would not only help maintain biodiversity but also contribute toward the creation of carbon sinks under the Kyoto Protocol.

What is interesting about the link between climate change and biodiversity is the fact that actors have not only identified this link in scientific terms, but have also started to connect the two policy areas in very practical political and economic terms. One obvious example would be the adoption of economic policy tools such as permit trading, which is already tried and tested in the climate change area and is now being envisaged for biodiversity. A certain degree of lesson learning is evident; this can also be seen in the fact that the UN is planning to introduce a group of biodiversity experts similar to that of the Intergovernmental Panel on Climate Change (IPCC).

AN ASSESSMENT OF BIODIVERSITY POLICY IN INTERNATIONAL RELATIONS

The above chapter has produced a number of findings that are typical for environmental regime-building. At the outset, there is an international consensus that biodiversity loss is real (i.e. scientifically proven) and predominantly a man-made problem which requires collective action. International actors also share some common ground on good intentions, such as helping "poor" indigenous communities maintain their habitats and species. However, once negotiators venture into policy details, there are many areas where styles, priorities, and interests depart from each other or simply clash. The result of the compromise-seeking process is that the biodiversity regime is somewhat watered down and—from an environmentalist's point of view—rather disappointing. In this sense, the biodiversity regime is not much different from other international environmental regimes, such as that of climate change.

As has been mentioned, the protection of species and habitats is not a simple and straightforward environmental task. Biodiversity is complex, not just in scientific but also in political and economic terms. For a start, it affects all levels of governance—from the global to the local. Consequently, its regime covers questions of self-determination vis-a-vis collective action and international responsibility (the "commons dilemma" springs to mind). In addition, it deals with economic interests, particularly those of transnational corporations which have an interest in accessing genetic and natural resources in order to feed their new technology and pharmaceutical industries. On top of that, biodiversity has *intra-* as well as intergenerational implications and generates debates over justice, equity, and responsibility: we have learned that natural resources are distributed unevenly between populations, as are their uses and benefits, and that the current generation has a responsibility toward future generations in taking effective and collective action to ensure that endangered species and habitats do not disappear forever. This is perhaps where biodiversity departs from other environmental policy areas. It is this sense of irreversibility and finality that makes biodiversity so special. As effective *and* collective action is not forthcoming, it is unlikely that the problem of biodiversity loss will be tackled in the foreseeable future. Biodiversity therefore touches upon a number of wider and fundamental discourses in political economics, global governance, and environmental justice, be it of an intergenerational or *intra*generational nature; these may not be evident to such an extent in other policy areas.

One of the main problems of biodiversity policy has been the "horse-trading" between actors, which in turn has distracted them away from the actual environmental objective (i.e. to stem the trend of biodiversity loss) and instead turned their attention toward the economic interests associated with natural or genetic resources. Since the convention's adoption, the parties have tried to live up to the official policy goal of a "significant reduction in the current rate of biodiversity loss by 2010." However, with the 2010 deadline looming, there is no scientific evidence to suggest that the goal has been achieved. The trend of biodiversity loss has not been halted or even slowed down; indeed, it continues to accelerate at an alarming rate. Strategies and plans at national levels may well exist alongside the convention, but they are evidently insufficient in stopping the trend of biodiversity loss. Alongside other government priorities and day-to-day business (such as addressing international terrorism and the global financial crisis), these strategies appear to be too small and too insignificant to make a difference. Biodiversity is positioned at the policy periphery and is not integrated properly into the political and economic paradigms of societies.

Having noted at the outset that biodiversity is a neglected research area, this chapter has (hopefully) demonstrated that this neglect is unjustified. In many respects, regime-building in biodiversity has proven to be a case of environmental regime "meddling." And yet biodiversity has also proven to be an interesting, complex, and pressing policy matter that has affected international relations. One just needs to look at the "coalitions" of states that were formed during negotiations and transnational actor groups such as the Global Industry Coalition and their increased involvement in international regime-building.

From the perspective of international relations theory, insights from biodiversity regime-building are useful for a number of reasons: biodiversity is a good example in that it highlights the involvement of different government levels (from local communities to the UN); it demonstrates how environmental and economic spheres are interconnected; and it is a policy area that illustrates how transnational actors become polarized on policy details, which in turn can have an effect on regime-building and problem-solving.

By the same token, international relations theory can help us make sense of biodiversity regime-building. At first sight, the problem of biodiversity loss may be clear to all and a solution may appear to be straightforward. However, in order to understand the discrepancy between policy ideal and reality, we need to apply an *actor-centered* approach which focuses on transnational actors, their (environmental or economic) interests, and their places in a complex global setting. One finding of this chapter is that transnational actors' commitments, policy approaches, and policy measures on biodiversity are too varied to allow for decisive collective action. An international regime may exist on paper but, when it comes to effective problem-solving, it is too weak to impose collective discipline across the board. As it stands, the biodiversity regime is merely a reflection of the lack of proper commitment on the part of global civil society. Similarly, a *policy tool* approach can provide an understanding of recent policy changes.

The chapter has described how the UN has sought to give biodiversity a new impetus by establishing a causal link between climate change and biodiversity loss. This deliberate link was established partly because of obvious scientific evidence, but also in order to raise the biodiversity profile in a highly competitive global governance setting. In this setting a multitude of policy issues compete for "ear-time" in international relations. Whether or not this strategy will bring about the desired effects remains to be seen. Another International Day for Biological Diversity was marked on 22 May 2010. However, in the light of the global financial crisis, the international community and the media hardly took note of this event. The success of a global biodiversity policy will depend on the wider international context, which includes not only economic factors but also new scientific evidence and societal changes.

NOTES

1. For the full text of the Biodiversity Convention, visit: www.cbd.int. Thematic programs focus on the following habitat categories—marine and coastal areas, forests, agricultural land, inland waters, dry and subhumid lands, and mountain regions.
2. Spangenberg (2007) points out that quantifying biodiversity is a difficult if not an impossible task. The estimate of 1.75 million identified species is based on UN and EU literature.
3. For a comparison of variations in sustainability commitments, see, for instance, Baker (1997).
4. See Hardin (1968). Vogler (2000) offers a more modern version.
5. Banerjee (2003) highlights this paradox.
6. For further information on how indigenous communities seek to improve their negotiating powers, see Pachamama (2008). Also visit the Indigenous Peoples Council on Biocolonialism at: www. ipcb.org.
7. Le Prestre (2002) investigates the regime's effectiveness by focusing on policy learning, capacity-building, and norm changes. He notes that its development has been uneven, which is partly owing to a lack of proper monitoring and the slow development of common indicators that would help measure its impact.
8. This initiative is interesting because the US is one of the very few nations that have, to date, not ratified the UN convention and is therefore not party to it.
9. The term "genetic robbery" has been used by Vandana Shiva on several occasions and has been cited by many. See, for instance, Shiva et al. (1997) and Chamerik (2003).
10. It should be noted that the patent was issued by a non-party state, the US, three years after the adoption of the convention. The patent was later revoked. See Baker (2008) for further information.
11. The difficulty of this exercise in economic valuation of biodiversity can be seen in Nunes and van den Bergh (2001).
12. Pretty and Smith (2004) investigate the notion of environmental, economic, and social capital in greater detail.
13. For the Asian Development Bank and specifically the Indonesian carbon sequestration project, visit: www.adb.org. For information on the Global Forest Coalition, visit: www.globalforest coalition.org.

RECOMMENDED READING

Le Prestre, P. (2002.) The CBD at ten: the long road to effectiveness. *Journal of International Wildlife Law and Policy* 5: 269–85.

Pretty, J. and D. Smith. (2004). Social capital in biodiversity conservation and management. *Conservation Biology* 18(3): 631–8.

Vogler, J. (2000). *The Global Commons: Environmental and Technological Governance*. Chichester: Wiley.

ONLINE RESOURCES

Congo Basin Forest Partnership: www.cbfp.org
Convention on Biological Diversity: www.cbd.int
Convention on International Trade in Endangered Species of Wild Fauna and Flora: www.cites.org
Global Forest Coalition: www.globalforestcoalition.org
Indigenous Peoples Council on Biocolonialism: www.ipcb.org
UN Environmental Programme: www.unep.org

REFERENCES

Baker, L. (2008). Turf battles: politics interfere with species identification. *Scientific American* 299(6): 22–4.

Baker, S. (1997). *The Politics of Sustainable Development.* London: Routledge.

Banerjee, S. B. (2003). Who sustains whose development? Sustainable development and the reinvention of nature. *Organization Studies* 24(1): 143–80.

Chamerik, S. (2003). Community rights in global perspective, in X. Jianchu and S. Mikesell, (eds.). *Landscapes of Diversity.* Kunming: Yunnan Science and Technology Press.

Chasek, P., D.L. Downie, and J.W. Brown. (2006). *Global Environmental Politics, 4th ed.* Boulder, CO: Westview Press.

Hardin, G. (1968). The tragedy of the commons. *Science* 162(3859): 1243–8.

Le Prestre, P. (2002). The CBD at ten: the long road to effectiveness. *Journal of International Wildlife Law and Policy* 5(3): 269–85.

Nunes, P. and J. van den Bergh. (2001). Economic valuation of biodiversity: sense or nonsense? *Ecological Economics* 39(2): 203–22.

Pachamama, (2008), "Pachamama Newsletter," 2(2,). Available: www.cbd.int/doc/newsletters/news-8j-02-02-low-en.pdf.

Pretty, J. and D. Smith. (2004). Social capital in biodiversity conservation and management. *Conservation Biology* 18(3): 631–8.

Shiva, V., et al. (1997). *The Enclosure and Recovery of the Commons: Biodiversity, Indigenous Knowledge and Intellectual Property Rights.* New Delhi: Research Foundation for Science, Technology and Ecology.

Spangenberg, J. H. (2007). Biodiversity pressure and the driving forces behind it. *Ecological Economics* 61(1): 146–58.

UNEP (United Nations Environmental Programme) (2007), "Global Environmental Outlook: Environment for Development (GEO-4)," Malta: Progress Press, Available: www.unep.org/geo/geo4/report/GEO-4_Report_Full_en.pdf (accessed 17 November 2009).

(2008). *Carbon and Biodiversity: A Demonstration Atlas.* Cambridge: UNEP World Conservation

Monitoring Centre, Available: www.unep.org/pdf/carbon_biodiversity.pdf.

Vogler, J. (2000). *The Global Commons: Environmental and Technological Governance,* Chichester: Wiley.

Tourism, Biodiversity and Global Environmental Change

By C. Michael Hall

Biodiversity (biological diversity) refers to the total sum of biotic variation, ranging from the genetic level, through the species level and on to the ecosystem level. The concept therefore indicates diversity within and between species as well as the diversity of ecosystems. The extent or quantity of diversity can be expressed in terms of the size of a population, the abundance of different species, as well as the size of an ecosystem (area) and the number of ecosystems in a given area. The integrity or quality of biodiversity can be expressed in terms of the extent of diversity at the genetic level, and the resilience at the species and ecosystem level (Martens *et al.* 2003).

Biodiversity loss is a major policy issue and, as with climate change with which it intersects in both environmental and regime terms, is the subject of an international convention (McNeely 1990; Rosendal 2001; Kim 2004). The extinction of species is a natural process (Lande 1993, 1998). However, species and ecosystem loss has accelerated as a result of human activity. The United Nations Environment Programme (UNEP) estimates that almost 4,000 mammal, bird, reptile, amphibia and fish species are threatened with extinction, while about 600 species of animals are on a critically endangered species list (UNEP 2002; Nielsen 2005). Pitman and Jorgensen (2002) estimate that between 24 and 48 percent of the world's plant species are presently faced with extinction. Wilson (1992) estimated that one species was being lost every 20 minutes, with approximately 27,000 species being lost per year. In contrast, Pimm *etal.* (1995) estimated that human-induced extinction of species was as high as 140,000 per year. At that rate, half of the existing species will be extinct in 70 years.

Although the exact rate of biodiversity loss is disputed (Purvis and Hector 2000) there is no doubt that human domination of the natural environment has led to a decline in biodiversity at all levels, with an acceleration in the rate of species extinction in recent years. According to Martens *et al.* (2003):

> The current speed of extinction of species through human intervention is approximately 100–1,000 times faster than the natural speed of extinction. In many groups of organisms 5–20% of all species are already extinct.

Also critical is recognition of endemic biodiversity versus the total amount of biodiversity for a given region. Endemic biodiversity refers to the biodiversity that is indigenous (native) or endemic for a given region as opposed to introduced or alien biodiversity, which is that biodiversity which is present in a specific environment because of human interference in natural systems and human mobility. The focus on biodiversity conservation is nearly always on the maintenance of endemic biodiversity, with the exception being when an introduced species or variety is endangered or extinct in its naturally occurring range. There are essentially six main tenets to be found in the desire for conserving biodiversity (see Soule 1985; Callicot 1990; Wilson 1992):

1. The diversity of organisms and habitats on different scales (for example, genetic, species, ecosystem) is positive.
2. The untimely extinction of organisms and habitats on different scales is negative.
3. Ecological complexity is good.
4. Allowing evolutionary processes to occur is positive.
5. Biodiversity has extrinsic or anthropocentric value in terms of the goods and services it provides humankind.
6. Biodiversity has intrinsic or biocentric value.

There are three main mechanisms by which biodiversity is being lost:

- reduction in the size and fragmentation of natural areas
- changes in ecosystem conditions
- deliberate extinction of species.

Conversion to agriculture, forest clearance and urbanisation are the main causes of the loss of natural areas on the global scale. For example, in 2004 the United Nations Environment Programme (UNEP) warned the governments of the Congo, Rwanda and Uganda that, according to satellite studies, the Virunga National Park was being colonised by farmers at the rate of 2km^2 a day. The park is home to half of the world's population of mountain gorillas of which there are only an estimated 700 left (Radford 2004) The most important causes of changes to ecosystem conditions are fragmentation, disruption and isolation of natural areas, eutrophication, pollution, climate change, erosion and the introduction of diseases and species (Martens *et al.* 2003). Species are also being deliberately extinguished. Not only through poaching of high profile species such as tigers, rhinoceros and elephant, but also though the use of herbicides and biocides and hunting, fishing and farming practices. As Andrew Purvis, a conservation biologist, commented:

> Other species generally have their numbers limited by competitors, predators, parasites and pathogens. ... Any competitors, we get rid of those pretty quickly, even if they are just competing with things like crop plants, or our livestock, or our golf courses. We are also doing things to eliminate parasites and pathogens.
>
> (quoted in Radford 2004)

Historic and current loss of biodiversity is related to growth in human population and consumption. Pressure on natural resources is occurring not only because of existing consumption in developed countries but also because of increased consumption levels in the less developed countries.

> Each human needs roughly two hectares of land to provide food, water, shelter, fibre, currency, fuel, medicine and a rubbish tip to sustain a lifespan. So the more land humans take, the less that is available for all other mammals, birds, reptiles and amphibians. ... humans and their livestock now consume 40 percent of the planet's primary production, while the planet's other 7 million species must scramble for the rest. No other single species on the planet—except possibly some termites and the Antarctic krill—can match human numbers. People are having such an impact: we are sharply reducing the numbers of other things and very quickly you can go from large numbers to nothing.
>
> (Radford 2004)

Tourism represents a significant part of the consumption practices that impact biodiversity (e.g. German Federal Agency for Nature Conservation 1997; Gossling 2002; Christ *et al.* 2003). However, the impact of tourism development, for example, through tourism urbanisation, and habitat and species disturbance is not all negative. In many locations, tourism provides an economic justification to establish conservation areas, such as national parks and private reserves, as an alternative to other land uses such as logging, clearance for agriculture, mining or urbanisation. Often such tourism is described as ecotourism, safari, wildlife or nature-based tourism or even sustainable tourism (e.g. Cater and Lowman 1994; Hall and Lew 1998; Fennell 1999; Newsome, et al. 2002; Hall and Boyd 2005). Regardless of the name that is used, it is apparent that charismatic mega-fauna, for example, such animals as dolphins, elephants, giraffes, gorillas, lions, orangutan, rhinoceros, tigers and whales, do serve as a significant basis for tourism in a number of parts of the world, while national parks and reserves can also be significant tourist attractions in their own right. Indeed, tourism is seen as a mechanism to directly benefit biodiversity and the maintenance of natural capital through several means (Brandon 1996; Christ, et al. 2003; Hall and Boyd 2005), including:

- an economic justification for biodiversity conservation practices, including the establishment of national parks and reserves (public and private)
- a source of financial support for biodiversity maintenance and conservation
- an economic alternative to other forms of development that may negatively impact biodiversity and to inappropriate exploitation or harvesting of wildlife, such as poaching
- a mechanism for educating people about the benefits of biodiversity conservation
- potentially involving local people in the maintenance of biodiversity and incorporating local ecological knowledge in biodiversity management practices.

In 2003 Conservation International, in collaboration with the United Nations Environment Programme (UNEP), produced a report on the relationships between tourism and biodiversity that focused on the potential role of tourism in biodiversity "hotspots"—"priority areas for urgent conservation on a global scale" (Christ *et al.* 2003: vi). Hotspots are areas that both support a high diversity of endemic species and have been significantly impacted by human activities. Plant diversity is the biological basis for designation as a biodiversity hotspot, according to Christ *et al.* (2003: 3): "a hotspot must have lost 70 percent or more of its original habitat. Overall, the hotspots have lost nearly 90 percent of their original natural vegetation." The biodiversity hotspots identified by Conservation International "contain 44 percent of all known endemic plant species and 35 percent of all known endemic species of birds, mammals, reptiles, and amphibians in only 1.4 percent of the planet's land area" (Christ, et al. 2003: 3). The report highlighted several key issues:

- Although most biodiversity is concentrated in less developed countries, five tourism destination regions in the developed world were also identified as biodiversity hotspots—the Mediterranean Basin, the California floristic province, the Florida Keys, southwest Australia and New Zealand.
- An increasing number of biodiversity hotspot countries in the less developed world are experiencing rapid tourism growth: 23 of them record over 100 percent growth in the last 10 years, and more than 50 percent of these receive over one million international tourists per year; 13 percent of biodiversity hotspot countries receive over five million international tourists per year.
- Although receiving fewer tourists overall than the developed countries, many biodiversity-rich countries in the less developed world receive substantial numbers of international tourists. Thirteen of them—Argentina, Brazil, Cyprus, the Dominican Republic, India, Indonesia, Macao, Malaysia, Mexico, Morocco, South Africa, Thailand and Vietnam—receive over two million foreign visitors per year, while domestic tourism is also of growing significance in some of these countries.
- More than half of the world's poorest 15 countries fall within the biodiversity hotspots and, in all of these, tourism has some economic significance or is forecast to increase according to the World Tourism Organization and the World Travel and Tourism Council.
- In several biodiversity hotspots in less developed countries—for example, Madagascar, Costa Rica, Belize, Rwanda, South Africa—biodiversity or elements of biodiversity, such as specific wildlife, is the major international tourism attraction.
- Forecast increases in international and domestic tourism suggest that pressures from tourism development will become increasingly important in other biodiversity hotspot countries, for example, in South and Southeast Asia.

The Conservation International report highlighted some of the key relationships between tourism and biodiversity and stated that:

> Biodiversity is essential for the continued development of the tourism industry, yet this study indicates an apparent lack of awareness of the links—positive and negative—between tourism development and biodiversity conservation.
>
> (Christ, et al. 2003: 41)

Indeed, it went on to note that while many ecosystems serve to attract tourists, for example, coral reefs, rainforest and alpine areas, many of the factors linked to the loss of biodiversity, such as land clearance, pollution and climate change, are also linked to tourism development. Unfortunately, the report failed to adequately emphasise what some of the strategies by which tourism could both contribute to diversity and economic development might be or to state the broader ramifications of those strategies. For example, while the concept of scarcity rent that underlay much of earlier thinking with respect to the value of ecotourism—reduce access to desirable wildlife in the face of high demand and charge more for the experience while reducing the stress on animals and the environment—sounds sensible, it has often foundered on cultural and political values that have historically favoured access. Indeed, for most of their history, national parks agencies have often sought to encourage visitation so as to meet the recreational component of their mandate and to create a political environment supportive of national parks. Unfortunately, in the face of growing populations and increasing personal mobility the access issue is becoming increasingly problematic for many conservation authorities who seek to conserve biodiversity (Budowski 1976; Runte 1987; Hall 1992; Cater and Lowman 1994; Butler and Boyd 2000; Hall and Boyd 2005).

Despite the growth of research and publications on tourism in natural areas, our understanding of the role and effects of tourism in natural areas is also surprisingly limited. Arguably, the majority of studies have examined the impacts of tourism and recreation on a particular environment or component of the environment rather than over a range of environments (Holden 2000; Weaver 2001; Hall and Boyd 2005). There is substantial research undertaken on tourism with respect to rainforest, reefs and dolphins and whales, for example, but very limited research undertaken on what are arguably less attractive environments, such as wetlands, or animal species that are not the charismatic mega-fauna that are a key component of wildlife viewing tourism but which are just as important a part of the ecosystem (Newsome, et al. 2002; Hall and Boyd 2005). Moreover, the scale on which interactions between tourism and biodiversity are examined is also critical. The Conservation International report on tourism and biodiversity (Christ, et al. 2003) can only serve to highlight relationships at the macroscopic level of biological provinces, it does not serve as a useful management tool at the level of ecosystems, let alone individual species. Such a comment is not to denigrate the report because it serves an extremely useful function in terms of policy debate, but the harsh reality is that knowledge of the structure and dynamics of the geographic range of species in terms of abundance, size and limits is extremely limited even before the implications of human impact, including tourism, on range and abundance is considered (Gaston 2003).

As noted above, species extinction is a natural process. It has long been recognised that extinction and colonisation of habitats is an ongoing process. But just as importantly it has also been recognised that without human interference such processes lead to equilibrium between extinction and immigration (e.g. see MacArthur and Wilson 1967; Whitehead and Jones 1969). Human impact is changing this natural balance with respect to both extinction and immigration of new species at a rate that is making it extremely difficult, if not impossible, for new equilibria to be established. Tourism's contribution to the present human-induced mass extinction of species (May, et al. 1995; Pimm, et al. 1995; Hilton-Taylor 2000) is several-fold and will be examined in the following sections. Although tourism rarely directly kills off species, tourism-related developments and land use contributes to species range contraction and extinctions through habitat loss and fragmentation. Tourism and other forms of human mobility also introduce alien organisms into areas beyond the natural limits of their geographical range, thereby creating new competition among species. Tourism also affects biodiversity through its contribution to climate change. Finally, we can raise issues over the extent to which national parks and other protected areas can be used to conserve biodiversity.

HABITAT LOSS AND FRAGMENTATION

Tourism directly affects habitat through processes of tourism urbanisation. As Chapter 8 notes, such processes are spatially and geographically distinct and are often related to high natural amenity areas such as the coast, where coastal ecosystems are subject to urbanisation, land clearance and the draining and clearance of wetlands. Tourism also contributes to habitat loss and fragmentation via its ecological footprint in terms of resource requirements and pollution and waste.

The loss of endemic biodiversity through species extinction can be expressed in relation to changes in the size of the geographic range of a species and its total population size as the total number of individuals in a species declines. Four different idealised forms of these relationships can be presented (Wilcove and Terborgh 1984; Schonewald-Cox and Buechner 1991; Lawton 1993; Gaston 1994, 2003: 168-74), although, as Gaston notes:

> Declines in extinction are often likely to walk much more varied paths through abundance-range space than these simple models might imply, given the complexities of the abundance structure of species' geographic ranges and of the processes causing reductions in overall population size.
>
> (2003: 174)

- The geographic range size remains approximately constant as the number of individuals declines and overall density declines with time. Activities such as hunting, pollution or climatic change can all lead to declines in the number of individuals of a species without affecting the overall geographic range in which they are found.
- The number of individuals and range size decline simultaneously so that the density remains constant. Two circumstances can be identified in which such a situation may occur. First, reduction of losses of individual members of a species may be balanced by losses of lower density areas. Gaston (2003: 168), for example, notes: "declines in the local abundances of persistent populations accompanied by the loss of small, and often peripheral, populations appear to be a widespread phenomena." Second, the total habitat area may be eroded without loss in the quality of the remaining habitat. One of the best examples of this situation that is often related to coastal tourism development is the loss of individual wetlands which are drained, leaving other wetlands as yet undeveloped.
- The number of individuals of a species and the size of their range decline simultaneously such that the density of individuals declines with time. Gaston (2003) recognises three cases in which such a situation might exist. First, with respect to broad-scale environmental change, including the effects of an increase in the proportion of "edge" as a habitat area is fragmented therefore leading to micro-climatic changes as well as other changes to patterns of predation and species invasion (Laurance 2000; Laurance, et al. 1997, 2000). Such a situation is consistent with the development of edge effects when recreational access is unmanaged in habitat fragments and people do not keep to trails, thereby creating further edges as new walking paths are created. Second, in areas that undergo differential exploitation through, for example, timber extraction. Third, if there is a causal link between abundance of a species and occupancy of a given habitat (see also Gates and Donald 2000; Lawton 1993, 2000).
- The number of individuals of a species and range size decline simultaneously such that the density of individuals increases with time. Such a pattern is most likely to occur when habitat is lost with no compensating increase in density of individuals in other available habitat. This particular model of extinction has substantial implications for conserving biodiversity "hot spots" which tends to assume (see Christ, et al. 2003) that protecting such areas in the face of the loss of other areas where the species are present should increase the density of individuals of the species. Yet such density increases may be only short-term if populations in the "hot spots" were dependent on their relationship with other populations of the species that have been made extinct. As Gaston highlights: "This emphasizes the need for a regional rather than a site- by-site approach to conservation planning and action, albeit this is at odds with the methodology

embodied in some international agreements, for example Ramsar Convention on Wetlands of International Importance ... and espoused by some conservation agencies" (2003: 174).

INTRODUCTION OF FOREIGN ORGANISMS

It is estimated that approximately 400,000 species have been accidentally or deliberately introduced to locations that lie beyond the natural limits of their geographic range (Pimentel 2001). The introduction of alien species into an environment is a major influence on biodiversity that is associated with tourism because of the capacity of tourists and the infrastructure of tourism to act as carriers of exotic species. Many introductions have no apparent adverse effects (Williamson 1996), although some introductions, such as deer, rabbits and possum in New Zealand, and cane toads and rabbits in Australia, have caused massive ecological damage and harm to native species (Fox and Adamson 1979).

In the nineteenth and early twentieth centuries many species were deliberately introduced from one part of the world to another by the European colonial powers as a means of economic development and the Europeanisation of the natural environment (Crosby 1986). Although agricultural development and the creation of an ideal environment were the primary motives for such introductions, leisure and tourism were also significant. For example, in New Zealand a number of Australian and European animal species were introduced for hunting purposes but it was not until the 1920s that substantial opposition emerged to the introduction of new species, including widespread indignation and opposition to Lady Liverpool's efforts to introduce grouse into Tongariro Park, New Zealand's oldest national park. Prof. H.B. Kirk, one of New Zealand's leading natural historians, sent an angry letter to the *Evening Post*, which had earlier applauded Lady Liverpool's efforts as likely to "give added attractions to sportsmen coming to New Zealand from the Old Country: No other country would do so ludicrous a thing as to convert the most distinctive of its national parks into a game preserve. This thing is an insult to the Maori donors and to all lovers of New Zealand as New Zealand." Kirk's letter appeared to find a supportive response among a wide range of individuals and authorities. By the end of 1924 the New Zealand Legislative Council had "pushed through a resolution condemning all introduction and proclaiming that the park should be held inviolate" (in Harris 1974: 109-110).

International trade has also served to introduce alien species through accidental carriage in shipping containers and on ships and aircraft (Drake and Mooney 1988; Carlton and Geller 1993). Nevertheless, travellers remain a major source of accidental and intentional species introductions to the point where they are a focus of biosecurity concerns at both international and regional levels (Timmins and Williams 1991; Hodgkinson and Thompson 1997; Hall 2003; Jay *et al.* 2003).

Much of the concern over the introduction of alien species lies in their potential economic damage. Pimentel *et al.* (2000) estimated that the approximately 50,000 exotic species in the US have an economic impact of US$137 billion per annum in terms of their economic damage and costs of control. For example, since 1998 the State of California has provided US$65.2 million for a statewide management programme and research to combat the glassy-winged sharpshooter and the deadly Pierce's disease (a bacterium, *Xylella fastidiosa)* that it carries. Accidentally introduced in 1989, 15 counties in California have been identified as being infested (Wine Institute of California 2002).

For many grape diseases, humans are a significant vector (Pearson and Goheen 1998), the most notable of which is grape phylloxera, an aphid *Daktulosphaira vitifoliae*, which wreaked havoc on the world's vineyards in the late nineteenth century (Ordish 1987). The economic impact of a phylloxera outbreak on the modern wine industry would be substantial. In Western Australia, it is estimated that phylloxera could cost affected growers A$20,000/ha in the first five years in lost production and replanting costs (Agriculture Western Australia 2000). Increased personal mobility, particularly through wine and food tourism, is a potential threat to the wine industry because of the potential for the relocation and introduction of pests. Yet, despite recognition of the potential role of humans in conveying grape pests, there is only limited awareness of the biosecurity risks of wine tourism (Hall 2003). However, it is likely that in the future concerns over the risks associated with the introduction of exotic species for endemic biodiversity can only increase as rates of international travel continue to grow and the climate change leads to the creation of environmental conditions conducive to the establishment of alien species.

CLIMATE CHANGE

Climate change sets particular challenges for conservation. One of the most significant long-term issues for the global network of protected areas that serve to help maintain biodiversity is how can they "be established and developed in such a way that it can accommodate the changes in species distributions that will follow from climate change" (Gaston 2003: 181). Substantial research has been undertaken on the implications of climate change for species' geographic ranges which has typically sought to model the relationships between climate and distribution in relation to such issues as habitat fragmentation (e.g. Nakano, et al. 1996) and loss (e.g. Keleher and Rahel 1996; Travis 2003), species distribution (e.g. Jeffree and Jeffree 1994, 1996), pests (e.g. Baker, et al. 1996) and disease (e.g. Rogers and Randolf 2000; Lieshout, et al. 2004). Undoubtedly, the distribution of species is affected by temperature. However, studies of the relationship between species distribution and future climate change scenarios tend to make a number of critical assumptions (Gaston 2003):

- Correlations between climate and species occurrence reflect causal relationships.
- Any influence of other factors on observed relationships between climate and the occurrence of a species, such as competitors, diseases, predators, parasites and resources will remain constant.
- Temporally generalised climatic conditions—for example, seasonal means, annual means, medians—are more important influences on the distribution of species than rates of climatic change and extreme events.
- Spatially generalised climatic conditions derived or interpolated from the nearest climate stations sufficiently characterise the conditions that individuals of a species actually experience.
- Climate change will be relatively simple, in that its influence on species distributions can be summarised in terms of the projected changes in one or a few variables.
- There is no physiological capacity to withstand environmental conditions which are not components of those existing conditions in areas in which a species is presently distributed.
- Range shifts, expansions, or contractions are not accompanied by physiological changes, other than local non-genetic acclimatisation.
- Dispersal limit is unimportant in the determination of the present distribution of species and in their ability to respond to changes in climate.

As Gaston (2003: 185) points out, the reality is that a number of these assumptions will be, and already have been, "severely violated" (see Lawton

1995, 2000; Spicer and Gaston 1999; Bradshaw and Holzapfel 2001). For example:

> Human activities impose a marked influence on the distribution of species, and how these alter with changes in climate is alone likely to be extremely complicated, and dependent on social pressures and technological developments.
>
> (Gaston 2003: 183)

The above observations are not to deny that climate change will affect the geographic range of species, it clearly will in the future just has it has in the past (e.g. Boer and de Groot 1990; Hengeveld 1990; Huntley 1991,1994; Huntley and Birks 1983; Huntley *et al.* 1989, 1995). However, the use of relatively simple models based on climate matching approaches is likely to prove misleading in terms of planning conservation regimes that can accommodate future climate change.

CONSERVING BIODIVERSITY

At the start of the twenty-first century the world's biodiversity is threatened as never before—as noted above, many species become extinct each year and the number is growing. This section looks at the role of national parks and reserves as present and future refugia. Tourism, and ecotourism in particular, has become a major economic rationale for the establishment of national parks and reserves that serve to conserve and present charismatic mega-fauna and habitats.

The global conservation estate has grown enormously since the first UN List of Protected Areas was published in 1962 with just over 1,000 protected areas. In 1997 there were over 12,754 sites listed. The 2003 edition listed 102,102 sites covering 18.8 million km^2. "This figure is equivalent to 12.65 percent of the Earth's land surface, or an area greater than the combined land area of China, South Asia and Southeast Asia" (Chape *et al.* 2003: 21). Of the total area protected, it is estimated that 17.1 million km^2 constitute terrestrial protected areas, or 11.5 percent of the global land surface, although some biomes, including Lake Systems and Temperate Grasslands, remain poorly represented. Marine areas are significantly under-represented in the global protected area system. Approximately 1.64 million km^2 comprise marine protected areas—an estimated 0.5 percent of the world's oceans and less than one tenth of the overall extent of protected areas worldwide (Chape *et al.* 2003).

The size of the global conservation estate raises the question of just how large the global network of protected areas needs to be (Rodrigues and Gaston 2001). The present size of global conservation estate exceeds the IUCN's earlier target of at least 10 percent of the total land area being set aside for conservation purposes, although there is clearly substantial variation between both countries and biomes in terms of the actual area set aside (Chape *et al.* 2003). Yet commentators such as Soule and Sanjayan (1998) have noted that the IUCN's target has been dictated more by political considerations than biological science. Rodrigues and Gaston (2001, 2002) observed that the minimum area needed to represent all species within a region increases with the number of targeted species, the level of endemism and the size of the selection units. They concluded that:

- No global target for the size of a network is appropriate because those regions with higher levels of endemism and/or higher diversity will correspondingly require larger areas to protect such characteristics.
- A minimum size conservation network sufficient for capturing the diversity of vertebrates will not be sufficient for biodiversity in general, because other groups are known to have higher levels of endemism (Gaston 2003).

- The 10 percent target is likely to be grossly inadequate to meet biodiversity conservation needs. Instead, Rodrigues and Gaston (2001) estimated that for a selection unit of 1° x 1° (approximately12,000 km²) 74.3 percent of the global land area and 92.7 percent of the global rain forest would be required to represent every plant species and 7.7 percent and 17.8 percent respectively to represent the higher vertebrates. However, Gaston (2003) also notes that even reserves of 12,000 km² may not be large enough for maintaining populations of many species, citing examples from the national parks of Africa (Newmark 1996; Nicholls, et al. 1996) and the USA (Newmark 1984; Mattson and Reid 1991). Indeed, it must be noted that while there has been well-considered literature on the size and shape of conservation reserves since the 1970s (e.g. Main and Yadav 1971; Diamond 1975; Slatyer 1975; MacMahon 1976) there has been inadequate utilisation of such knowledge with respect to park and reserve establishment and design and their dual role conservation and tourism roles.

A further concern in terms of biodiversity conservation is the capacity of a national park and reserve system to cope with the impact of global environmental change (GEC) including climate change (Dockerty, et al. 2003), surrounding land-use change and anthropogenic pressure (Cardillo, et al. 2004). Given the migration of species as a result of climate change, present reserves may not be suitable for conservation of target species and ecosystems (Huntley 1994, 1999). Given the potential scale of GEC it is therefore important that sites are identified that can act as refugia from future change. A refuge is a region in which certain species are able to persist during a period in which most of the original geographic range becomes uninhabitable because of environmental change. Historically, such changes have been climatic although in terms of contemporary biodiversity conservation anthropogenic environmental change is just as significant. Although some present national parks and reserves are likely to fill this role, it is also important that sites that are available which have attributes that may potentially fulfil the role of refugia for endangered species in the future also be identified, conserved and managed so as to reduce the impacts of GEC on populations. Such "future refugia" may then become locations from which future species migration can occur should climate become stabilised.

CONCLUSION

The loss of biodiversity is one of the most significant aspects of GEC given the extent to which it underpins the global economy and human welfare (Martens, et al. 2003). Biodiversity, or at least the existence of certain charismatic species (usually mega-fauna) and ecosystems is also significant as an attraction for 'ecotourism' and 'nature-based tourism'. Nevertheless, the interrelationships between tourism and biodiversity are poorly understood in terms of empirical data, although the potential impacts of the loss of some charismatic species such as the polar bear (The Age 2005), or African wildlife, or even entire ecosystems, such as the Great Barrier Reef (Fyfe 2005), on tourism would be dramatic.

The extent to which tourism contributes towards biodiversity loss through tourism urbanisation, habitat loss and fragmentation and contribution to climate change is also dramatic and, arguably, makes a lie out of attempts to paint a picture of tourism as a benign industry. Undoubtedly, tourism can make a contribution to the conservation and maintenance of biodiversity. However, in reality the success stories are few and far between and are generally isolated to individual species and relatively small areas of habitat (e.g. see Newsome *et al.* 2002) rather than a comprehensive contribution to conservation. Such a comment is not to belittle the efforts that have been made with respect to developing a positive contribution from tourism toward biological conservation. Instead, it is to highlight the fact that while tourism has led to biodiversity maintenance at a local level in some instances, the global picture is one in which tourism, like many other industries that have a large ecological footprint and lead to clearance of natural areas, is not a net contributor to biodiversity.

REFERENCES

Agriculture Western Australia (2000). *Grape Phylloxera: Exotic Threat to Western Australia, Factsheet no.0002-2000.* Perth: Department of Agriculture.

Baker, R.H.A., R.J.C Cannon, and K.F.A. Walters. (1996). An assessment of the risks posed by selected non-indigenous pests to UK crops under climate change. *Aspects of Applied Biology* 45: 323–30.

Boer, M.M. and R.S. de Groot, (eds.). (1990). *Landscape-Ecological Impact of Climate Changes.* Amsterdam: IOS Press.

Bradshaw, W.E. and C.M. Holzapfel. (2001). Genetic shift in photoperiodic response correlated with global warming. *Proceedings of the National Academy of Sciences of the USA* 98: 14509–11.

Brandon, K. (1996). Ecotourism and Conservation: A Review of Key Issues. *Environment Department Paper no.33*. Washington, DC: World Bank.

Budowski, G. (1976). Tourism and conservation: conflict, coexistence or symbiosis. *Environmental Conservation* 3(1): 27–31.

Butler, R. and S. Boyd, (eds.). (2000). *Tourism and National Parks.* Chichester: John Wiley.

Callicott, J.B. (1990). Whither conservation ethics? *Conservation Biology* 4: 15–20.

Cardillo, M., A. Purvis, W. Sechrest, J.L. Gittleman, J. Bielby, and G.M. Mace. (2004). Human population density and extinction risk in the world's carnivores. *PloSBiology* 2(7): 0909–0914

Carlton, J.T. and J.B. Geller. (1993). Ecological roulette: the global transport of non-indigenous marine organisms. *Science* 261: 78–82.

Cater, E. and G. Lowman, (eds.). (1994). *Ecotourism: A Sustainable Option?* Chichester: John Wiley.

Chape, S., S. Blyth, L. Fish, P. Fox, and M. Spalding, (compilers). (2003). *2003 United Nations List of Protected Areas.* IUCN, Gland and Cambridge: IUCN and UNEP-WCMC.

Christ, C., O. Hilel, S. Matus, and J. Sweeting. (2003). *Tourism and Biodiversity: Mapping Tourism's Global Footprint.* Washington, DC: Conservation International.

Crosby, A.W. (1986). *Ecological Imperialism: The Biological Expansion of Europe, 900–1900.* Cambridge: Cambridge University Press.

Diamond, J.M. (1975). The island dilemma: lessons of modern biogeographic studies for the design of natural reserves. *Biological Conservation* 7: 129–46.

Dockerty, T., A. Lovett, and A. Watkinson. (2003). Climate change and nature reserves: examining the potential impacts, with examples from Great Britain. *Global Environmental Change* 13: 125–35.

Drake, J.A. and H.A. Mooney, (eds.). (1988). *Biological Invasions: A Global Perspective.* New York: John Wiley.

Fennell, D. (1999). *Ecotourism: An Introduction.* London: Routledge.

Fox, M.D. and D. Adamson. (1979). The ecology of invasions, in H. Recher, D. Lunney and I. Dunn (eds.) *A Natural Legacy: Ecology in Australia.* Rushcutter's Bay: Pergamon Press.

Fyfe, M. (2005). Too late to save the reef. *The Age* February 12.

Gaston, K.J. (1994). Geographic range sizes and trajectories to extinction. *Biodiversity Letters* 2: 163–70.

Gaston, K.J. (2003). *The Structure and Dynamics of Geographic Ranges.* Oxford: Oxford University Press.

Gates, S. and P.F. Donald. (2000). Local extinction of British farmland birds and the prediction of further loss. *Journal of Applied Ecology* 37: 806–20.

German Federal Agency for Nature Conservation. (1997). *Biodiversity and Tourism: Conflicts on the World's Seacoasts and Strategies for Their Solution.* Berlin: Springer Verlag.

Gossling, S. (2002). Global environmental consequences of tourism. *Global Environmental Change* 12: 283–302.

Hall, C.M. (1992.) *Wasteland to World Heritage: Preserving Australia's Wilderness.* Carlton: Melbourne University Press.

Hall, C.M. (2003). Biosecurity and wine tourism: Is a vineyard a farm? *Journal of Wine Research* 14(2–3): 121–6.

Hall, C.M. and S. Boyd. (2005). Nature-based tourism and regional development in peripheral areas: Introduction, in C.M. Hall and S. Boyd, (eds.) *Tourism and Nature-based Tourism in Peripheral Areas: Development or Disaster,* Clevedon: Channelview Publications, pp.3–17.

Hall, C.M. and A. Lew. (1998). The geography of sustainable tourism development: an introduction, In C.M. Hall and A. Lew, (eds.) *Sustainable Tourism: A Geographical Perspective.* London: Addison-Wesley Longman, pp.1–12.

Harris, W.W. (1974). Three parks: An analysis of the origins and evolution of the national parks movement, unpublished MA Thesis, Christchurch: Department of Geography, University of Canterbury.

Hengeveld, R. (1990). *Dynamic Biogeography.* Cambridge: Cambridge University Press.

Hilton-Taylor, C. (2000). *The 2000 IUCN Red List of Threatened Species.* Gland: IUCN.

Hodgkinson, D.J. and K. Thompson. (1997). Plant dispersal; the role of man., *Journal of Applied Ecology* 34: 1484–96.

Holden, A. (2000). *Environment and Tourism.* London: Routledge.

Huntley, B. (1991.) How plants respond to climate change: migration rates, individualism and the consequences for plant communities. *Annals of Botany* 67 (Supplement 1): 15–22.

Huntley, B. (1994). Plant species' responses to climate change: implications for the conservation of European birds. *Ibis* 137: S127–S138.

Huntley, B. (1999). Species distribution and environmental change: considerations from the site to the landscape scale, in E. Maltby, M. Holdgate, M. Acreman and A. Weir, (eds.) *Ecosystem Management: Questions for Science and Society*. Virginia Water: Royal Holloway Institute for Environmental Research. pp.115–29.

Huntley, B. and H.J.B. Birk. (1983) *An Atlas of Past and Present Pollen Maps for Europe: 13000 Years Ago.*, Cambridge: Cambridge University Press.

Huntley, B., P.J. Bartlein, and I.C. Prentice. (1989) Climatic control of the distribution and abundance of beech *(Fagus L.)* in Europe and North America. *Journal of Biogeography* 16: 551–60.

Huntley, B., P.M. Berry, W. Cramer, and A.P. McDonald. (1995). Modelling present and potential future ranges of some European higher plants using climate response surfaces. *Journal of Biogeography* 16: 551–60.

Jay, M., M. Morad, and A. Bell. (2003). Biosecurity—A policy dilemma for New Zealand. *Land Use Policy* 20(2): 121–29.

Jeffree, C.E. and E.P. Jeffree. (1996). Redistribution of the potential geographical ranges of Mistletoe and Colorado Beetle in Europe in response to the temperature component of climate change. *Functional Ecology* 10: 562–77.

Jeffree, E.P. and C.E. Jeffree. (1994). Temperature and the biogeographical distributions of species. *Functional Ecology* 8: 640–50.

Keleher, C.J. and F.J. Rahel. (1996). Thermal limits to salmonoid distributions in the Rocky Mountain Region and potential habitat loss due to global warming: a geographic information system (GIS) approach. *Transactions of the American Fisheries Society* 125:13.

Kim, J.A. (2004). Regime interplay: the case of biodiversity and climate change. *Global Environmental Change* 14: 314–24.

Lande, R. (1993). Risks of population extinction from demographic and environmental stochasticity and random catastrophes. *American Naturalist* 142: 911–27.

Lande, R. (1998). Anthropogenic, ecological and genetic factors in extinction, in G.M. Mace, A. Balmford and J.R. Ginsberg (eds.) *Conservation in a Changing World.* Cambridge: Cambridge University Press, pp.29–51.

Laurance, W.F. (2000). Do edge effects occur over large spatial scales. *Trends in Ecology and Evolution* 15: 134–5.

Laurance, W.F., S.G. Laurance, L.V. Ferreira, J.M. Rankin-de Merona, C. Gascon, and T.E. Lovejoy. (1997). Biomass collapse in Amazonian forest fragments. *Science* 278: 1117–8.

Laurance, W.F., P. Delamonica, S.G. Laurance, H.L. Vasconcelos, and T.E. Lovejoy. (2000). Rainforest fragmentation kills big trees. *Nature* 404: 836.

Lawton, J.H. (1993). Range, population abundance and conservation. *Trends in Ecology and Evolution* 8: 409–13.

Lawton, J.H. (1995). The response of insects to environmental change, in R. Harrington and N.E. Stork (eds.) *Insects in a Changing Environment.* London: Academic Press, pp.3–26.

Lawton, J.H. (2000). *Community Ecology in a Changing World.* Oldendorf: Ecology Institute.

Lieshout, M. van, R.S. Kovats, M.T.J. Livermore, and P. Martens. (2004). Climate change andmalaria: Analysis of the SRES climate and socio-economic scenarios. *Global Environmental Change* 14: 87–99.

MacArthur, R.H. and E.O. Wilson. (1967). *The Theory of Island Biogeography*. Princeton, NJ: Princeton University Press.

MacMahon, J.A. (1976). Thoughts on the optimum size of natural reserves based on ecological principles, in J. Franklin and S. Krugman, (eds.) Selection, Management and Utilization of Biosphere Reserves. *Proceedings of the USA–USSR Symposium on Biosphere Reserves*. Moscow, May 1976, Corvallis: Pacific Northwest Forest and Range Experiment Station, United States Department of Agriculture, Forest Service, pp.128–34.

McNeely, J.A., (1990). Climate change and biological diversity: policy implications, in M. Boer and R.S. de Groot, (eds.) *Landscape-Ecological Impact of Climate Change*. Amsterdam: IOS Press, pp.406–29.

Main, A.R. and M. Yadav. (1971). Conservation of macropods in reserves in Western Australia. *Biological Conservation* 3: 123–33.

Martens, P., J. Rotmans, and D. de Groot. (2003). Biodiversity: luxury or necessity? *Global Environmental Change* 13: 75–81.

Mattson, D.J. and M.M. Reid. (1991). Conservation of the Yellowstone grizzly bear. *Conservation Biology* 5: 364–72.

May, R.M., J.H. Lawton, and N.E. Stork. (1995). Assessing extinction rates, in J.H. Lawton and R.M. May (eds.) *Extinction Rates*. Oxford: Oxford University Press, pp.1–24.

Nakano, S., F. Kitano, and K. Maekawa. (1996). Potential fragmentation and loss of thermal habitats for charrs in the Japanese archipelago due to climatic warming. *Freshwater Biology* 36: 711–22.

Newmark, W.D. (1987). A land-bridge perspective on mammalian extinctions in western North American parks. *Nature* 325: 430–2.

Newmark, W.D. (1996). Insularization of Tanzanian national parks and the local extinction of large mammals. *Conservation Biology* 10; 1549–56.

Newsome, D., S.A. Moore, and R.K. Dowling. (2002). *Natural Area Tourism: Ecology, Impacts and Management*. Clevedon: Channelview Publications.

Nicholls, A.O., P.C. Viljoen, M.H. Knight, and A.S. van Jaarsveld. (1996). Evaluating population persistence of censused and unmanaged herbivore populations from the Kruger National Park, South Africa. *Biological Conservation* 76: 57–67.

Nielsen, R. (2005). *The Little Green Handbook: A Guide to Critical Global Trends*. Carlton North: Scribe Publications.

Ordish, G. (1987). *The Great Wine Blight, 2nd ed.* London: Sedgwick and Jackson.

Pearson, R.C. and A.C. Goheen, (eds.). (1998). *Compendium of Grape Diseases*. Saint Paul: The American Phytopathological Society.

Pimentel, D. (2001). Agricultural invasions, in S.A. Levin (ed.) *Encyclopedia of Biodiversity.* San Diego: Academic Press, pp.71-85.

Pimentel, D., L. Lach, R. Zuniga, and D. Morrison. (2000). Environmental and economic cost of nonindigenous species in the United States. *BioScience,* 50: 53–65.

Pimm, S.L., G.J. Russell, J.L. Gittleman, and T.M. Brooks. (1995). The future of biodiversity. *Science* 269: 347–50.

Pitman, N.A. and P.M. Jorgensen. (2002). Estimating the size of the world's threatened flora. *Science* 298: 989.

Polar bears' days may be numbered. (2005). *The Age* February 3.

Purvis, A. and A. Hector. (2000). Getting the measure of biodiversity. *Nature* 405: 212–19.

Radford, T. (2004). Life on the brink. *The Age* 11 August.

Rodrigues, A.S.L. and K.J. Gaston. (2001). How large do reserve networks need to be? *Ecological Letters* 4: 602–9.

Rodrigues, A.S.L. and K.J. Gaston. (2002). Rarity and conservation planning across geopolitical units. *Conservation Biology* 16: 674–82.

Rogers, D.J. and S.E. Randol. (2000). The global spread of malaria in a future, warmer world. *Science* 289: 1763–6.

Rosendal, G.K., (2001). Overlapping international regimes: the case of the inter-governmental forum on forests (IFF) between climate change and biodiversity. *International Environmental Agreements: Politics, Law and Economics* 1(4): 447–68.

Runte, A., (1987). *National Parks: The American Experience 2nd ed.* Lincoln, NE: University of Nebraska Press.

Schonewald-Cox, C. and M. Buechner. (1991). Housing viable populations in protected habitats: the value of a coarse-grained geographic analysis of density patterns and available habitat, in A. Seitz and Loeschke (eds.) *Species Conservation: A Population- Biological Approach.* Basel: Birkhauser Verlag, pp.213–26.

Slatyer, R.O. (1975). Ecological reserves: size, structure and management, in F. Fenner (ed.) *A National System of Ecological Reserves in Australia.* Canberra, Australian Academy of Science, pp.22–38.

Soule, M. (1985). What is conservation biology? *Bioscience* 35: 727–34.

Soule, M. and M.A. Sanjayan. (1988). Ecology-conservation targets: do they help? *Science* 279: 2060–1.

Spicer, J.J. and K.J. Gaston. (1999). *Physiological Diversity and its Ecological Implications.* Oxford: Blackwell Science.

Timmins, S.M. and P.A. Williams. (1991). Factors affecting weed numbers in New Zealand's forest and scrub reserves. *New Zealand Journal of Ecology* 15: 153–62.

Travis, J. (2003). Climate change and habitat destruction: A deadly anthropogenic cocktail. *Proceedings of the British Royal Society B,* 270: 467–73.

United Nations Environment Programme (UNEP) (2002.) *Global Environmental Outlook Three Past, Present and Future Perspectives.* Nairobi: UNEP.

Weaver, D. (2001). *The Encyclopedia of Ecotourism.* Oxford: CABI Publishing.

Whitehead, D.R. and Jones, C.E. (1969) 'Small islands and the equilibrium theory of insular biogeography', *Evolution*, 23: 171-9.

Wilcove, D.S. and Terborgh, J.W. (1984) 'Patterns of population decline in birds', *American Birds*, 38: 10-13.

Williamson, M. (1996) *Biological Invasions.* London: Chapman and Hall.

Wilson, E.O. (1992) *The Diversity of Life.* New York: Norton.

Wine Institute of California (2002) *Pierce's Disease Update.* San Francisco: Wine Institute of California.

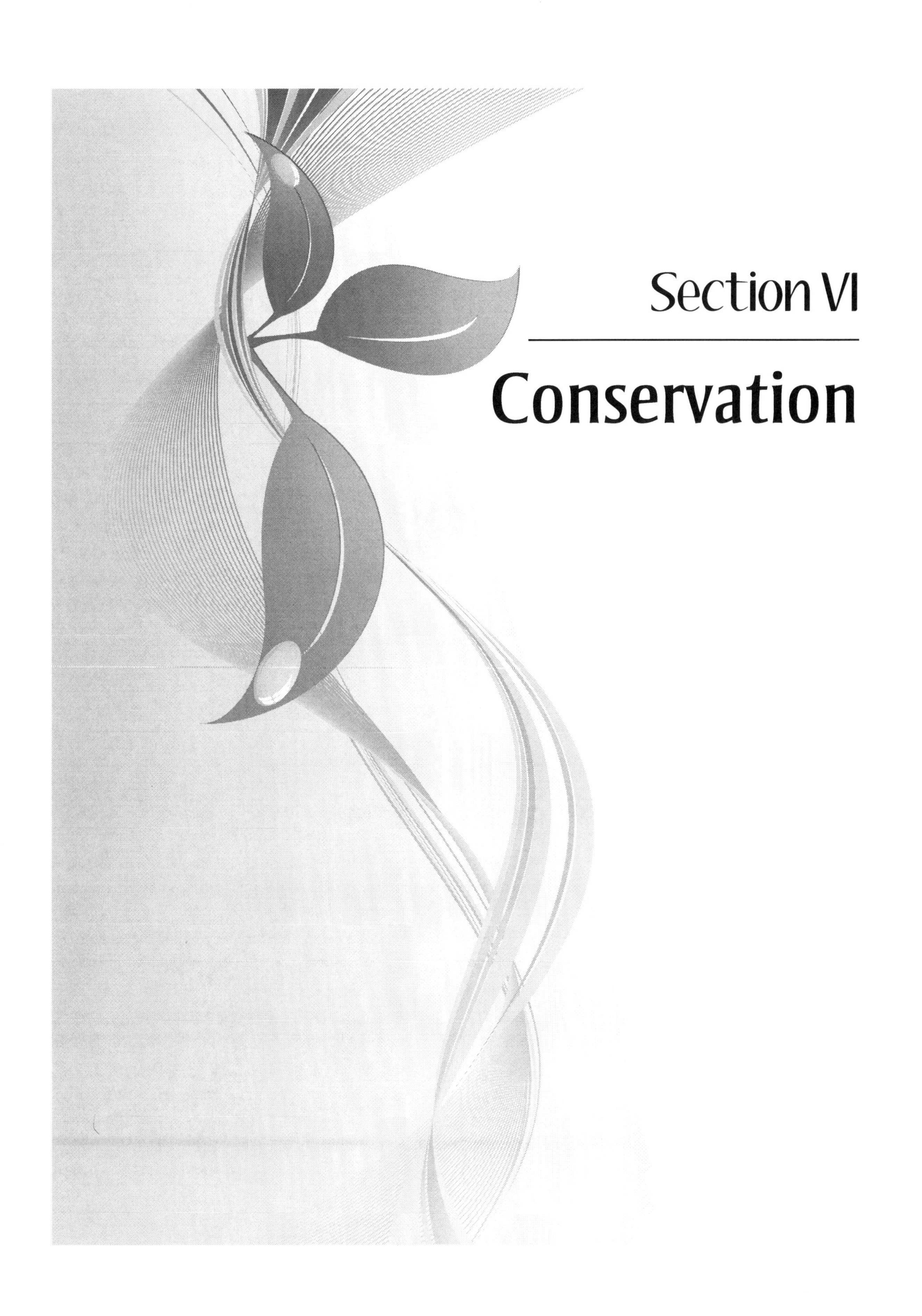

Section VI

Conservation

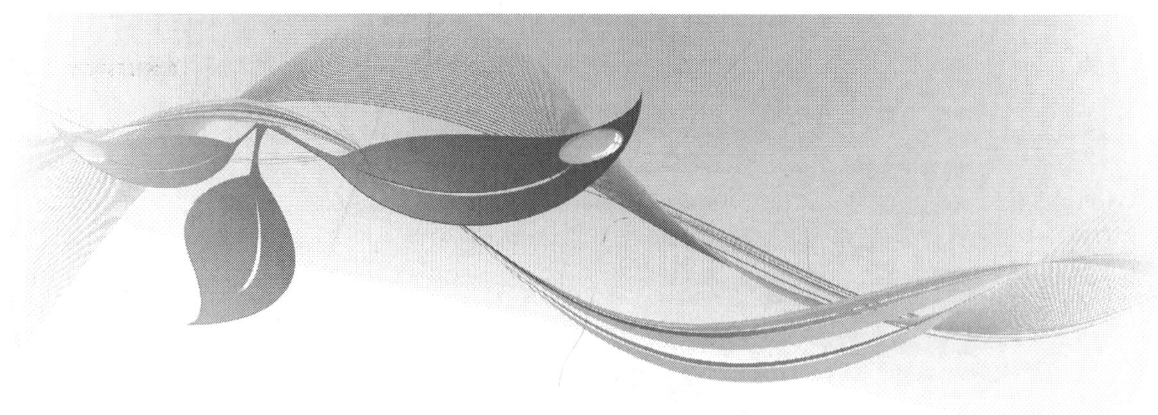

By Anne Marie Zimeri

This chapter will discuss conservation of forest resources and grasslands, as well as conservation efforts that have led to national parks and preserves.

Conservation Biology

This branch of conservation seeks to preserve biodiversity across a range of biomes. The modern era of conservation biology originated at a conference in 1978 when the term was coined. The ultimate goal of conservation biology is to prevent extinction of species, especially those extinctions caused by human activities. There have certainly been several mass extinctions in the distant past, but none was caused by human interference such as the events of today (Table 6.1).

Current extinction events now are considered part of the "sixth extinction," which is caused by humans. It has been estimated that the global extinction rate is 100 times higher than it would be without humans. Sadly, within the past 50,000 years, the hardest hit for extinctions were mammals weighing 100 pounds or more. These large mammals are important as predator species, and as a food source.

What are these human activities that threaten animals with extinction? First, humans are causing harm to the biosphere by burning fossil fuels and releasing greenhouse gases that have led to climate change. Secondly, humans have destroyed the natural habitats for many species, both plant and animal. We have razed forests to farm for exotic hardwood, increased the need for the acquisition of fuelwood, and expanded the geographical areas where we live, work, and shop.

Forests

Humans have been chopping down trees throughout history, but never at the rate at which we are currently deforesting our land. In the past, most deforestation has occurred in temperate forests; however, in the last century, deforestation has mostly occurred in tropical regions, and mostly in the rain forests of Brazil (Figure 6.1).

Why are humans deforesting these biodiverse areas on such a grand scale? Mostly the deforested area is converted to agricultural fields for crop plants, such as soy, or for grazing land to raise beef cattle. Deforesting these areas sets up many of these agricultural areas for failure because most of the nutrients in the rain forest are in the canopy. The topsoil is

Mass Extinction Event	Time Frame (mya)	Types of Life Affected
Late/End Ordovician	443 million years ago	Many species of Trilobites, Brachiopods, Graptolites, Echinoderms and Corals
Late/End Davonian	354 million years ago	Many marine familes on tropical reefs, Corals, Brachiopods, Bivalves, Sponges
Late/End Permian	248 million years ago	57% of all marine families, Trilobites, Eurypterids, Mollusca devastated along with Brachiopods, Many vertebrates
Late/End Triassic	206 million years ago	Mollusea phyla, Sponges, marine vertebrates, large Amphibians, many Mammal-like Reptiles
Late/End Cretaceous	65 million years ago	Ammonites, Marine Reptiles, Dinosaurs, Pterosaurs, microscopic marine plankton, Brachiopods, Bivalves and Echninoderms

Table 6.1. Over the past 540 million years, five mass-extinction events have occurred through naturally induced events.

quite thin and acidic and cannot support large-scale agriculture for more than a few years without massive soil amendments, which may make those endeavors unprofitable.

Rangelands

Rangelands are present in arid or semiarid areas and typically support the growth of native grasses and shrubs. Most of the time rangelands would not be able to grow agricultural crops due to soil quality and low rainfall levels, though there are exceptions. Nonetheless, rangelands are important for livestock rearing and grazing, and the maintenance of rangelands in a sustainable manner is key to supporting that industry. More than half of the land on the planet is rangeland, which represents more land than any other type of ecosystem.

The western states house most of the rangeland in the United States. Much of this land is under preservation law in the national parks such as Yellowstone National Park. In the United States as well as the rest of the world, the majority of the ranching industry occurs on rangeland. Livestock raised on rangeland in a conventional way can seriously degrade the quality of the land. When too many animals graze on a small plot of land, the overgrazing can lead to **desertification**. Ranching cannot be successful sustainably if overgrazing occurs. As a result, ranchers have had to employ new methods to preserve the health of their land. Most of these methods require that areas of land not be grazed for a period of time. An extreme case of this management practice occurred on 1,500 acres of rangeland in San Mateo County, California,

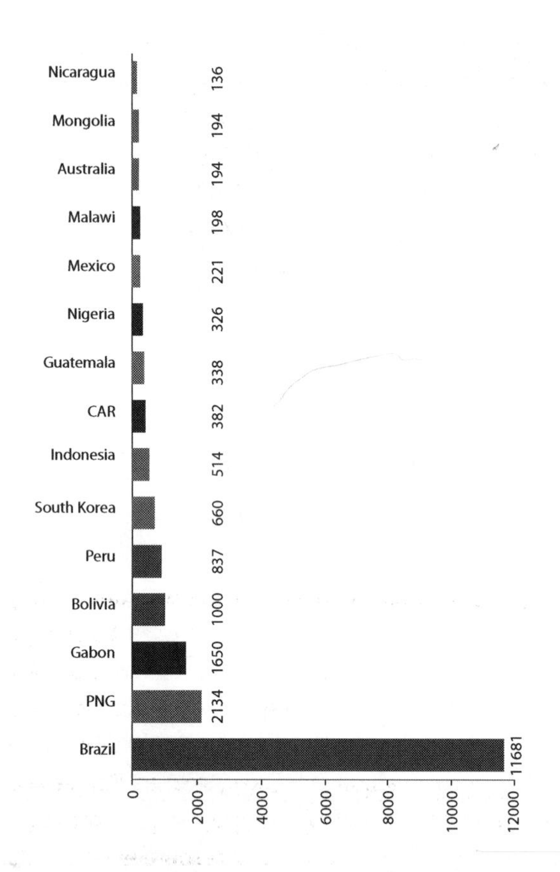

Figure 6.1. Deforestation primarily occurs in tropical nations, and mostly in Brazil.

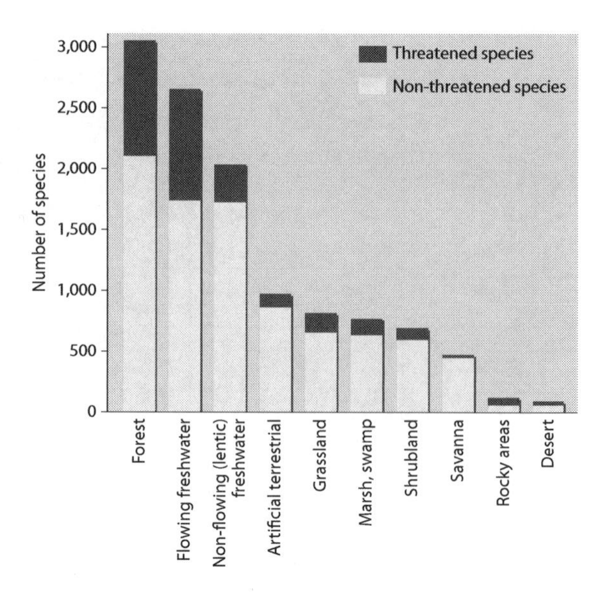

Figure 6.2. Besides the ecological service wetlands provide in cleaning our water, they also house a number of species, many of which are threatened. Source: Global Amphibian Assessment.

where grazing by livestock was not allowed for almost a decade, starting in the 1980s. More common, however, is **rotational grazing**. Rotational grazing is a process by which livestock are moved or rotated to fresh pastures, or pasture areas, to allow vegetation in previously grazed pastures to regenerate.

Wetlands

Wetlands are natural areas that hold on to water for a period of time before it continues on to a river, lake, stream, or under the ground to a groundwater source. Configurations that are included as wetlands include, but are not limited to, swamps, marshes, peat lands, estuaries, and tidal flats. While residing in a wetland, toxins are removed from the water either by degradation/mineralization from microbes, or adsorption of heavy metals to soil particles. Wetlands are also excellent at mitigating excess nutrients such as phosphorus and nitrogen from agricultural runoff. The destruction of wetlands has severely impacted water quality around the world. Wetlands are also important in food production in some areas of the world where crop plants, such as rice, are grown in paddies.

Loss of Natural Habitats and Biodiversity

By Robert W. Hastings

LAND CLEARING AND URBAN DEVELOPMENT

Most if not all environmental problems begin with human modification of the natural environment, such as clearing of land, changing water flow characteristics, introduction of new species, and release of wastes. As long as the human population is small and functioning in harmony with the natural environment, the environmental problems will be minimal. Such was apparently the case with the indigenous peoples who occupied the Pontchartrain basin. Although there was some land clearing for agriculture or construction of lodges and setting fires to clear forest undergrowth, their activities were mostly in harmony with nature and relatively minimal compared to modern human impacts. However, Kidder (1998) considered the concept that the Native Americans lived in harmony with the land a "myth" and their alteration of the Mississippi River delta area as substantial. In any case, European settlement brought with it a change in philosophy and an increased incentive to modify the environment. The change from a subsistence existence to a market-based economy meant increased harvesting of natural resources, such as timber and fur, and introduction of new crop species for commercial production. Export of resources or profits back to the mother country (initially France) meant greater pressure on residents to cut more trees, kill more game, and plant more crops. As the human population grew, so did these modifications of the environment. One tragic result of European settlement of the region was the almost total extinction of the indigenous peoples within about 200 years. The natural environment was to suffer as well.

Even in the late 1700s, cypress trees were considered rare in the immediate vicinity of New Orleans and were said to have been "wasted so imprudently" (Le Page du Pratz 1774). In addition, the supply of wild game near the city was severely depleted

(Surrey 1916; Kniffen 1990). Throughout the basin, loss of natural habitat was a significant and continuing concern, and with the loss of natural habitat, there was also a loss of wildlife and biodiversity. Such waste with little regard for conservation of natural areas continued until the present, when only a small remnant of these natural areas remains.

CYPRESS LOGGING AND CANAL DREDGING

Extensive logging of cypress *(Taxodium distichum)* goes back to the 1700s, but continued to increase to reach a peak in the early to mid 1900s. In the late 1800s, the "industrial cypress logging" period began and virtually all of the virgin cypress forests were destroyed within about fifty years. The logging of swamp trees has had a profound effect on the Pontchartrain basin wetlands, with its cypress and tupelo swamps changed dramatically in the past 100 years. Forests of massive cypress trees typically 400–600 years old (Sharitz and Mitsch 1993), but thousands of years old in some cases, have been destroyed, in many places never recovering but converting to open marsh habitat or even open water. Where regeneration did occur, only small, relatively young trees of only about a hundred years of age are now present. Those few large ancient trees that do survive are virtually all hollow, having been spared the saw because of their limited potential for board-feet of lumber. According to Conner et al. (1986), four requirements are necessary for successful regeneration and reestablishment of cypress forests: an abundant supply of seeds, abundant moisture during germination, lack of flooding for a period sufficient for the seedlings to grow tall enough to stay above subsequent flooding, and lack of predators. Cypress seeds will not germinate if submerged, and seedlings must grow above flood levels to survive their first year (Demaree 1932). The abundance of nutria *(Myocastor coypus)* now creates excessive predation on any cypress that do germinate. Increased flooding and predation appear to be the major factors preventing regeneration of cypress in these areas, although saltwater intrusion may also be significant.

To gain access to the dense swamps, the loggers dug canals and ditches, and additional canals have been dug to facilitate boating activity or provide access to wetlands areas for oil and gas exploration drilling. The canals expedited boating activity into once remote areas, but they also contributed to wetland loss and erosion. Some canals have increased in size through shoreline erosion and subsidence and have become conduits for increased flooding of the interior swamps.

Such hydrological changes often involve waters of increased salinity, which can have additional detrimental impacts on the cypress and other swamp vegetation. Another new threat to cypress in recent years has been the cutting of second-growth cypress to be shredded for garden mulch, a tragic waste of hundred-year-old trees and wildlife habitat.

The loss of virgin cypress swamps with numerous large mature trees must have eliminated important habitat for many animals and greatly reduced their numbers. The classic examples are the ivory-billed woodpecker *(Campephilus principalis)*, Bachman's warbler *(Vermivora bachmanii)*, Louisiana black bear *(Ursus americanus luteolus)*, and cougar *(Puma concolor)*, all of which are now extirpated from the Pontchartrain basin.

PINE FOREST LOGGING

Hickman (1966) estimated that originally more than 75% of the upland terrace area in the Florida Parishes was covered by pure longleaf pine *(Pinus palustris)* forests, with almost no shrubbery or undergrowth.

These open pine savannas were important habitats for many wild species and were used for grazing cattle and sheep, harvesting resin for naval stores, and for lumbering. By the early 1920s, the upland longleaf pine forests had been virtually eliminated, either from lumbering or clearing for agriculture and residential development (Ellis 1981). Forest lands that were replanted mostly became monoculture loblolly pine *(Pinus taeda)* plantations to supply the pulpwood industry (Hickman 1966). The southern pine beetle *(Dendroctonus frontalis)* became a more serious pest because loblolly pines are more susceptible to their attack than are longleaf pines, and this insect has been the cause of much forest mortality in recent years. Exclusion of fire from the region has also contributed to a replacement of the former fire-maintained longleaf pine forests with a southern mixed hardwood forest (Ware et al. 1993). Today the upland terrace area is a mix of urban and suburban development, cleared agricultural lands, and a few remnant reforested areas.

IMPERILED SPECIES AND ECOLOGICAL COMMUNITIES

Major changes of many natural areas occurred with human development of the Pontchartrain basin, and today they are quite different from what was present when European explorers first entered the area. Undisturbed natural areas are unfortunately rare in the Pontchartrain basin. Those that do remain in most cases now support species assemblages quite different from those originally present.

The estuarine communities of the lake have been impacted by water quality changes, including increased turbidity and nutrients. Substrates have been disturbed by dredging, trawling, and other human activities, and shorelines have been modified by filling and armoring projects. Most wetland areas have been modified by the digging of canals and ditches, and many have been drained and leveed or filled to create new land for development. Almost the entire south shore of Lake Pontchartrain is now occupied by the metropolitan area of New Orleans, Metairie, and Kenner, and urban sprawl is rampant on the north shore in Slidell and Mandeville on the east and Baton Rouge and Denham Springs on the west, as well as other communities in between. Most forested lands have been logged at least once, and many of the upland areas have been permanently cleared for agriculture or construction.

Correlated with this natural habitat destruction and environmental abuse, as well as overhunting and overfishing of some species, many species and ecological communities have become imperiled. Of fifty ecological communities identified by the Louisiana Natural Heritage Program (1988) as occurring in the Pontchartrain basin, at least twenty are regarded as being of conservation concern (Table 4). Smith (1999) listed twenty-two historic vegetation types occurring in the Florida parishes of the basin and noted that eleven had become rare or extirpated from the area (Table 5). These included the dominant forest types of longleaf pine flatwoods/savannahs and upland longleaf pine forests (dominated by longleaf pine). In recent years there has been increased concern for protecting the few remaining stands of longleaf pine, as well as increased restoration of some sites with this species.

Some species have become extinct or extirpated from the area, and many others are stressed and could be in danger of extinction if not properly protected. Some 141 species of plants and sixty-six species of animals occurring in the Pontchartrain basin are listed as being of conservation concern by the Louisiana Natural Heritage Program (www.wlf.louisiana.gov/experience/naturalheritage). Fourteen species officially listed as threatened or endangered by the U.S. Fish and Wildlife Service occur in the Pontchartrain basin (Table 6). Extinct species include the once abundant passenger pigeon *(Ectopistes*

Table 4. The natural communities of the Pontchartrain basin

COMMUNITY SYSTEM	COMMUNITY TYPE	NATURAL COMMUNITY
Estuarine	A. Intertidal emergent vegetation	1. Salt marsh
		2. Brackish marsh
		3. Intermediate marsh
	B. Subtidal aquatic bed	4. Submergent algal vegetation
		5. Submergent vascular vegetation
	C. Intertidal flat	6. Intertidal sand/shell flat
		7. Intertidal mud/organic flat
		8. Intertidal mollusk reef
	D. Subtidal open water	9. Bay
		10. Tidal channel/creek
		11. Tidal pass
Lacustrine	A. Limnetic open water	12. Upland lake
	B. Littoral open water	13. Marsh lake
		14. Swamp lake
Palustrine	A. Aquatic bed	15. Submergent algal bed
		16. Submerged/floating vascular vegetation
	B. Emergent vegetation	17. Freshwater marsh
		18. Hillside bog
	C. Scrub/shrub wetland vegetation	19. Scrub/shrub swamp
		20. Shrub swamp
	D. Forested wetland	21. Bald cypress-tupelo swamp
		22. Bald cypress swamp
		23. Tupelo-black gum swamp
		24. Batture
		25–29. Bottomland forest
		25. Overcup oak-water hickory
		26. Sugarberry-American elm-green ash
		27. Sycamore-sweetgum-American elm
		28. Sweetgum-water oak
		29. Live oak forest
		30. Wooded seep

		31. Bayhead swamp
		32. Slash pine-cypress/hardwood forest
		33. Pine flatwoods
		34. Wet hardwood flatwoods
		35. Wet/mesic spruce pine/hardwood flatwoods
		36. Pine savannah
		37. Riparian forest
Riverine	A. Riverine subtidal channel	38. Tidal mud flat
		39. Subtidal open water
	B. Riverine lower perennial channel	40. Sand/gravel beach/bar
		41. Mud bar
		42. Lower perennial open water
	C. Aquatic bed	43. Submerged/floating vascular vegetation
Terrestrial	A. Deciduous forests	44. Hardwood slope forest
	B. Mixed evergreen/deciduous forests	45. Shortleaf pine/oak-hickory forest
		46. Mixed hardwood-loblolly forest
		47. Slash pine/post oak forest
		48. Live oak-pine-magnolia forest
	C. Evergreen forests	49. Upland longleaf pine forest
		50. Sandy woodland
	D. Woodland	

Modified from Louisiana Natural Heritage Program (1988)

migratorius), which was last recorded in Louisiana in about 1903 (Lowery 1974b). The Carolina parakeet *(Conuropsis carolinensis),* described by Du Ru in the 1700s as being present by the thousands at the bayou portage near New Orleans (Butler 1934), had not been seen in the area for many years prior to 1900 (Kopman 1900). It was last recorded in Louisiana in 1880 (Lowery 1974b). The ivory-billed woodpecker *(Campephilus principalis)* was also virtually extirpated, having been dependent upon the large expanses of mature bottomland swamp forests for its habitat. Recent reports of its possible occurrence in the Pearl River swamps of the Pontchartrain basin have yet to be verified, although it was reliably sighted in the Big Woods of Arkansas in 2004 (Fitzpatrick et al. 2005; Gallagher 2005) and in the mature swamp forests of the Choctawhatchee River in western Florida in 2005 (Hill et al. 2006), more than sixty years after its previous confirmed sighting. Bachman's warbler *(Vermivora bachmanii)* is apparently extinct, as a result of destruction of its old-growth bottomland forest habitat.

Other species once common but now extirpated from the basin include American bison *(Bison bison),* cougar *(Puma concolor),* and black bear *(Ursus americanus luteolus).* Those that have become rare include Gulf sturgeon *(Acipenser oxyrinchus desotoi),* striped bass *(Morone saxatilis),* Alabama shad

Table 5. Historic vegetation types of the Florida parishes in the Pontchartrain basin

Vegetation Type	EBR	EF	Liv	SH	ST	Tang	Wash
Brackish/intermediate marsh					S		
Submergent estuarine grass beds (r)					ss*		
Fresh marsh					S	S	
Hillside seepage bog (r)					ss*	ss*	ss*
Bald cypress swamp	S	ss	P	ss	S	P	S
Pond cypress/black gum swamp					S	S	
Gum swamp		ss	ss	ss	ss	ss	ss
Bottomland hardwood forest	S	ss	S	ss	S	S	S
Live oak forest (r)	ss*						
Forested (wooded) seep					ss	ss	ss
Bayhead swamp		ss*	ss*	ss*	S	S	S
Slash pine-pond cypress/ hardwood forest (r)					S*		
Longleaf pine flatwoods/ savannahs (r)		(ss)	(P)	(ss)	P*	P*	S*
Wet hardwood flatwoods	P						
Spruce pine/ Hardwood flatwoods (r)	P*		P*				
Small stream (riparian) forest	S	S	S	S	S	S	S
Hardwood slope forest (r)	ss	ss*	ss*	ss*	ss*	ss*	ss*
Shortleaf pine/oak-hickory forest (r)	ss*	P*	ss*	ss*	ss*	ss*	ss*
Mixed hardwood-loblolly pine forest	ss	S	S	S	S	S	S
Upland longleaf pine forest (r)	(S)	(P)	(S)	P*	P*	P*	P*
Prairie terrace loess forest (r)	P*						
Saline prairie (extirpated from Pontchartrain basin)	(ss)	(ss)	(ss)				

Modified from Smith (1999)
EBR, East Baton Rouge Parish; EF, East Feliciana Parish; Liv, Livingston Parish; SH, St. Helena Parish; ST, St. Tammany Parish; Tang, Tangipahoa Parish; Wash, Washington Parish; r, now rare in the Pontchartrain basin; P, primary vegetation type; S, secondary vegetation type; ss, historic small-scale vegetation type; ( ), now extirpated from parish; *, now rare in parish

(Alosa alabamae), gopher frog *(Rana sevosa)*, and red-cockaded woodpecker *(Picoides borealis)*. Gulf sturgeon, striped bass, and Alabama shad are anadromous species once common in the Pontchartrain basin, spending most of the year and feeding in the estuary and coastal Gulf waters and then migrating up tributary rivers to spawn in early spring (March through May). Critical habitat was designated for Gulf sturgeon in 2003 and included the eastern half of Lake Pontchartrain and all of Lake Borgne, although the species has also been recorded in lesser numbers in the upper parts of the system, including several of

Table 6. Federally listed threatened and endangered species occurring in the Pontchartrain basin

SPECIES	STATUS	NOTES
Plants		
Louisiana quillwort, *Isoetes louisianensis*	endangered	Occurs on sand and gravel beach bars in the Bogue Chitto River drainage and the Bogue Falaya and Abita Rivers
Animals		
Inflated heelsplitter mussel, *Potamilus inflatus*	threatened	Currently found in the Amite River; once also occurred in the Pearl and Tangipahoa Rivers
Gulf sturgeon, *Acipenser oxyrinchus desotoi*	threatened	An anadromous marine species that moves up into coastal rivers to spawn, including the Pearl and Tchefuncte Rivers
Gopher tortoise, *Gopherus polyphemus*	threatened	Occurs in upland pine forests with sandy soils, including Tangipahoa, Washington, and St. Tammany Parishes
Ringed map turtle, *Graptemys oculifera*	threatened	Occurs only in the Pearl River drainage, including the Bogue Chitto River
Loggerhead sea turtle, *Caretta caretta*	threatened	A marine turtle usually found along the Gulf coast but may occasionally stray into Lake Pontchartrain
Bald eagle, *Haliaeetus leucocephalus*	threatened	Once rare in the state, but now becoming more common; several nests occur in the Pontchartrain basin
Louisiana black bear, *Ursus americanus luteolus*	threatened	Once common in the basin but now mostly characteristic of heavily wooded bottomland hardwood forests of the Tensas and Atchafalaya River basins; apparent strays may occasionally be seen in the Pearl River swamp
Mississippi gopher frog, *Rana sevosa*	endangered	Now known to survive only in a few isolated ponds in southern Mississippi
Kemp's ridley sea turtle, *Lepidochelys kempii*	endangered	A marine turtle usually found along the Gulf coast but may occasionally stray into Lake Pontchartrain
Brown pelican, *Pelecanus occidentalis*	endangered	A shore bird extirpated from Louisiana by the early 1960s, but reintroduced and now common on Lake Pontchartrain
Red-cockaded woodpecker, *Picoides borealis*	endangered	Nests only in mature pine forests
Ivory-billed woodpecker, *Campephilus principalis*	endangered	Possibly extirpated but supposedly sighted in the Pearl River swamp in 1999; sighted in Arkansas in 2004
West Indian manatee, *Trichechus manatus*	endangered	Primarily a resident of more southern coastal waters but occasionally sighted in Louisiana waters, apparently becoming more frequent in recent years

the rivers (U.S. Fish and Wildlife Service 2003). Ross (2001) reported that the species may be fairly common in the lower Pearl River.

Striped bass abundance in the Gulf of Mexico was never as great as on the Atlantic coast, and the Pontchartrain basin was apparently the western limit of its natural range in the Gulf. However, large schools were once recorded in the Tangipahoa River near Osyka, Mississippi (Bean 1884; McIlwain 1968; Ross 2001). Most Gulf coast populations were extirpated in the early 1900s as a result of dam building and water quality degradation, and individuals seen today in the Pontchartrain area are most likely the Atlantic coast strain present as a result of restocking efforts. There does not appear to have been any natural spawning of striped bass in the Pontchartrain basin in recent years.

Alabama shad was once abundant in the Mississippi River and other rivers of the northern Gulf coast, and it occurred in rivers of the Pontchartrain basin. This species has experienced a significant decline since the 1960s, apparently as a result of dams that block migration and increased siltation (Gunning and Suttkus 1990).

Several species have been nearly extirpated in the past but with proper management and protection have once again become common. Among these are the American alligator *(Alligator mississippiensis)*, brown pelican *(Pelecanus occidentalis)*, osprey *(Pandion haliaetus)*, bald eagle *(Haliaeetus leucocephalus)*, wild turkey *(Meleagris gallopavo)*, and beaver *(Castor canadensis)*.

Recovery of the alligator populations in Louisiana and other states is a true success story in wildlife conservation. The species was hunted almost to extinction in the late 1800s and early 1900s. There are numerous reports of alligator hunters who had killed thousands of individuals (Walker 1885). According to McIlhenny (1935), an estimated 3.0-3.5 million alligators were killed in southern Louisiana between 1880 and 1933. Populations reached a low point about 1960, and the species was placed on the federal endangered species list in 1967 (Newsom et al. 1987). By 1972 with protection and proper management, alligator populations in Louisiana had recovered sufficiently that the state implemented a controlled harvest program. Alligators are now common throughout the Pontchartrain basin in appropriate habitats.

The restoration of Louisiana's state bird, the brown pelican, is another impressive conservation success story (Norman and Purrington 1970). The species was abundant in coastal Louisiana prior to the late 1950s, but then suffered a precipitous decline in numbers until there were no brown pelicans in Louisiana by the mid-1960s. The cause of this dramatic loss was unknown initially, but was later attributed to the poisonous effects of DDT and other chlorinated hydrocarbon pesticides. Nesting colonies were subsequently reestablished in coastal Louisiana beginning in 1968, with birds imported from Florida, and the populations have steadily increased since then (McNease et al. 1992). Brown pelicans began reappearing on Lake Pontchartrain in 1987 and have since become quite common (Brantley 1998). Ospreys, bald eagles, and double-crested cormorants *(Phalacrocorax auritus)* experienced a similar decline and subsequent increase during the same period, apparently associated with the same causes. Their recovery in the Pontchartrain basin, however, has occurred without the benefit of human reintroductions.

The recovery and restoration of such species demonstrates what can be done with a reasonable measure of environmental protection, proper wildlife management, and species protection. The ultimate concern in protecting any species must be adequate protection for its habitat. Fortunately large areas of natural habitat for some species still remain within the Pontchartrain basin, although these are continually threatened by human impacts. Hopefully sufficient natural environments can be protected and restored in perpetuity so that future generations can experience and enjoy such natural biodiversity.

INVASIVE EXOTIC SPECIES

Many introduced exotic species have become established in the Pontchartrain basin, and in some cases these have become among the most abundant species, usually with detrimental effects on native biota. In their atlas of vascular plants in Louisiana, Thomas and Allen (1993, 1996, 1998) listed 2423 native species and 826 exotic species (25%) statewide. Not all of these exotic species are established in the Pontchartrain basin and not all are invasive, but the number certainly reflects the enormity of the problem. The number of nonnative animals established in the state is unknown (especially among the invertebrates), but their numbers are substantial. Exotic species usually compete with native species and can cause serious environmental problems by displacing native species or by becoming significant pests. Among the most serious pests are various aquatic plants, such as water lettuce *(Pistia stratiotes)*, water hyacinth *(Eichhornia crassipes;* see Figure 7.2), alligatorweed *(Alternanthera philoxeroides)*, Eurasian watermilfoil *(Myriophyllum spicatum)*, and water spangle *(Salvinia minima)*. Many of these exotic plant species have become so abundant in basin waterways that they impede or prevent boat traffic. Their abundance also inhibits the growth of native aquatic plants and can reduce aquatic habitat or dissolved oxygen levels and thus reduce the populations of aquatic animals.

Water lettuce, thought to be native to South America or Africa, was first reported in Florida in 1765 and has been established so long that it is sometimes listed as a native species. Water hyacinths from Venezuela were on display at the World's Industrial and Cotton Centennial Exposition held in New Orleans in the 1880s (Roberts 1946). After being released into local waterways, it became one of the most abundant aquatic plants in much of southeastern North America. Alligatorweed, another native of South America, was introduced to Louisiana in the early 1900s. Montz and Cherubini (1973) noted that a dense undergrowth of alligatorweed covered most of the soil surface in a cypress swamp of the LaBranche Wetlands area. Eurasian watermilfoil was introduced to North America in the 1880s, but the plant was first recorded in Louisiana in 1966 (in False River, Pointe Coupe Parish; Reed 1977). It probably soon spread to other waterways. Water spangle is a fairly recent invader, having been first reported from southeastern Louisiana and the Pontchartrain basin in the 1980s. It has spread rapidly and now occurs in most enclosed waterways tributary to Lakes Pontchartrain and Maurepas. Several terrestrial exotic plant species have also become invasive in the Pontchartrain basin, including Chinese tallow tree *(Triadica* or *Sapium sebiferum)*, Chinese privet *(Ligustrum sinense)*, Japanese privet *(Ligustrum japonicum)*, Japanese honeysuckle *(Lonicera japonica)*, kudzu *(Pueraria montana)*, cogon grass *(Imperata cylindrica)*, and Japanese climbing fern *(Lygodium japonicum)*. These also tend to compete with and displace native vegetation.

The introduced nutria *(Myocastor coypus)* has greatly affected the wetlands of coastal Louisiana (Kidder 1998). This South American rodent's consumption of wetlands vegetation, and especially seedling cypress trees, has contributed to the changed nature of many wetland sites. In some areas, such as the Manchac Wildlife Management Area, there is virtually no survival of seedling cypress because they are quickly destroyed by nutria. Nutria also appear to have caused a decline in muskrat *(Ondatra zibethicus)* populations in coastal Louisiana. They were first introduced to Louisiana in the 1930s, but quickly spread throughout the state. Tradition has blamed Edward Avery McIlhenny of Avery Island, Iberia Parish, for introducing nutria to Louisiana (Lowery 1974a), but this has been challenged by Bernard (2002). However, McIlhenny certainly was instrumental in the propagation and release of large numbers of nutria in Louisiana. The first importation of nutria to the United States (in California) was in 1899. The first recorded occurrence in Louisiana was in 1933, when

Susan and Conrad Brote established a nutria farm in Abita Springs, which closed after only four years, having sold some nutria and "turned the rest out." Another nutria farm in St. Bernard Parish provided McIlhenny with his first nutria in 1938. Subsequently McIlhenny's captive population increased to more than 500, and many were deliberately released into the wild. All of his remaining nutria were released in late 1945. During the 1945–1946 trapping season, at least 8784 nutria were captured statewide and during the following season 18,015. By 1970 nutria numbers had reached such a level that more than 1.5 million were trapped for fur (Lowery 1974a).

This introduction of an exotic fur-bearing animal was encouraged and promoted by then-director of Louisiana Department of Conservation's Fur and Wildlife Division Armand P. Daspit, in spite of warnings from the U.S. Bureau of Biological Survey: "It may be highly objectionable to turn them loose ... Numerous examples exist in this and foreign countries of the introduction of species from one part of the world to another with very disastrous results" (Bernard 2002). Those disastrous results were to become clearly evident within about forty years. Soon after its initial introduction to Louisiana in the 1930s, nutria had replaced muskrat as the dominant furbearer being trapped. By 1962–1963, the statewide harvest of nutria was more than 1.3 million skins valued at $1.35 per skin (compared to 300,000 muskrat skins valued at $1.60 per skin; Davis, Donald W. 1978). By 1976–1977, the 1.9 million nutria harvested in the state were valued at about $8.00 per skin, for a total of over $14 million. However, with declining markets for furs in the 1980s, the price of furs became too low to provide a profit, and trapping of nutria declined dramatically and remained low through the 1990s. With the decrease in trapping, the population of nutria in the marshes increased to a level that nutria damage to the marsh vegetation became apparent in many areas. In recent years, cash incentives paid to nutria trappers to increase their profit margin have been attempted as a means of controlling the nutria population. In addition, human consumption of nutria meat has been encouraged, but with only limited success. In Metairie, sheriff's department sharpshooters were employed to reduce nutria populations along the urban canals, where they have also become an abundant nuisance.

The common carp *(Cyprinus carpio)* is another exotic species present in the Pontchartrain basin, although it is not especially common. Very large individuals are occasionally seen in the lakes. The species was first introduced to North America in 1831 and subsequently widely distributed, but its first occurrence in Louisiana is unknown. Fuentes and Cashner (2002) documented the establishment of the Rio Grande cichlid *(Cichlasoma cyanoguttatum)* in the canals of New Orleans and along the southern shore of Lake Pontchartrain near canal pumping stations. It was first reported in 1996, but now seems to be well established in most of the urban canals of Orleans and Jefferson Parishes and may be expected to spread to other areas. It will potentially compete with native sunfishes (*Centrarchids*) and other species. Two other nonnative fish species, the goldfish *(Carassius auratus)* and the fathead minnow *(Pimephalespromelas)*, have been reported from the Pearl River (Ross 2001), but are not known to occur elsewhere in the Pontchartrain basin.

Large numbers of the Florida strain (subspecies) of largemouth bass *(Micropterus salmoides floridanus)* have been released into waters of Louisiana, including the Pontchartrain basin, in recent years. Initial reports of trophy-sized bass being caught have encouraged additional introductions. The potential impacts of this exotic genetic strain on the native subspecies (M. *s. salmoides)* are unknown, but hybridization could result in an inferior strain not well adapted to this environment (Ross 2001).

The exotic Asian clam *Corbicula fluminea* was introduced to North America in 1938 and has occurred in Louisiana at least since 1961 (Counts 1986, 1991). It is now present in all freshwater tributaries of the

Pontchartrain basin. *Corbicula fluminea* is the most common species of bivalve mollusk in the upper Tangipahoa River, where it represents almost 88% of the total bivalve fauna (Miller et al. 1986).

Another mollusk of special concern is the zebra mussel *(Dreissena polymorpha)*, a native of western Asia, which has become established in the Mississippi River. It was first recorded in the Great Lakes in 1988, and by 1993 this species had become abundant in the lower Mississippi River. Thus far no records of zebra mussel have been documented from Lake Pontchartrain, although the planktonic veliger larvae have been found in the Bonnet Carre Spillway (Bruce Thompson, personal communication). Research suggests that the species should be able to survive throughout most of the Pontchartrain basin, except possibly the more saline parts of Lake Borgne, so why it has not become established is not known. If it ever does become abundant in Lake Pontchartrain, it could profoundly affect food webs by filtering vast quantities of phytoplankton and outcompeting native species of filter-feeders.

The South American fire ant *(Solenopsis invicta)*, introduced to North America in 1918, is now one of the most abundant, and hated, species in southeastern Louisiana, as well as throughout the southeastern United States. It is common in both upland and wetland habitats, including the swamps and marshes surrounding Lake Pontchartrain, where the nests occupy any slight elevation, including mounds, cypress stumps and logs, canal spoil banks, and alligator nests. In addition to being a dangerous pest to humans, fire ants are thought to be detrimental to numerous ground-nesting species, including other insects, reptiles, birds, and mammals. They kill and consume many other animals and displace others with their aggressive and painful stings. An interesting characteristic of the fire ants is its ability to survive flood waters in wetland areas by congregating in floating masses. Thousands of ants will hold on to one another in a writhing mass, apparently alternating those who are submerged with those on top of the mass. This behavior allows the colony to survive flood waters of hurricanes and other storms.

Most introduced exotic species can have profound effects on native species and significantly modify the environment. Although any introduction of an exotic species is potentially harmful and should be avoided, some introductions have been considered either beneficial or neutral. The honey bee *(Apis mellifera)*, first introduced to North America in about 1638, is often cited as an example of a beneficial introduction, but its impact upon native species of bees is unknown. Other exotic species established in the Pontchartrain basin that are considered beneficial or neutral include Mediterranean gecko *(Hemidactylus turcicus)*, greenhouse frog *(Eleutherodactylus recordi)*, and Queen Anne's lace *(Daucus carota*; Thomas 1996). Additional exotic wetland plants that do not appear to cause significant harm include wild taro ("elephant ear," *Colocasia esculenta)*, which is fairly common along much of the lakeshore, and Timothy canary grass *(Phalaris angusta)*, which is also well established in much of the basin, including the Manchac marshes. Such species, however, may have subtle but significant effects upon native species, and they may also introduce exotic parasites and diseases that can be detrimental to native species. Once a nonnative becomes established, it is too late to become concerned about the consequences and virtually impossible to eradicate the invader. The best approach is to be satisfied with our native flora and fauna and not experiment with nature under uncontrolled conditions.

SHORELINE EROSION AND WETLAND LOSS

Coastal erosion and wetland loss is a significant environmental problem throughout coastal Louisiana, including the Pontchartrain basin (Templet and

Meyer-Arendt 1988; Penland et al. 1990; Boesch et al. 1994; Turner 1997). Louisiana has 40% of the nation's coastal wetlands but experiences 80% of the nation's coastal wetland loss, at a current estimated rate of about 25–35 sq. mi. (65–90 km^2) per year (Louisiana Coastal Wetlands Conservation and Restoration Task Force 1993). The total amount of coastal land in Louisiana in the early 1930s was 8511 sq. mi. (22,043 km^2) and in 1990, 6985 sq. mi. (18,091 km^2), for a loss of 1526 sq. mi. (3952 km^2) or 17.9% (Britsch and Dunbar 1993). The greatest land loss occurred in the period from 1956 to 1974, at a rate of approximately 42 sq. mi. (109 km^2) per year. The rate decreased to about 25 sq. mi. (65 km^2) per year by 1990. The concerted efforts to reduce the rate of land loss and to restore many eroded areas have been only slightly successful. A more realistic explanation of the reduction in land loss rate is that the more sensitive lands were lost prior to the 1980s, and lands remaining were somewhat more resistant to erosion.

Land loss rates within the Pontchartrain basin (Table 7) are generally lower than those along coastal Louisiana, but still are highest for the period of 1954–1974 (Britsch and Dunbar 1993). As might be expected, rates were higher in the areas nearest the open coast (that is, the Lake Borgne area, including the Chef Menteur quadrangle) than within Lakes Pontchartrain and Maurepas. Relatively high rates were recorded for the Bonnet Carre quadrangle, apparently related to land loss in the LaBranche Wetlands.

Penland et al. (1990) identified two major types of coastal land loss: coastal erosion (or retreat of shorelines) and wetland loss (development or expansion of water bodies within wetlands). Shoreline erosion had apparently been a problem for structures built on the unstable soils of the lake shoreline since the early 1700s, but it became an even greater problem in the 1900s. Shoreline armoring had been employed at least since 1852 to reduce this erosion and now covers some 40% of the Lake Pontchartrain shoreline (Lopez 2003). There was significant shoreline erosion on the northern shore near Mandeville in 1895 (Mugnier photograph in Kemp and King 1975). Parts of the lake shoreline in Orleans Parish had eroded as much as 500 ft. (152 m) at the turn of the century, providing justification for the lakeshore reclamation project and seawall between West End and the Industrial Canal. Along many eroded shorelines living cypress trees are now standing in the lake, and other trees once growing on land now lie prostrate along the shore (Figure 9.1).

Steinmayer (1939) reported shoreline recession between 1870 and 1917 at several locations on the eastern side of Lake Pontchartrain, such as between Point aux Herbes and Chef Menteur Pass (annual rate of about 14 ft. or 4.3 m) and between Bayou St. John and New Basin Canal (about 8 ft. or 2.4 m). He noted shorelines on the northeastern side of the lake where some areas had receded while others showed a slight advance into the lake (at Grand Lagoon [now Howze Beach and Eden Isles] and between Big Point and Point Platte).

The western side of the lake also was experiencing significant shoreline erosion during the period from 1930 to 1974 (Britsch and Dunbar 1996). Pearson et al. (1993) estimated that shorelines in the LaBranche area had receded almost 1000 ft. (305 m) during the past 100 years. In the Ruddock area, a wooden seawall was built at some early date along the shoreline at a site known as the "washout," which threatened the Illinois-Central Railroad. Erosion apparently continued, and only remnants of this seawall now remain. The shoreline was later reinforced with rocks, many of which contain crinoid fossils, a unique occurrence in the Lake Pontchartrain area where rocks, and thus fossils, do not naturally occur.

The Manchac lighthouse (see Figure 5.1) also illustrates the problems of shoreline erosion and wetland loss affecting coastal Louisiana. The existing lighthouse was built on land in 1857 approximately 1000 ft. (305 m) from the current shoreline. Sea level rise, land subsidence, and storm surges from the lake have combined to erode the land, currently at a rate

Table 7. Land loss rates for the Pontchartrain basin

	Annual Land Loss Rate by Time Period			
Quadrangle	1932–1958	1954–1974	1974–1983	1983–1990
North Shore				
Springfield	0.01 sq. mi. (0.03 km^2)	0.01 sq. mi. (0.03 km^2)	0.03 sq. mi. (0.08 km^2)	0.003 sq. mi. (0.01 km^2)
Ponchatoula	0.07 sq. mi. (0.18 km^2)	0.09 sq. mi. (0.23 km^2)	0.08 sq. mi. (0.21 km^2)	0.05 sq. mi. (0.13 km^2)
Covington	0.02 sq. mi. (0.05 km^2)	0.18 sq. mi. (0.47 km^2)	0.02 sq. mi. (0.05 km^2)	0.18 sq. mi. (0.47 km^2)
Slidell	0.06 sq. mi. (0.16 km^2)	0.15 sq. mi. (0.39 km^2)	0.05 sq. mi. (0.13 km^2)	0.04 sq. mi. (0.10 km^2)
South Shore				
Mount Airy	0.05 sq. mi. (0.13 km^2)	0.08 sq. mi. (0.21 km^2)	0.08 sq. mi. (0.21 km^2)	0.12 sq. mi. (0.31 km^2)
Bonnet Carre	0.10 sq. mi. (0.26 km^2)	0.44 sq. mi. (1.14 km^2)	0.19 sq. mi. (0.49 km^2)	0.07 sq. mi. (0.18 km^2)
Spanish Fort	0.03 sq. mi. (0.08 km^2)	0.01 sq. mi. (0.03 km^2)	0.003 sq. mi. (0.01 km^2)	0.01 sq. mi. (0.03 km^2)
Chef Menteur	0.49 sq. mi. (1.27 km^2)	0.41 sq. mi. (1.06 km^2)	0.28 sq. mi. (0.73 km^2)	0.28 sq. mi. (0.73 km^2)
Lake Borgne				
Rigolets	0.11 sq. mi. (0.29 km^2)	0.24 sq. mi. (0.62 km^2)	0.26 sq. mi. (0.67 km^2)	0.12 sq. mi. (0.31 km^2)
St. Bernard	0.29 sq. mi. (0.75 km^2)	1.23 sq. mi. (3.19 km^2)	0.70 sq. mi. (1.81 km^2)	0.26 sq. mi. (0.67 km^2)
Yscloskey	0.12 sq. mi. (0.31 km^2)	0.60 sq. mi. (1.55 km^2)	0.53 sq. mi. (1.37 km^2)	0.14 sq. mi. (0.36 km^2)
Total	1.35 sq. mi. (0.94 km^2)	3.44 sq. mi. (8.91 km^2)	2.22 sq. mi. (5.75 km^2)	1.27 sq. mi. (3.29 km^2)

Data from Britsch and Dunbar (1993)

of 12 ft. (3.7 m) per year in this area, and threaten the future existence of the lighthouse. The lighthouse tower is only a remnant of the once more extensive structure that included a lighthouse keeper's residence, several outbuildings, and a dock and boat shed. Today only the lighthouse tower and a jumble of broken brickwork remain in the lake, about 1000 ft. from shore.

By comparing shorelines on recent topographic maps with those of the 1800s, Zganjar et al. (2001) estimated the rates of shoreline movement in the basin (Table 8). Virtually all areas have experienced some shoreline erosion. Overall rates were highest in Lake Borgne and lowest in Lake Maurepas. In Lake Pontchartrain, erosion rates on the northern shore exceeded those on the southern shore because of the higher level of armoring and land creation along the New Orleans lakefront. The unprotected shorelines on the western side, such as near Pass Manchac, and on the eastern side near Point aux Herbes have experienced the highest rates of erosion.

In addition to the loss of land from shoreline erosion, there has also been substantial loss within the wetlands of the Pontchartrain basin. The total land loss within the basin (but including Chandeleur and Breton Sounds) between 1932 and 1990 has been estimated to be 188,356 acres (76,226 ha), of which 127,000 acres (51,396 ha) or 67% was interior loss (Penland et al. 2001b). Marsh areas in the basin between Lake Maurepas and Lake Borgne (excluding the Plaquemines Wetland and Birdfoot Delta Areas) decreased from 303,555 acres (122,847 ha) in 1932 to 231,370 acres (93,634 ha) in 1990, a loss of 72,185 acres (29,213 ha) or 24% (Penland et al. 2001c).

The Louisiana Coastal Wetlands Conservation and Restoration Task Force (1993) identified four critical wetland loss problems facing the Pontchartrain basin: (1) increased salinity and reduced sediment and nutrient input; (2) erosion along the Mississippi River Gulf Outlet (MRGO); (3) potential loss of the land bridges separating Lakes Pontchartrain and Borgne and Pontchartrain and Maurepas; and (4) potential rapid erosion of especially vulnerable wetlands, such as those separated from the lakes by just a narrow rim of shore (such as the Prairie at the Manchac Wildlife Management Area, the LaBranche marshes, and the shoreline from Goose Point to Green Point).

Land loss in the Pontchartrain basin, as in all of coastal Louisiana, results from a complex mixture of both natural processes and human activities, but the human-induced losses far surpass those caused by natural factors (61.2% vs. 38.8%; Penland et al. 2001b). Human activities in the basin have greatly exacerbated a natural process that would have occurred at a much slower rate in the absence of human influences. Thirteen factors contributing to land loss have been identified, which can be classified in three primary categories (Table 9): erosion (mechanical removal and transport of land by water action, 34.6%), submergence (increase of water level relative to ground surface elevation, 52.7%), and direct removal (physical removal of land by actions other than water, 12.7%).

Hurricane Katrina resulted in major losses of marsh habitat in the Pontchartrain basin, including 9–14 sq. mi. (23–36 km²) in the middle sub-basin, especially in the north shore marshes between Green Point and North Shore, in the LaBranche Wetlands, and in the East Orleans Land Bridge (LPBF 2006a). Even greater losses (estimated to be 40.9 sq. mi. or 106 km²) occurred in the lower sub-basin southwest of Lake Borgne. These preliminary estimates of the one day loss from Katrina exceeded the total land lost in the previous decade (1990–2000).

SEA LEVEL RISE AND SUBSIDENCE

The natural processes contributing to land loss in the Pontchartrain basin, as in all of coastal Louisiana, are in general correlated with sea level rise and subsidence. Mean global sea level has risen about 4.7 in. (12 cm) in the past 100 years or about 0.05 in. (1–2 mm) per year (Gornitz et al. 1982; Gornitz 1995), and the sea level has risen about 0.09 in. (2.3 mm) per year in the Gulf of Mexico. During the next 100 years as global warming continues, the rates of sea level rise are projected to increase four to seven times over current rates, or an additional 19 in. (48 cm) by the year 2100 (Gornitz 1995; Twilley et al. 2001). In coastal Louisiana, relative rates of sea level rise are considerably higher than global sea level rates because of the combined effects of global sea level rise and land subsidence (Ramsey and Penland 1989).

Figure 9.1. Shoreline erosion in Lake Pontchartrain near (A) the mouth of the Tangipahoa River and (B) Ruddock (Photographs by the author)

Table 8. Annual rates of shoreline change in the Pontchartrain basin

Location	1850–1995*	1930–1995	1960–1995
Lake Maurepas	-3.15 ft./year (-0.96 m/year)	-1.97 ft./year (-0.60 m/year)	-2.59 ft./year (-0.79 m/year)
Lake Pontchartrain, northern side	-4.10 ft./year (-1.25 m/year)	-6.23 ft./year (-1.90 m/year)	-3.90 ft./year (-1.19 m/year)
Lake Pontchartrain, southern side	-2.43 ft./year (-0.74 m/year)	-4.07 ft./year (-1.24 m/year)	-3.81 ft./year (-1.16 m/year)
Lake Borgne	-7.87 ft./year (-2.40 m/year)	-8.83 ft./year (-2.69 m/year)	-7.12 ft./year (-2.17 m/year)

Data from Zganjar et al. (2001)
*Rates for 1899–1995 for Lake Maurepas

Table 9. Causes of land loss in the Pontchartrain basin (in order of their significance and percent contribution to total land loss)

Causes of Land Loss	Source	Acres Lost(ha)	Percent Effect
Erosion by natural (wind-generated) waves	N	55,603 (22,502)	29.5%
Submergence due to altered hydrology with multiple causes	H	54,514 (22,061)	28.9%
Submergence due to altered hydrology from oil/gas channels	H	16,715 (6764)	8.9%
Direct removal of land to form oil/gas channels	H	12,781 (5172)	6.8%
Submergence due to natural water-logging or subsidence	N	11,188 (4528)	5.9%
Submergence due to failed land reclamation	H	7091 (2870)	3.8%
Direct removal of land to form navigation channels	H	6787 (2747)	3.6%
Erosion by channel flow	N?	6334 (2563)	3.4%
Submergence due to altered hydrology from roads	H	4767 (1929)	2.5%
Submergence due to altered hydrology from impoundments	H	4480 (1813)	2.4%
Erosion by navigation waves or boat wakes	H	3139 (1270)	1.7%
Direct removal of land to form borrow pits	H	3117 (1261)	1.6%
Direct removal of land to form access channels	H	1280 (518)	0.7%
Submergence due to substrate collapse following excessive herbivory or overgrazing	H?	561 (227)	0.3%
Direct removal to form drainage channels	H	0.13 (0.05)	<0.01%

Data from Penland et al. (2001b) N, natural; H, human-induced cause

Subsidence, resulting from consolidation and compaction of Holocene (Recent), Pleistocene, and Tertiary sediments of the Mississippi River deltas, is at least partly a natural process in that alluvial sediments tend to become more consolidated and compacted with time, and thus subside. If new sediments are continually or regularly deposited, then the subsidence is not noticeable. Without those new sediments, the land sinks, and low-elevation wetlands may become submerged. Loss of sedimentation from the Mississippi River is partly a natural process in the Pontchartrain basin correlated with Mississippi River delta shifts, which have occurred about every 1000 years. The Pontchartrain area has not been the primary building-delta of the Mississippi River since the formation of the St. Bernard delta (and the lake) some 1000–4000 years ago (Saucier 1994). However, some significant sedimentation and land building probably occurred in the basin during the formation of the Modern (Plaquemines or Balize) delta and even up through relatively modern times before Bayou Manchac was closed (in 1812) and the Mississippi River levees were finally completed (in 1930). Most of the sediments now carried by the river are retained within the river levees and lost to deeper waters of the Gulf of Mexico. With the loss of sedimentation or severe reduction in its amount, the effects of delta subsidence have become evident. The combined effects of subsidence and sea level rise produce a dramatic rate of land loss in coastal Louisiana.

Subsidence rates have been estimated at several sites in the Pontchartrain basin to be about 0.06–0.19 in. (0.15–0.47 cm) per year (Ramsey et al. 2001), based upon a contribution of about 43% to the relative rate of water level rise (Ramsey and Penland 1989). This combined effect of subsidence and global sea level rise resulted in a relative sea level rise for sites in the basin of 0.40 in. (1.01 cm) per year at South Point, 0.43 in. (1.09 cm) per year at Little Woods, 0.16 in. (0.40 cm) per year at West End, 0.14 in. (0.36 cm) per year at Frenier, and 0.18 in. (0.45 cm) per year at Mandeville since 1931.

Storms, and especially hurricanes, can have a much more profound effect in coastal areas because of the continuing subsidence and loss of coastal buffer lands. The storms further contribute to the erosion of coastal areas, resulting in accelerated shoreline retreat, destruction of beaches, destruction or modification of vegetation, and either scouring or filling of channels (Saucier 1963). Although some storm tides can deposit silt in swamp and marsh areas, thus building land, the net effect is not nearly sufficient to offset the loss of sediments from the Mississippi River.

A major continuing threat to New Orleans has been potential flooding, a long-standing problem exacerbated by the subsidence of most of the city below sea level (or lake level). Many had warned that it was just a matter of time before New Orleans was hit by a devastating hurricane with extensive flooding, property damage, and loss of life (McQuaid and Schleifstein 2002). That devastation finally came in August 2005, with Hurricane Katrina, and flooding of 80% of the city.

LAND RECLAMATION

Another significant factor in the loss of wetlands in some locations is land reclamation projects, where wetlands have been drained for agriculture or drained and filled for residential developments or other construction. Wetlands were once considered almost worthless, and landowners were encouraged to drain or fill them to make more profitable land. The extensive marshes surrounding Lakes Pontchartrain, Maurepas, and Borgne, which early explorers described as a wide "prairie" between the lake and the tall trees on the shore, have been greatly reduced in extent and dissected by numerous canals and ditches. Most of the marshes on the southern shore have been destroyed by the dredging, filling, and bulkheading of the New Orleans lakeshore. Others, such as the lakeshore areas near Slidell, have been filled for construction of housing developments including North Shore, Northshore Beach, Howze Beach, Eden Isles, Oak Harbor, Lakeshore Estates, Treasure Island, and Rigolets Estates, and Venetian Isles on Chef Menteur Pass. Although such developments create high ground and choice waterfront property for residents who desire convenient recreational opportunities on the lake, they also contribute to its demise, by increasing surface runoff, adding pollutants, and destroying wetlands or water bottom habitat. The developments also become prime targets for storm damage and flooding. Most suffered severe damage from Hurricane Katrina, as well as from previous storms.

The wetland reclamation projects for agriculture, such as the LaBranche Wetlands and the Madisonville rice fields, also contributed to wetland loss. Such sites were apparently productive for a few years, but eventually failed because of increased subsidence and flooding. Today they are mostly open-water ponds. The Bayou LaBranche Wetlands Restoration Project in 1994 was an attempt to return one of these areas to a productive marsh wetland. It resulted in the filling of 342 acres (138 ha) of eroded wetlands to create new marsh by dredging sediments from the adjacent lake bottom (Pearson et al. 1993; Louisiana Coastal Wetlands Conservation and Restoration Task Force 1997). Additional information on the LaBranche Wetlands is available in an excellent teacher's guide, "LaBranche: Lessons of a Wetland Paradise" (Maygarden 1996), and in a videotape "Bayou of the Lost: The Legacy of the LaBranche Wetlands" (Tyler 1996).

One marsh type that seems to have been virtually eliminated from the Pontchartrain basin is the immense canebrakes (monotypic stands of giant cane or switch cane, *Arundinaria gigantea)* described by Darby (1816) as once occurring along waterways such as the Amite, Comite, and New Rivers, on lands "not liable to annual submersion." Even at the time of Darby, much of this community type had been destroyed by clearing of the land for cultivation, the presence of cane being a sign of soil fertility (Platt and Brantley 1997). It is difficult to imagine the vast canebrakes and massive sizes of this native bamboo that once occurred along waterways in the Pontchartrain area and reportedly grew as tall as 40 ft. (12 m) and 4 in. (10 cm) in diameter; the canebrakes are now an ecological community type considered critically endangered (Platt and Brantley 1997; Brantley and Platt 2001; Platt, et al. 2001). Canebrakes have also been suggested as an important site for feeding of the apparently extinct Bachman's warbler *(Vermivora bachmanii).*

SALTWATER INTRUSION

Saltwater intrusion continues to be a problem in much of coastal Louisiana, including the Pontchartrain basin. The increased salinity has caused mortality of cypress and other primarily freshwater wetland species, and standing forests of dead cypress can be seen in areas severely stressed by saltwater intrusion. These standing dead trees may remain for years as reminders of the negative impacts that human activities can have even in remote locations. Studies have indicated that salinities in Lake Pontchartrain have continued to increase at least since the 1960s, when the most significant cypress mortality seems to have occurred. The MRGO, which provided a direct route for high-salinity Gulf water to move into the lake, has been cited as the most significant cause of this increase (Overton, et al. 1986).

Sikora and Kjerfve (1985) demonstrated a mean annual salinity increase of about 2 ppt in Lake Pontchartrain between 1963 and 1982, thought to be correlated with the opening of the MRGO (in 1963). However, statistical significance of such a correlation was not possible because of the extreme variation in the salinity records.

Salinity stratification has been demonstrated in the lake adjacent to the Inner Harbor Navigation Canal (Poirrier 1978b; Junot et al. 1983), with more dense, high-salinity water flowing into the lake below the lower salinity lake water. Such salt wedges can occur when fresh waters leaving the estuary move along the surface and high-salinity waters entering the estuary remain near the bottom. This is especially true in the MRGO and Inner Harbor Navigation Canal because of the greater salinity difference between water entering at Breton Sound and water flowing out of Lakes Pontchartrain and Borgne (Swenson 1980b).

Because the dense, saline water from the Inner Harbor Navigation Canal is also low in oxygen, it has caused "dead zones" in the adjacent part of the lake. The saltwater intrusion creates a plume of higher salinity water in the lake adjacent to the canal, and the stratified bottom water in such areas can become hypoxic and devoid of life, with dissolved oxygen levels of 1–4 ppm (Poirrier 1978b; Sikora and Sikora 1982; Junot et al. 1983; Overton et al. 1986; McCorquodale et al. 2001b). The increased salinity has also caused dramatic changes in wetlands vegetation along the course of the MRGO. Its margins have eroded very rapidly so that today it has increased in width from 500 ft. (51 m) to more than 2000 ft. (600 m). This erosion had destroyed 3000 acres (1241 ha) of marsh by the mid-1970s (Shallat 2001).

The MRGO was also blamed for much of the storm surge that entered adjacent wetlands and Lake Pontchartrain as a result of Hurricane Katrina (Van Heerden and Bryan 2006). Subsequently the MRGO was deauthorized as a deep draft ship channel and may eventually be closed (LPBF 2006b).

Several major hurricanes and storms since 1960 have produced significant flooding of the Pontchartrain basin by saline waters (including Betsy, 1965; Camille, 1969; Bob, 1979; Juan, 1985; Tropical Storms Beryl and Florence, 1988; Andrew, 1992; Tropical Storm Frances, 1998; Georges, 1998; and Katrina and Rita, 2005). In addition to the human consequences of hurricanes and other storms, the flood waters generated may contribute additional stresses to adjacent marsh vegetation, and cause rapid coastal erosion, wildlife mortality, and increased pollution levels. Such flooding now has a more significant impact on basin wetlands, because many areas are lower as a result of subsidence and numerous canals and ditches allow flood waters to more rapidly penetrate deep into these wetland areas. Once such saline water enters the swamps and marshes, evaporation may further concentrate the salt in ponds and ditches.

On a positive note, Froomer (1982) suggested that increased salinity can decrease the erodibility of marsh soils in the basin. In addition, the higher salinity water may be clearer and can result in increased biodiversity as more marine species move into the lake. However, these minor benefits cannot compensate for their substantial negative impacts.

Subsidence, increased flooding, and saltwater intrusion have all contributed to additional land loss and vegetational changes in places such as the Prairie, an open grassy area in the Manchac Wildlife Management Area adjacent to Lake Pontchartrain. This area has been popular for many years as a significant duck-hunting site. Aerial photographs taken in 1953 show the Prairie to be almost entirely covered with marsh vegetation, which was most likely fresh floating marsh of maiden cane or paille fine *(Panicum hemitomon).* By 1970 the area was 75% open water, and by 1983, 92% open water. There has been significant concern that eventually the narrow strip of land separating the Prairie from Lake Pontchartrain (at a small shoreline indentation called Turtle Cove) would be eroded away and the Prairie would become part of the lake. Because of this concern, the shoreline along this part of the lake was armored in 1994 with rock-filled gabions. A few years later, additional shoreline protection was added by the U.S. Army Corps of Engineers in the form of rock breakwaters constructed at intervals along 4.5 mi. (7.2 km) of shoreline. Other shoreline stabilization projects around the lake, such as bulkheads and seawalls, have effectively stopped shoreline erosion, but have also usually eliminated the adjacent wetlands.

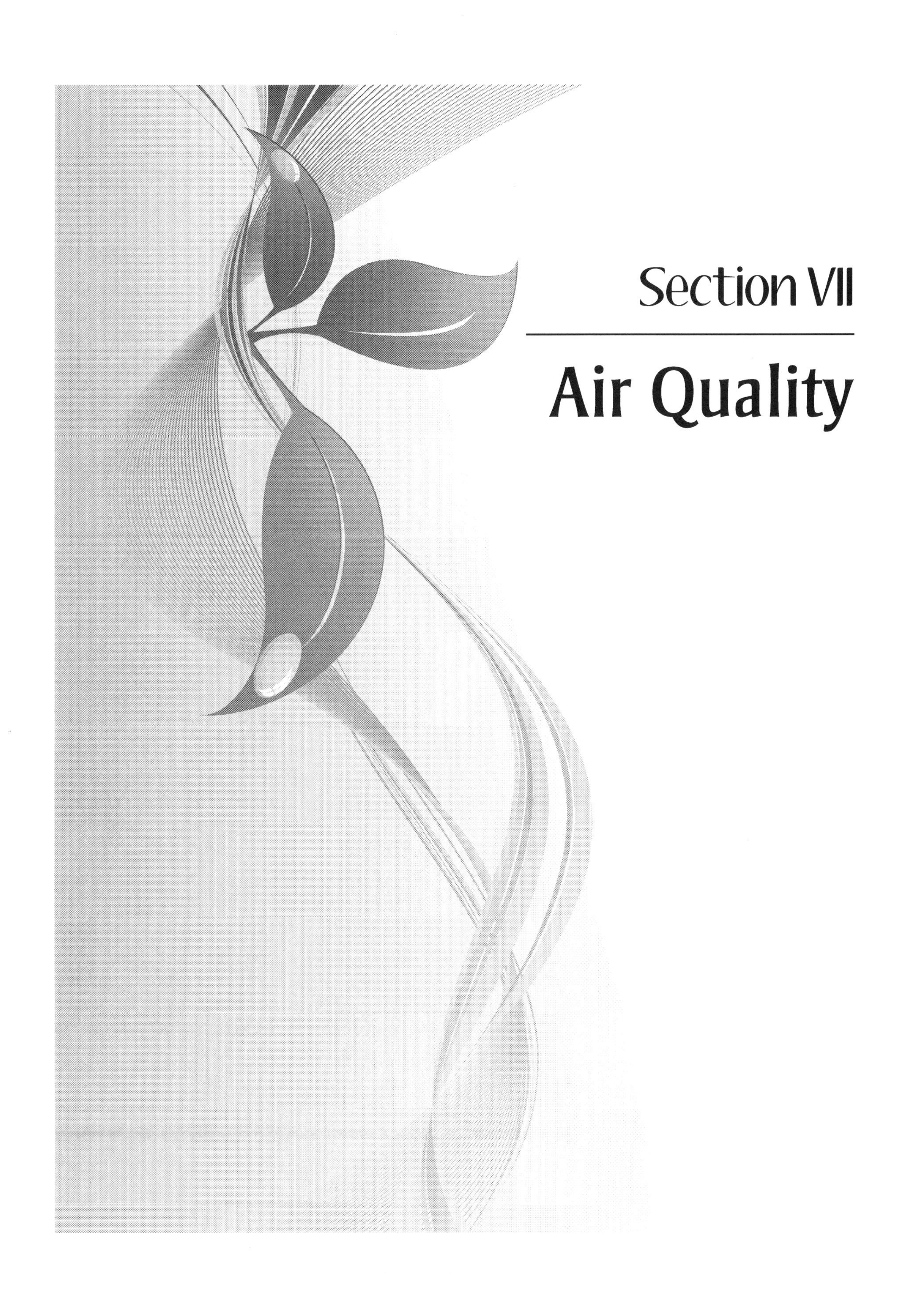

Section VII

Air Quality

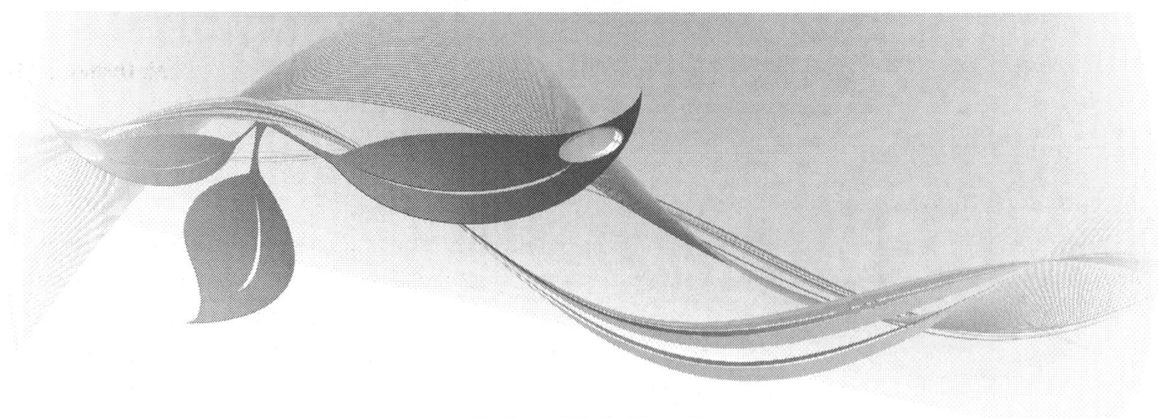

By Anne Marie Zimeri

The atmosphere and climate are extremely important to human health and the health of most organisms on the planet. Industrial emissions can affect both human health and the overall health of the planet. Air pollution emitted from one area can disperse around the planet such that ecosystems far from the source of the pollution can be affected. Therefore, air quality must be addressed on a global scale.

Our atmosphere is mostly nitrogen gas (78%), followed by oxygen (21%). There are a variety of other gaseous materials that make up a small percentage of what we breathe. For example, mercury vapor and argon can be detected at low levels well below one percent. The composition of our atmosphere has remained the same for at least 10^9 years, and organisms that are currently on the planet are well adapted to these ratios. If the composition of the atmosphere changes faster than organisms can adapt, we may experience a severe loss of species. We may also sicken humans and other species by emitting toxic gases and particulate matter into the air we breathe.

The troposphere is the area of the atmosphere that reaches from sea level to 12 km, in altitude. All terrestrial life on earth lives in the troposphere. Wind currents mix the air in this area and disperse pollution around the planet. Because of the great mass of the earth and the force of gravity, atmospheric components decrease in density as altitude increases. Ninety percent of the atmospheric mass is below 12 km. Temperature also changes as altitude increases. In the troposphere, temperature decreases at an average rate of 6.5C per km. This is called the normal lapse rate. Some areas stray from the normal lapse rate based on moisture and particulate matter in that specific area. If the lapse rate for a specific site is measured, that is the environmental lapse rate.

Above the troposphere lies the stratosphere (from 12 km to about 46 km above sea level). The atmosphere in the stratosphere is less dense and mixes less than the troposphere. Also in the stratosphere is the protective layer of ozone (O_3), which prevents most UV radiation from hitting the earth. Preserving the ozone layer in the stratosphere has become of global importance since it was discovered that it was being degraded by chemical releases. Chlorofluorocarbons have been used as propellants in the past to such a degree that they formed a hole in the ozone layer. This was discovered by Professor Paul Crutzen, Max-Planck-Institute for Chemistry, Mainz, Germany (Dutch citizen); Professor Mario Molina, Department of Earth, Atmospheric and Planetary Sciences and Department of Chemistry, MIT, Cambridge, MA, USA; and Professor F. Sherwood Rowland, Department of Chemistry, University of California, Irvine, CA, USA, who, for their work in atmospheric

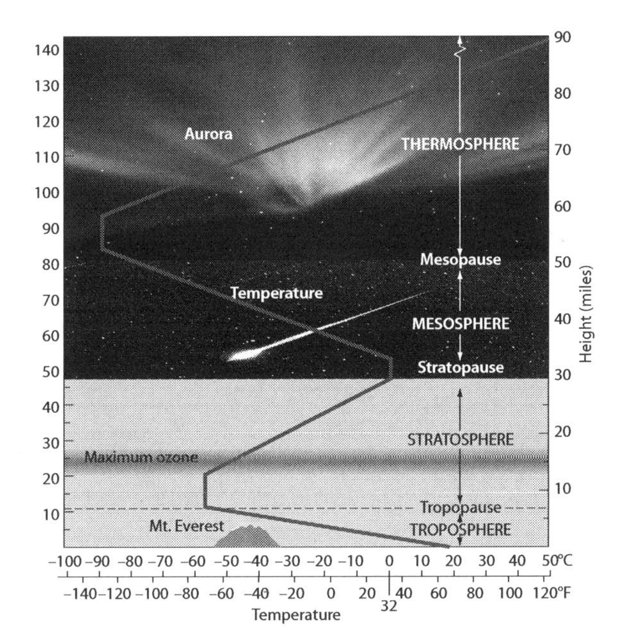

Figure 7.1. Each layer of the atmosphere has a different temperature and density profile.

chemistry, particularly concerning the formation and decomposition of ozone, won the Nobel Prize in 1995.

Ozone in the Troposphere

While ozone in the stratosphere is protective, ozone in the troposphere, or at "nose level" is harmful to human health. When inhaled, ozone can release free radicals in cells that can cause oxidative damage. Ground-level ozone is regulated by federal and state clean air legislation. The state and federal standards are supported by documented health effects of ozone measured in human and animal studies. Low-level ozone exposures can irritate the eyes, nose, throat, and lungs, and have even been shown to cause significant temporary decreases in lung capacity in healthy adults.

Coriolus Effect

There are large cells of air movement on the earth that disperse pollutants but also move moisture to areas that may not otherwise have access to water. These cells move based on the lower density of warm air rising and more dense, cool air, falling toward the earth. As warm moist air rises, it pulls cooler, denser air into its place (Figure 7.2). Once warm, moist air rises, it begins to cool, and the moisture condenses and falls to the earth as precipitation. When the air cools, it then falls to the earth. The Coriolis effect is caused by the movement of the air and the rotation of the earth. These predominant winds correlate with Hadley cells (Figure 7.3).

Atmospheric Inversions

Under normal conditions, warmer air masses can rise unimpeded, and the Coriolus effect and Hadley cell mechanisms work well. However, a couple of circumstances cause warm moist air to get trapped beneath a descending layer of cool air in an atmospheric inversion. If a cool air mass descends over a valley or area and traps the warmer air for several days, it is

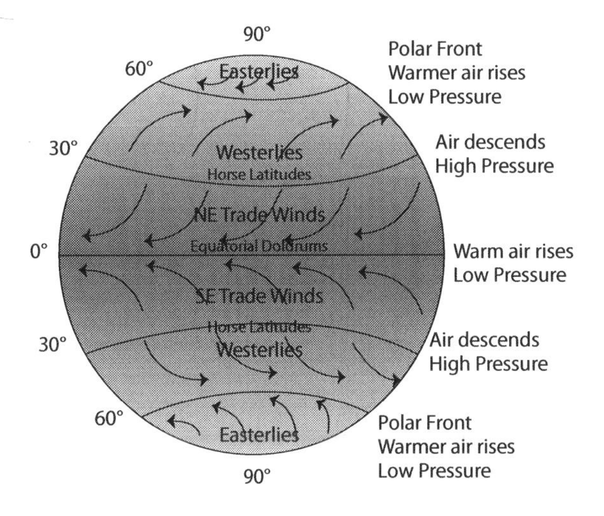

Figure 7.2. The predominant winds are caused by the Coriolis effect.

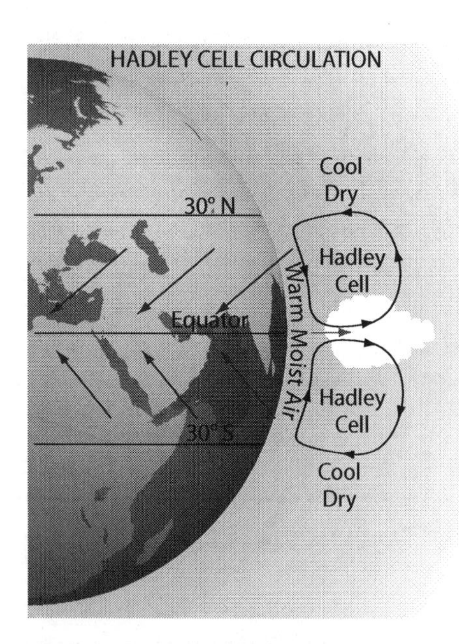

Figure 7.3. The earth is encircled by several broad prevailing wind belts. The direction and location of these wind belts are determined by solar radiation and the rotation of the earth. The three primary circulation cells at the equator are known as the Hadley cell.

called a subsidence inversion. These inversions trap the warm air along with pollution from industry and emission from automobiles. When this occurs over a large city, it can cause dangerous levels of smog.

When an atmospheric inversion occurs in the evening because the sun has set and cool air begins to descend, it is called a radiation inversion. Radiation inversions are remedied in the morning when the sun rises and the warm air can begin to ascend as usual.

Clean Air Act

The Clean Air Act is the law that defines the EPA's responsibilities for protecting and improving the nation's air quality and the stratospheric ozone layer. The last major change in the law, the Clean Air Act Amendments of 1990, was enacted by Congress in 1990. The 1990 amendments included a provision for National Ambient Air Quality Standards (NAAQS). These standards were based on the identification of several criteria pollutants and standards below which human health or infrastructure would not be impeded. Each standard has a level that must be measured and averaged over a specific period of time. For example, carbon monoxide must be measured and averaged over an eight-hour period and remain below nine parts per million (ppm) to attain the primary standard set by the Clean Air Act. Primary standards are set for human health. Secondary standards take into account damage to plants and abiotic things such as infrastructure (buildings and bridges) and artwork. Hydrocarbons, such as carbon monoxide, primarily are emitted by transportation vehicles and must be measured and averaged over the three hours from six to nine a.m. when most traffic is on the road. Nitrogen dioxide and sulfur dioxide, the two main components that cause acid rain, must also be measured.

Indoor Air Pollution

On average, people in the United States spend 90% of their time indoors. Some people, the ill, the elderly, and infants, often spend all of their time indoors. There are more concentrated levels of pollutants indoors as a result of volatile organic compounds being emitted from carpet and furniture, vapors from cleaning products, biological contamination such as mold and mildew, and secondhand tobacco smoke, to name a few. In addition, volatile organic compounds (VOCs) are emitted as gases from certain solids or liquids. Examples include paints and lacquers, paint strippers, cleaning supplies, pesticides, building materials and furnishings, office equipment such as copiers and printers, correction fluids and carbonless copy paper, graphics and craft materials including glues and adhesives.

Pollutant (final rule cite)		Primary/ Secondary	Averaging Time	Level	Form
Carbon Monoxide [76 FR 54294, Aug 31, 2011]		primary	8-hour	9 ppm	Not to be exceeded more than once per year
			1-hour	35 ppm	
Lead [73 FR 66964, Nov 12, 2008]		primary and secondary	Rolling 3 month average	0.15 $\mu g/m^3$ [1]	Not to be exceeded
Nitrogen Dioxide [75 FR 6474, Feb 9, 2010] [61 FR 52852, Oct 8, 1996]		primary	1-hour	100 ppb	98th percentile, averaged over 3 years
		primary and secondary	Annual	53 ppb [2]	Annual Mean
Ozone [73 FR 16436, Mar 27, 2008]		primary and secondary	8-hour	0.075 ppm [2]	Annual fourth-highest daily maximum 8-hr concentration, averaged over 3 years
Particle Pollution [71 FR 61144, Oct 17, 2006]	$PM_{2.5}$	primary and secondary	Annual	15 $\mu g/m^3$	annual mean, averaged over 3 years
			24-hour	35 $\mu g/m^3$	98th percentile, averaged over 3 years
	PM_{10}	primary and secondary	24-hour	150 $\mu g/m^3$	Not to be exceeded more than once per year on average over 3 years
Sulfur Dioxide [75 FR 35520, Jun 22, 2010] [38 FR 25678, Sept 14, 1973]		primary	1-hour	75 ppb [4]	99th percentile of 1-hour daily maximum concentrations, averaged over 3 years
		secondary	3-hour	0.5 ppm	Not to be exceeded more than once per year

as of October 2011

Table 7.1. The EPA has set National Ambient Air Quality Standards for six principal pollutants, which are called "criteria" pollutants. Units of measure for the standards are parts per million (ppm) by volume, parts per billion (ppb) by volume, and micrograms per cubic meter of air ($\mu g/m^3$).

The best remedy for indoor air pollution is ventilation. Opening doors and windows or installing attic fans can turn the air over a period of time to prevent the accumulation of pollutants.

Health indications upon exposure to indoor air pollution can include reactions such as eye, nose, and throat irritation; headaches, loss of coordination, nausea; and damage to the liver, kidneys, and the central nervous system. Some organics can cause cancer in animals; some are suspected or known to cause cancer in humans. Key signs or symptoms associated with exposure to VOCs include conjunctival irritation, nose and throat discomfort, headache, allergic skin reaction, dyspnea, declines in serum cholinesterase levels, nausea, emesis, epistaxis, fatigue, and dizziness, according to the U.S. EPA.

Radon

Radon is a colorless, odorless, tasteless gas that occurs naturally from the decay of uranium-rich rocks. As the gas seeps from the rocks, it can collect in low-lying, unventilated areas such as basements, crawl spaces, or lower levels of a home. This can be hazardous because radon has a radioactive isotope that can cause lung cancer.

To reduce the risk of lung cancer from domestic exposure to radon, ventilating areas susceptible to radon accumulation is key.

How do you know if you have radon in your home? Radon alpha track tests are available at many home improvement stores. These tests are placed in the home for a period of time, and then sent to the company for radon concentration testing. Radon is interesting because though you may have it in your home, your next-door neighbors may not. Therefore, as part of being a responsible homeowner, you should determine whether radon is present.

Air Pollution and Your Lungs

By Harvey Blatt

> There's so much pollution in the air now that if it weren't for our lungs there'd be no place to put it all.
>
> —Robert Orben

Long, long ago, on a planet far, far away, the air was pure. It contained only nitrogen (78 percent), oxygen (21 percent), argon (1 percent), and trace amounts of other gases necessary to make complex life possible, such as water vapor (0–6 percent), carbon dioxide (0.034 percent), ozone, nitrous oxide, and a few others. But then the planet's inhabitants, believing themselves of superior intelligence and recognizing that about 10,000 quarts of air and several billion dust particles enter a person's nose and mouth every day, decided to not only increase the amounts of some of the trace gases but also to add other, more creative things. They increased the amounts of carbon dioxide, ozone, sulfur dioxide, nitrogen dioxide, and carbon monoxide. Some gaseous organic compounds were thrown in as well. Then, to spice up the mixture, they added to the air they breathed particulate materials such as microscopic soot (black smoke), lead, asbestos, rubber, arsenic, cadmium, mercury, and other interesting substances.

The results were predictable. There were great increases in respiratory diseases such as bacterial infections, bronchitis, allergies, and asthma. The incidence of eye infections also rose. Babies born in the areas where the air had undergone the most change had smaller heads, lower weight at birth, damaged DNA, and increased rates of birth defects. Many died. Children had smaller lungs. Respiratory disease became the greatest killer of children on the planet. Elderly people also died prematurely, often from heart attacks because their weakened lungs could not process enough oxygen. Bus drivers in urban areas had increased cancer rates, chromosomal abnormalities, and DNA damage from breathing diesel fumes all day.

Crop yields were reduced as pollutant haze reduced the penetration of sunlight and decreased photosynthesis, the process by which plants use sunlight to convert carbon dioxide and water to food and plant fiber. Fruit size and weight decreased. Constituents in the haze affected plant metabolism. Market value was reduced because of spotting on leaves and fruit. Plant death in the field increased as the vegetation became more vulnerable to injury from diseases and pests.

Clearly, experimentation with the atmosphere in which they had evolved was not a good idea for the people of this planet. Because of their superior intelligence they decided to scale down or end the experiment. But it would not be easy.

WHERE IS THIS PLACE?

> We all live downwind.
>
> —Bumper sticker

As you no doubt have surmised, the long, long ago, far, far away planet is earth in the twenty-first century. Everything described is true. Air pollution has harmed crop production and causes more deaths in the United States than traffic accidents. Traffic fatalities total just over 40,000 per year, while air pollution claims 70,000 lives annually.

Air circulates, so Americans do not live in an atmospheric cocoon. We are not immune to the effects of air pollution arising in other parts of the world. And in many other parts of the world the air is so bad that people walk around wearing surgical masks to filter out some of the particles. Most urban children in the "developing countries" inhale the equivalent of two packs of cigarettes each day just by breathing. "Oxygen bars" have opened in Los Angeles, New York, Mexico City, Beijing, London, and Tokyo. In Los Angeles, for $15 you can breathe lemon- and lime-scented clean air briefly to let your lungs know what they are missing ($13 without the citrus). In Beijing a slug of good air is only $6, presumably because of lower production costs. In Mexico City the cost for a supplement of commercial oxygen is $2 per minute. A gambling casino in Cripple Creek, Colorado, offers oxygen hits in eight delicious flavors, including peppermint and tangerine. And air pollution is getting worse worldwide as impoverished nations such as India and China attempt coal-based rapid industrialization in their increasingly urbanized and car-using societies. The world's worst air is in Dacca, Bangladesh. It replaced the former leader, Mexico City, in 1997.

What are the major pollutants, how did they get into our air, and what can we do about getting them out? What do American political leaders, both federal and local, propose to do to clean our air? Have they made much progress? What can we hope for in the near future? Do we want to encourage enterprising entrepreneurs to increase the number of oxygen bars in our major cities? Should we impose a graduated tax on people who breathe cleaner air to pay for cleaning up the air in dirty cities? It could work like the federal income tax. If the air where you live is unusually clean (as it is for penguins in the Antarctic) your tax bill would be high. If your air is unusually dirty (Houston, Los Angeles, New Orleans) you would be exempt from paying the tax. Clean air is already traded on stock exchanges in Australia and at the Chicago Climate Exchange.

POLLUTANT GASES IN THE AIR

> When you can't breathe, nothing else matters.
>
> —American Lung Association

Each of us takes a breath every 4 seconds, about 8.5 million breaths per year. When resting, you and I inhale about 2,500 gallons of air per day, 7 quarts per minute. That's 1,500 trillion (1.5,000,000,000,000) gallons per day for the world's population of 6 billion. The molecules of air we breathe were at one time in the lungs of Beethoven, Napoleon, Attila the Hun, Jesus, and Moses. When doing heavy work—that is, huffing and puffing—the daily amount for each person rises from 2,500 to 15,000 gallons. An average American breathes 3,400 gallons. We use a lot of air. It would be nice if it were clean. Should a "Right to Breathe" be added to our Bill of Rights?

The noxious gases in our air that are monitored by urban air-quality authorities are sulfur dioxide, nitrogen oxide, ozone, and carbon monoxide. In 2004, more than half of all Americans lived in counties that did not meet the EPA standard for at least one of these pollutants. The problem is worldwide. Twenty of the 24 megacities of the world (more than 10 million people) do not pass muster for all of these gases and lead. A few cities have succeeded in improving their air somewhat during the past few years, but in most cities the air has continued to worsen. In Athens the death rate jumps by 12 percent when the level of sulfur dioxide exceeds a critical threshold. In New Delhi, 20 percent of the traffic police at busy intersections need regular medical attention for lung problems. Soon the number of coughs per hour in these cities may rival the number of words spoken.

Perhaps more distressing than increases in coughs is the effect of traffic fumes on male sperm and, therefore, reproductive success. In a study of male toll-booth attendants on Italian motorways, a recent investigation found the men had poorer-quality sperm than other young and middle-aged workers in the same area.

SULFUR DIOXIDE

Two-thirds of the noxious sulfur compounds in the air of North America have been produced by human activities, mostly by the combustion of coal and oil. Coal burning in America's electric power-generating plants produces 80–85 percent of America's annual emissions of sulfur dioxide. The other 15–20 percent comes from oil refining, commercial ships, and smelting of sulfide ores. When the sulfur dioxide leaves the smokestacks hundreds of feet above the ground it interacts chemically with the moisture and oxygen in the air to produce sulfuric acid, a major component of "acid rain." The problem is worldwide. China, because of its massive use of coal, leads the world in sulfur emissions.

How strong is the acid in acid rain? How does it compare, for example with Coca-Cola, lemon juice, or battery acid? The measuring stick used to determine the level of acidity is the pH scale (Figure 7.1). The scale extends from 0 to 14. Values greater than 7 are basic; values less than 7 are acidic. The scale is logarithmic, meaning that each unit change is an increase or decrease of ten times. Hence, a value of 5 is ten times more acidic than a value of 6, pH 4 is 100 times more acidic that 6, and pH 3 is 1,000 times more acidic than 6. Although some acid rains and acid fogs (Los Angeles) have been found to have the pH of lemon juice, average acid rain has a pH between 5 and 4, more acid than urine but less acid than tomatoes.

Unpolluted rainwater has a pH of 5.6, so it is acidic. When we speak of acid rain we mean rain more acidic (lower pH value) than 5.6, enhanced acidity rather than normal acidity (just as when we speak of the normal greenhouse effect versus the enhanced greenhouse effect). The acidity of rain over the Eastern United States, where most of our population and most of our coal-burning industries are located, is much lower than 5.6.

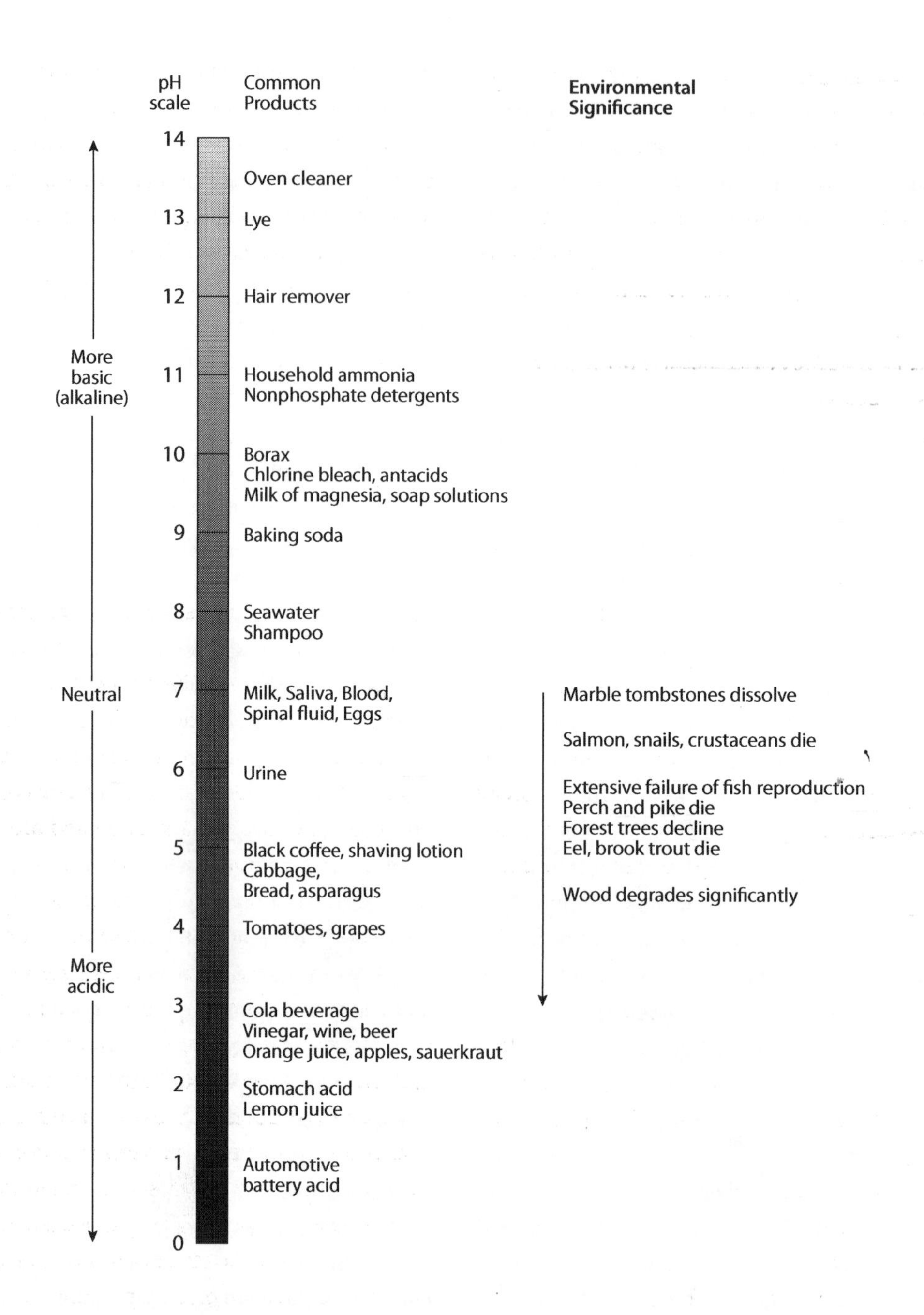

Figure 7.1 The pH scale, the pH of some common liquids and solids, and the environmental significance of acid pH values.

Recall that each unit change is a change of ten times, so pH 4.6, typical of rain in the Eastern United States, is ten times more acidic than 5.6. Clearly, there is a problem with rainfall in the Eastern United States.

Well, so what difference does more acid rain make? There are several differences.

Air with enhanced acid content kills fish in lakes. In a third of the lakes in New York's Adirondack

Mountains the pH is commonly 4.3 and all the fish have died. Winds from the Upper Midwest blow toward the northeast, so that some of America's air pollution rains on eastern Canada. As a result, in Nova Scotia the pH has fallen so low in some rivers that salmon cannot live in them. Acid rain harms plant life and crops as well. It removes nutrients from soil, destroys leaves on trees, and mobilizes both nutrients and toxic metals in the soil, which end up in nearby lake waters.

Acid precipitation in polluted cities causes statuary and building surfaces to deteriorate two to three times faster than in rural areas. Do you remember the supposed midnight ride of Paul Revere in 1775 to warn the American colonists that the British were coming? His tombstone doesn't. Acid precipitation has erased the inscription. The Parthenon, Taj Mahal, and Michelangelo's statues are dissolving under the onslaught of the acid pouring out of the skies.

Half a century ago, if you stood on a hilltop on a clear day just about anywhere east of the Rocky Mountains, you could have seen things 70 miles away. Now, average visibility—even far from cities—is about 15 miles. There is a permanent haze. A 10-year study of visibility in 12 U.S. National Parks found continually decreasing visibility. According to the researchers the cause is mostly sulfur compounds. At times in the East "it is almost pure dilute sulfuric acid," according to one investigator. Between 1982 and 1992, summer sulfate hazes in the Great Smokey Mountains National Park, Tennessee, soared almost 40 percent. The most improved site was Chiracahua National Monument, Arizona. Sulfur levels dropped by one-third because of the closing of, or emission controls on, nearby copper smelters.

Fortunately, there has been progress in recent years in slowing the emission of sulfur from smokestacks, thanks to the Clean Air Act acid-rain amendments passed by Congress in 1990. Between 1981 and 2000 sulfur emissions dropped by 50 percent. Industry was required to cut its sulfur emissions. In the first phase, which began in 1995, 445 power plants cut their sulfur emissions by 50 percent. In the second phase, which began in 2000, another 700 plants were required to start reducing their emissions. A 2003 study found that more than half of our artificially acidified lakes are bouncing back, their acidity reduced from previous lows, but recovery is slow. The acidity took decades to develop and will take decades to reverse.

NITROGEN DIOXIDE

Approximately 90 percent of nitrogen dioxide emissions (and those of other nitrogen compounds) are caused by human activities, about one- third from motor vehicles. Thus, more than half the emissions of nitrogen dioxide are made at ground level, in contrast to sulfur emissions. Nitrogen emissions grew 4 percent between 1981 and 2000 and are one- third greater than sulfur emissions. However, because nitrogen compounds are a major ingredient in the formation of smog (see below) near ground level, they are removed from the air very quickly near their source and are not a major cause of acid rain.

Until the mid-1990s, the catalytic converters in cars driven in the United States increased emissions of nitrogen oxides from the car. Since then, the catalyst in the converters has been changed and emissions of nitrogenous gases have been halted. The nitrogenous compounds are converted to nitrogen and water, Under ideal conditions, the converters can reduce nitrogen oxide emissions by 95 percent. However, catalytic converters can only clean up vehicle exhausts if the engine is warmed up, and many trips taken by walk-avoiding Americans are too short or involve lots of stopping and starting, so the engine does not get hot enough for the catalytic converter to do its job.

OZONE AND SMOG

As I will explain in chapter 8, ozone in the stratosphere (more than 10 miles above our heads) is essential for our well-being. But ozone near ground level is bad news. It is the chief component of urban smog and is most severe in areas with little wind, much sunlight, and many motor vehicles. Ground-level ozone is a secondary pollutant, because it does not come belching out of smokestacks or car exhausts. Most ozone forms when sunlight stimulates a chemical reaction between the nitrogen dioxide and hydrocarbon fumes coming out of car and truck exhausts. The main function of catalytic converters is to convert the fumes into carbon dioxide and water. In short, cars produce smog. In 2000 half of all Americans (more than 141 million) lived in communities that had severe smog pollution during the May-through-September period when higher temperatures and increased sunlight combine with stagnant air to produce ozone. The number of sufferers was up from 132 million in 1999. The air in nearly all the biggest cities, including New York, Los Angeles, Houston, Chicago, and Dallas, fell short of federal standards for ozone pollution.

Houston, because of its unusually high concentration of petrochemical plants, and the Los Angeles Basin, because of too many cars, have the nation's worst smog problems. With 10 million cars, Los Angeles has the highest ratio of cars to people in the world. Californians use 5 percent of the *world's* gasoline. Its 24,000 public school buses are among the most polluting in the nation. Examination of the lungs of young accident and homicide victims in Southern California found 75 percent had airspace inflammation and 27 percent had severe damage. That's one-quarter of *young* people. For these reasons, California is commonly at the forefront of the push for cleaner-burning gasoline-powered cars and hybrid cars.

Environmental efforts to reduce smog in the Los Angeles area have been paying off. In 1977 the number of stage 1 smog alerts, in which avoidance of vigorous outdoor exercise is recommended and those with health problems are advised to stay indoors, was 121. In 1996 there were only 7 such alerts. In 1999 the area had the best smog record ever recorded there. It was the first summer in 50 years without a single stage 1 warning. Sunburn advisories may soon become more common than smog alerts.

Smog is a particularly serious problem for asthma sufferers, who number 17 million in the United States, 6 percent of the population. Every day 14 people die from asthma attacks. A study in France determined that asthma attacks increased by 30 percent on smoggy days. In Paris, 42 percent of visits to pediatricians are for respiratory problems. A study in the Netherlands found that people living near major roadways are twice as likely to die from heart and lung diseases as those who live farther away. Smog in Europe kills more than twice as many people as car accidents.

Pregnant women exposed to the high concentrations of smog and carbon monoxide (see below) characteristic of large cities have triple the risk of having a child with heart malformations, the rate jumping from the normal two per thousand to six per thousand.

CARBON MONOXIDE 

Carbon monoxide interferes with the ability of the blood to carry oxygen throughout your body. The hemoglobin in your blood is the oxygen transporter, but it likes carbon monoxide 200 times better than it likes oxygen. So it preferentially transports the monoxide to your organs instead of the oxygen your organs would rather have. Unfortunately, this gas is colorless and odorless, so it cannot be detected without special instruments. Symptoms of carbon monoxide poisoning include headache, nausea, vomiting, dizziness, coma, and even death. The amount of carbon monoxide in the air we breathe has risen over time in the Northern Hemisphere but the degree of increase is uncertain.

Guess where 60 percent of carbon monoxide pollution comes from? Why, motor vehicles of course. Between 1990 and 1999 carbon monoxide emissions decreased 2 percent and are at their lowest level since 1981. Catalytic converters now present in most American cars remove the carbon monoxide from car exhausts (assuming the engine has warmed up), changing it into carbon dioxide, which won't kill you but instead contributes to global warming. Sometimes you win, sometimes you lose, and sometimes you just break even.

PARTICULATE MATTER

> It is much easier to remove the olive from a martini than it is to remove the vermouth. ... We must rely on prevention rather than decontamination.
>
> —Ivan L. Bennett, Jr.

There is a wide variety of microscopic particles floating around in the air we breathe. Some, like soil (dirt), have always been there and most of us are able to tolerate them if the amounts are not too great. But much microscopic airborne stuff today consists of human-made materials such as lead, rubber, soot from coal-burning power plants and motor-vehicle emissions, and poisonous chemical elements and compounds emitted from smokestacks. Some particulate materials, such as diesel soot, carry toxic compounds like benzene and dioxin that can increase cancer risk. The 15 percent of lung cancers not attributable to cigarette smoking are caused by particulate air pollution. And it has been found that minuscule particles in the air pose a greater risk to the heart than to the lungs. According to the EPA, particles in the air may be responsible for 60,000 deaths annually in the United States.

LEAD

The decline in atmospheric lead is perhaps the brightest spot on the air-pollution scene. Its dominant source was tetraethyl lead, formerly added routinely to gasoline to increase engine performance. Lead additions started in the mid-1920s and continued until 1986, when the sale of leaded gas was finally banned in the United States. Lead in the air has decreased 93 percent since 1981. The blood of Americans now contains 50 percent less lead then it used to. Lake and reservoir waters have seen decreases of as much as 70 percent. Lead poisoning causes antisocial behavior and learning disabilities and may be one cause of criminal activity.

Because the major industrial countries have most of the cars and are phasing out leaded gasoline, the lead added to the world's gasoline dropped 75 percent between 1970 and 1993. But in Third World countries leaded gas still dominates, although there has been some progress. In 1997 the Chinese government banned the sale of leaded gasoline in Shanghai, one of the country's most polluted cities. Manila, capital of the Philippines, banned leaded gas in 2000. Many African countries have either banned leaded gas or are phasing it out. About half of the countries in Central and South America have introduced unleaded gasoline.

In China, two-thirds of the gasoline sold is leaded and studies from different parts of that country indicate lead poisoning of children living in industrial and heavy-traffic areas ranges between 65 and 99.5 percent, based on EPA standards. Perhaps 50 percent of children living outside such areas have lead poisoning. Of the 10 cities with the worst air pollution, 7 are in China. The microscopic lead particles emitted from motor vehicles can remain airborne for weeks and circulate around the globe. Of course, the amount in the air decreases as the distance from the source increases and repeated rains wash the lead particles from the air.

RUBBER

Everyone who drives knows that tires wear and need to be replaced. Ever think about the microscopic pieces of rubber that are worn from the tire treads? Where do they go? Well, they don't disappear. They get into the air we breathe and adorn our lungs. Do your lungs need retreading? In addition to rubber particles, wear and tear of tires releases large amounts of chemicals called polycyclic aromatic hydrocarbons into the air and these are suspected of being carcinogenic. About 15 percent of lung cancers are caused by factors other than cigarette smoking.

THE PARTICLE-SIZE CONTROVERSY

What size of particle in the air is dangerous? Is it the larger ones or the smaller ones? Or perhaps it makes no difference. These questions are debated with great ferocity among industrial representatives, the federal government, and environmental groups. Until mid-1997 federal law mandated that the particles with diameters less than 10 micrometers (1/2,500 of an inch, one-seventh the diameter of a human hair) be below a certain threshold amount. Otherwise the air would be considered polluted with particulate matter and must be cleansed. The size of 10 micrometers was chosen by the EPA to limit regulation to those particles small enough to penetrate beyond the upper airways of your body's defenses.

However, studies have indicated that the most damage to human lungs is done by particles less than 2.5 micrometers in diameter (1/10,000 of an inch, less than 1/40th the width of a human hair). These are the particles that penetrate deepest into the lungs and cause the most damage. They appear to be responsible for most of lung-cancer deaths among nonsmokers. So in 1997 the EPA changed the law to emphasize the amount of particles smaller than 2.5 micrometers in diameter. Industry objects to this change because most of these tiny particles result from the combustion of coal and oil (soot). The larger particles tend to be road dust and soil clay. The change from 10 to 2.5 micrometers cost industry a lot of money in new pollution controls.

Those opposed to the 2.5 standard also point out that the material smaller than 2.5 micrometers is so tiny and stays suspended in the air for so long before settling that much of it comes from sources outside any particular region. It is not within the control of those who would be penalized. For example, several states on the East Coast receive large amounts of this cryptodust from the Sahara Desert. Florida will be in noncompliance much of the time. The Western United States receives cryptodust from the Gobi Desert in China. Critics also note that the size designation ignores the nature of the particle. The particles smaller than 2.5 micrometers in midsummer in the farm belt differ dramatically from those in the streets of a major city. One is mostly soil dirt, the other dangerous chemicals. The complexity of particle types is enormous but is not taken into account by considering size alone. It is perhaps worth noting that 73 percent of the particles less than 2.5 micrometers in diameter have diameters less than 0.1 micrometers. These may be even more dangerous than their larger brothers. Should they be regulated? What price safety? The battle continues.

PARTICLE POLLUTION AND RAINFALL

Recent research indicates that particles suspended in the air lower the amount of precipitation by preventing large droplets of water from forming.

They have a dampening effect on rainfall, which can have effects that extend far beyond suppressing local precipitation. For example, tropical rain produces much of the energy needed for worldwide movement of air. Any change in rainfall in the tropics will affect global climate in uncertain ways. As environmentalists often note, everything is connected to everything else.

WHAT PRICE CLEAN AIR?

> When someone is chronically ill, the cost of pollution to him is almost infinite.
>
> —Anonymous congressional staff member

Should the EPA disregard cost in establishing clean-air standards? A health-only approach is the historical norm in federal environmental policy. In recent years, however, talk of cost-benefit analysis has crept into most regulatory discussions. What is the average American willing to pay for what degree of health? Those who favor a cost-benefit approach say that a health-only standard is simply naive. If populations are going to be protected with a margin of safety, then *no* level of air pollution is acceptable. Some degree of balance is needed. A dollar's worth of benefit should be generated for a dollar's worth of expense. What are *you* willing to pay? Ten dollars a month for lowering the particle-size standard to 2.5 micrometers? How can you know how much health benefit the change from 10 to 2.5 micrometers will bring you? Tough questions.

PLANES AND POLLUTION

> As in a quiet backwater, pollution collects in the stratosphere and no rain washes it away.
>
> —Louise Young, *Earth's Aura*

Commercial airlines are handling more passengers each year. Air traffic in the United States is expected to double between 1997 and 2017. Like cars, planes burn gasoline (jet fuel) and currently account for 10 percent of America's oil consumption. A jumbo jet burns 3,250 gallons *per hour* of flight; the now-retired Concorde, 6,500, spewing large amounts of pollutants into the atmosphere. And thousands of planes are flying every day of the year. Air transport is the fastest growing source of greenhouse-gas emissions, currently accounting for 3.5 percent of them. The newest and largest oceangoing ship, the *Queen Mary II*, burns 13,000 gallons of diesel fuel per hour. A car burns 2 gallons of refined petroleum per hour.

According to the Natural Resources Defense Council, a single 747 airliner landing and departing generates gaseous organic compounds equal to those of a car traveling 5,600 miles and nitrous oxide equivalent to a car driving 26,500 miles. This suggests that people living downwind of large airports should suffer more health problems than other people. This expectation is borne out by a study of health records in Seattle. Those living downwind of the city's airport had significantly higher rates of respiratory diseases, pregnancy complications, infant mortality, and other health problems.

Nevertheless, airports are exempt from many air-pollution regulations. In many cases they are among the top polluters in major metropolitan areas. The two major airports in New York City are among the ten largest sources of smog in the city. Los Angeles International Airport is second only to Chevron Corporation as a source of smog in the City of Angels. Chicago O'Hare International Airport is the fifth largest source of pollution in the Windy City area. The two airports in the Washington, D.C., area ranked between two garbage incinerators as the fourth and sixth largest sources of smog in the nation's capital. Clearly, airports need more pollution controls than they are now subjected to by the EPA.

In 1997 Zurich airport became the first in the world to charge stiffer landing fees for aircraft that emit more pollution. The cleanest aircraft pay 5 percent less than before, while the dirtiest pay 40 percent more. Is this a model the United States should adopt?

SHIPS AND POLLUTION

> Man has lost the capacity to foresee and to forestall. He will end by destroying the earth.
>
> —Albert Schweitzer

Commercial airlines are not the only polluting nonautomotive means of transport that is growing rapidly. Ocean-going ships are also of increasing concern. During the past 15 years, as international trade has exploded and shipping capacity has grown by 50 percent, cargo ships have become one of the nation's leading sources of air pollution, threatening the health of millions of people living in port cities. These ships burn the dirtiest grades of fuel, literally the dregs of the oil barrel after refiners have removed cleaner fuels like gasoline and jet fuel. These low-grade hydrocarbons have the consistency of mud, with sulfur levels 3,000 times that of gasoline. They must be heated simply to allow them to move through pipes to enter the engine cylinders. A single cargo ship coming into New York harbor can release in an hour as much pollution as 350,000 2004-model cars.

Satellite photos show that trails of pollution thousands of miles long are causing semipermanent clouds above oceanic shipping routes. Scientists who study climate change are concerned about the effect on global warming as well as on atmospheric pollution. Foreign-flagged ships are responsible for almost 90 percent of the pollution in American ports. So far our government has refused to demand stricter regulations for these ships that dock at American ports.

PROGRESS IN CLEANING THE AIR

> Every American in every city in America will breathe clean air (by early in the next century).
>
> —President George H. W. Bush, 1989

The marked decrease in air quality since the end of World War II provoked angry complaints from environmental groups in the 1960s, culminating in passage by Congress of the Clean Air Act in 1970 (amended in 1997 and 1990), despite vociferous opposition from industrial polluters. Within a few years air quality improved nationwide. Between 1970 and 2000 aggregate emissions of the six principal pollutants tracked nationally decreased 29 percent, despite a 45 percent increase in energy consumption. Nevertheless, half of the people in the United States still breathe air that sometimes fails to meet the standards set by the Clean Air Act. Most of the problem is (ozone) smog generated by motor vehicles.

The greatest improvement in air quality has been in the amount of particulate lead. It has dropped like a, well, like a hunk of lead because of the phasing out of leaded gasoline. Sulfur dioxide levels have also shown marked improvement, although some of the decrease is illusory, resulting from an increased height of smokestacks in existing and new industrial facilities. Pollution measurements are made in cities, and the taller smokestacks simply transport their pollutant sulfur further away from the measuring stations. Clever, huh? The pollutant showing the least improvement is the nitrogen oxides that are an ingredient in ozone formation (smog), in large part because of the increase in motor-vehicle use.

INDOOR AIR POLLUTION

> Mother Nature says: "Clean up your room."
>
> —Anonymous T-shirt slogan

Studies indicate that residents of highly developed countries such as the United States spend little time outdoors. This raises the question of air quality indoors, not only at home but also in the sealed office buildings where many American work. It is common for some air pollutants to be two to five times more concentrated inside homes than outdoors. Sources include air fresheners, hair sprays, and oven cleaners. At work the ventilation system in the building can be a source of both germs and noxious chemicals. For some people, the least healthy air they breathe all day is indoor air.

The pollutants that lurk indoors can come from a wide variety of sources, including cooking appliances, furnishings, household products, and pets. Because modern, energy-efficient buildings tend to be tightly sealed, with very little fresh (?) air entering from outdoors, pollutants can reach high levels inside. Most of the time, however, air-pollutant levels indoors are low and have not been shown to pose a serious health threat for most Americans.

CIGARETTE SMOKE

The most common and serious indoor air hazard faced by Americans is one that is produced deliberately, secondhand cigarette smoke. People in a room with a smoker breathe in cigarette smoke that contains about 4,000 different chemicals, including more than 40 cancer-causing agents. Among the chemicals in the smoke are arsenic, methanol (rocket fuel), and toluene (banned for use in nail polish), a bit of lead, formaldehyde (embalming fluid), and hydrogen cyanide (gas-chamber poison), naphthalene (used in moth-repellent balls), a smidgen of DDT (the classic pesticide), a bit of butane (the stuff we use to start fires quicker), cadmium (a heavy metal found in batteries, which is implicated in cancers that have struck down thousands of industrial workers exposed to it), and traces of the highly carcinogenic polonium-210 (a radioactive substance).

Continuous involuntary smoking increases your risk of lung cancer, high blood pressure, rheumatoid arthritis, heart disease (by 20–70 percent), respiratory infections, and other health maladies known to be associated with the cigarette smoker. The National Research Council and Surgeon General of the United States have both come out strongly against cigarette smoke, no matter how you encounter it. As stated by the World Health Organization in an unusual worldwide release, "Passive smoking does cause lung cancer. Don't let them (the cigarette companies) fool you." As former Surgeon General Charles Everett Koop said, "The cigarette is the only product in the world that kills if you use it according to the manufacturer's instructions."

The risk of being affected by secondhand cigarette smoke is particularly high for children, house pets, spouses of smokers, pop musicians, and bartenders, people continually exposed to high levels of air pollution. The unborn children of pregnant women are affected by the chemicals in the mothers' blood, and children of smoking mothers have significantly lower

IQ scores and are 27 percent more likely to have respiratory illnesses during their first six years. Passive cigarette smoking damages your arteries about 40 percent as much as it does to the smokers themselves. Women occasionally exposed to cigarette smoke are 58 percent more likely to develop coronary heart disease; those regularly exposed are 91 percent more likely. Secondhand cigarette smoke is also implicated in cervical cancer and hearing loss. Anyone who has been to indoor music festivals or bars where smoking is permitted is well aware of the air pollution in such places. In a remarkable, precedent-setting decision, the U.S. Supreme Court ruled in February 2004 that an airline can be held liable for the death of a passenger from a severe asthma attack caused by exposure to secondhand smoke.

On June 16, 2003, the World Health Assembly drafted a tobacco treaty that was signed that same day by 28 nations. The United States was not among them. The document calls on each country's politicians to pass laws protecting its citizens from exposure to tobacco smoke. The treaty requires each ratifying nation to ban all tobacco advertising, promotion, and sponsorship. I can only suggest that you hold your breath when in the presence of tobacco smoke, but don't try to hold it until the United States signs this treaty.

RADON GAS

Radon is a colorless, tasteless, odorless, radioactive gas that is produced by the natural disintegration of uranium in rocks and soils. Although it originates outdoors, it is heavily diluted in the air and poses no threat. Radon gas is threatening only indoors, when it accumulates to high levels in poorly ventilated areas such as the basement of tightly sealed homes with cracked floors.

A few decades ago, one of those famous "faceless bureaucrats" in Washington established a "safe limit" for exposure to radioactivity from radon, based not on health considerations but on his estimate of what an average American could afford to pay for cleanup. His estimate of your financial resources for radon cleanup was $500–$2,500. Using this criterion, the EPA estimated that about 7 percent of American homes were at risk and that "radon is a national health problem." Incredible but true. Both the EPA and the Surgeon General recommend that all houses be tested for radon. They recommend taking remedial action if radon levels of more than 4 picocuries per liter of air are found. An average American home has 1.6 picocuries. In other nations the average is much higher: Finland, 4.6; Sweden, 3.8; Norway, 2.2; France and Germany, 2.9; Denmark, 1.9. Most of these nations have set higher levels because their uranium-bearing rocks are widespread and they feel that a level of only 4 picocuries is an unrealistic goal. Finland's recommended action level is 7.4.

Suppose the radon level in your basement exceeds the EPA's upper limit. Should you panic? It's a judgment call. Research indicates that people who have lived for 20 years in houses that have 27 picocuries of radon per cubic meter (35 cubic feet) face an additional 2–3 percent chance of contracting lung cancer. But the average American moves 10–11 times during a lifetime. If only 7 percent of homes are in danger (1 home in 14), the chance of moving to another home that is over the radon limit is small. In summary, the radon threat is negligible for all but a very few people who may live their entire lives on rock or soil with an extremely high uranium content, and who have an unventilated basement in which they spend most of the day.

AIR POLLUTION IN THE OFFICE

In July 1976, a faulty cooling tower allowed the pathogenic bacterium *Legionella* to be dispersed through the air-conditioning system in a Philadelphia hotel, causing 182 of the guests there to become ill with Legionnaires' disease. Before the cause of the sickness had been discovered, 29 people had died. This is perhaps the best-known example of how the indoor human environment, where we spend about three-quarters of the 168 hours of each week, can become deadly.

In the United States, up to 21 million employees are exposed to poor indoor air quality. Several major office buildings have recently made headlines by being diagnosed as "sick." A new disease called "sick-building syndrome" (SBS) has arisen. The disease had its origin in 1973 when the energy crisis caused by the Arab oil embargo dictated a cut in air-handling costs. The standard for the minimum amount of outdoor air brought into buildings was reduced by 70 percent.

The outdoor-air cutback was accompanied by a gradual rise in the use of photocopiers, laser printers, personal computers, and other equipment that may release chemical fumes. What's more, architectural designs changed, and sealed windows, wall-to-wall carpeting, and fiberglass or particle-board materials that may also contribute to the problem were increasingly used in buildings. The lower rate of air exchange (ventilation rate) combined with increased exposure to indoor pollutants probably explains the rise in SBS illnesses. Studies show that a doubling of ventilation rates does decrease SBS symptoms.

In addition to chemical irritants, microzoos can arise in building ventilation systems. Fungi, bacteria, viruses, algae, and other microbes lurk inside air ducts, grow around ceiling tiles, and thrive on almost any damp surface. Bacteria and fungi can produce airborne particles such as spores that leave employees with symptoms like coughing, headaches, and other allergic reactions. In exceptionally severe cases, biologically polluted air can lead to serious illness or even death, as in the case of Legionnaires' disease in 1976.

CONCLUSION

Based on the lung diseases, heart problems, and cancers attributable to air pollution in the United States, the quality of the air Americans breathe is bad. According to the Harvard School of Public Health, air pollution causes 60,000 deaths each year. But air quality has improved markedly (29 percent decrease in total pollutants) since passage of the Clean Air Act in 1970, over the prostrate bodies of the major polluters. The noxious materials in the air include sulfur dioxide, nitrogen dioxide, ozone, carbon monoxide, and soot. All of these materials arise from the combustion of coal and oil and cannot be eliminated without a major decrease in their use. However, improvements in catching the pollutants before they leave smokestacks and the exhaust pipes of motor vehicles have had a beneficial effect. An unchecked and rapidly growing source of air pollution is airplanes. They are currently exempt from most pollution rules, and this needs to be changed.

Indoor air is normally safe, although there have been some highly publicized exceptions. Few people are in serious danger from noxious chemicals or biological agents, either at home or where they work.

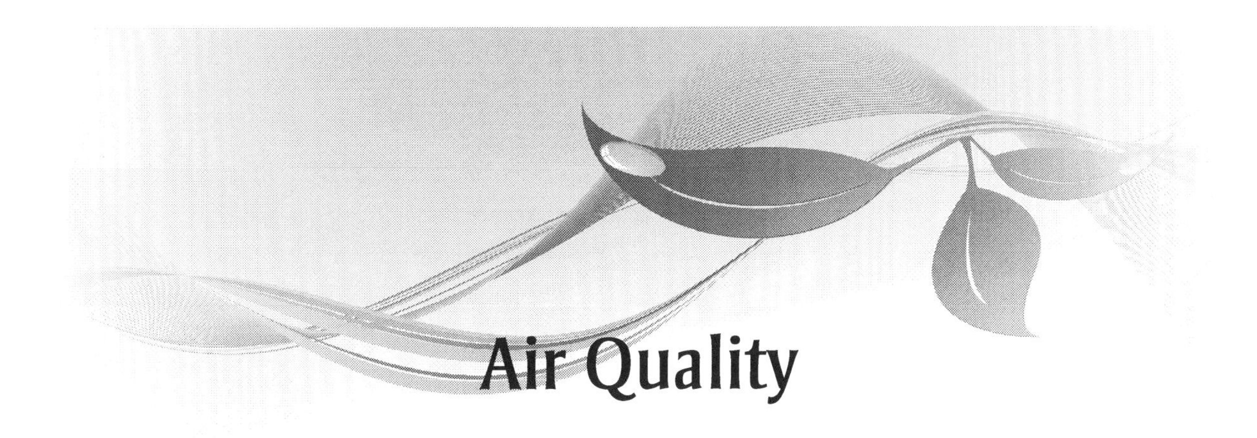

Air Quality

By Monroe T. Morgan

A person may survive many days without food, or a few days without water, but without air a person could not exist long enough to walk 100 feet. The air that humans must have is an odorless, colorless mixture of natural gases, roughly 78% nitrogen and 21% oxygen. The remaining 1% is mostly argon (0.93%), carbon dioxide (0.032%), and traces of neon, helium, ozone, xenon, hydrogen, methane, krypton, and varying amounts of water vapor. When anything else is added, it becomes air pollution.

HISTORICAL PERSPECTIVE

Until the 19th century and the industrial revolution, air pollution was not a problem because pollution was readily diluted in the atmosphere and did not build up over densely populated areas. When humans learned to burn fuel (wood, coal, and others) to convert water into a stream to turn a turbine, they started creating air pollution problems. The belching smokestacks brought new wealth to the industrialized nations, but grimy effluents became the price for the desired affluence. Industrialization raised the standard of living while lowering the visibility and causing disease. People sought affluence without regard for the effluent from the affluent society. Only after air pollution disasters were obviously responsible for multiple deaths did people become concerned about air pollution.

During the last week of October 1948, a high concentration of pollutants—then called smog by Harold Antoine Des Voeuy at a London Public Health Congress—settled down over the air surrounding Donora, Pennsylvania, and the surrounding area. This particular smog encompassed the Donora area on the morning of Wednesday, October 27, reducing visibility to the extent that native Donorians became lost. On Saturday morning at 2:00 a.m. the first death occurred. The deaths continued until, by Sunday night, 19 people had died, and one became ill and died a week later.

In London, 4 years later, in 1952 an air pollution episode gripped the city for 5 days. The thick yellow smog was so dense that people walked with handkerchiefs over their noses, and visibility was about 4 yards. People walked into each other, and only the blind knew where they were going. The "pea souper" of December 1952 caused the death of 4,000 people in London. These deaths, attributed to the smog, were far in excess of those normally expected to occur during that time of year.

New York City also has had air pollution episodes. The worst, in 1965, caused the death of 400 people. These episodes are not limited to large cities or cities downwind from large cities. Secluded Muse Valley, Belgium, underwent an air pollution episode in 1930 that resulted in 63 deaths and 6,000 illnesses. These and other pollution disasters are listed in Table 14.1. The episodes in the table represented cases where the "dumping ground" (the atmosphere) could not disperse the materials being emitted from natural and manmade sources. As populations grew, power demands to operate machinery, provide transportation, heat homes and other buildings, prepare food, and so forth, increased, and the ability of the atmosphere to dilute or disperse the pollutant was overcome, causing the air pollution disasters.

SOURCES OF AIR POLLUTION

Some natural sources of air pollution are forest fires, dust storms, and volcanic eruptions. Plants such as ragweed contaminate the air with pollen. Decaying leaves and other forms of vegetation release gases that contribute to air pollution and cause a haze.

Anthropogenic air pollution, that produced by humans, also may affect human health adversely. Some sources are smoke from chimneys, gases from septic tanks and house sewer systems vents, odors from cooking food, and fumes, gases, vapors and particles released from paint, household cleaner, hair sprays, and so on. Industrial pollution is created by the release of gases, vapors, fumes, and the like, where industry is making the cars, clothing, cleaning agents, furniture, and other products we purchase.

Air pollutants are created in agriculture where our food is grown. For example, crop yields are increased greatly when insecticides and herbicides are used to rid the crops of pests.

At the same time, these insecticides and herbicides add to the air pollution problem.

Transportation contributes to the pollution problem and, according to some sources, accounts for approximately 50% of all air pollution. Carbon monoxide (CO) is a major source of air pollution generated by transportation. In 1983, 70% of the nonnatural emissions of carbon monoxide were from highway vehicles. Now catalytic converters installed on automobiles have reduced CO emissions from this source significantly.

CO is the result of incomplete combustion of products; in contrast, complete combustion produces carbon dioxide (CO_2). Nitrogen oxides and hydrocarbons are additional byproducts of the combustion of petroleum products. They undergo photochemical reactions to produce what is called photochemical smog, a major problem in large cities.

Energy use and production are the major contributors to deterioration of our air quality. When coal or wood is burned to produce electricity or heat, the combustion process releases air pollutants: CO, CO_2, sulfur dioxide (SO_2), nitrogen oxides, heat and particulate matter, depending on the fuel—to name a few. Of particular importance is SO_2. This gas is emitted into air by burning oil and coal, which contain sulfur impurities ($S + O_2 = SO_2$) In the United States, 15% of SO_2 emissions are from industrial plants and 68% is from coal and oil-burning electric power plants. Refuse often is burned to generate heat as electricity,

Table 14.1. Air Pollution Disasters Since 1930 and Associated Death Rates

Date	Location	Mortality Deaths: Normal Predictions — Exceeding
1930	Meuse Valley, Belgium	63
1948	Donora Valley, PA	20
1950	Poza Rica, Mexico	22
1952	London	4,000
1953	New York City	250
1956	London	1,000
1957	London	700–800
1962	London	700
1963	New York City	200–400
1966	New York City	168

a process called waste-heat recovery, which also generates a small percentage of air pollution.

EFFECTS OF AIR POLLUTION

Health Effects

Epidemiological studies indicate that high levels of air pollutants contribute to or cause a number of respiratory conditions. A Harvard study estimated that as many as 60,000 people die annually from particulate air pollution. A phenomenon called thermal inversion traps pollutants in layer of cool air that cannot rise to disperse the pollutants (see Figure 14.1). Twenty-eight million Americans with chronic respiratory problems are exposed regularly to harmful levels of smog that worsen their illnesses. Some of these respiratory ailments are the following.

1. **Asthma,** an irritation of the bronchial passages that leads to severe difficulties in breathing, is a growing public health problem nationwide. From 1983 to 1993, its prevalence increased 34%, according to the National Institutes of Health. The nation's urban areas, especially those with high levels of air pollutants, seem to be the most affected. Particulates and SO_2 are among the air pollutants that seem to be linked to asthma.
2. **Chronic bronchitis** occurs when an excessive amount of mucus is produced in the bronchi, which results in a lasting cough. There seems to be a significant correlation between death rates from chronic bronchitis and SO_2 concentrations. Sulfur dioxide may irritate the nasopharynx (mucous membrane) and the bronchi. Repeated exposure to high levels of SO_2 over time may cause the body to produce excessive mucus as a defense.
3. **Pulmonary emphysema** is characterized by weakening of the walls of the alveoli, the tiny air sacs in the lungs. As the disease progresses, the alveoli become enlarged, lose their resilience, and their walls disintegrate. Shortness of breath is a primary symptom. Nitrogen dioxide has been identified as one of the air pollutants that may contribute to emphysema.

Lung cancer and heart disorders also may be caused or exacerbated by exposure to air pollutants.

Other Effects

Sulfur dioxide, carbon monoxide, nitrogen oxides and other contaminants not only adversely affect our health, but they also affect our property. Some pollutants damage vegetation, thus affecting the landscape. Near Los Angeles, smog is destroying pine trees around the city. Some forms of air pollution directly damage leaves of crops and trees when these gases enter leaf pores (stomata). Chronic exposure to these air pollutants (including NO_2, SO_2, and ozone) breaks down the waxy coating, allowing excessive water loss and damage from disease, pests, drought, and frost. In addition, acid deposition can leach vital plant nutrients such as calcium from the soil and kill essential microorganisms such as the decomposers.

Each year air pollutants cause millions of dollars in damage to various materials. Ozone causes rubber to crack and lose of strength. Sulfur dioxide is responsible for loss of strength and surface deterioration of leather and other natural fabrics. Pollutants can cause the corrosion, erosion, discoloration, and soiling of stone, metals, paint, paper, and glass. Table 14.2 notes the significant impacts of some air pollutants.

AIR POLLUTION CONTROL

Because air pollution affects our health, crops, buildings, and the natural environment, efforts are made to reduce air pollution. Catalytic converters are used to improve the burning of petroleum products so as to reduce the amount of carbon monoxide, nitric oxides, and hydrocarbons in the air. More fuel-efficient cars not only save energy but also emit less exhaust.

Coal gasification and sulfur removal of sulfur dioxide are being practiced to reduce the amount of sulfur dioxide in the air, especially during thermal inversions. Coal that is low in sulfur is being burned to help lower the SO_2 in the air. That practice, however, can hurt the economy in coal mining areas that have coal with a higher sulfur content.

Emission control efforts concentrate on the settling chamber, the afterburner—which requires a high temperature to ignite and burn particles—and the electrostatic precipitator. First used for fly ash control in 1925, the electrostatic precipitator now is used widely in the United States. Operation consists of attaching electric charges to particles, achieved by high-voltage discharges of the particles that give them a charge. The charged particles then are attracted to metal plates called collection electrodes. The particles are intermittently cleaned from the plates.

The bag house method, used for years by the grain mills, cement factories, and the like utilizes fabric bags to capture particles in the air. This method works like a large vacuum cleaner. Now electric utilities are trying to use the bag house method to reduce air pollution in large, power-generating plants.

Pollution control methods are approached from two aspects: input control and output control. Input control concentrates on preventing or reducing the severity of the problem, whereas output control treats the symptoms, by attempting to remove pollutants once they have entered the environment.

Input Control

Some input control methods for reducing the amount of pollution before it reaches the environment are to:

1. Control population growth.
2. Reduce the need for energy.
3. Enhance fuel-dependent units such as gas engines.
4. Recycle resources and prevent loss of metals and chemicals into the environment.
5. Emphasize quality in products (e.g., cars) so they will last longer.
6. Encourage repair of products rather than supporting remove-and-replace practices.

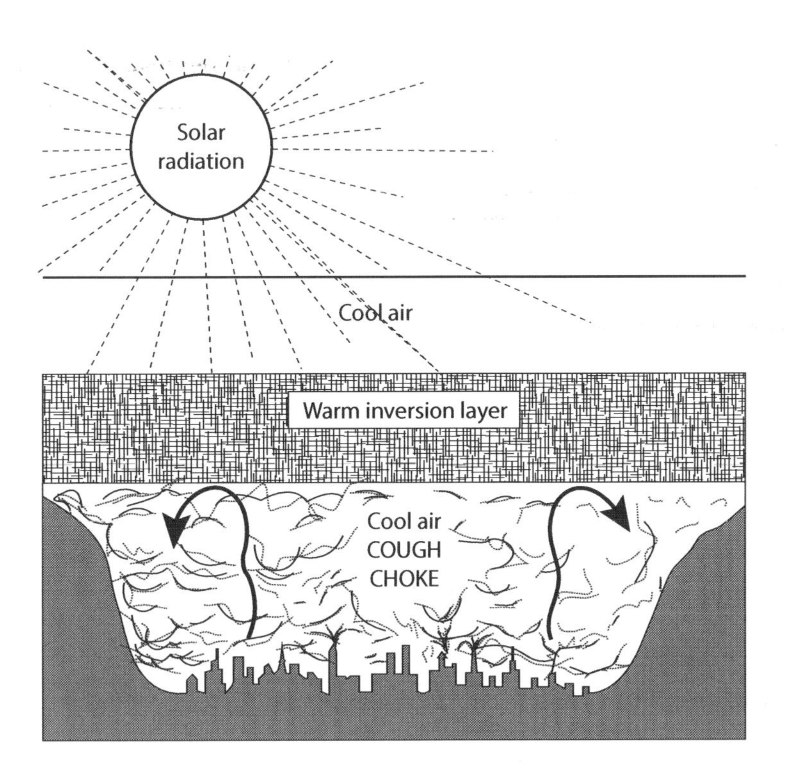

Figure 14.1. Thermal inversion.

7. Reduce our dependency upon conveniences and our desire for affluence.
8. Reduce our dependency upon fossil fuels.
9. Find new nonpolluting sources of energy (e.g., wind, tidal, and solar energy).
10. Reduce our dependency upon electricity by reducing the use of electric toothbrushes, electric knives, and electric can openers and instead emphasizing simplicity—hand- operated toothbrushes, knives, and can openers.

Output Control

Strong emphasis on input control methods reduces the need for output control, which consists of the following practices:

1. Remove pollutants after combustion by using scrubbers and electrostatic precipitators.
2. Add lime and other materials to raise the pH of lakes, streams, and the soil damaged by acid rain.
3. Support improved methods of emission control.
4. Research new methods of removing pollutants in emissions.
5. Find ways to convert pollutants to a resource.
6. Improve catalytic conversion in automobiles.

Legislation

Four years after the London episode of 1952, Great Britain passed the Clean Air Act, aimed at improving the air by banning the burning of soft coal at home and in industries. The Londoners were unhappy at first, but after seeing the improvement, they welcomed the absence of the thick, yellow smog that had caused some 4,000 deaths.

Federal statutory law addressing air pollution began in the United States with the 1963 and 1967 Clean Air Acts. Although these laws provided broad clean air goals and money for air research, they did not provide for air pollution control throughout the entire United States. In 1970 the Clean Air Act was amended to cover the entire United States, and the U. S. Environmental Protection Agency was created to promulgate the 1970 amendments.

In accordance with a 1977 amendment to the Clean Air Act, the Environmental Protection Agency (EPA), and the Council on Environmental Quality, along with other agencies, developed a Pollutant Standard Index (PSI). The PSI is a national air-monitoring network using a uniform air quality index. The EPA believed it was necessary to devise a method of conveying air quality data to the public in a way that would give people a good understanding of how daily levels of air pollution might be affecting their health. The PSI is given in Table 14.3.

The act was revised most recently in 1990. Under the Clean Air Act, most enforcement power is concentrated at the federal level and is delegated to the states by the EPA. The states must show the EPA that they can clean up the air to the levels required by the National Ambient Air Quality Standards (NAAQS), given in Table 14.4. The main intent of NAAQS is to protect public health and to "protect welfare." Air quality levels are determined as those needed to protect health and welfare (see Table 14.4). The states must have an Air Quality Implementation Plan (AQIP), containing all of the state's regulations governing air pollution control, including local regulations within the state. The EPA must approve the AQIP. Once approved, it has the force of federal law.

Table 14.2. Characteristics of Some Air Pollutants

Name	Formula	Properties of Importance	Significance as Air Pollutant
Sulfur Dioxide	SO_2	Colorless gas, intense choking, odor, somewhat soluble in water to form sulfurous acid (H_2SOA_3)	Damage to vegetation, property, and health
Hydrogen Sulfide	H_2S	Rotten egg odor at low concentrations, odorless at high concentrations	Highly toxic
Nitric Oxide	NO	Colorless gas	Produced during high-temperature, high-pressure combustion. Oxidizes to No_2.
Nitrogen Dioxide	NO_2	Colored gas, used as carrier	Relatively inert. Not greatly produced in combustion.
Carbon Monoxide	CO	Colorless and odorless	Product of incomplete combustion. Poisonous.
Carbon Dioxide	CO_2	Colorless and odorless	Formed during complete combustion. Possible effects in producing changes in global climate.
Ozone	O_3	Highly reactive	Damage to vegetation and property. Produced mainly during the formation of photochemical smog.

SOME AIR QUALITY PHENOMENA

Acid Precipitation

All rainfall is somewhat acidic. Decomposing organic matter, the movement of the sea, and volcanic eruptions all contribute to accumulation of acidic chemicals in the atmosphere. The main contributor is atmospheric carbon dioxide. Manmade pollutants, too, accelerate the acidification of rainfall. Emission of sulfur dioxide (SO_2) and nitrogen oxides transform into acids in the atmosphere. The acid deposition more commonly called acid rain is a misleading term because these acids and acid-forming substances are deposited not only in rain but also in snow, sleet, fog, and dew.

Acid precipitation has become a worldwide problem, first in the Scandinavian countries, then in the northeastern United States and Southeastern Canada, then Europe, Japan and Taiwan. Studies are revealing that what was considered pure rainwater is now highly acidic. This precipitation is the product of sulfur oxides and nitrogen oxides produced in the burning process. The sulfur oxides come from burning coal and other fuels that contain sulfur. Under certain conditions, the sulfur oxides convert to sulfuric acid in the atmosphere and fall to earth in precipitation. The nitrogen oxides are produced from the high-temperature combustion of fossil fuels, such as in cars, in which the nitric oxides are oxidized to nitrogen dioxide, which further oxidizes and dissolves in water droplets to form nitric acid. These acids depress the pH of the soil, lakes, rivers, and other natural resources after precipitation such as rain or snow.

In the northeastern United States the culprit in most of the acid precipitation is sulfuric acid coming from coal-fueled power plants. Prevailing winds blowing from the Southwest to the Northeast dump the sulfuric acid on the northeastern United States and Canada (see Figure 14.2).

In Copperhill, Tennessee, coal was used to smelt the sulfur-containing ore and extract pure copper in copper-smelting plants of that area. The high-sulfur coal was burned for years with much So_2 coming from the smokestacks and falling out around Copperhill and Ducktown, Tennessee. The pollution killed trees, grass, and other forms of vegetation.

With time the pollution spread to Georgia, where the Georgia legislature contacted the U. S. Public Health Service, the controlling federal agency at the time. The PHS contacted the Tennessee government and required that the pollution stop, as Tennessee did not have a right to pollute Georgia. Tennessee subsequently contacted the industry and relayed the information.

At that point, the industry started looking for ways to stop or reduce the pollution. It began removing So_2 from the stack emissions, converting it into a resource, sulfuric acid, and then related products, which the industry sold.

Only precipitation that has a pH of 5.6 and below is considered acid precipitation. In some parts of the world, the acidity of rainfall has fallen well below 5.6. In the northeastern United States, for example, the average pH of rainfall is 4.6, and rainfall with a pH of 4.0—1,000 times more acidic than distilled water—is not unusual. Figure 14.3 compares acid rain to other products in terms of pH.

The extent of damage caused by acid deposition in an area depends on several factors. For example, an area with acid-neutralizing compounds in the soil does not reveal problems as quickly as an area without the neutralizing compounds.

Aquatic ecosystems reveal the effects of acid precipitation more clearly. The acids lower the pH, creating an unfavorable environment for aquatic life. When exposed to acid water, female fish, frogs, salamanders, and other sea creatures fail to produce eggs or may produce eggs that do not develop normally. Some scientists believe that acid water can kill fish and other aquatic life, using as evidence some lakes in areas of high acid deposition, which have been found to be highly acidic and lifeless.

The effect of acid precipitation on land, crops, forest, and other vegetation is unknown. Some believe that "dieback" (unexplained death of whole sections of once-thriving forest) is caused by acid precipitation. Present concerns have been expressed in regard to trees dying in Germany and in the Appalachian Mountains of the United States.

We do not know the effect, if any, of acid precipitation on human health. We do know that acid water can leach out as well as other chemicals from pipes. The effects of inhaling air that contains sulfur dioxide and nitrogen oxides are not known at this time.

Some methods of controlling acid precipitation are the following:

1. Reduce population growth, and thus the number of people driving cars and needing energy. Prevent unwanted pregnancies and restrict immigration.
2. Reduce the need to travel by using communication methods such as phones, faxes, and e-mail.
3. Engage in carpooling to work, school, and other trips, whenever possible.
4. Use energy more efficiently. Over 50% of energy used in the United States each year is wasted. Use energy-efficient heat pumps to provide space heating rather than burning oil.
5. Convert coal to gaseous fuel or a liquid to remove sulfur and to reduce emissions of sulfur oxides from burning solid coal.
6. Shift from fossil fuels to a mix of energy sources such as solar, tidal, nuclear, geo- thermal, hydroelectric, refuse, and biomass energy.
7. Shift to low-sulfur coal (less than 1%) for power plants, homes, and so on, and convert high-sulfur coal to gasoline.

Table 14.3. Pollutant Standard Index

Index Value	Health Effect Descriptor	General Health Effects	Cautionary Statements
0–50	Good		
51–100	Moderate		
101–200	Unhealthful	Mild aggravation of symptoms in susceptible persons, with imitation symptoms in the healthy population.	Persons with existing heart or respiratory ailments should reduce physical exertion and outdoor activity.
201–300	Very unhealthful	Significant aggravation of symptoms and decreased exercise tolerance in persons with heart or lung disease, with widespread symptoms in the healthy population.	Elderly and persons with existing heart or lung disease should stay indoors and avoid physical exertion and outdoor activity.
301–400	Hazardous	Premature onset of certain diseases in addition to significant aggravation of symptoms and decreased exercise tolerance in healthy persons.	Elderly and persons with existing diseases should stay indoors and avoid physical exertion. General population should avoid outdoor activity.
401–500	Hazardous	Premature death of ill and elderly. Healthy people will experience adverse symptoms that affect their normal activity.	All persons should remain indoors, keeping windows and doors closed. All persons should minimize physical exertion and avoid traffic.

8. Remove sulfur from coal, even if expensive, rather than pay high prices for foreign oil.
9. Design better and more efficient automobiles, heating units, and coal- and oil-fueled power plants.
10. Increase the efficiency of scrubbers and electro-static precipitators.
11. Reduce the need for energy by emphasizing recycling and conservation and by deem- phasizing "keeping up with the Joneses."
12. Plant trees and other plants to remove carbon dioxide and produce more oxygen for better burning.

Table 14.4. National Ambient Air Quality Standards (MAAQ5) 1990

POLLUTANT	PRIMARY (HEALTH-RELATED) STANDARD LEVEL		SECONDARY (WELFARE-RELATED) STANDARD LEVEL	
	AVERAGING TIME	CONCENTRATION	AVERAGING TIME	CONCENTRATION
PM_{10} [b]	Annual Arithmetic Mean	50 $\mu g/m^3$	Same as Primary	
	24-hour	150 $\mu g/m^3$	Same as Primary	
TSP[b]	Annual Geometric Mean	75 $\mu g/m^3$	Annual Geometric Mean	60 $\mu g/m^3$
	24-hour	260 $\mu g/m^3$	24-hour	150 $\mu g/m^3$
SO_2	Annual Arithmetic Mean	80 $\mu g/m^3$ (0.03 ppm)		
	24-hour	365 $\mu g/m^3$ (0.14 ppm)	3-hour[c]	1300 $\mu g/m^3$ (0.50 ppm)
CO	8-hour[c]	9 ppm (10 $\mu g/m^3$)	No Secondary Standard	
	1-hour[c]	35 ppm (40 $\mu g/m^3$)	No Secondary Standard	
$NO\text{-}_2$	Annual Arithmetic Mean	0.053 ppm (100 $\mu xg/m^3$)	Same as Primary	
O_3	Maximum Daily 1-hour Average[d]	0.12 ppm (235 $\mu g/m^3$)	Same as Primary	
P[b]	Maximum Quarterly Average	1.5 μg/m3	Same as Primary	

a. The value in parentheses is an approximately equivalent concentration; the standard is in the first units shown.
b. Until July 1, 1987, total suspended particulate matter (TSP) was the indicator pollutant for the particulate matter standards. In 1987, EPA adopted the PM_{10} standard (for particles less than ten micrometers [$\mu g/m^3$] in diameter). Until attainment status of all Air Quality Control regions for PM_{10} is determined, and new plans submitted and approved, many State Implementation Plans (SIPs) will continue to address TSP. The PM_{10} annual standard is attained when the expected annual arithmetic mean concentration is less than or equal to 50 $\mu g/m^3$; the PM_{10} 24-hour standard is attained when the expected number of days per calendar year above 150 $\mu g/m^3$ is equal to or less than one.
c. These standards are not to be exceeded more than once per year.
d. The standard is achieved when the expected number of days per calendar year with maximum hourly average concentrations above 0.12 ppm is equal to or less than one.
Source: 40 CFR Part 50 (1989); U.S. EPA (1990, March), National Air Quality and Emissions Trends Report, 1988, EPA-450/4-90-002, Research Triangle Park, NC

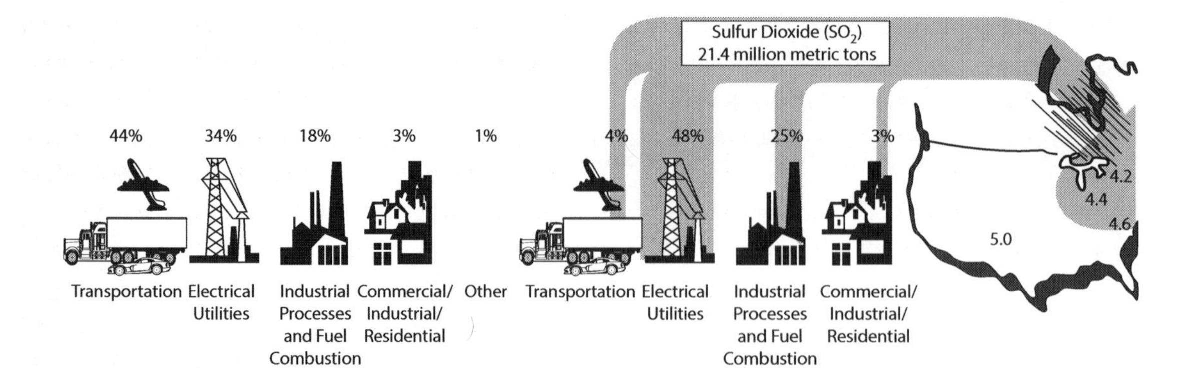

Source: U.S. Environmental Protection Agency

Figure 14.2. Precursors of acid precipitation in northeastern United States.

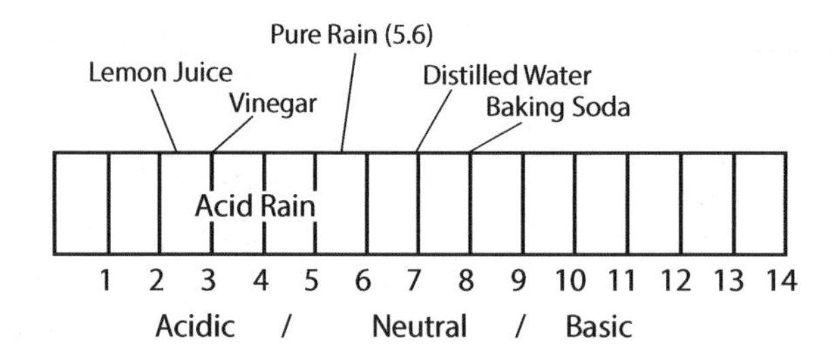

The pH scale ranges from 0 to 14. A value of 7.0 is neutral. Readings below 7.0 are acidic; readings above 7.0 are alkaline. The more pH decreases below 7.0, the more acidity increases. Because the pH scale is logarithmic, there is a tenfold difference between one number and the one next to it. Therefore, a drop in pH from 6.0 to 5.0 represents a tenfold increase in acidity, and a drop from 6.0 to 4.0 represents a hundredfold increase.

Source: U. S. Environmental Protection Agency

Figure 14.3. Comparative pH of add precipitation.

The Greenhouse Effect

The earth's temperature remains relatively constant because some of the solar energy absorbed by earth eventually is radiated back into space. Since the 1800s, however, the heat is not being radiated back into space as much as it once was, and thus the earth is undergoing a warming trend. This warming is caused by CO_2 and water vapor in the air, as well other substances.

With the discovery that coal could be burned to convert water into steam to rotate a turbine that turns wheels on trains, pulleys on machinery, paddles on river boats, and generators in electricity-producing plants, the potential to add CO_2 to the air increased greatly. This fossil-fuel burning began producing carbon dioxide in huge quantities.

CO_2 can be removed from the atmosphere in two ways:

LONDON'S AIR EXPERIENCE

London used to be a place where the sky was always dingy, where the only birds were pigeons, where people lived in dread of the thick yellow smog that sneaked up on the city and strangled thousands.

By 1972 Londoners lived in one of the cleanest atmospheres in the world. Hawks, wild ducks, and bullfinches returned to nest in the parks. The sun began to shine brightly again. Even the clothing on the line dried whiter. The reason: Britain's Clean Air Act, which deprived Englishmen of their traditional glowing coal fire for the privilege of breathing fresh air.

The 1956 Clean Air Act was the first legislation of its kind in the world. Its main feature was to ban the burning of soft coal—a move that meant homeowners had to board up their fireplaces and switch to electric or gas heaters. Only the well-to-do could afford the hard, smokeless coal the act required. Lord Kennet said:

> The hearth has always been the focal point of an Englishman's living room. The sight of a glowing coal fire is built very deep in him and people felt very lost at first without it.
>
> But it was the domestic hearth which was the killer. Three fourths of the terrible concentration of smoke in the air in 1956 came from the chimneys. They were a health horror.

The long-range advantages of the act far outweighed any temporary disgruntlement. Even the most home-loving Englishman conceded a preference for clean air over glowing hearths. The amount of smoke was reduced by 75% in London and 50% countrywide.

The number of chest diseases and heart complaints was reduced greatly. The hours of sunshine increased by 50%. People now could see the dome of St. Paul's Cathedral from Westminster Bridge, several miles away, on a clear morning—something they couldn't in the 1950s.

Whiter buildings and cleaner clothes were other benefits. Plants began to flourish in the parks, and a range of bird species returned to London. "They came back in the hundreds suddenly one spring in the early sixties—as if they had sniffed the air and said, 'It smells good; let's nest in Berkely Square,"' Lord Kennet said. Most important, the killer smogs that had hit London every winter completely disappeared.

1. By green plants using it in the process of photosynthesis
2. By dissolving in the oceans, where it may be converted to dolomite (calcium magnesium carbonate).

The capability of these two systems is severely limited, however. As humans remove forests and pollute oceans, the need for CO_2 from the air decreases. At the same time, the world produces more CO_2 each year as the population grows. Thus, more CO_2 remains in the air each year. Studies reveal that about half of the CO_2 now emitted into the air remains in the air.

The sun radiates heat into space in every direction and has been reaching earth ever since its genesis. Some of the radiation is reflected back into space or absorbed by the earth itself. Ozone (O_3) in the upper atmosphere absorbs some of the short-wavelength range of ultraviolet light, X-ray, and gamma radiation, preventing it from reaching the earth's surface. At levels lower than the ozone layer, dust particles suspended in the air, clouds, and the earth's surface reflect about 3% of the incoming radiation back into space. About 20% is absorbed by water droplets, water vapor, and dust in the air and some 50% reaches vegetation, seas, snow-covered land and ice-covered seas that can reflect it back into space.

Burning fossil fuels in automobiles, coal- fired electric plants, and the like produces more CO_2 and, consequently, more remains, in the atmosphere. Research has shown that water vapor and CO_2 absorb radiation. Hence, more and more radiation is absorbed and trapped in the atmosphere. This causes an increase in the earth's temperature, termed the greenhouse effect. A greenhouse lets in sunlight through its glass. The light warms the inside, and the glass prevents the heat from escaping. Like the glass, CO_2 and water vapor in the atmosphere absorb the long-wavelength heat radiated by the earth.

The greenhouse effect causes much concern because of the possible repercussions of global warming. A National Academy of Sciences committee of experts estimated a global temperature increase of 3°C with a doubling of the CO_2 content of the atmosphere. A normal sample of air contains approximately 0.03% CO_2 or 320 ppm. According to the committee, from 1880 to 1980 the CO_2 level increased by 10% to 12%, contributing to a mean global temperature increase of 0.4°C. The temperature increase resulted in melting of some of the polar icecaps and glaciers, and the global ocean level increased by 5.4 inches (14 cm) in that time span.

A 1983 report by the National Academy of Sciences indicated a general consensus among 70 atmospheric scientists that, based on computer models of atmospheric processes, a doubling of the 1980 carbon dioxide levels in the atmosphere would raise the average global temperature of the atmosphere between 1.5°C and 4.5°C (2.7° and 8.1°F) and two to three times this temperature increase at the earth's polar regions. A 1985 model of the atmosphere suggested that global warming from CO_2 buildup may be only half as great as these earlier projections because the denser and wetter clouds containing more CO_2 should reflect more sunlight into space. Other studies, however, indicate that dozens of other gases such as chlorofluorocarbons (CFCs) found in trace amounts in the atmosphere could produce a global warming at least as great as that caused by carbon dioxide alone.

Global warming has two possible harmful effects:

1. Distribution of rainfall and snowfall over much of the earth could change. This change would mean that the world's major food- growing regions (such as those in much of the United States) would shift northward to areas of Canada and other northern countries where the soils tend to be poorer and less productive.
2. Glaciers and icefields in polar regions would melt, causing a projected rise in the average sea level of about 2.4 meters (8 feet) by 2100—thus possibly causing flooding of coastal cities and industrial areas.

Ozone Layer Depletion

The main cause of the depletion of the ozone layer is chlorofluorocarbons (CFCs). The chlorofluorocarbons are propellants used in products such as hair spray, bathroom cleaners, and other aerosol products. Chlorofluorocarbons can be found in the production of foam coffee cups, egg cartons, furniture cushions, and building insulation. Hospitals use a nonflammable gas made of chlorofluorocarbons for sterilization of medical equipment.

Early in 1989, scientists were surprised that the ozone layer above the Antarctic had become thinner. David Hofman of the University of Wyoming discovered this by conducting tests with balloon-borne instruments. The Antarctic ozone "hole" in 1989 was almost twice the area of the Antarctic continent.

The Arctic ozone layer also has been researched. Scientists from Norway concluded that weather conditions over the Arctic are too dark for chlorine to destroy the ozone layer. They say, however, that destruction will come later because of an increase in sunlight. These scientists concluded that the "polar stratospheric clouds," together with the sun, promote chemical reactions that turn pollutants into ozone-depleting chemicals.

CFCs

Chlorofluorocarbons (CFCs) have become suspect in depletion of the ozone layer. Thomas Midgley developed CFCs in 1930 to replace the poisonous ammonia used in refrigerators during the 1920s. CFCs were considered safe because they did not react with other substances or break down easily. Although CFCs did not break down below the stratosphere, their inventor did not see their reaction in the stratosphere.

CFCs contain chlorine. When the CFCs reach 10 to 20 miles into the atmosphere, chlorine is released when the chloroflourocarbon's molecular bonds break down. Once released, the chlorine takes a molecule from an atom of ozone. Ordinary oxygen, which has no sun- blocking properties, is left.

Some researchers suspect that bromine also destroys the ozone layer. A study of bromine levels in the Arctic by Walter W. Berg, a scientist for the National Center for Atmospheric Research, and other scientists found a substantial amount of bromine in the Arctic atmosphere—a level 10 times higher than normal and about the same level found in a heavily polluted environment. Berg suggests that the two major contributors to depletion of the ozone layer are long-range manmade air pollution and red algae, which produces large quantities of bromine-contaminated compounds in the water under the Arctic ice. Other scientists agree that the major source of bromine is marine but question the mechanism for releasing the bromine into the atmosphere. The significance of the bromine is believed to be its synergism with the chlorine chains, which could aid in destruction of the ozone layer at lower altitudes in the stratosphere and under dark conditions—the opposite of what is done by chlorine.

The major concern with depletion of the ozone layer is that the increase in ultraviolet light will reach the surface, increasing the incidence of skin cancer and cataracts. It also could affect plants and the food chain in some undetermined way.

Not withstanding all the research, experts disagree as to the severity of ozone depletion.

Research must continue to understand the problems fully.

Solutions

According to many researchers, the best answer to the ozone problem is a new breed of CFCs that are less likely to harm the environment. One compound already on the market is 95% less destructive to the ozone layer than standard CFCs. Although this compound, called HCFC-22, costs up to 50% more than its predecessors, it is gaining popularity as a coolant for commercial and residential air-conditioning systems. In December 1987, the Food and Drug Administration approved HCFC-22 for use in containers used by the fast food industry.

One drawback to HCFC-22 is that it is less versatile than many of the ozone-depleting CFCs. It is a poor candidate for building insulation and for use in automobile air conditioning systems. It has poor insulating qualities and a high boiling point, and it requires higher pressure for air conditioning systems.

Chemical companies have developed other alternative CFCs. These have caused problems in the waste products they give off during the manufacturing process. Years of toxicity testing are necessary before these compounds can be marketed commercially.

The National Aeronautics and Space Administration (NASA), the National Oceanic and Atmospheric Administration, the Federal Aviation Administration (FAA), the World Meteorological Organization, and the United Nations Environment Program have studied ozone depletion using ground-based and satellite instrumentation. They concluded that from 1969 to 1986, the decline in annual average of ozone was from 1.7% to 3% in the Northern Hemisphere. They also found losses of 95% between altitudes of 15 and 20 kilometers.

By far the main focus of research seems to be developing a new form of CFCs that will not harm the environment, especially the ozone layer. After years of denying the adverse effects of CFCs, manufacturers now recognize the problem. They have begun using CFC replacements as well as curbing CFC production. Manufacturers of foam food containers also are involved in researching substitute compounds.

The control measures being taken are simple: an outright ban on CFCs, reduction in CFCs, and development of alternatives. Europe has recognized the CFC problem and has made a proposal to ban all CFCs by the end of the century. The United States seeks a ban on all CFCs by the year 2000 and already has banned CFCs in aerosols. Dupont, the largest producer of CFCs, has indicated that it will phase out these compounds by the end of the century if its replacement is ready by then.

DEFORESTATION EFFECTS

Although they may seem peaceful, forests are places of intense activity. Countless animals, plants, and microorganisms grow and reproduce there, and in the process they filter the air and water, regulate stream flow, store water, and reduce soil erosion. Removing trees changes the ecology in several ways. For example, if even a small plot of tropical forest is cut, the temperature of the region fluctuates from extremely high during the day to cool temperatures at night.

Deforestation can change weather patterns, as it has in Panama, where rainfall in areas where deforestation occurred 50 years ago has decreased by 1 cm every year (50 cm for the 50 years) compared to adjacent uncleared land. Forest soil filters the polluted rainfall and cleanses it before it reenters the surface or groundwater supply. Forest vegetation and soil also purify the air. Many environmental pollutants stick to leaves and branches and are removed. Environmental pollutants also are removed by leaves, detoxified by microorganisms, and taken up by plant roots. Hence, air is cleaner when it leaves the forest than when it enters.

Desertification is a problem related to deforestation. When trees are removed and the land is overgrazed or cultivated, it affects the ecosystem. That is evidenced by the drought that killed tens of thousands of people in the Sahel region of Africa in the early 1970s. In many developing countries, desertification is accelerated by deforestation where wood is burned for fuel and the land is overused. This was evidenced around Khartoum, Sudan, where the acacia tree no longer grows.

Soil erosion and flooding are direct results of deforestation. Where clear-cutting of trees (cutting everything) occurs, soil erosion follows. Also, when all of the trees are cut, more water runs off and flooding ensues. This happens because the trees that once held water on their trunks, limbs, and leaves are no longer there to hold water after a rain during peak runoff. Some of the water retained on the trees before they were cut reentered the hydrologic cycle by evaporation, thus not adding to the peak flow and causing floods. Also, forests have a "sponge effect"; they soak up rainfall during wet weather and release it during dry weather.

Another effect of deforestation is the loss of the oxygen generating "factories." By depleting the forest, we reduce the means of converting carbon dioxide at a time when there is a greater need for oxygen.

Finally, plants use carbon dioxide and give off oxygen. By depleting the forest, we reduce the means of converting carbon dioxide at a time when there is a greater need for oxygen.

SUMMARY

Health effects of air pollution can be serious, including asthma, chronic bronchitis, pulmonary emphysema, lung cancer, and heart disorders. Air pollution has been classified by its major sources. Examples of natural sources of air pollution are forest fires, dust storms, and volcanic eruptions, as well as pollen-laden plants that irritate the mucous membranes. Anthropogenic air pollution—that which is produced by humans—includes things such as smoke, paint vapors and particles, and gases from septic tanks and house sewer systems.

Industrial pollution is created wherever industry releases gases, vapors, fumes, and the like from the manufacture of cars, clothing, furniture, and so on. Agricultural pollutants are created where our food is grown, particularly where insecticides and herbicides are used.

Transportation is probably the major contributor to air pollution, accounting for approximately half of all air pollution, although catalytic converters in automobiles have drastically reduced pollution from carbon monoxide emissions. A final source is energy use and production.

Some air quality phenomena are acid precipitation, the greenhouse effect, and ozone layer depletion.

A 10-year study by EPA suggests that urban air quality is improving. The study shows that since 1984:

- smog (ground-level ozone) dropped 12%
- lead decreased 89%
- sulfur dioxide fell 26%
- carbon monoxide declined 37%
- nitrogen dioxide dropped 12%

In addition, particulate levels decreased 20% from 1998 to 1993.

These findings are encouraging and motivate us to continue to make improvements that will protect the health of millions of Americans who live, work, and play in areas where air pollutant levels are still too high.

REFERENCES

National Tuberculosis and Respiratory Disease Association. *Air Pollution Primer.* 1969. New York.

Chiras, Daniel D. 1988. *Environmental Science: A Framework for Decision Making*. Benjamin/Cummings Publishing, Menlo Park, CA.

Faith, W. L. and Arthur Atkisson, 1972. *Air Pollution.* John Wiley and Sons, New York.

Miller, G. Tyler, Jr. 1988. *Environmental Science: An Introduction*. Wadsworth Publishing, Belmont, CA.

Revelle, Penelope and Charles Revelle. 1988. *The Environment: Issues and Choices for Society*. Jones and Bartlett Publishing, Boston.

Sproull, Wayne T. 1970. *Air Pollution and Its Control.* Exposition Press, New York.

Turco, Harold P. 1997. *Earth Under Siege: From Air Pollution to Global Change*. Oxford University Press, NY.

Vesilind, Anne. 1996. *Environmental Engineering*. PWS Publishing, Boston.

Airborne Endotoxin Concentrations in Homes Burning Biomass Fuel

By Sean Semple, Delan Devakumar/Duncan G. Fullerton, Peter S. Thorne, Nervana Metwali, Anthony Costello, Stephen B. Gordon, Dharma S. Manandhar, and Jon G. Ayres

The use of solid or biomass fuels to cook and to heat homes is widespread in large parts of the developing world, with an estimated 3 billion people exposed to smoke from burning these fuels in their own home (International Energy Agency and Organisation for Economic Co-operation and Development 2004). The World Health Organization estimates that biomass fuel smoke exposure is responsible for about 1.5 million early deaths per year (Prüss-Ustün et al. 2008), with a global burden of disease of approximately 2.5% of all healthy life-years lost. Most of the burden of disease arises from respiratory infections, especially in children <5 years of age, with a disproportionate amount of health problems falling on women and children, who are more likely to be at home or to have responsibilities for cooking and heating activities (Rehfuess et al. 2006).

Research into indoor air pollution in homes burning biomass fuels has tended to focus on airborne concentrations of fine particulate matter (PM) (Albalak et al. 2001; Edwards et al. 2007; Ezzati et al. 2000; Fullerton et al. 2009; Kurmi et al. 2008), but airborne endotoxin may also play an important role.

Endotoxin or lipopolysaccharide is part of the cell wall of Gram-negative bacteria and has been measured in airborne PM in occupational settings, office buildings, households, and ambient air. Once inhaled, endotoxin stimulates an amplifying series of endotoxin-protein and protein-protein interactions, sequentially binding to a range of proteins and receptors, leading to production of chemotactic cytokines and chemokines (Hadina et al. 2008) and lung inflammation and resultant oxidative stress. Studies have shown associations between household endotoxin concentrations and diagnosed asthma, asthma medication use, and severity of asthma symptoms (Michel, et al. 1996; Thorne, et al. 2005). Respiratory illness in endoroxin-exposed working populations has been frequendy documented (Smit, et al. 2008; Thorne and Duchaine 2007). In asthma, endotoxin exposure has been shown to

Sean Semple, Delan Devakumar, and Duncan G. Fullerton, "Airborne Endotoxin Concentrations in Homes Burning Biomass Fuel," *Environmental Health Perspectives*, vol. 118, no. 7, pp. 988-991.

be protective of development of allergic disease at low levels while also producing nonallergic asthma and/ or aggravating symptoms of existing asthma (Douwes, et al. 2003; Smit, et al. 2009).

Thorne and Duchaine (2007) tabulated data on endotoxin levels in a wide variety of industries and home environments, which indicated geometric mean (GM), Inhalable fraction, personal exposures of 12–8,300 endotoxin units (EU)/m^3 in agricultural occupations and 5.8 EU/m^3 in homes of rural asthmatic children (n = 326).

Airborne endotoxin concentrations in homes in Boston, Massachusetts (USA), have been shown to be significantly associated with the presence of dogs, moisture sources in the home, and the amount of settled dust (Park et al. 2001). Endotoxin has also been identified in tobacco smoke (Larsson, et al. 2004) and in homes where smoking takes place, pets are present, and/or dampness or mold is found (Rennie et al. 2008; Tavernier et al. 2006). In the large U.S. National Survey of Endotoxin in Housing (Vojta et al. 2002), increased household endotoxin was most strongly associated with living in poverty, number of people in the home, pet ownership, and household cleanliness (Thorne et al. 2009). Most studies have used endotoxin levels in settled dust as a surrogate for personal exposure. There are very few studies that have measured airborne endotoxin concentrations in household settings.

It seems probable that the burning of common biomass fuels such as wood, charcoal, dried animal dung, and crop residues within small and poorly ventilated homes will produce high endotoxin exposures. The only available report in the scientific literature comes from a small study in the Ladakh region of India, where short-term sampling (<60 min) of two homes produced average endotoxin concentrations of 24 and 190 EU/m^3 (Rosati. et al. 2005). These concentrations are within the range of those found in occupations involved in the handling and processing of large volumes of biological material.

In this article we present results from a study to measure endotoxin levels within the main living area of 69 homes in Malawi and Nepal and to explore differences in these concentrations based on the fuel type being used.

MATERIALS AND METHODS

Study population and sampling strategy.

Samples of airborne PM were collected from homes in two studies that assessed indoor air pollution and health in Malawi and Nepal. In Nepal the Dhanusha district was selected. This is a flat, low-lying area of the country close to the border with India. Two villages were sampled: one in the south of the district (Lohana), where dried cow or buffalo dung is burned, and one in the north (Dhalkebar), where wood is burned. Fifteen homes were sampled in each village, during cooking time in the morning or evening in December 2008. After consent was given by an adult householder, air sampling equipment was placed in the main living area of the home and sampled air between 90 and 180 min. This study had ethical approval from the Nepal Health Research Council

For the Malawi study, details of methods used and results of PM concentrations measured have been previously published (Fullerton et al. 2009). In summary, a total of 75 homes were recruited from around Blantyre and rural Chikwawa villages during April 2008. Sampling equipment was placed in the main living area of each of these homes for a period of approximately 24 hr, except in six homes where short-term sampling similar to that used in Nepal (60–200 min duration around the time of a cooking event) was carried out (all respirable samples; n = 4 wood burning; n = 2 maize crop residue burning). Not all homes received an instrument capable of providing a sample for the measurement of endotoxin concentrations;

we therefore present a subsample of data from 38 (19 rural and 19 urban) of the 75 Malawian homes. This study had ethical approval from the Research Ethics Committee of the College of Medicine, University of Malawi, and the Liverpool School of Tropical Medicine.

Sample collection.

Air sampling was conducted by placing a small Apex air pump (Casella, Bedford, UK) attached to either a cyclone sampling head (2.2 L/min) or an Institute of Occupational Medicine sampling head (2.0 L/min) to sample the respirable (median aerodynamic diameter, 4 pm) or the total inhalable (defined as anything that can be breathed into the nose and mouth and is broadly particulate with an aerodynamic diameter <100 pm) particle size fraction of PM, respectively. Both types of sampling heads were loaded with preweighed 25-mm glass-fiber filters with a 0.7-pm pore size. All samples in Nepal were collected using an IOM sampling head, whereas 32 of the 38 Malawian samples were collected using a cyclone. Sampling was performed in accordance with Methods for Determination of Hazardous Substances (MDHS) no. 14/3 (Health and Safety Executive 2000). The equipment was placed in the main living area of the home at a height of approximately 1.0 m and, where possible, at about 1.0 m from the main stove or cooking area. After sampling, each filter was placed in a sealed metal tin and sent back to the United Kingdom for reweighing before being further transported to the United States for endotoxin analysis. Field blanks were used to correct the data for changes in filter weight associated with manipulation.

Endotoxin analysis.

Endotoxin concentrations of samples were measured using a modification of the kinetic chromogenic Limulus amebocyte lysate assay (Lonza, Inc., Waikersviile, MD, USA) (Thome et al. 2005). Briefly, air sampling filters were extracted in sterile, pyrogen-free water containing 0.05% Tween 20 for 1 hr at 22°C, with continuous shaking. Filter extracts were centrifuged 20 min at 600 x g. Two-fold serial dilutions of endotoxin standards (*Escherichia coli* O111:B4) and 5-fold serial dilutions of sample extracts were prepared using sterile, pyrogen-free water in heat-treated borosilicate glass tubes. A 13-point standard curve was generated ranging from 0.025 to 100 EU/mL R^2 >0.995), with absorbance measured at 405 nm (SpecrraMax 340; Molecular Devices, Inc., Sunnyvale, CA, USA). Endotoxin determinations were based upon the maximum slope of the absorbance versus time plot for each well.

The arithmetic mean (14.4 EU/sample) for the six Malawi filter field blanks was subtracted from each of the other Malawi filter results. The analytical limit of detection (LOD) was derived from using a value of three times the standard deviation (9.24 EU/filter) of the field blank measurements (Malawi filter analytical LOD = 27.7 EU/filter). Where corrected filter values were less than the LOD ($n = 12$), the filter was assigned a value of one-half the LOD (13.9 EU/filter). A similar process was applied to the Nepal filters based on results from four field blanks (arithmetic mean = 4.4; SD = 0.95 EU/filter; analytical LOD = 2.85 EU/filter). For the Nepal filters with corrected values less than the LOD ($n = 4$), a value of 1.43 EU/liker was assigned.

Statistical analysis.

Data are double entered to a Statistical Package for the Social Sciences (SPSS), version 17.0 file (SPSS Inc., Chicago, IL, USA), and summary statistics and box plots were generated directly. Mean endotoxin concentrations measured in Nepalese total inhalable dust samples from wood- and dung-burning homes were compared using a Mann-Whitney U-test. A similar test was used to test for differences in respirable endotoxin concentrations in Malawian charcoal- and wood-burning homes.

RESULTS

Tables 1 and 2 provide summary statistics of the measured total inhalable and respirable dust concentrations and the endotoxin concentrations measured in the homes. Data are subdivided by country, primary fuel type of the home, and measurement duration. The PM concentrations from the short-duration samples (Table 1) are generally about an order of magnitude higher than the 24-hr samples collected in Malawi, reflecting the much higher smoke concentrations during cooking events than at other times in the household. Total inhalable endotoxin concentrations during cooking-time sampling show much higher median values during dung burning in Nepal (365 EU/m^3) than during wood burning in Nepal (43 EU/m^3). For 24-hr samples, total inhalable endotoxin median values were higher in wood-burning (40 EU/m^3) than in charcoal-burning (24 EU/m^3) homes. Although values for respirable endotoxin concentrations are not directly comparable with total inhalable endotoxin concentrations and will be an underestimate of total inhalable levels, the respirable data are broadly supportive of the increasing gradient in endotoxin concentrations: charcoal < wood < cow dung < maize crop residues.

Figure 1 is a box plot of airborne endotoxin concentrations from the directly comparable samples taken during cooking from wood-burning and dung-burning homes in Nepal. There is a statistically significant difference in airborne endotoxin concentrations in Nepalese homes burning dung compared with those burning wood (Mann-Whitney (*U*-test, $z = 4.0$; $p < 0.01$). The much larger endotoxin concentrations in dung-burning homes do not appear to be simply a function of the increased PM produced in this type of fuel. Figure 2 illustrates the amount of endotoxin per mass of PM measured in the Nepalese villages and demonstrates that dung-generated smoke tends to contain much more endotoxin than does a similar mass of wood-generated smoke ($z = 2.2$; $p = 0.024$).

Figure 3 presents data on 24-hr respirable endotoxin concentrations from the Malawian homes burning charcoal or wood. The difference between fuel types is not statistically significant ($z = 0.46$; $p = 0.647$). Figure 4 shows the median endotoxin concentration per mass of respirable PM, again for the 24-hr samples collected in homes in Malawi. The median concentrations of endotoxin per mass of dust is higher in wood-burning homes than in charcoal-burning homes, although this is not statistically significant ($z = 0.243$; $p = 0.808$).

DISCUSSION

Endotoxin concentrations reported in this study are high and much higher than those found in a recent study measuring airborne endotoxin in 10 homes in northern California (Chen and Hildemann 2009), where mean concentrations were generally <1 EU/m^3, and in a study of homes of rural asthmatic children, where the GM inhalable endotoxin was 5.8 EU/m^3 ($n = 326$) (Thorne and Duchaine 2007). They were also considerably higher than those measured from a large study of the homes of 332 children in Canada (Dales, et al. 2006). The mean airborne endotoxin concentration in the Canadian study was 0.49 EU/m^3, almost 100 times less than the 24-hr average levels measured in this study for charcoal-burning homes and close to 1,000 times lower than the average level during cooking with dried dung in homes in Nepal. However, results from the Canadian study showed that even at the relatively low levels of exposure experienced by the Canadian study population, there was a statistically significant relationship between airborne endotoxin and respiratory illness in the first 2 years of life.

The only previous study of endotoxin concentrations in biomass-burning homes was carried out in two homes in the Ladakh region of India (Rosati et al. 2005), where endotoxin levels of 24 and 190 EU/m³ were found, broadly in line with our data. The Indian homes were small, portable tentike structures with little in the way of ventilation or extraction of smoke generated from burning dung and crop residues.

A health-based guidance limit of 50 EU/m³ has been recommended for occupational settings in the Netherlands (Heederik and Douwes 1997) for an 8-hr time-weighted average exposure. The median value of 24-hr samples collected from charcoal-burning homes (using respirable dust size selection and hence conservative compared with the total inhalable dust sampler used for the limits proposed in the Netherlands) was approximately 20 EU/m³. Scaling this to an 8-hr time-weighted average would produce levels of around 60 EU/m³, exceeding the concentration deemed to be acceptable for a healthy workforce. From our results, we would anticipate much higher 8-hr time-weighted average values from wood- and dung-burning homes, and it seems likely that many of these would approach or exceed the health-based guidance limit value.

The health effects of exposure to the endotoxin concentrations measured in the homes in this study may be considerable, particularly because exposure is sustained and occurs from birth in most homes. Personal exposures of women who carry out cooking and fire lighting have the potential to be even higher than the static or area measurements made in this study because of regular close proximity to the smoke plume. There is a need for personal exposure data in these settings.

We acknowledge that this study has several important weaknesses. We did not design the study to collect samples for analysis of endotoxin, but rather "piggy-backed" it onto two studies that set out to characterize PM concentrations in homes in Malawi and Nepal. As a consequence, our results present data from both short cooking periods and longer 24-hr samples and also a mixture of total inhalable and respirable PM size selection. In addition, there was an extended period between the collection of the filters and analysis for endotoxin, and we believe that this led to the high levels of contamination of some of the field blanks that we have reported. This is particularly evident for the Malawi samples, which were stored for the longest duration. We report our data separately by size fraction, sampling duration, country, and fuel type and used appropriate methods for blank correction to overcome these weaknesses where possible.

Further work should use a standard protocol for endotoxin measurement and should seek to standardize durations of sample collection. Optimally, personal exposure measurements should be considered, especially in the context of health-related exposure measurement. Our study design collected only two samples from homes burning crop residues, and any future study should seek to address this data gap.

Controlling and reducing exposure to bio- mass fuel smoke in homes in the developing world are complex and difficult areas with such options as

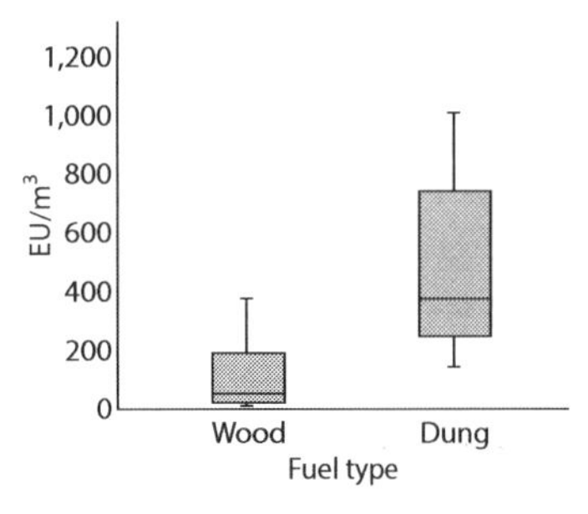

Figure 1. Box plot of airborne total inhalable endotoxin concentrations by fuel type during cooking in Nepalese homes. The line inside the box represents the median value, the lower and upper box lines represent the limits of the interquartile range (25th and 75th percentiles), and the "whiskers" represent the 5th and 95th percentiles of the distribution. Difference in means $p < 0.01$.

Table 1. Summary statistics of PM and endotoxin concentrations by fuel type, sampling fraction, and country for short (<200 min) cooking-interval samples.

	PM (MG/m³)				ENDOTOXIN (EU/m³)			
FUEL/PARTICLE SIZE	n	RANGE	MEAN±SD	MEDIAN	RANGE	MEAN±SD	MEDIAN	< LCD (n)
Nepal, TIP								
Wood	16	0.22–4.08	1.14(1.12)	0.68	6–371	100(113)	43	4[a]
Dung	15	0.91–12.8	3.78(3.51)	2.33	133–1,002	498(291)	365	0
Malawi, Resp								
Wood	4	1.40–9.65	4.73(3.97)	3.94	63–520	202(217)	113	1[b]
Maize crop residue	2	0.65–9.56	5.11(6.30)	5.11	45–3,172	1,609(2,211)	1,609	1[c]

Abbreviations: LOD, analytical limit of detection (Malawi, 27.7 EU/filter; Nepal, 2,85 EU/filter); Resp, respirable PM; TIP, total inhalable PM.
[a]Concentrations generated from filters assigned values of LOD/2 were 6.18, 6.23, 7.02, and 7.96 EU/m³. [b]Concentrations generated from filters assigned values of LOD/2 was 63.3 EU/m³. [c]Concentrations generated from filters assigned values of LOD/2 was 45.1 EU/m³.

Table 2. Summary statistics of PM and endotoxin concentrations by fuel type and sampling fraction for 24-hr Malawi samples.

	PM (MG/m³)				ENDOTOXIN (EU/m³)			
FUEL/PARTICLE SIZE	n	RANGE	MEAN ± SD	MEDIAN	RANGE	MEAN ± SD	MEDIAN	< LOD
Wood								
TIP	4	0.43-0.81	0.65 (0.18)	0.68	34-141	64 (52)	40	0
Resp	9	0.03-0.70	0.32(0.25)	0.24	5-106	31 (34)	26	4[a]
Charcoal								
TIP	2	0.20-0.32	0.26 (0.09)	0.26	21-26	24(3.6)	24	0
Resp	17	0.04-0.72	0.25(0.171	0.23	4-256	35 (59]	21	6[b]

Abbreviations: LOD, analytical limit of detection (Malawi, 27.7 EU/filter; Nepal, 2.85 EU/filter); Resp, respirable PM; TIP, total Inhalable PM.
[a]Concentrations generated from filters assigned values of LOD/2 were 4.88, 5.22, 5.24, and 5.40 EU/m³. [b]Concentrations generated from filters assigned values of LoD/2 were 4.08, 4.26, 4.29, 4.60 ,5.20, and 5.31 EU/m³.

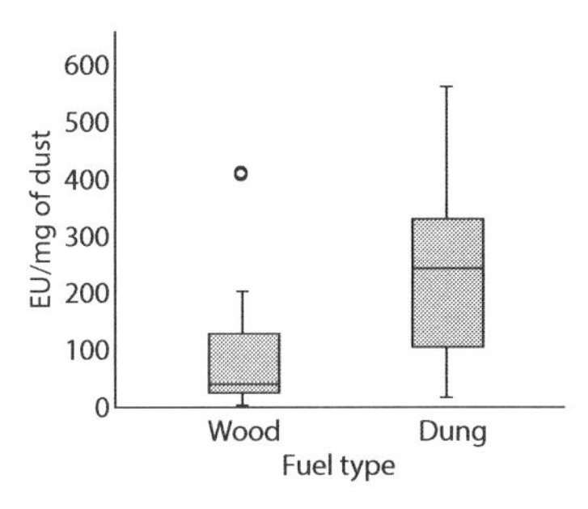

Figure 2. Box plot of airborne endotoxin by fuel type during cooking per PM mass on the filter in Nepalese homes. The line inside the box represents the median value, the lower and upper box lines represent the limits of the interquartile range (25th and 75th percentiles), and the "whiskers" represent the 5th and 95th percentiles of the distribution. Circles indicate outlier observations with values 1.5-3.0 times the interquartile range from the 25th or 75th percentile, Difference in means $p = 0.024$.

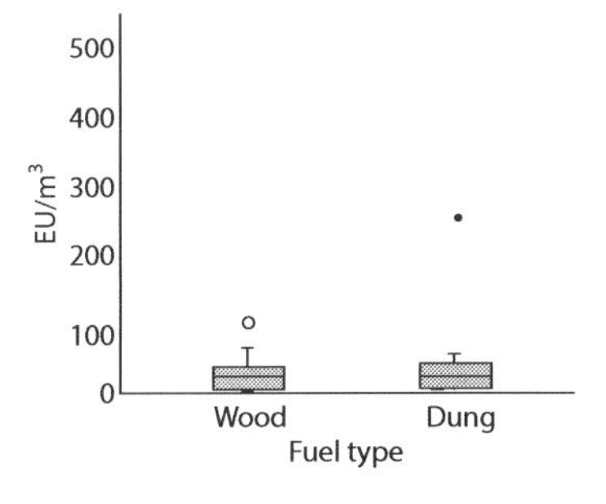

Figure 3. Box plot of 24-hr airborne respirable endotoxin concentrations by fuel type in Malawian homes. The line inside the box represents the median value, the lower and upper box lines represent the limits of the interquartile range (25th and 75th percentiles], and the "whiskers" represent the 5th and 95th percentiles of the distribution. The circle indicates an outlier observation as described in Figure 2; the asterisk indicates an observation more than three times the interquartile range from the 25th or 75th percentile. Difference in means $p = 0.647$.

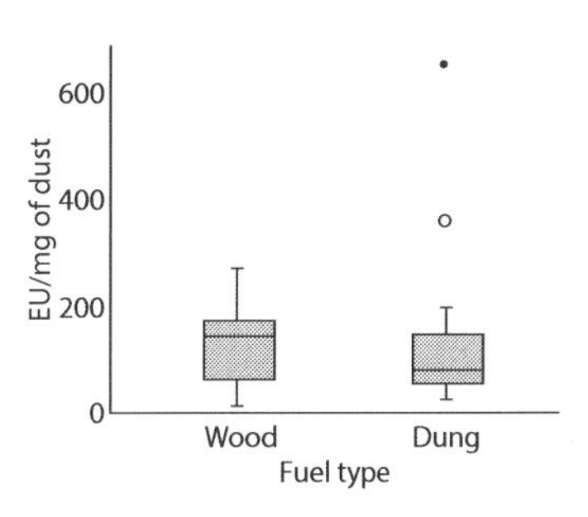

Figure 4. Box plot of 24-hr airborne respirable endotoxin by fuel type per PM mass on the filter in Malawian homes. The line inside the box represents the median value, the lower and upper box lines represent the limits of the interquartile range 125th and 75th percentiles), and the "whiskers" represent the 5th and 95th percentiles of the distribution. The circle indicates an outlier observation as described in Figure 2; the asterisk indicates an observation more than three times the interquartile range from the 25th or 75th percentile. Difference in means $p = 0.808$.

modifications of behavior, introduction of better and more efficient stoves, and improved household ventilation (Zhang and Smith 2007). Methods of reducing airborne endotoxin concentrations will be broadly similar, but there may also be opportunities to reduce bacterial and endotoxin content of the source fuel via harvesting and/or production methods and changes to how fuel is stored. Higher cooking temperatures are likely to degrade endotoxin, and more efficient cooking using improved stove technologies can also reduce the generation of PM-bound endotoxin. A recent study has also suggested that outdoor storage of wood chips increased endotoxin content (Sebastian et al. 2006), so dry, indoor storage areas for fuel may reduce the airborne endotoxin levels when burning eventually takes place.

Our study raises the possibility of an important new risk factor, and preventive strategies, for respiratory morbidity and mortality in the developing world. The mechanism for the association between biomass smoke exposure and infections of the lower respiratory tract in children remains unclear but is likely to be multifactorial and influenced by housing conditions, nutritional status, and other coexposures. It is possible that inhaled endotoxin, being proinflammatory, may be one contributory factor in this mechanistic pathway. Pneumonia remains one of the largest contributors to under-five mortality, and exposure to high concentrations of airborne endotoxin may be an important risk factor for the severity of illness (Dales, et al. 2006). From a public health perspective, interventions to reduce PM and endotoxin exposures generated from household combustion of solid fuels should be implemented as a matter of urgency.

CONCLUSIONS

This study has shown that airborne endotoxin concentrations in homes burning biomass fuels are considerably higher than those found in homes in the developed world and at levels comparable to agricultural-related occupations. Some homes recorded cooking period concentrations >1,000 EU/m³, more than 20 times the health-based occupational guidance limit suggested in the Netherlands. There is a need for a larger study using a standard protocol that allows further identification of the determinants of exposure in these homes. This would increase our understanding of which fuels produce the high levels. Methods to separate the influence of endotoxin concentrations from those of high airborne PM levels are also required, as are epidemiologic and intervention studies to determine the health effects of reducing exposure to these high endotoxin levels.

REFERENCES

Albalak R, N. Bruce, J.P. McCracken, K.R. Smith, and T. De Gallardo. 2001 Indoor respirable particulate matter concentrations from an open fire, improved cookstove, and LPG/open fire combination in a rural Guatemalan community. *Environ Sci Techno* 35:2650–2655.

Chen O. and L.M. Hildemann. 2009. The effects of human activities on exposure to particulate matter and bioaerosois in residential homes. *Environ SciTechno* 143:4541–4546.

Dales R, D. Miller, K. Ruest, M. Quay, and S. Judek. 2006. Airborne endotoxin is associated with respiratory illness in the first 2 years of life. *Environ Health Perspect* 114:610–614.

J. Douwes, P.S. Thome, N. Pearce, and D. Heederik. 2003. Bioaerosol health effects and exposure assessment: progress and prospects. *Ann Occup Hyg* 47:187–200.

Edwards RD, Y. Liu, G. He, Z. Yin, J. Sinton, and J. Peabody, et al. 2007. Household CO and PM measured as part of a review of China's National Improved Stove Program. *Indoor Air* 17:189–203.

Ezzati M, H. Saleh, and D.M. Kammen. 2000. The contributions of emissions and spatiaf microenvironments to exposure to indoor air pollution from biomass combustion in Kenya. *Environ Health Perspect* 108:833—839.

Fullerton DG, S. Semple, F. Kalarnbo, R. Malamba, A. Suseno, and G. Henderson, et al. 2009. Biomass fuef use and indoor air pollution in homes in Malawi. *Occup Environ Med* 66:777–783.

Hadina S, J.P. Weiss, P.B. McCray Jr, K. Kulhankova, and P.S. Thome. 2008. MD-2-dependent pulmonary immune responses to inhaled lipooligosaccharides: effect of acylation state. *Am J Respir Cell Mol Biol* 38:647–664.

Hearth and Safety Executive, 2000, "General Methods for Sampling and Gravimetric Analysis of Respirable and Inhalable Dust Methods for Determination of Hazardous Substances no.

14/3," Available: http://www.hse.gov.ulc/pubns/mdhs/pdfs/ mdhsl4-3.pdf (accessed 14 September 2509).

Heederik D and J. Douwes. 1997. Towards occupational exposure limit for endotoxins? *Ann Agric Environ Med* 4:17–19.

International Energy Agency and Organisation for Economic Cooperation and Development 2004. *World Energy Outlook 2004*. Paris:International Energy Agency and Organisation for Economic Co-operation and Development.

Kurmi O.P., S. Sample, M. Steiner, G.D. Henderson, and J.G. Ayres. Particulate matter exposure during domestic work in HepeS. *Ann Occup Hyg* 52:509–517.

Larsson L, B. Szpona, and C. Pehrson. 2004. Tobacco smoking increases dramatically air concentrations of endotoxin. *Indoor Air* 14:421–424.

Michel O., J. Kips J., J. Duchateau, F. Vertongen, L. Robert Land, and H. Collet, et al. 1996. Severity of asthma is related to endotoxin in house dust *Am J Respir Crit Care Med* 154:1641–1646.

Park J.H., D.L. Spiegelman, D.R. Gold, H.A. Surge, and D.K. Milton. 2001. Predictors of airborne endotoxin in the home. *Environ Health Perspect* 109:859–864.

Prüss-Ustün A., S. Bonjour, and C. Corvalán. 2008. The impact of the environment on health by country: a meta-synthesis. *Environ Health* 7:7, doi: 10.1186/1476-069X-7-7 (Online 25 February 2007).

Rehfuess E., S. Mehta S, and A. Prüss-Ustün. 2006. Assessing household solid fuel use: multiple implications for the Millennium Development Goals. *Environ Health Perspect* 114:373–378.

Rennie D.C., J.A. Lawsort, S.P. Kirychuk, C. Paterson, P.J. Willson, and A. Senthilseivan, et al. 2008. Assessment of endotoxin levels in the home and current asthma and wheeze in school-age children. *Indoor Air* 18:447–453.

Rosati J.A., K.Y. Yoneda, S. Yasmeen, S. Wood, and M.W. Eldridge. 2005. Respiratory health and indoor air pollution at high elevation. *Arch Environ Occup Health* 60:96–105.

Sebastian A., A.M. Madsen, L. Martensson, D. Pomorska, and L. Larsson. 2006. Assessment of microbial exposure risks from handling of biofuel wood chips and straw-effect of outdoor storage. *Ann Agric Environ Med* 13:139–145.

Smit L.A., D. Heederik, G. Doekes, C. Blom, I. van Zweden, and I.M. Wouters. 2008. Exposure-response analysis of allergy and respiratory symptoms in endotoxin-exposed adults. *Eur Respir J* 31:1241–1248.

Smit L.A., O. Heederik, G. Doekes. E.J. Krop, G.T. Rijkers, and I.M. Wouters, *Ex vivo* cytokine release reflects sensitivity to occupational endotoxin exposure. *Eur Respir J* 34(4):795–802.

Tavernier G., G. Fletcher, I. Gee, A. Watson, G. Blacklock, and H. Francis, et al. 2006. IPEADAM study: indoor endotoxin exposure, family status, and some housing characteristics in English children. *J Allergy Clin Immunol* 117:656–662.

Thorne P.S., R. Cohn, O. Mav, S.J. Arbes Jr., and D.C. Zeldin. 2009. Predictors of endotoxin levels in U.S. housing. *Environ Health Perspect* 117:763–771.

Thorne P.S.and C. Duchaine. 2007. Airborne bacteria and endotoxin. In: *Manual of Environmental Microbiology* (Hurst C.J., R.L. Crawford. J.L. Garland, D.A. Lipson, A.L. Mills, and L.D. Stetzenbach, edsl. 3rd ed. Washington, DC:ASM Press, 989–1004.

Thorne P.S.,K. Kulhankova, M. Yin, R. Cohn, S.J. Arbes Jr., and D.C. Zeldin. 2005. Endotoxin exposure is a risk factor for asthma: the National Survey of Endotoxin in United States Housing. *Am J Respir Grit Care Med* 172:1371–1377.

Vojta P.J., W. Friedman, D.A. Marker, R. Clickner, J.W. Rogers, and S.M. Viet, et al. 2001. First National Survey of Lead and Allergens in Housing: survey design and methods for the allergen and endotoxin components. *Environ Health Perspect* 110:527–532.

Zhang J.J. and K.R. Smith. 2007. Household air pollution fTpm coal and biomass fuels in China: measurements, health impacts, and interventions. *Environ Health Perspect* 115:848–855.

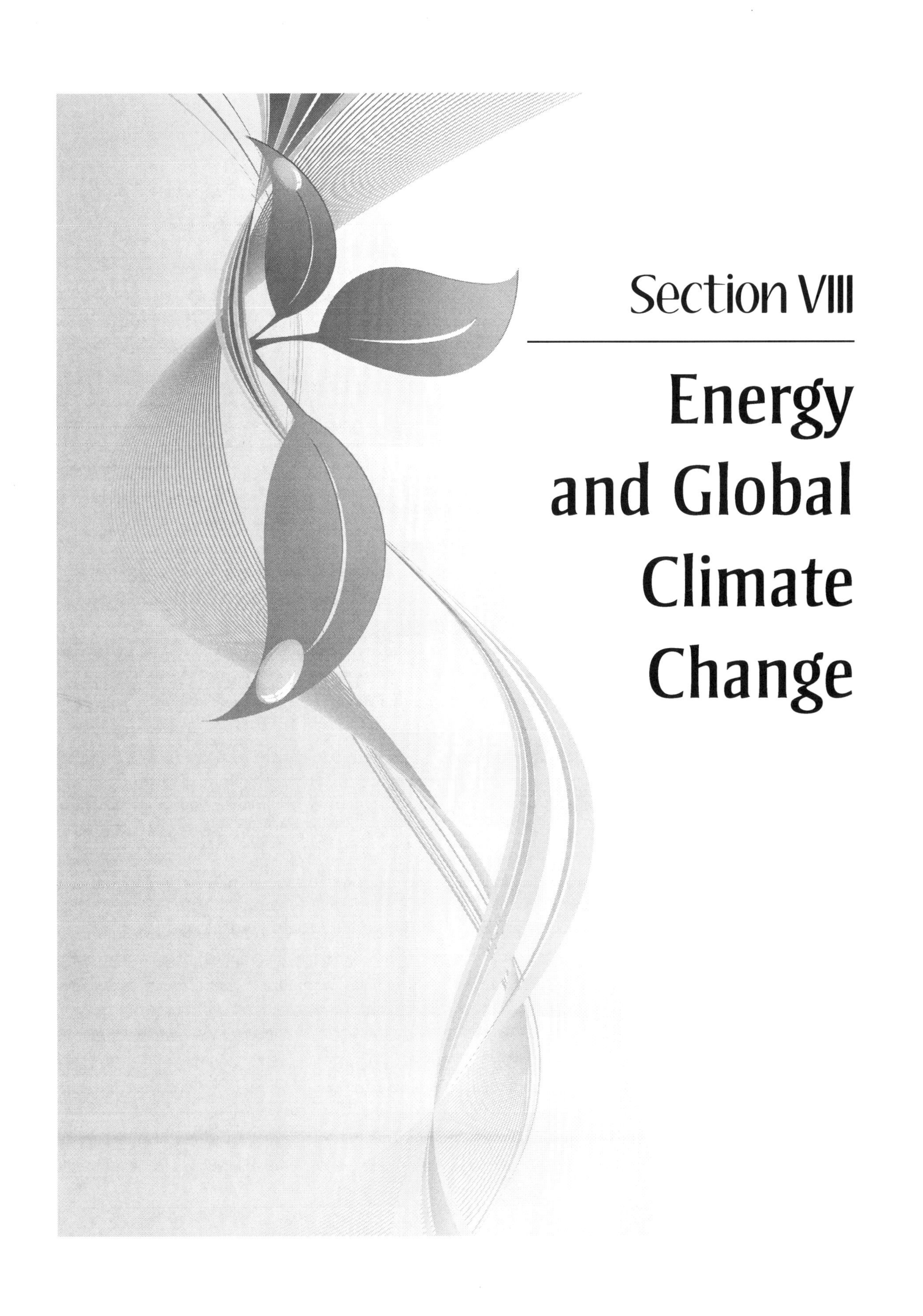

Section VIII

Energy and Global Climate Change

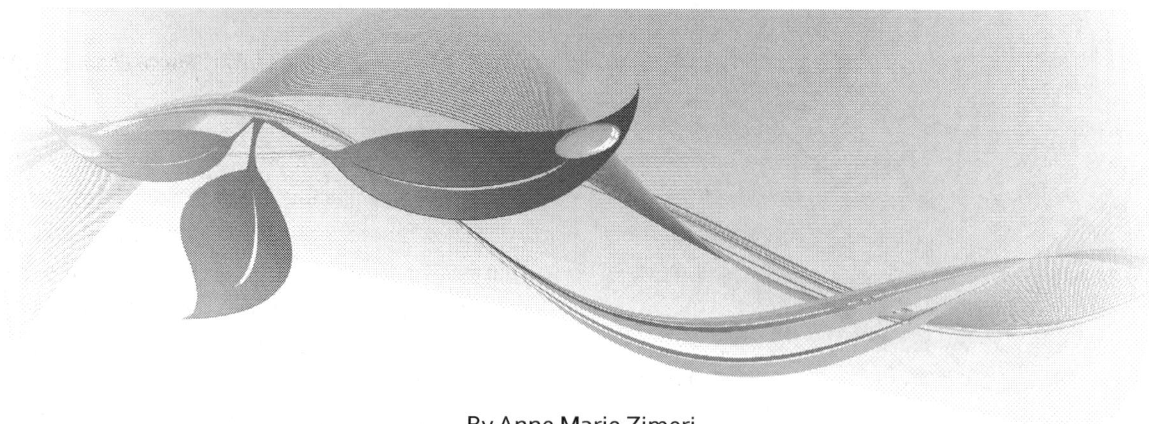

By Anne Marie Zimeri

The previous chapters have focused on issues and activities that rely on resources on the earth's surface. This chapter will delve beneath the earth's crust so that we can understand the environmental geology of the planet and the valuable resources it holds. It will also discuss energy sources, both fossil and sustainable.

The earth was formed billions of years ago in an extremely hot reaction that has been cooling ever since. In fact, the center of the earth is still quite hot at its core. The core is mostly made of iron and is intensely hot. This iron generates the magnetic field that surrounds the earth. Between the core and the earth's outer crust is the mantle. The mantle begins approximately 200 km. below the surface of the earth and extends 2900 km the core. It is less dense than the core and contains a slow-moving magmatic convection current. The relatively fragile earth's crust lies on top of the mantle in a series of plates, called tectonic plates. As these plates slide slowly across the earth's surface, some are moving away from one another. When this is occurring, we see the formation of deep ocean basins. As magma seeps upward from the mantle, it cools at the site of these basins and can form deep ocean ridges. When plates collide directly, they can force both edges upward to form mountain ranges, or one plate can subduct (go beneath) another. Because the plates are dynamic, geologists have suggested that several times in the earth's long history the plates were arranged such that all the continents were together to form one supercontinent, Pangea (Figure 8.1).

The earth's crust contains similar components as the earth below the surface, but in different proportions. If we take the earth as a whole, we see that it is mostly made of iron (33%) followed by oxygen (30%) and silicon (15%). The crust alone, however, is mostly oxygen (45%). Iron makes up only 6% of the crust.

The crust is made up of several rocks and minerals. The most common type of rock on the planet was formed as a result of cooling magma. These **igneous** rocks can form from rapid cooling (basalt) or slow cooling (granite). The study of this and of other types of rocks and minerals that can be important for humans is called economic mineralogy. Economic mineralogists work hard to identify **ore,** rock rich enough in one type of component to make it profitable enough to mine. Minerals and ores are not distributed evenly around the planet, making them strategic, i.e., one country or region is beholden to another in order to obtain something it needs. For example, the United States is rich in copper and lead, but it needs to import tin and nickel.

Extracting and mining for these resources once they are located can cause an enormous amount

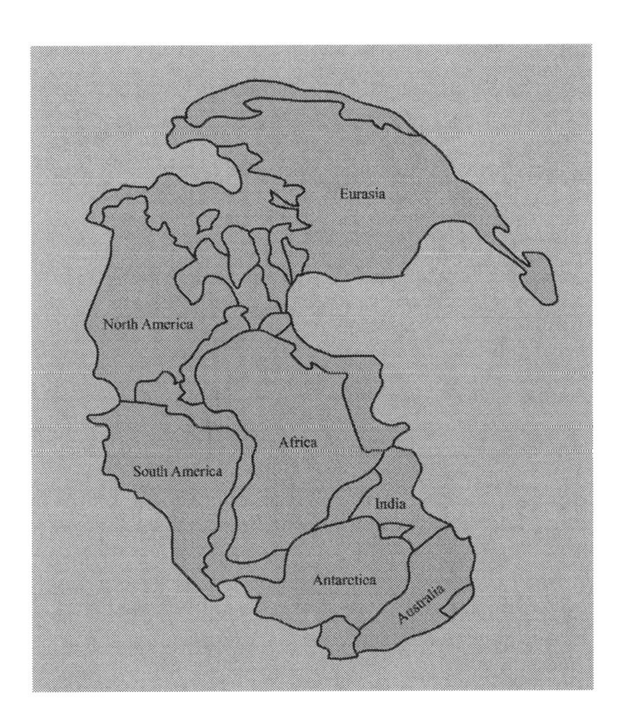

Figure 8.1. Pangea, the supercontinent, was last assembled around 200 million years ago.

of environmental degradation. **Open-pit** mines are those set up to extract or dig for ore or building materials (in the case of quarries) and remove large portions of earth in the process. Oftentimes the pit is dug below the water table, so water must be pumped out of the pit. After the pit is spent and abandoned, it will fill with water. The water in the pit moves across rock faces that are newly exposed to oxygen and can therefore release toxic heavy metals into the water. Other types of mining activities can be used to extract ore from the ground with a variety of environmental and public health consequences. Tunnel mining is more expensive than open-pit mining, but, if the ore is located deep enough that the waste rock on top cannot be removed or managed cheaply, then shafts or tunnels are dug to get to the valuable items. Underground tunnel mines are extremely dangerous places to work because of exposure to particulate matter, methane gas, and potential tunnel collapse. It has been estimated that up to one hundred mine workers die each year in tunnel mines.

In recent news there are many stories concerning mountaintop mining operations. These types of removals target coal seams under the peaks of mountains. In order to get to the coal, mining companies completely remove the top of the mountain and push the rubble and tailing into the adjoining valley. Mountaintop coal-mining operations are mostly found in eastern Kentucky, southern West Virginia, western Virginia, and eastern Tennessee. Environmental impacts of mountaintop mining, based on studies of more than 1,200 stream segments impacted by this type of mining practice, are increases of metals and minerals in the water, the abolishment of some smaller streams, fragmentation of forests, and general habitat destruction.

Placer mining is a method employed to extract valuable resources, often gold and platinum, from alluvial (or placer) deposits. Alluvial deposits are dredged and sifted in enormous rockers called cradles, to separate out the ore. Streams and rivers that are being mined have their bottom layers removed; therefore the benthic or bottom layer habitats are destroyed. In addition, the sandy silt that flows down into other streams from the mine can obstruct lower layers from receiving sunlight necessary for photosynthetic organisms that often make the lowest trophic layers of the food web.

To alleviate some of the destruction of the ecology at a mining site, the United States government passed the Surface Mining Control and Reclamation Act in 1977. This act required that mining operations restore the land to at least prime farmland once the mining operation is complete. This adds to the cost of the recovered products from mining because it can cost more than $5,000 per acre to restore.

There has been a real movement to conserve geologic resources in order to minimize ecological destruction and in many cases to reduce energy costs. There are many examples where using raw materials or mining for virgin materials is more costly when it comes to energy than reusing previously mined materials. For example, mining bauxite to produce aluminum uses more than thirty-one times as much energy as recycling aluminum from beverage containers. Recycling copper uses eight times less energy than mining for new copper, based on current market prices.

Fossil Energy

Worldwide energy production is overwhelmingly dominated by the fossil fuels: oil, coal, and natural gas. These fuels are deemed fossil because we are extracting them at a faster rate than they can be formed. It takes millions of years for these fuels to form; consequently, some would argue that fossil fuel supplies are essentially finite. Of the three main fossil fuels, coal represents 37% of worldwide commercial energy production, followed by oil (26%), and natural gas (24%). Alternative fuels and energy, such as nuclear power, hydropower, wind power and all other sources combined, make up only 13% of worldwide **energy** production.

What is **energy**? Energy is the capacity to do work. The work we ask our fossil fuels to do is to provide electricity for our homes and commercial buildings. We use these fuels in the transportation industry as well. In the United States, we typically measure energy from the power company in BTUs (British Thermal Units). One BTU is the amount of energy required to heat one pound of water one degree Fahrenheit. Worldwide, it is more common to measure energy in terms of small calories, which is the amount of energy required to heat one gram of water one degree centigrade.

If we look at how energy is used, we find that it can be divided into three major categories: industry, residential and commercial building use, and transportation. Each of these categories has room

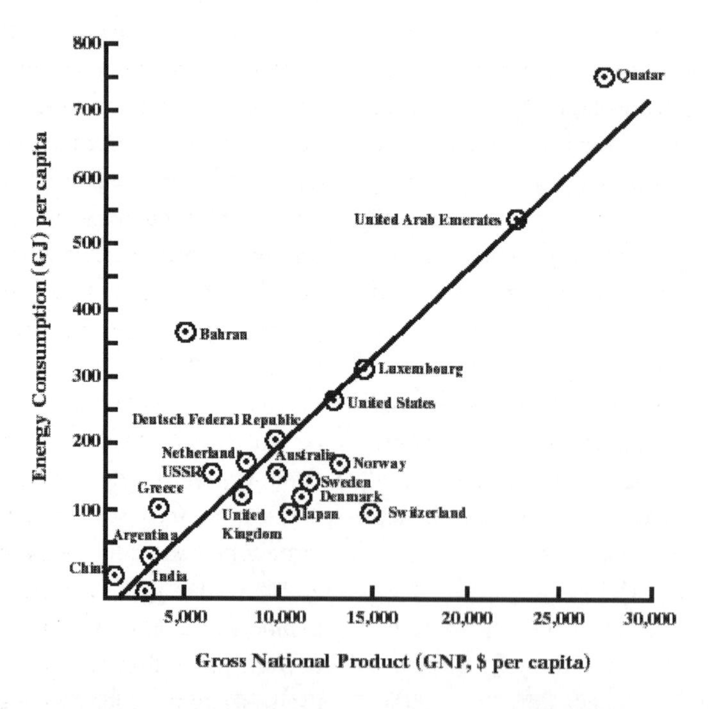

Figure 8.2. In general, more affluent countries use more energy per capita than less affluent countries. There are several exceptions to this, however. The United States should aspire to model energy use after some of the countries that have a higher standard of living, yet use less energy per capita. As the graph shows, Switzerland, even though having a higher standard of living, or GNP, than the United States, uses less energy per capita. Source: www.hendrix2.uoregon.edu/~dlivelyb/phys161/L17.html.

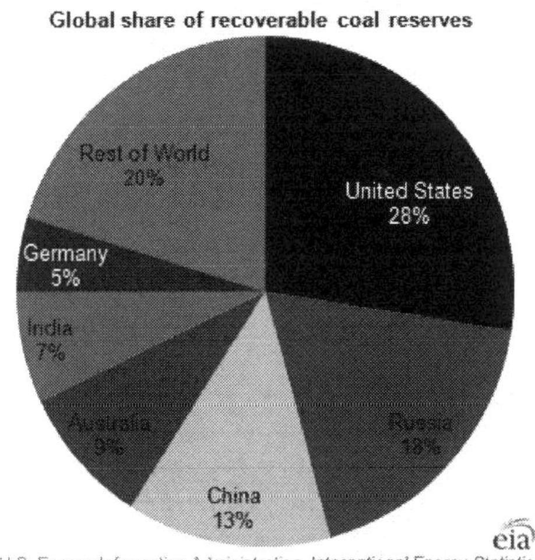

Source: U.S. Energy Information Administration, International Energy Statistics.
Note: Latest data available for the U.S. are 2009, international data are 2008. Coal reserves are relatively stable from year to year.

Figure 8.3. Global coal distribution

for improvement when it comes to efficiency of use. And all three are susceptible to energy losses that occur while converting fossil fuels to useful energy. For example, more than half of the energy in coal can be lost during thermal conversion in traditional coal-fired power plants. Moving electricity across wires, no matter how well insulated, allows for up to a 10% energy loss.

Like most topics in this text, we find that energy use and generation are not evenly distributed around the world. More developed countries use most of the energy on the planet. A typical person in a developed nation uses several hundred times more energy per year than a person in a lesser developed nation. Data also show a close correlation with standard of living (as demonstrated by income level) and energy consumption (Figure 8.2).

Coal Extraction and Use

Coal can be mined in several ways, which have been previously introduced in this chapter: mountaintop mining, strip mining, and open-pit mining, all of which leave large scars on the land and destroy the local ecology. Mine tailings, the waste and rubble associated with mining activities, easily leach heavy metals out into the soil and into the local water table. Coal burning expels millions of tons of waste each year from the almost one billion tons of coal burned each year in the United States. Two of the main culprits are sulfur dioxide and nitrogen dioxide, which are two of the main components in acid rain. Where is coal found? Coal is mostly found in the United States and Russia (Figure 8.3).

Oil Extraction and Use

The majority of proven-in-place oil reserves are located in the Middle East. Proven-in-place reserves are those that are profitable to extract with current technologies. When mining companies drill for oil, there is quite a bit of water pollution caused from the brine pumped into the wells to force the crude oil to the surface. The majority of waste from the oil industry comes from the production and refining business; however, use of oil also expels waste. More than 90% of oil is used in the transportation industry and is therefore one of the most widely used sources of energy worldwide. Like coal, oil is not distributed around the world equally, and some countries are richer in this resource than others. One country in particular, Saudi Arabia, sits above more than 25% of the world's proven-in-place oil reserves. Many other oil-rich countries in the Middle East have a volatile relationship with the United States and have power over the United States in this regard. The United States currently has about 3% of the world's proven-in-place reserves.

Natural Gas Extraction and Use

Natural gas has recently gained favor because it has been shown to release half the amount of CO_2 into the atmosphere as compared to an equivalent amount of coal when burned. Natural gas, which is primarily methane, is difficult to ship and store, and it is converted to liquid natural gas (LNG) in order to do so. LNG takes up only 1/600th of the volume of gas in its natural state and is transported in this compressed form in large sea vessels or gas pipelines. At present usage rates, and based on proven-in-place-reserve data, natural gas supplies will be depleted in less than 70 years.

Hydraulic fracturing, or "fracking," is an increasingly common technique used to release natural gas reserves that are trapped in shale formations. There are several shale gas reserves in the United States (Figure 8.4).

Figure 8.4. U.S. Shale gas reserves with volumes in trillion cubic feet (tcf).

In order to release the natural gas from the shale bed, fracturing fluid is pumped into the shale to create fissures that allow for the release of the natural gas. Fracking fluid can consist of water, sand, and a combination of up to several hundred chemicals to better release the gas. It is controversial because the fracking fluid can be hazardous to human health and may expose humans upon its migration into underground water supplies or through poor management and collection of waste water from fracking operations.

Nuclear Power

Nuclear power is responsible for around 7% of U.S. energy production and is expected to rise with the approval of new construction for several reactors around the country, including in Augusta, Georgia. A nuclear facility uses a chain reaction to generate heat so that the heat can be used to generate steam and turn a turbine to then generate electricity. The advantage of using a nuclear reaction to generate heat over burning coal is that there are no gaseous emissions from the reaction if it is handled properly.

What fuels a nuclear reaction? In order to start the chain reaction that generates heat in a nuclear power plant, uranium must be mined for fuel. Most uranium is of the isotope U238, which is not useful to the reaction, but a small fraction (less than 1%) is U235, which can drive the reaction. First, the mined uranium must be enriched so that the percent of U235 is at least 3%. This enriched uranium is pressed into pellets that are placed into long rods, which, when bundled, form a nuclear core. The U235 is then bombarded with neutrons. When this happens, an extremely unstable isotope of uranium U236 forms and then releases heat, and more neutrons that can react with another U235 atom; thus the chain reaction. To control the reaction, the core is surrounded

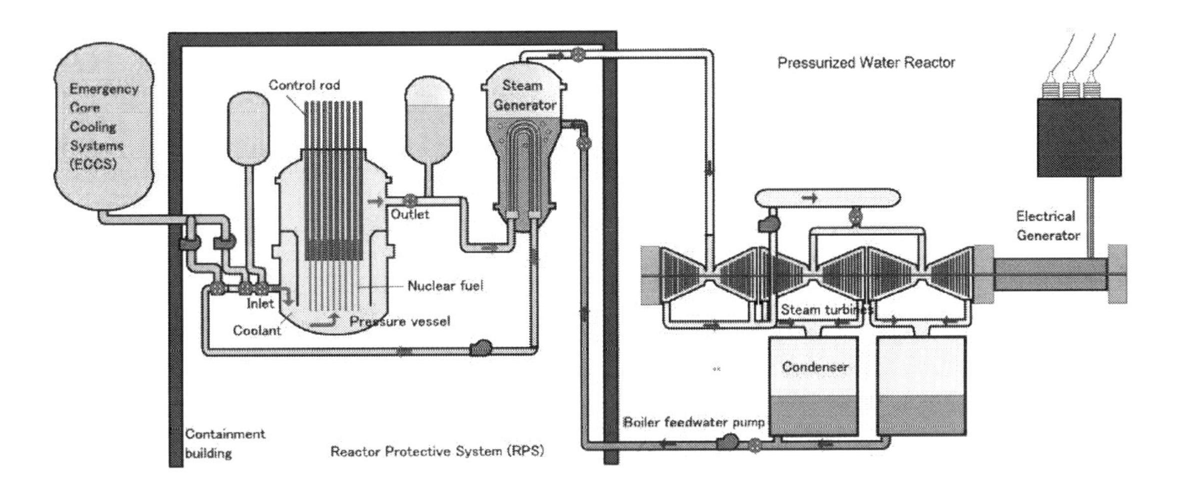

Figure 8.5. A pressurized water reactor uses two closed loops of water to generate heat and steam that will ultimately turn a turbine that generates electricity.

with neutron-absorbing material and kept. Most nuclear reactors house the chain reaction in one building and the power generation in another. These are known as pressurized water reactors.

Nuclear power is not without its environmental impact. Mining for uranium scars the land and creates tailings not unlike mining for coal. There is also some radioactive waste left after the nuclear fuel rods are spent. This fuel has been stored in the past in underground pools but is now typically in an above-ground cask that can be monitored from all sides. Potentially to open in the future is a centralized nuclear waste repository in Yucca Mountain, Nevada. This site has a partially constructed area for nuclear waste, though its continued development has been removed from the 2011 budget.

RENEWABLE ENERGY

Solar Energy

Photovoltaic Energy

Photovoltaic cells capture solar energy and convert it directly to electrical current instead of generating heat that will eventually be converted to electricity. You may be familiar with these cells already if you have a solar-powered calculator or watch.

Why even bother with generating heat from solar energy if we can use it directly with photovoltaic cells? When the cells were first invented in Bell Laboratories in the late 1950s, the price per watt was an exorbitant two thousand dollars. Technical advances have lowered the price extensively to a current average of five dollars per watt. However, since current utility company processes are nearly six cents per watt, photovoltaic cells are cost prohibitive.

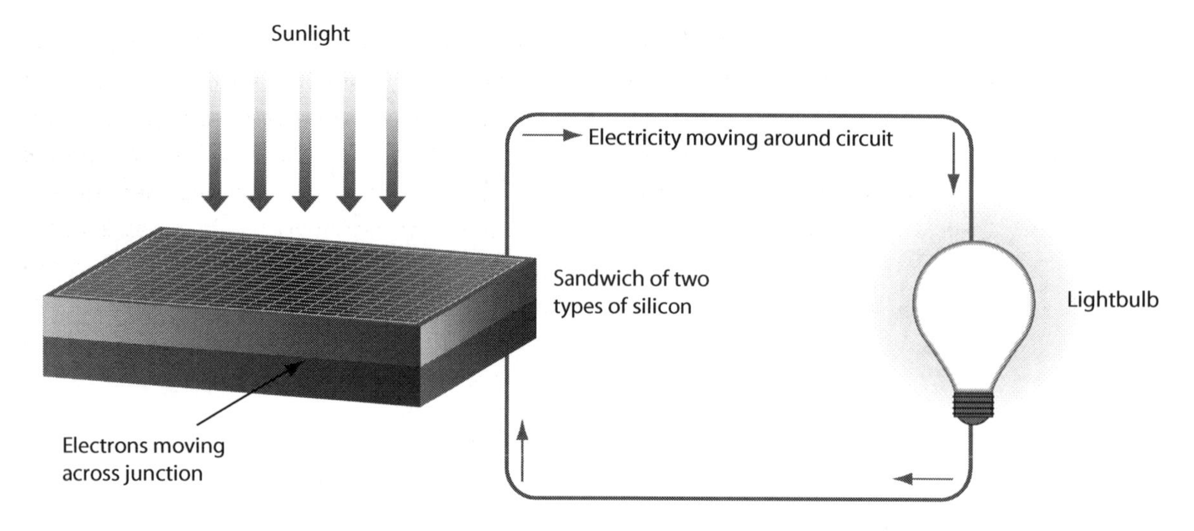

Figure 8.6. Photovoltaic cells are made with layers of amorphous silicon that allow for the movement of electrons when exposed to light.

Storing the Sun's Energy

Both solar and photovoltaic energy require light from the sun, but what can we do in the evening or in the winter when the sun is not shining or is shining less intensely because of the tilt of the earth? Energy storage in the past has relied heavily on the battery, though to power a home with a battery large enough to store energy for low-sunlight months would cost thousands of dollars and likely be the size of a Dumpster. More eco-friendly and efficient is **pumped hydro-storage**. In this scenario, during daylight hours, when sunlight is plentiful, energy is generated for use and for pumping water against gravity into a cistern or tower. Once the sun goes down, the water is released so that it passes by a paddle wheel that will turn a turbine that will generate electricity. Another option for energy storage is using daylight energy to turn a flywheel in as frictionless an environment as possible such that once energy is no longer applied to the wheel, it continues to spin for a period of time while generating electricity.

Biomass/Fuelwood

Half of the worldwide annual wood harvest is used as fuel, primarily fuel to heat homes and for cooking. Most of this fuelwood is harvested and used in LDCs, and it is thought that if fuelwood consumption continues at its current rate, there will simply not be enough to satisfy the demand within only a few decades. Besides wood, however, other types of fast-growing, more sustainable plants may be used as biomass for fuel. Switchgrass (*Panicum virgatum)* is an example of a plant that holds promise for the future.

Hydropower

Much of the increase in hydropower as of late has been due to the construction of extremely large dams such as the Three Gorges Dam that spans the Yangtze River in China. These structures hold water behind a dam wall as potential energy. When water is released, it turns large turbines that generate electricity. The main advantage to hydropower is that there is no combustion involved; therefore it does not release hydrocarbons and CO_2 into the atmosphere. There are, however, several ecological disadvantages associated with dams. When Three Gorges Dam was built, it displaced more than one million residents who lived along the river. It changed ecosystems and habitats and altered the flow of sediment and nutrients downriver.

Geothermal Energy

Geothermal energy exploits the natural heat from below the surface of the earth. The amount of heat within 10,000 meters (about 33,000 feet) of the earth's surface contains 50,000 times more energy than all the oil and natural gas resources in the world. This energy can be accessed in many areas of the world and can be used to run large power plants, or to simply stabilize the ambient temperature in a private home.

Wind Energy

Wind turbines can be constructed in areas of the country with plentiful winds in order to generate electricity. There are even wind farms just offshore to harness wind energy off the coast. Like hydropower and some other renewable energy sources, wind energy does not release greenhouse gases. Is has the added advantage of using little land when compared to solar fields or biomass generation. There are aesthetic disadvantages; many communities fight the installation of wind farms because they can obstruct the natural view from an area, but beyond such obstructions and some cases of injured birds, wind power has little environmental impact.

Tidal Energy

The tides are extremely regular and predictable and completely renewable when it comes to energy generation. Tidal energy can be harnessed in a couple of ways. It can be used in a manner similar to traditional hydropower in that when the tide comes in, it fills a large storage area with water. This full storage area can then be drained when the tide goes out such that the water, upon release, turns a turbine. Tidal energy can also be harnessed by using structures that undulate with the tide and generate energy just offshore. All tidal energy is subjected to the constraints of how building materials resist corrosion from salt water. It may also disturb marine life or even kill some fish if they get caught in the wheels that turn the turbines.

Climate Change

By William H. Schlesinger, The Cary Institute of Ecosystem Studies. Millbrook, New York

INTRODUCTION

It is early fall as I write this in Millbrook, New York, which just finished its hottest summer on record. The National Weather Service reports that such record temperatures extended across the United States during 2010. Satellite views of our planet show rising sea levels and melting polar ice at the highest rates seen during the past thirty years. And on our annual summer vacation to Maine, I heard the locals talk of the first appearance of Lyme disease, a tick-borne illness, normally found only in southern New England.

Meanwhile, the leadership of the U.S. Senate scuttled a bill that might have curbed our emissions of carbon dioxide and other "greenhouse gases" to the atmosphere, reflecting the mood of the electorate across our nation. Why is there such a disconnect between our understanding of climate change, the early symptoms of a planetary fever, and the public's willingness to believe, let alone act, as if global warming is real?

THE GREENHOUSE EFFECT

The physics of the greenhouse effect have been known for more than 150 years, since John Tyndall put different types of gases in laboratory chambers and noted how some of them absorbed heat (infrared) radiation, while they passed visible light without incident.[1] Indeed, even early naturalists were witness to Earth's greenhouse effect when they noted that a night without clouds was likely to be much colder than a humid or cloudy night preceded by the same

1 John Tyndall. *Heat: A Mode of Motion.* New York: D. Appleton and Co., 1865.

late afternoon temperature.[2] Water vapor absorbs heat that is reradiated from the Earth's surface, and it is easiest to sense this effect at night, when it is not overwhelmed by incoming radiation from the sun.

The term, "greenhouse effect," stems from the observation that sunlight passes through glass relatively easily, warming the inside of glass houses, whereas infrared or heat radiation is unable to pass so readily in the opposite direction. The inside of a greenhouse, an automobile, or a passive solar home gets warm on a sunny day and retains the heat for a long time after sunset.

The Earth has a natural greenhouse effect that derives from water vapor, carbon dioxide (CO_2), methane, and a few other "radiatively active" gases in its atmosphere.[3] And we should be thankful for it! Without its natural greenhouse effect, the temperature of our planet would be 33° centigrade cooler than today, and all water on Earth's surface would be frozen. Look at Mars—a planet with a thin atmosphere and not much CO_2—and you see a cold planet without much greenhouse effect.[4] Carbon dioxide freezes on Mars' south pole to form dry ice. In comparison, Venus, with a thick atmosphere largely composed of CO_2, has a surface temperature of 474°C, much hotter than it would be (54° C) without its huge greenhouse effect

2 For example, reflecting on the formation of dew and mist in his journal, November 20, 1853, Henry David Thoreau notes, "there is most dew in clear nights, because clouds prevent the cooling down of the air; they radiate the heat of the Earth back to it." (ed. Bradford Torrey; *The Writings of Henry David Thoreau: Journal,* March 5–November 30, 1853. Boston, Houghton Mifflin and Company, 1906, 513.) The effect of a clear night is also seen in the Christmas carol, Good King Wenceslas, in the stanza, "Brightly shone the moon that night, tho' the frost was cruel" (1853).

3 Gavin Schmidt, et al., Attribution of the Present-day Total Greenhouse Effect. *Journal of Geophysical Research* 115 (2010): doi:10.1029/2010JD014287.

4 The dominant gas in the Martian atmosphere is CO_2 (95%), but the atmosphere is less than 1% of that on Earth.

THE CARBON CYCLE

Studies of Earth's greenhouse effect are closely tied to our understanding of the movement of carbon between the land, oceans, and atmosphere on Earth.[5] Most of this carbon moves as carbon dioxide; for instance, the transport of carbon in soils eroded by rivers to the sea (500 million tons per year) is dwarfed by the release of carbon dioxide from soils to the atmosphere by the action of decomposing organisms (75,000 million metric tons per year). Atmospheric scientists focus on CO_2 because it is a powerful greenhouse gas, which has shown remarkable variations in its concentration in the atmosphere through Earth's history. Water vapor is also a greenhouse gas, but with so much liquid water exposed on Earth, the concentration of water vapor in the atmosphere varies largely as a function of Earth's temperature, not vice-versa.[6] Moreover, the average molecule of CO_2 spends about five years in the atmosphere, versus about nine days for water vapor.

The important greenhouse gases spend enough time in the atmosphere to have a significant effect.

Every year, CO_2 moves in and out of the oceans and in and out of land vegetation, mostly forests. In the ocean, CO_2 dissolves in cold, dense waters, which tend to sink to the deep ocean near the poles of the Earth. Carbon dioxide returns to the atmosphere when the deep waters upwell to the surface at warm tropical latitudes, such as off the coast of Peru, where nutrient-rich deep waters support an important historical fishery. Carbon dioxide dissolves in seawater in proportion to its concentration in the atmosphere, so as atmospheric CO_2 concentrations rise, more will enter the ocean's waters. Carbon dioxide also enters the ocean as a result of its uptake by marine

5 William H. Schlesinger. *Biogeochemistry*. San Diego: Academic Press, 1997..

6 Andrew A. Lacis, et al. Atmospheric CO_2: Principal Control Knob Governing Earth's Temperature. *Science* 330 (2010): 356–59.

algae—phytoplankton—some of which die and sink to the deep sea.

Land plants take up CO_2 in photosynthesis and return CO_2 to the atmosphere via their own metabolism or that of fungi and bacteria, which decompose the vast amount of vegetative material that is produced by land plants each year. Fires also return CO_2 to the atmosphere. The small amount of plant production that escapes decomposition is stored in soils, peat bogs, and sediments. During periods of Earth's history when there have been vast interior swamplands, large amounts of plant material escaped decomposition and formed coal. Similar sedimentary deposits of organic matter in the oceans formed petroleum. Today, we drill for oil in old sedimentary rocks exposed on land and buried in shallow seas.

Through geologic time, there have been large variations in the concentration of CO_2 in Earth's atmosphere. Of course, no one was here to measure it or take samples for later analysis, so we study these changes in atmospheric CO_2 indirectly, using proxies. A proxy is something that leaves a record of the changes in Earth's prior condition. The relative content of boron isotopes in certain marine sediments, largely limestone, is a proxy for changes in the acidity of seawater as a result of changes in atmospheric CO_2 concentrations. With such proxies, we are able to reconstruct the past variations in atmospheric CO_2 and ascertain what may have caused them. Extensive volcanic activity is one such cause; CO_2 rose during the Eocene and Oligocene, reaching levels >1300 parts-per-million[7] (ppm) 33 million years ago, as a result of widespread volcanic emissions. Other records show that the Earth's climate was also very warm at that time and cooled as CO_2 declined.[8]

Using the same boron-isotope proxy, we know that the concentration of CO_2 in Earth's atmosphere has not risen above 400 ppm during the past 20 million years.[9]

Our understanding of fluctuations in atmospheric CO_2 during the past 800,000 years is more straightforward than for earlier times, because we are able to analyze the bubbles of gas buried in layers of snow and now trapped in the Antarctic ice pact This record shows that CO_2 has varied between 170 and 290 ppm, in regular oscillations that correspond to glacial intervals of the Pleistocene.[10] These oscillations in Earth's climate are linked to variations in the Earth's orbit around the sun and variations in its inclination on its axis. (The Earth wobbles on its axis like a spinning top.) Interestingly, at the end of each ice age, the temperature rose before the rise in CO_2. Carbon dioxide may have reinforced the return to warmer temperatures, but it apparently did not initiate the change.

The last ice age of the Pleistocene ended about 12,000 years ago, when we entered the relatively stable climate conditions of the modern epoch, the Holocene. This is not to say that the ice ages are over, only that with an anthropocentric view, we have given a different name to the stable conditions of Earth's recent, recorded history, including all human civilizations beginning about 8,000 years ago. From the start of the Holocene until the Industrial Revolution, the Earth's temperature and CO_2 have been remarkably constat, with CO_2 ranging only from about 260 to 285 ppm.[11] The recent rise in CO_2 to today's value of 388

7 Environmental scientists use the unit parts-per-million or ppm, to express the concentration of trace substances in whole numbers. For instance, today, the concentration of CO_2 in Earth's atmosphere is close to 0.0388% or 388 ppm, meaning that for every 1 million liters of gas one might collect in a tank, 388 liters will be CO_2.

8 Pearson, Paul N., G. L. Foster, and B. S. Wade. Atmospheric Carbon Dioxide through the Eocene-Oligocene Climate Transition. *Nature* 461 (2009): 1110–13; Owen, R.M. and D. K. Rea. Sea-floor Hydrothermal Activity Links Climate to Tectonics: The Eocene Carbon Dioxide Greenhouse. *Science* 227 (1985): 166–69.

9 Pearson, Paul N. and M. R. Palmer .Atmospheric Carbon Dioxide Concentrations over the Past 60 Million Years. *Nature* 406 (2000): 695–99.

10 Luthi, Dieter, et al. High-resolution Carbon Dioxide Concentration Record 650,000-800,000 Years before Present. *Nature* 453 (2009): 379–82.

11 Fluckiger, Jacqueline, et al. High-resolution Holocene N2O Ice Core Record and its Relationship with CH4

ppm is unprecedented, and the increase continues at a rate of about 1.9 ppm per year.

The economic activity of the Industrial Revolution has been powered by fossil fuels—coal, oil, and natural gas. In a very real sense, humans are extracting the products of past plant growth on Earth, bringing fossil organic materials to the surface and burning them— releasing captured sunbeams! By one estimate, each year we are burning the organic materials that represent plant growth over 400 years of Earth's history—perhaps deposited 300 million years ago.[12]

If we were burning small amounts of fossil fuels, then the Earth's carbon cycle could compensate for the CO_2 added to the atmosphere. More CO_2 would dissolve in the ocean's waters and more CO_2 might be assimilated (and a little bit stored) by land plants. The current rapid rise in CO_2 in Earth's atmosphere is a rate problem. By burning fossil fuels, we are adding CO_2 more rapidly than the natural cycles can accommodate it, and we have not created a counterbalancing force that takes CO_2 out of the atmosphere. Indeed, another major human impact stems from tropical deforestation, which releases CO_2 to the atmosphere when tropical forests are burned and their soils are cultivated. As much as 20% of the current human emissions of CO_2 to the atmosphere may stem from forest destruction.

CLIMATE CHANGE

Physicists tell us that higher CO_2 in Earth's atmosphere should warm our planet—the fundamentals of the greenhouse effect. Records from tree rings[13] and ice cores[14] show a rise in temperature during the past 150 years, corresponding to the rise in CO_2. Historical records of the extent of sea ice, the spring break-up of ice on rivers and lakes, the spring flowering of plants and bird migrations, all indicate unprecedented warm temperature during the past few decades. Satellite measurements show an increase in Earth's surface temperature, particularly in northern latitudes, over the past 30 years.[15] Even the oceans' temperature has warmed over the past 40 years.[16]

But, beyond human activities, how do we know that there are not other factors that might be responsible for the current warming or that might reverse it sometime in the near future? How can we separate the human effects on Earth's greenhouse effect from natural variations? For this, we need to consider the Earth's radiation budget in slightly more detail.

Incoming solar radiation delivers about 340 W/m^2 (watts per square meter) to the Earth. The natural greenhouse effect warms the planet about 33° C by trapping 153 W/m^2 of outgoing radiation.[17] For the past 30 years or so, there has been a small increase in the sun's luminosity (+0.12 to 0.16 W/m^2).[18] By com-

and CO2. *Global biogeochemical Cycles* 16 (2002): doi; 10.1029/2001GB001417; C. MacFarling Meure. et al., "Law Dome CO2, CH< and N2O Ice Core Records Extended to 2000 BY," *Geophysical Research Letters* 33 (2006): doi: 10.1029/2006 GL026152.

12 Jeffrey S. Dukes. Burning Buried Sunshine: Human Consumption of Ancient Solar Energy. *Climatic Change* 61 (2003): 31–44.

13 Michael E. Mann, R. S. Bradley, and M. K. Hughes, Global-scale Temperature Patterns and Climate Forcing over the Past Six Centuries *Nature* 392 (1998): 779–87.

14 Lonnie G. Thompson, et al. A High-resolution Millennial Record of the South Asian Monsoon from Himalayan Ice Cores. *Science* 289 (2000): 1916–19.

15 *National Research Council, Reconciling Observations of Global Temperature Change* Washington: National Academy Press, 2000.

16 Tim P. Barnett, et al. Tenetration of Human-induced Warming into the World's Oceans. *Science* 309 (2005): 284–87.

17 IPCC (Intergovernmental Panel on Climate Change), *Climate Change 2007: The Physical Science Basis* (Cambridge: Cambridge University Press, 2007). W/m2 is a timeless expression of the receipt of energy. If you place a 100-Watt bulb on a square meter of land, it adds 100 W/m2 to that unit of the Earth's surface. The total energy received in a day would be 2400 W or 2.4 kW/hr.

18 P. Foukal, et al. Variations in Solar Luminosity and their Effects on Earth's Climate. *Nature* 443 (2006): 161-66; R. T. Pinker, B. Zhang, and E. G. Dutton, "Do Satellites

parison, the human impact to radiative forcing due to increasing concentrations of greenhouse gases currently adds about 2.3 W/m^2 to the natural greenhouse effect—about 20 times the change in solar forcing. A substantial component of that warming derives from higher concentrations of water vapor in Earth's atmosphere, as a result of warmer temperatures.

Aerosols and some clouds tend to cool the atmosphere by increasing the reflectivity of the Earth to incoming radiation. We see the effect of aerosols in the first few years after major volcanic eruptions, such as Mount Pinatubo, which add sulfate aerosols to the stratosphere. This cooling effect disappears within a few years, because aerosols drop out of the atmosphere fairly quickly. The overall reflectivity or ***albedo*** of the Earth, as measured by changes in "earthshine" seen on a New Moon, is about 30%.[19] Earth's ***albedo*** has apparently increased slightly in recent years (i.e., global dimming), presumably due to particulate air pollutants.[20] Increases in aerosols due to human activities are thought to reduce the current global radiative forcing by about 1.2 W/m^2. It is interesting to note that aerosol concentrations were higher and temperatures were lower during the last glacial period, but the cause and effect relation of these observations is unknown.

Besides CO_2 and water vapor, a variety of other gases contribute to Earth's greenhouse effect, including methane (CH_4, natural gas) and nitrous oxide (N_2O), which is best known from the dentist's office. The majority of the annual emission of these gases is natural, although both have human sources as well. Methane is released from the leakage of natural gas from wells and pipelines, and from an expanding cultivation of rice and cattle. Nitrous oxide is released from fertilized soils. The atmospheric lifetimes of methane (12 years) and nitrous oxide (>100 years) are such that they contribute substantially to Earth's greenhouse effect, versus other potential greenhouse gases, such as ammonia (NH_3) and ozone (O_3), which would also be important if it were not for their short persistence in the atmosphere. Various industrial gases, such as chloroflurocarbons (CFCs) and nitrogen trifluoride (NF_3, a solvent in the computer industry), are also potent greenhouse gases, with long atmospheric lifetimes. Methane, nitrous oxide, and these industrial gases have all increased in concentration in Earth's atmosphere since the beginning of the Industrial Revolution. Nitrous oxide is particularly worrisome, since it is rising at a rate of 0.3%/year, and destined to rise even faster as fertilized agriculture expands globally.

CLIMATE CHANGE IMPACTS

All evidence suggests that humans are having an impact on the radiation budget of the Earth, and that human-induced global warming exceeds our effects on processes that might cool the planet The human impacts on the Earth's radiation budget are also well in excess of all known natural variations, such as in the sun's luminosity, although it is possible that a future gigantic volcanic eruption might drastically cool the planet for a few years.

How are the human effects on climate predicted? In the same way that atmospheric scientists have built models to predict our daily weather—with much success relative to a few decades ago—they can build models that predict the behavior of the Earth's atmosphere over longer periods. General circulation models for the atmosphere consider parcels of air in vertical stacks to the top of the atmosphere and in horizontal divisions extending from pole to pole. Each of these cubes contains a parcel of the

Detect Trends in Surface Solar Radiation?" *Science* 308 (2005): 850-54.

19 P.R. Goode, et al. Earthshine Observations of Earth's Reflectance. *Geophysical Research Letters* 28 (2001): 1671–74.

20 Kaicun Wang, R. E. Dickinson, and S. Liang. Clear Sky Visibility Has Decreased over Land Globally from 1973 to 2007. *Science* 323 (2009): 1468–70.

atmosphere that reflects, passes, or absorbs radiation entering from outside the Earth or re-radiated from its surface.

The parcel is heated and circulates, potentially exchanging gas and heat with adjacent parcels, as the Earth undergoes its seasons. If clouds form and become supersaturated at the predicted temperature, rain or snow falls from that parcel to the Earth's surface. These models are immensely complicated, often requiring supercomputers to operate. Their performance is best validated by comparing their predictions for past climate, for periods when we have historical records or proxies to check the results. A large number of these models have been built, which allow different teams of atmospheric scientists working independendy to derive and check predictions. Disagreements typically produce heated discussion, but they have helped to refine the current generation of general circulation models and achieve some consensus about future climate trends.

Nearly all climate models predict that a substantial warming of the atmosphere (2° to 4.5° C during this century) will accompany increasing concentrations of greenhouse gases in the atmosphere. The predicted warming of future climate is greatest near the poles, where there is normally the greatest net loss of infrared radiation. For the same reason, the models predict that future nighttime and wintertime temperatures will be most likely to show the greatest changes relative to today's conditions. Presumably, the oceans will warm more slowly than the atmosphere, but eventually, warmer ocean waters will allow greater rates of evaporation, increasing the circulation of water through the atmosphere. Water vapor also absorbs infrared radiation, so it is likely to accelerate further the potential greenhouse effect. Thus, most models predict that higher concentrations of CO_2 and other trace gases in the atmosphere will make the Earth a warmer and more humid planet.

The effects of rising CO_2 include a fertilizing of plant growth and increasing acidity of seawater. As a reactant for photosynthesis, CO_2 stimulates plant growth by 15–18% when the concentration rises from 350 to 550 ppm, as seen in long-term field experiments with forests and agricultural crops.[21] Unfortunately, high CO_2 also stimulates the growth of weeds, poison ivy, and other plant allergens.[22] Higher levels of CO_2 also cause additional CO_2 to dissolve in seawater, where it forms carbonic acid. Recent measures indicate a greater level of acidity in seawaters, as indicated by a drop of 0.06 unit in the pH of seawater in the North Pacific since 1991 [23] In the face of higher acidity, which dissolves their carbonate skeletons, many coral reef ecosystems are threatened by rising CO_2.

The greatest concerns from rising CO_2 stem from its effects on climate. Mean global temperatures are anticipated to rise 2° to 4.5° centigrade by the end of the twenty-first century. Already, dramatic losses of sea ice are recorded in the Arctic[24] as well as a net

---

21 Long, Stephen P., et al. Food for Thought: Lower-than-expected Crop Yield Stimulation with Rising CO_2 Concentrations. *Science* 312 (2006): 1918–21; Norby, Richard, et al. Forest Response to Elevated CO_2 is Conserved Across a Broad Range of Productivity. *Proceedings of the National Academy of Sciences, U.S.* 102 (2005): 18052-56.

22 Ziska, Lewis H., P. R. Epstein, and W H. Schlesinger, Rising CO_2, Climate Change, and Public Health: Exploring the Links to Plant Biology. *Environmental Health Perspectives* 117 (2009): 155–58; Mohan, Jacqueline E., et al. Biomass and Toxicity Responses of Poison Ivy (Toxicodendron Radicans) to Elevated Atmospheric CO_2. *Proceedings of the National Academy of Sciences, U.S.* 103 (2006): 9086–89; LaDeau, Shannon L. and J. S. Clark. Pollen Production by Pinus Taeda Growing in Elevated Atmospheric CO_2. *Functional Ecology* 20 (2006): 541–47.

23 Byrne, Robert H., et al. Direct Observations of Basin-wide Acidification of the North Pacific Ocean. *Geophysical Research Letters* 37 (2010):doi: 10.1029/2009GL040999; Doney, Scott C. The Growing Human Footprint on Coastal and Open-ocean Biogeochemistry. *Science* 328 (2010): 1512–16.

24 Serreze, Mark C., et al. Perspectives on the Arctic's Shrinking Sea-ice Cover. *Science* 315 (2007): 1533–36; Kwok, R. and D.A. Rothrock. Decline in Arctic Sea Ice Thickness from Submarine and ICESat Records: 1958-2008. *Geophysical Research Letters* 36 (2009): doi: 10.1029/2009GL039035.

loss of ice from the Antarctic ice pack[25] While melting sea ice, which is floating on the ocean surface, does not affect sea level, melting continental glaciers contribute to the current 3.5 millimeters per year (mm/yr) global rise in sea level—up from 0.8 mm/yr 100 years ago.[26] Warmer arctic conditions also lead to the loss of permafrost, which has retained huge stores of dead organic matter and peat during the Holocene. As these soils thaw, the organic matter decomposes, releasing more CO_2 to the atmosphere, exacerbating global warming.[27] Another positive feedback to global warming will follow the loss of polar ice itself. Normally, this ice contributes significantly to Earth's ***albedo,*** but when it melts, it exposes the ocean surface or land, which is normally less reflective than ice to incoming solar radiation.

Rapid changes in climate associated with global warming have several indirect effects on human health and well-being. Many diseases that are transmitted by insects, especially mosquitoes, occur in climatic regions that are defined by conditions of temperature and moisture. A warmer, wetter world in the future is likely to allow an expansion of the occurrence of malaria, dengue fever, and other insect-borne diseases, or require a substantial human investment to prevent it.[28] Anticipated effects on plant diseases are similar. Already, a northward expansion of the hemlock woolly adelgid, due to warmer winters, is thought to be responsible for the loss of hemlock from northeastern forests.[29] Noah Diffenbaugh and his colleagues show potential expansions in the range of the corn-borer and other insect pests of major crops, which could threaten the breadbasket of major foods in the Great Plains.[30] While some crops may grow better in warmer conditions, many of the world's major crops show lower yields.[31] Even wine growers should expect a shift in the optimal range for wine production from California to points northward.[32] Many models for future climate indicate a substantial drying in the

25 Chen, J. L., et al. Accelerated Antarctic Ice Loss from Satellite Gravity Measurements. *Nature Geoscience* 2 (2009): 859–62.

26 Church, J. A. and N. J. White. A 20th-century Acceleration in Global Sea-level Rise. *Geophysical Research Letters* 33 (2005): doi.10.1029/2005GL024826; Merrifield, M. A., et al. An Anomalous Recent Acceleration of Global Sea Level Rise. *Journal of Climate* 22 (2009): 5772–81.

27 Dorrepaal, Ellen, et al.Carbon Respiration from Subsurface Peat Accelerated by Climate Warming in the Subarctic. *Nature* 460 (2009): 616–19; Schuur, Edward A. G., et al. The Effect of Permafrost Thaw on Old Carbon Release and Net Carbon Exchange from Tundra *Nature* 459 (2009): 556–59; Oechel, Wlater G., et al. Acclimation of Ecosystem CO_2 Exchange in the Alaskan Arctic in Response to Decadal Climate Warming. *Nature* 406 (2000): 978–81.

28 Pascual, Mercedes, et al. Malaria Resurgence in the East African Highlands: Temperature Trends Revisited. *Proceedings of the National Academy of Sciences, V.S.* 103 (2006): 5829–34; Hales, Simon, et al. Potential Effect of Population and Climate Change on Global Distribution of Dengue Fever: An Empirical Model. *Lancet* 360 (2002): 830–34. For alternative views, see Rogers, David J. and S. E. Randolph. The Global Spread of Malaria in a Future, Warmer World. *Science* 289 (2000): 1763–66, who suggest that the spread of malaria with global warming will be rather modest, and Gething, Peter W., et al. Climate Change and the Global Malaria Recession. *Nature* 465 (2010): 342–45, who argue that in the past increased economic development has reduced the incidence of malaria and will do so in the future. Neither of the latter papers offers an indication of the costs involved and the ability of the developing world to pay the cost to prevent a greater global incident of malaria for humans.

29 Skinner, Margaret, et al. Regional Responses of Hemlock Woolly Adelgid (Homoptera: Adelgidae) to Low Temperatures. *Environmental Entomology* 32 (2003): 523–28; Dukes, Jeffrey S.,et al. Responses of Insect Pests, Pathogens, and Invasive Plant Species to Climate Change in the Forests of Northeastern North America: What Can We Predict? *Canadian Journal of Forest Research* 39 (2009): 231–48.

30 Diffenbaugh, Noah S., et al. Global Warming Presents New Challenges for Maize Pest Management. *Environmental Research Letters* 3 (2008): doi: 10.1088/1748–9326/3/4/044007.

31 Lobell, David B. and C. B. Field. Global Scale Climate-crop Yield Relationships and the Impacts of Recent Warming. *Environmental Research Letters* 2 (2007): doi: 10.1088/1748–9326/2/1/014002; Lobell, David B., et al. Prioritizing Climate Change Adaptation Needs for Food Security in 2030. *Science* 319 (2008): 607–10.

32 White, M. A., et al. Extreme Heat Reduces and Shifts United States Premium Wine Production in the 21st

southwestern United States, an area of rapid current population growth and limited water supply.[33]

In the eastern United States, predictions of the effects of climate change on the distribution of forest species show sugar maple being eliminated from most of its present range, persisting only in Canada.[34] There are substantial changes in the predicted range of southern pine species, which should be of major concern to all those who depend on the current forest products industry of that region. The range of many bird species in New York State has already shifted northward during the past several decades,[35] and in many areas of the eastern United States, springtime migrating birds are arriving earlier from the South[36] Simultaneous, but disconnected, shifts in insects, birds, and plant species threaten a reconfiguration of the major components of nature in many areas. It is likely that some species will lose the entire envelope of climate that now supports their existence.[37] Chris Thomas and colleagues predict a loss of 18 to 35% of species with the global warming expected in this century.[38]

TIPPING POINTS

Some global change scientists speculate that the human impacts on climate may carry us to a planetary threshold or tipping point, beyond which our impacts on the Earth will have been so large that a return to prior conditions will be impossible, even if we were to cease our actions.[39] For instance, a complete loss of Arctic ice may so increase the absorption of solar radiation and warming in the northern latitudes that the frozen conditions would not return, even if emissions of CO_2 and the amount accumulated in the atmosphere were to decline substantially. A warming of the Arctic ocean may lead to the degassing of methane from the ocean sediments, where it is now held frozen in sediments known as methane hydrates. This release of methane would dramatically increase the concentration of this greenhouse gas in the atmosphere, exacerbating global warming, and minimizing that chance that the planet could ever return to something resembling the conditions of the Holocene. The geologic record shows evidence of past, catastrophic degassing of methane from ocean sediments, with large effects on Earth's climate and its biota.[40]

Of course, tipping points are speculative, but even remote probabilities of their occurrence have garnered the attention of security and defense communities in our government More than one past

Century. *Proceedings of the National Academy of Sciences, U.S.* 103 (2006): 11217–22.

33 Milly, P. C. D., K. A. Dunne, and A. V. Vecchia. Global Pattern of Trends in Streamflow and Water Availability in a Changing Climate. *Nature* 438 (2005): 347–50; Seager, Richard, et al. Model Projections of an Imminent Transition to a More Arid Climate in Southwestern North America. *Science* 316 (2007): 1181–84.

34 Louis R. Iverson and A. M. Prasad, "Predicting Abundance of 80 Tree Species Following Climate Change in the Eastern United States," *Ecological Monographs* 68 (1998): 465-85.

35 Benjamin Zuckerberg, A. M. Woods, and W F. Porter, "Poleward Shifts in Breeding Bird Distribution in New York State," *Global Change Biology* 15 (2009): 1866-83.

36 Jessica Vitale and W. H. Schlesinger, "Historical Analysis of the Spring Arrival of Migratory Birds to Dutchess County, New York—A 123-Year Record," *Northeastern Naturalist* 18 (2011), in press.

37 John W Williams, S. T. Jackson, and J. E. Kutzbach, "Projected Distributions of Novel and Disappearing Climates by 2100 A.D.," *Proceedings of the National Academy cf Sciences, U.S.* 104 (2007): 5738-42; Terry L. Root et al., "Fingerprints of Global Warming on Wild Animals and Plants," Nature 421 (2003): 57-60.

38 Chris D. Thomas et al., "Extinction Risk from Climate Change," *Nature* 427 (2004): 145^18.

39 Timothy M. Lenton et al., "Tipping Elements in the Earth's Climate System," *Proceedings of the National Academy of Sciences, U.S.* 105 (2008): 1786-93.

40 Miriam E. Katz et al., "The Source and Fate of Massive Carbon Input During the Latest Paleocene Thermal Maximum," *Science* 286 (1999): 1531-33.

civilization is thought to have perished as a result of past climate change and drought.[41]

DENIAL

In the face of so much evidence that humans are changing the composition of the atmosphere, that the changes will warm the planet, and that the warming could be costly to human health, economics, and welfare, why do we see so much denial? I believe the reasons are many. First, it is difficult for people to grasp that an odorless, colorless, unreactive gas like CO_2 could be this harmful, especially when it is measured in parts-per-million. No one wakes up feeling like the CO_2 levels are awfully high this morning. Except at exceptional levels, CO_2 does not produce direct impacts on human health, like mercury, ozone, or urban aerosols.

Second, people have difficulty separating the concepts of weather and climate. As Mark Twain put it: "Climate is what you expect, weather is what you get."[42] For instance, even though I know that the average temperature will be warmer in July than in January, the path from January to July will not be a uniform, gradual rise, especially in Millbrook, New York. I could easily imagine a warm spell in February, and at least a few days when it seems like winter has returned in April. When I dress for work each morning, I focus on the weather. It is difficult to perceive a change of a few degrees in mean annual temperature per decade, when the fluctuations in daily values are often much larger than the expected change in climate. Those who deny climate change like to point out that there has been little change in the global temperature since 1998, even though the first decade of this century (2001–2010) is the warmest on record.n fact, the mean decadal temperature has increased every decade since 1960.[43]

Third, even while they accept that the climate has warmed about 1° centigrade during the past century and may warm further in the coming years, many believe that these changes are part of the natural cycle of things over which we have little control. This belief persists in the face of unquestionable evidence that variations in the sun's luminosity have been minor over this same interval and that the human impact on atmospheric aerosols is adequately included in current global climate models and likely to decline in the future as we reduce air pollution emissions.[44] Waiting for the next glacial epoch is no way to respond to rapid climate change induced by human activities.

Fourth, the impacts of global warming are often seen as a future problem—something that will appear slowly—so we can deal with it later. I hope that is the case, but the evidence for past rapid climate change and future tipping points does not give much consolation that this belief is right. Careful studies of the climate change at the end of the last glacial age show short intervals (decades) in which the mean annual temperature in Greenland rose as much as 9° centigrade over a couple of decades—a greater rate of change than we predict for the ongoing global warming.[45] Growing up in the stable conditions of the Holocene, we forget that the Earth's climate can change dramatically in short periods.[46]

41 Peter B. deMenocal, "Cultural Response to Climate Change During the Late Holocene," *Science* 292 (2001): 667-73.

42 Hayhoe, Katharine and Andrew Farley. *A Climate for Change.* New York: Faith Words, 2009.

43 James Hansen, *The Storms of My Grandchildren* (New York: Bloomsbury, 2009).

44 Peter A. Stott, et al., "External Control of 20th-century Temperature by Natural and Anthropogenic Forcings," *Science* 290 (2000): 2133-37.

45 ICendrick Taylor, "Rapid Climate Change," *American Scientist* 87 (1999): 320-27; Jeffrey P Sveringhaus and E. J. Brook, "Abrupt Climate Change at the End of the Last Glacial Period Inferred from Trapped Air in Polar Ice," *Science* 286 (1999): 930-34.

46 Richard B. Alley et al., "Abrupt Climate Change," *Science* 299 (2003): 2005-10.

Finally, the human impacts on the climate are a classic case of the tragedy of the commons.[47] Using fossil fuels, like gasoline, is convenient. We each feel that our small daily contribution to the rise of CO_2 in Earth's atmosphere cannot possibly have much effect. If your neighbor is concerned about climate change, then let him cut back. But, the collective inputs from nearly seven billion of us, over 365 days a year, over decades is likely to result in a doubling of CO_2 in the atmosphere in less than 200 years. Fossil fuels have simultaneously allowed more people than ever before to live at higher levels of nutrition, health, and mobility and with the amenities of light, heat, and clean water, but also allowed the human population to grow beyond the carrying capacity of the planet.[48] How and when we adjust to reality will be an interesting time in history to be alive.

In sum, we know that rising concentrations of CO_2 will raise the temperature of Earth, and we are fairly certain that this will have effects on our food supply, our health, the flooding of major cities, and the persistence of species of plants and animals that share the planet with us. Why are we reluctant to provide planetary stewardship?

WHAT TO DO?

The most direct way for us to avoid the consequences of global warming is to reduce dramatically our use of fossil fuels. When we mine coal and extract oil and gas, we bring carbon from the crust to the surface of the Earth, where it is burned, returning ancient CO_2 to the modern global carbon cycle. Several recent accounts suggest that we must limit the total emissions of carbon dioxide (and other greenhouse gases), not to exceed 1 trillion tons, if we are to limit global warming to less than 2° C, which many believe is the maximum temperature change that will not produce huge impacts to life on Earth.[49] To date, we have emitted about 45% of that target, and we are poised to emit the remainder by 2050 or sooner.[50] Alternatively, if we were to change our current trajectory and replace the existing fossil fuel facilities, at the end of their useful lifetimes, with alternative energy technologies, we could potentially limit atmospheric CO_2 concentrations to 430 ppm and avoid the most costly of the global warming effects.[51] It may not be easy, but the choice is clearly defined.

Fossil fuels are so integral to our current social and economic systems that it is difficult to envision how we will implement the necessary reductions without major societal disruption and increases in the cost of energy. It is tempting to place today's economy over tomorrow's uncertainties. Nevertheless, the costs of inaction and the burden left for future generations are too large to ignore. Those who are impoverished today are likely to suffer the most from changes in climate in the near future.[52] Through no fault of their own, 17 million citizens of Bangladesh may lose their homeland as a result of rising sea level.

Reductions in the use of fossil fuels are best achieved by a combination of increased efficiency and transitions to alternative energy sources (wind,

47 Garrett Hardin, "Tragedy of the Commons," *Science* 162 (1968): 1243-48.

48 Marcus M. Wagernackel et al., "Tracking the Ecological Overshoot of the Human Economy," *Proceedings of the National Academy of Sciences, U.S.* 99 (2002): 9266-71. Humans exceed the carrying capacity of the planet when their emissions of carbon dioxide to the atmosphere exceed the capacity of nature to absorb that CO2 and store it in wood, soil, or ocean sediments.

49 Myles R. Allen et al., "Warming Caused by Cumulative Carbon Emissions Towards the Trillionth Ton," ***Nature*** 458 (2009): 1163-66.

50 *Scitor Corporation,* Progress Towards the Two-degree Cap. Science and Impacts of Climate Change Technical Note *(McLean, Va: Scitor Corporation, 2010).*

51 Steven J. Davis, K. Caldeira, and H. D. Matthews, "Future **CO2** Emissions and Climate Change from Existing Energy Infrastructure," ***Science*** 329 (2010): 1330-35.

52 Miranda, Marie Lynn, et al. The Environmental Justice Dimensions of Climate Change. *Environmental Justice* 4 (2011): 17–25.

solar, geothermal) that are not based on fossil carbon. For the latter, time is short Until recently, the availability of cheap sources of fossil fuels has inhibited incentives to look for alternatives and motivated the existing energy supply chain to work hard to codify its continued dominance of our economy. While elaborate schemes have been proposed to limit the emissions of CO2,I believe that a tax on emissions of fossil carbon will be the simplest, fairest, and most effective way for us to reduce our impacts on climate. The revenue from a tax on carbon could be used to reduce personal income taxes, so that we shift from taxing productivity to taxing resource use. The point is to motivate a lower use of fossil fuels in favor of alternative, carbon-free energy sources.

Various schemes have been proposed to use geoengineering to solve the global warming problem. Many geoengineering schemes, such as the capture of CO_2 from power plants and its deep injection into the ground, are fairly expensive, but straightforward and potentially without many negative impacts. Others, such as seeding the oceans with iron to stimulate the growth and carbon uptake by phytoplankton or adding sulfate aerosols to the stratosphere, where they might reflect incoming solar radiation, are potentially more problematic. In both cases, we know very little about the ancillary impacts of these actions on the biosphere. I believe that it will be much less risky to curb CO_2 emissions at the source than to try to gather up the CO_2 or mitigate its impacts at a later time.

The point is this: time is short, and we must get on with a program of action. We are failing in the planetary stewardship expected of us.

Children Are Likely to Suffer Most from Our Fossil Fuel Addiction

By Frederica P. Perera

CHILDREN AT RISK FROM THE TOXIC AND CARCINOGENIC EFFECTS OF AIR POLLUTION FROM FOSSIL FUEL COMBUSTION

As environmental health scientists, we have seen the direct damage inflicted on children in the United States and worldwide by our society's addiction to fossil fuel. Fine particles, polycyclic aromatic hydrocarbons (PAHs), sulfur and nitrogen oxides, benzene and mercury emitted by coal-burning power plants, and diesel and gasoline-powered vehicles have been variously linked to infant mortality, lower birth weight, deficits in lung function, respiratory symptoms, childhood asthma, developmental disorders, and cancer (Bobak and Leon 1992; Gauderman, er al. 2004; Grandjean and Landrigan 2006; Ha, et al. 2003; Miller, et al. 2004; Perera,et al. 2006b; Sram, et al. 2005; Woodruff, et al. 1997). The many observed adverse effects are not surprising, given the diversity of fossil fuel combustion products (Bernard, et al. 2001); moreover, the same pollutant can exert multiple toxic effects. For example, *in utero* exposure to PAHs as a result of mothers breathing polluted air during pregnancy has been associated with lower birth weight, reduced birth head circumference, preterm birth, and small size for gestational age (Choi, et al. 2006, 2008; Perera, et al. 2003; Srdm, et al. 2005). The same air pollutants have also been linked to developmental delay in U.S. and Chinese children (Perera, et al. 2006b; Tang, et al. 2006). Air pollution is not only an established trigger of asthma in children; but there is evidence that prenatal exposure to PAHs may be an early risk factor for the development of asthma (Miller, et al. 2004). There is also a suggested link between PAHs and cancer (Bocskay, et al. 2005).

These health effects represent a major societal and public health burden. A significant proportion of U.S. children 6–17 years of age are reported to have developmental problems including learning disabilities (11.5%), attention-deficit/hyperactiviry disorder

(8.8%), and behavioral problems (6.3%) (Blanchard, et al. 2006). Asthma affects as many as 25% of children in certain inner-city communities in the United States (Nicholas, et al. 2005), and the prevalence of asthma has increased throughout the developed world over the past 30 years (Beasley, et al. 2003; National Institutes of Health 2001). Approximately 10,400 U.S. children under the age of 15 years were diagnosed with cancer in 2007 (American Cancer Society 2007). Although data are lacking on attributable risk of specific pollutants and relationships between trends in pollution and rates of disease, air pollutants such as lead and mercury are known to contribute to the burden of neurobehavioral disorders (Cheuk and Wong 2006; Lanphear, et al. 2005; Stewart, et al. 2006), and fine particles, ozone, diesel emissions, and PAHs are known or suspected contributors to childhood asthma (Etzel 2003; Strachan 2000).

Insults sustained early in development can have lifelong consequences. Some adult diseases can be launched *in utero* or in childhood. For example, exposure to air pollution in childhood may result in a reduction in lung function and ultimately to increased risk of chronic respiratory illness (Gauderman, et al. 2000; Shea 2003) and greater susceptibility to cardiovascular disease in adulthood (Shea 2003). Similarly, several studies have indicated that genetic damage in the form of DNA adducts or chromosomal abnormalities can be acquired *in utero* as a result of air pollution exposure (Bocskay, et al. 2005; Perera, et al. 2005). Such types of genetic damage have been associated in prospective studies with increased risk of cancer and are considered biomarkers of increased cancer risk (Bonassi, et al. 1995; Hagmar, et al. 1994, 1998; Tang, et al. 2002).

Epigenetic effects of developmental exposure to air pollutants have been less well studied. However, exposure to PAHs has been associated with epigenetic effects experimentally (Santangelo, et al. 2002; Shin, et al. 2005; Vercelli 2004; Wilson and Jones 1983; Wojciechowski and Meehan 1984), and prenatal exposure to PAHs in humans was shown to alter methylation status of a number of genes with known or suspected roles in asthma development (Perera, et al. 2007). In addition, experimental studies, in some cases supported by human evidence, have demonstrated that epigenetic dysregulation resulting from *in utero* environmental exposures can lead to reproductive disorders and adult onset diseases such as cancer (Adam, et al. 1985; Anway, et al. 2005; Dolinoy, et al. 2007; Feinberg and Tycko 2004; Ho, et al. 2006).

A recent report from the American Lung Association noted that, although ozone levels have decreased in the United States since 2002, particle pollution has increased over that period, and coal-fired power plants are responsible for much of the increase in particle pollution in the eastern United States (American Lung Association 2007). The authors estimated that nearly half of the U.S. population (136 million) lives in counties that have unhealthful levels of either ozone (including 25 million children) or particle pollution (including 14 million children).

CHILDREN AT RISK FROM THE EFFECTS OF GLOBAL WARMING DUE TO CARBON DIOXIDE FROM FOSSIL FUEL COMBUSTION

Children are also particularly vulnerable to the effects of global warming (Bunyavanich, et al. 2003; Shea 2003). Anthropogenic carbon dioxide from fossil fuel burning is the most important climate-altering greenhouse gas (GHG) (Intergovernmental Panel on Climate Change [IPCC] 2007). Fossil fuel use has been the primary source of CO_2 concentrations since the preindustrial period (IPCC 2007). In the United States, energy-related activities account for three-quarters of human-generated GHG emissions, mostly in the form of CO_2 emissions from burning fossil fuels (U.S. Environmental Protection Agency [EPA] 2008). More than half the energy-related emissions come from large stationary sources such as power plants, while about a third (in the United States) comes from transportation (U.S. EPA 2008). In contrast, the livestock activities sector is responsible for approximately 18% of total anthropogenic GHG emissions measured in Co_2 equivalent (Steinfeld, et al. 2006). The average temperature of the earth is predicted to rise by 2–4°C (3.1–7.2°F) in this century (IPCC 2007). A temperature increase of this magnitude will bring more heat waves, flooding of coastal areas, famine, and forced migration (Bunyavanich, et al. 2003; Haines and Patz 2004; IPCC 2001; Shea 2003). As a result of these changes, children are more at risk of heat stroke, drowning, malnutrition, diarrhea, allergies, infectious disease such as malaria and encephalitis, and psychological trauma (Bunyavanich, et al. 2003; Haines and Patz 2004; IPCC 2001; Shea 2003).

Global warming also compounds the direct toxicity of fossil-fuel pollutants such as ozone, an important trigger of childhood asthma (Bernard, et al. 2001). Ozone formation from volatile organic chemicals and nitrogen dioxide is accelerated at higher temperatures (Bernard, et al. 2001). Another consequence of a warmer climate is increased plant growth and pollen production, and thus higher levels of natural allergens leading to more allergy and asthma in children (Bunyavanich, et al. 2003).

HEIGHTENED VULNERABILITY OF THE FETUS AND CHILD: POVERTY AND RACISM AS COMPOUNDING FACTORS

The fetus and child are especially susceptible to air pollution and many other environmental contaminants because of their rapid development and immature defense systems; thus, they may be affected by levels of exposure that have no apparent effects in adults (Bearer 1995; Etzel and Balk 1999; Grandjean and Landrigan 2006; Perera, et al. 2002, 2006a). For example, several studies have demonstrated the heightened susceptibility of the fetus to genetic damage in the form of carcinogen-DNA adducts (specifically PAH-DNA adducts) measured in white blood cells (Perera, et al. 2004). Comparison of levels of adducts in paired maternal and umbilical cord white blood cells has found that, despite the estimated 10-fold lower PAH exposure of the fetus compared with the mother, the levels of PAH-DNA adducts were comparable (Perera, et al. 2004). Moreover, although adolescence and old age are also periods of susceptibility to epigenetic reprogramming (Dolinoy, et al. 2007), the epigenome is particularly susceptible to dysregulation from environmental factors during embryogenesis, when the elaborate DNA methylation patterning and chromatin structure required for normal tissue development are established (Dolinoy, et al. 2007). Considering both their inherent biologic susceptibility and their long future lifetimes over which early insults can be manifest as chronic disease or cognitive impairment, the fetus and young child can be considered especially vulnerable and at risk of the multiple, cumulative, and long-term health effects of air pollution.

Poverty and racism compound the susceptibility of the fetus and child. Poor children, especially those in urban areas and developing countries, are most at risk, because the effects of toxic exposures are magnified by inadequate nutrition and psychosocial stress due to poverty or racism (Wood 2003). The shocking inequalities that now exist in children's health within and between countries (Marmot 2006; Marmot, et al. 1991; Waterston and Lenton 2000) will be exacerbated by global warming. The World Health Organization estimates that one-third of the global burden of disease is caused by environmental factors and that children <5 years of age already bear >40% of that burden, even though they represent only 10% of the world's population (Prüss-Ustün and Corvalán 2006). That inequality will only get worse. Finally, perpetuation of fossil fuel burning violates the principle of intergenerational equity that no significant environmental burden should be inherited by future generations (World Commission on Environment and Development 1987).

Although more has been written about the heightened susceptibility of the fetus and child to toxic exposures, children also are likely to be especially susceptible to dehydration and heat stroke, malnutrition, diarrhea, allergies, malaria and encephalitis, and psychological trauma (Bunyavanich, et al. 2003; Committee on Environmental Health 2007; Haines and Patz 2004; Shea 2003, 2007). The American Academy of Pediatrics Committee on Environmental Health noted: "Children represent a particularly vulnerable group that is likely to suffer disproportionately from both direct and indirect adverse health effects of climate change," (American Academy of Pediatrics 2007). For example, infants and young children are a high-risk group for heat-related deaths and hospitalizations, along with the elderly (Anonymous 2002). Children spend more time outdoors, particularly playing sports, which puts them at increased risk of heat stroke and heat exhaustion, as well as ultraviolet radiation—related basal cell carcinoma and malignant melanoma (American Academy of Pediatrics 2000). Because they lack specific immunity, children also experience disproportionately high levels of both morbidity and mortality from malaria; 75% of malaria deaths occur in children <5 years of age. The young are also more susceptible to cerebral malaria, which can lead to lifelong neurologic damage in those who survive (Shea 2007). Once again, health and psychological damage occurring early in life can play out over the lifetime, manifesting as adult chronic disease or impairment.

SOLUTIONS EXIST

The most recent IPCC concluded that significant progress toward stabilizing or reducing global warming emissions can be achieved at relatively low cost using known technologies and practices currently available (IPCC 2007). A recent McKinsey report concluded that the United States could reduce GHG emissions in 2030 by 3.0–4.5 gigatons of CO_2 equivalents using tested approaches and high-potential emerging technologies. The report stated: "Our research suggests that the net cost of achieving these levels of GHG abatement could be quite low on a societal basis" (McKinsey 2007).

These reports indicate that the cost of acting now to make power generation, transport, buildings, and appliances more efficient and to invest in alternative fuels and technologies would be minimal compared with the cost of doing nothing. The benefits of reducing air pollution and global warming would offset a substantial fraction of mitigation costs. These benefits include the individual and societal benefits of health and security extending multigenerationally and the monetary savings from fewer cases of children with asthma, developmental delay, cancer, heat stroke, drowning, malnutrition, diarrhea, allergies, and infectious disease.

CONCLUSION

Summarizing the recent series of articles in *Lancet*, Richard Horton notes that "Policies to improve access to affordable clean energy should be pro-poor," and that "Policies to reduce the progress and impact of climate change should explicitly aim to maximise health benefits and minimise health risks" (Horton 2007). Based on the present review, environmental and energy policies must also explicitly account for *all* the impacts of fossil fuel combustion on child health and development and maximize the health benefits to this susceptible population. Our addiction can be cured. We do not have to leave our children a double legacy of ill health and ecologic disaster.

REFERENCES

Adam, E., R.H. Kaufman, K. Adler-Storth, J.L. Melnick, and G.R. Dreesman. 1985. A prospective study of the association of herpes simplex virus and human papillomavirus infection with cervical neoplasia in women exposed to diethylstilbestrol in utero. *Int J Cancer* 35:19–26.

American Academy of Pediatrics, Committee on Environmental Health. 2007. Global Climate Change and Children's Health. *Pediatrics* 120:1149–1152.

American Academy of Pediatrics, Committee on Sports Medicine and Fitness. 2000. Climatic heat stress and the exercising child and adolescent. *Pediatrics* 106:158–159.

American Cancer Society, 2007, "Cancer Facts and Figures 2007," Atlanta, GA, American Cancer Society, Available: http:// www.cancer.org/downloadsySTT/CAFF2007PWSecured.pdf (accessed 26 December 2007).

American Lung Association. 2007. *State of the Air 2007*. New York:American Lung Association.

(Anonymous.) 2002. Heat-related deaths: four states, July–August 2001, and United States, 1979–1999. *MMWR Morb Mortal Wkly Rep* 51:567–570.

Anway, M.D., A.S. Cupp, M. Uzumcu, and M.K. Skinner. 2005. Epigenetic transgenerational actions of endocrine disruptors and male fertility. *Science* 308:1466–1469.

Bearer, C.F. 1995. Environmental health hazards: how children are different from adults. *Future Child* 5:11–26.

Beasley, R., P. Ellwood, and I. Asher. 2003. International patterns of the prevalence of pediatric asthma: the ISAAC program. *Peadiatr Clin North Am* 50:539–553.

Bernard, S.M., J.M. Samet, A. Grambsch, K.L. Ebiand I. Romieu. 2001. The potential impacts of climate variability and change on air pollution-related health effects in the United States. *Environ Health Perspect* 109(suppl 2):199–209.

Blanchard, L., M. Gurka, and J. Blackman. 2006. Emotional, developmental, and behavioral health of American children and their families: a report from the 2003 National Survey of Children's Health Emotional, Developmental, and Behavioral Health of American Children. *Pediatrics* 117:e1202–e1212.

Bobak, M. and D.A. Leon. 1992. Air pollution and infant mortality in the Czech Republic, 1986–88. *Lancet* 340:1010–1014.

Bocskay, K.A., D. Tang, M.A. Grjuela, L. Xinhua, D.P. Warburton, and F.P. Perera. 2005. Chromosomal aberrations in cord blood are associated with prenatal exposure to carcinogenic polycyclic aromatic hydrocarbons. *Cancer Epidemiol Biomarker Prev* 14:506–511.

Bonassi, S., A. Abbondandolo, L. Camurri, L. Dal Pra, M De Ferrari, and F. Degrassi, et al. 1995. Are chromosome aberrations in circulating lymphocytes predictive of future cancer onset in humans? Preliminary results of an Italian cohort study. *Cancer Genet Cytogenet* 79:133–135,

Bunyavanich, S., C.P. Landrigan, A.J. McMichael, and P.R. Epstein. 2003. The impact of climate change on child health. *Ambul Pediatr* 3:44–52.

Cheuk, D. and V. Wong. 2006. Attention-deficit hyperactivity disorder and blood mercury level: a case-control study in Chinese children. *Neuropediatrics* 37:234–240.

Choi, H., W. Jedrychowski, J. Spengler, D.E. Camann, R.M. Whyatt, and Rauh V, et al. 2006. International studies of prenatal exposure to polycyclic aromatic hydrocarbons and fetal growth. *Environ Health Perspect* 114:1744–1750.

Choi, H., V. Rauh, R. Garfinkel, Y. Tu, and F.P. Perera. 2008. Prenatal exposure to airborne polycyclic aromatic hydrocarbons and risk of intrauterine growth restriction. *Environ Health Perspect* 116:658–665.

Dolinoy, D.C., J.R. Weidman, and R.L. Jirtle. 2007. Epigenetic gene regulation: linking early developmental environment to adult disease. *Heprod Toxicol* 23:297–307.

Etzel, R.A. 2003. How environmental exposures influence the development and exacerbation of asthma. *Pediatrics* 112:233–239.

Etzel, R.A. and S.J. Balk. 1999. *Handbook of Pediatric Environmental Health*. Elk Grove Village, IL:American Academy of Pediatrics.

Feinberg, A.P. and B. Tycko. 2004. The history of cancer epigenetics. *Nat Rev Cancer* 4:143–153.

Gauderman, W.J., E. Avol, F. Gilliland, H. Vora, D. Thomas, and K. Berhane, et al. 2004. The effect of air pollution on lung development from 10 to 18 years of age. *N Engl J Med* 351:1057–1067.

Gauderman, W.J., R. McCorinell, F. Gilliland, S. London, D. Thomas, and E. Avol, et al. 2000. Association between air pollution and lung function growth in southern California children. *Am J Respir Crit Care Med* 162:1383–1390.

Grandjean, P. and P.J. Landrigan. 2006. Developmental neurotoxicity of industrial chemicals. *Lancet* 368:2167–2178.

Ha, E.H., J.T. Lee, H. Kim, Y.C. Hong, B.E. Lee, and H.S. Park, et al. 2003. Infant susceptibility of mortality to air pollution in Seoul, South Korea. *Pediatrics* 111:284–290.

Hagmar, L., S. Bonassi, U. Stromberg, A. Brogge, L.E. Knudsen, and H. Norppa, etal. 1998. Chromosomal aberrations in lymphocytes predict human cancer: a report from the European Study Group on Cytogenetic Biomarkers and Health (EOCH). *Cancer Res* 58:4117–4121.

Hagmar, L, A. Brogge, I.L. Hansteen, S. Heim, B. Hogstedt, and L. Knudsen, et al. 1994. Cancer risk in humans predicted by increased levels of chromosomal aberrations in lymphocytes: Nordic Study group on the health risk of chromosome damage. *Cancer Res* 54:2919–2922.

Haines, A. and J.A. Patz. 2004. Health effects of climate change. *JAMA* 291:99–103.

Ho, S.M., W.Y. Tang, J. Belmonte de Frausto, and G.S. Prins. 2006. Developmental exposure to estradiol and bisphenol A increases susceptibility to prostate carcinogenesis and epigenetically regulates phosphodiesterase type 4 variant 4. *Cancer Res* 66:5624--5632.

Horton, R. 2007. Righting the balance: energy for health. Lancet 370:921-921.

IPCC. 2001. *IPCC Third Assessment Report.* Geneva:Intergovern mental Panel on Climate Change, World Meteorological Organization.

IPCC. 2007. *Fourth Assessment Report.* Geneva:Intergovernmental Panel on Climate Change, World Meteorological Organization.

Lanphear, B.P., R. Hornung, J. Khoury, K. Yolton, P. Baghurst, and D.C. Bellinger, et al. 2005. Low-level environmental lead exposure and children's intellectual function: an international pooled analysis. *Environ Health Perspect* 113:894–899.

Marmot, M. 2006. Harveian Oration: Health in an unequal world. *Lancet* 368:2081.

Marmot, M.G., G.D. Smith, S. Stansfeld, C. Patel, F. North, and J. Head, et al. 1991. Health inequalities among British civil servants: The Whitehall II Study. *Lancet* 337:1387.

McKinsey C.A., 2007, "Reducing U.S. Greenhouse Gas Emissions: How much at what cost?" *Greenhouse Gas Report*, Available: http://www.mckinsey.com/clientservice/ccsi/greenhousegas.asp (accessed 16 June 2008).

Miller, R.L., R. Garfinkel, M. Horton, D. Camann, F.P. Perera, and R.M. Whyatt, et al. 2004. Polycyclic aromatic hydrocarbons, environmental tobacco smoke, and respiratory symptoms in an inner-city birth cohort. *Chest* 126:1071–1078.

National Institutes of Health, 2001, "NHLBI Reports New Asthma Data for World Asthma Day 2001: Asthma Still a Problem but More Groups Fighting It" (Press Release), Bethesda, MD:National Institutes of Health. Available: http://www. nhlbi.nih.gov/new/press/01-Q5-03.htm (accessed 31 March 2001).

Nicholas, S.W., B. Jean-Louis, B. Ortiz, M. Northridge, K. Shoemaker, and R. Vaughan, et al. 2005. Addressing the childhood asthma crisis in Harlem: the Harlem Children's Zone Asthma Initiative. *Am J Public Health* 95:245–249.

Perera, F., D. Tang, R. Whyatt, S.A. Lederman, and W. Jedrychowski. 2005. DNA damage from polycyclic aromatic hydrocarbons measured by benzopyrene-DNA adducts in mothers and newborns from northern Manhattan, the World Trade Center area, Poland, and China. *Cancer Epidemiol Biomarkers Prev* 14:709–714.

Perera, F., S. Viswanathan, R. Whyatt, D. Tang, R.L. Miller, and V. Rauh. 2006a. Children's environmental health research—highlights from the Columbia Center for Children's Environmental Health. *Ann NY Acad Sci* 1076:15–28.

Perera, F.P., S.M. Illman, P.L. Kinney, R.M. Whyatt, E.A. Kelvin, and P. Shepard, et al. 2002. The challenge of preventing environmentally related disease in young children: community-based research in New York City. *Environ Health Perspect* 110:197–204.

Perera F.P., V. Rauh, W.Y. Tsai, P. Kinney, D. Camann, and D. Barr, et al. Effects of transplacental exposure to environmental pollutants on birth outcomes in a multi-ethnic population. *Environ Health Perspect* 111:201–205.

Perera F.P., V. Rauh, R.M. Whyatt, W.Y. Tsai, D. Tang, and D. Diaz, et al. 2006b. Effect of prenatal exposure to airborne polycyclic aromatic hydrocarbons on neurodevelopment in the first 3 years of life among inner-city children. *Environ Health Perspect* 114:1287–1292.

Perera F.P., D. Tang, R.M. Whyatt, S.A. Lederman, and W. Jedrychowski. Comparison of PAH-DNA adducts in four populations of mothers and newborns in the U.S., Poland and China. In: *AACR 94th Annual Meeting* 27–31 March 2004, Orlando, FL American Association for Cancer Research, 454.

Perera, F.P., W. Tang, J. Herbstman, S.C.Edwards, D. Tang, and S.M. Ho. 2007. Prenatal exposure to airborne polycyclic aromatic hydrocarbons and alterations in DNA methylation in cord blood. In: *19th Annual Conference of the International Society for Environmental Epidemiology (USEE)* 5–9 September 2007. Mexico City, Mexico International Society for Environmental Epidemiology, 330.

Prüss-Ustün, A. and C. Corvalán C. 2006. *Preventing Disease through Healthy Environments: Towards an Estimate of the Environmental Burden of Disease*. Geneva:World Health Organization.

Santangelo, S., D.J. Cousins, N.E. Winkelmann, and D.Z. Staynov. 2002, DHA methylation changes at human TH2 cytokine genes coincide with DNase I hypersensitive site formation during CD4(+) T cell differentiation. *J Immunol* 169:1893–1903.

Shea, K.M. 2003. Global environmental change and children's health: understanding the challenges and finding solutions. *J Pediatr* 143:149–154.

Shea, K.M. 2007. Global climate change and children's health. *Pediatrics* 120:E1359–E1367.

Shin, H.J., H.Y. Park, S.J. Jeong, H.W. Park, Y.K. Kim, and S.H. Cho, et al. STAT4 expression in human T cells is regulated by DNA methylation but not by promoter polymorphism. *J Immunol* 175:7143–7150.

Sram, R.J., B. Binkova, J. Dejmek, and M. Bobak. 2005. Ambient air pollution and pregnancy outcomes: a review of the literature. *Environ Health Perspect* 113:375–382.

Steinfeld, H., P. Gerber, T. Wassenaar, V. Castel, M. Rosales, and C. de Haan. 2006. *Livestock's Long Shadow: Environmental Issues and Options*. Rome, Italy:Food and Agriculture Organization of the United Nations.

Stewart, P.W., D.M. Sargent, J. Reihman, B.B. Gump, Lonky E, and T. Darvill, et al. 2006. Response inhibition during differential reinforcement of low rates (DRL) schedules may be sensitive to low-level polychlorinated biphenyl, methylmercury, and lead exposure in children. *Environ Health Perspect* 114:1923–1929.

Strachan, D.P. 2000. The role of environmental factors in asthma. *Br Med Bull* 56:865–882.

Tang, D., S. Cho, A. Rundle, S. Chen, D. Phillips, and J. Zhou, et al. 2002. Polymorphisms in the DNA repair enzyme XPD are associated with increased levels of PAH-DNA adducts in a case-control study of breast cancer. *Breast Cancer Res Treat* 75:159–166.

Tang, D., T.Y. Li, J.J. Liu, Y.H. Chen, L. Qu, and F.P. Perera. 2006. PAH-DNA Adducts in cord blood and fetal and child development in a Chinese cohort. *Environ Health Perspect* 114:1297–1300.

U.S. EPA, 2008, "Basic Information," Washington, DC:U.S. Environmental Protection Agency.

Available: http://wwvv.epa. gov/climatechange/basicinfo.html (accessed 8 April 2008).

Vercelli, D. 2004. Genetics, epigenetics, and the environment: switching, buffering, releasing. J *Allergy Clin Immunol* 113:381–386.

Waterston, T. and S. Lenton. 2000, Public health: sustainable development, human induced global climate change, and the health of children. *Arch Dis Child* 82:95–97.

Wilkinson, P., K. Smith, S. Beever, C. Tonne, and T. Oreszczyn. 2007, Energy, energy efficiency, and the built environment. *Lancet* 370:1175–1187.

Wilson, V.L. and P.A. Jones. 1983. Inhibition of DNA methylation by chemical carcinogens in vitro. *Cell* 32:239–246.

Wojciechowsk,i M. and T. Meehan. 1984. Inhibition of DNA methyltransferases in vitro by benzo(a) pyrene diol epoxide-modified substrates. *J Biol Chem* 259:9711–9716.

Wood, D. 2003. Effect of child and family poverty on child health in the United States. *Pediatrics* 112:707–711.

Woodruff, T.J., J. Grille, and K.C. Schoendorf. 1997. The relationship between selected causes of postneonatal infant mortality and particulate air pollution in the United States. *Environ Health Perspect* 105:608–612.

World Commission on Environment and Development. 1987. *Our Common Future (Brundtland Report).* Oxford, UK:oxford University Press.

Other Unconventional Fuels

By James T. Bartis

Presently, petroleum demand in the United States stands at between 20 million and 21 million bpd. Imports meet 60 percent of this demand: ten million bpd of imported crude oil and 2.5 million bpd of imported petroleum products. Analyses of U.S. energy requirements for the next 20 to 25 years generally show slightly growing demand for liquid fuels and continued high dependence on imported petroleum. Looking at global trends, we anticipate that rapidly growing energy demand from large developing nations, such as China and India, will raise global petroleum consumption by 20 to 50 percent beyond current levels by 2030.[1] These trends strongly suggest that, without significant additional changes in the energy policies of the United States, we should anticipate petroleum imports in the next few decades to range between 10 million and 12 million bpd (EIA, 2008c). Moreover, unless nations with large or rapidly growing economies make significant changes to their energy policies, we should anticipate growing world demand for petroleum, a long-term trend toward higher prices, and increased dependence on supplies from the Middle East.

In this chapter, we review other approaches for using domestic resources to produce transportation fuels that can substitute for conventional petroleum and lessen U.S. dependence on imports. Our focus is on understanding whether and to what extent other unconventional-fuel options are available to produce liquid fuels that can substitute for conventional petroleum-derived products.[2]

1 Current world demand for liquid fuels is about 85 million bpd (including crude and natural-gas plant liquids) (EIA, 2008b, Table 1.7). Examples of projected 2030 petroleum demand are the International Energy Agency's reference-case projection—116 million bpd (IEA, 2007)—and EIA's 2008 reference- and high-oil-price-case projections, 113 million and 98 million bpd (EIA, 2008c, Table C6).

2 Heavy government promotion of certain of these fuels has raised important public policy issues, but we do not address these issues here.

James T. Bartis, Frank Camm, and David S. Ortiz, "Other Unconventional Fuels," *Producing Liquid Fuels from Coal*, pp. 49-57

We excluded certain domestic petroleum resources that might qualify as unconventional but were beyond the scope of our study. These include heavy oils and oil deposits that require advanced enhanced oil recovery methods. We also excluded consideration of long-range, high-risk concepts (such as using hydrogen as a transportation fuel and all-electric cars) and advanced biomass concepts (such as genetically engineered algae). While research on these approaches may be worthy of support, insufficient information is available to speculate on what, if any, contribution to the U.S. fuel supply they offer for the next few decades.

The more we learned about the benefits of reducing dependence on conventional sources of petroleum, the more we appreciated the importance of improved energy conservation and energy efficiency as ways to accomplish this goal. Understanding the opportunities in these areas, as well as exploring such interesting concepts as plug-in hybrid vehicles, is extremely important but beyond the scope of our study.

COMMERCIALLY READY UNCONVENTIONAL FUELS

In addition to FT CTL and MTG CTL, two other approaches are commercially available for producing liquid fuels that can substitute for petroleum: fermenting food crops to produce alcohols and deriving fuels from renewable oils, such as soybean oil.

Food-Crop Alcohols

In the United States, fuel ethanol is fermented primarily from corn and is used as a blendstock for gasoline. Production in 2007 averaged 423,000 bpd (EIA, 2008a), which is the energy equivalent of about 280,000 bpd of gasoline. This production is driven by federal requirements to include oxygenates in automotive gasoline and by a federal subsidy of $0.51 per gallon of ethanol, which corresponds to $0.76 for the amount of energy in a gallon of gasoline or about $35 for the amount of energy in one barrel of crude oil.

The Energy Independence and Security Act of 2007 (P.L. 110–140) extended and increased the renewable-fuel standard originally set in the Energy Policy Act of 2005 (P.L. 109–58). The law now requires a minimum of nine billion gallons of ethanol in transportation fuels used in the United States during 2008. This is the energy equivalent of roughly 400,000 bpd of gasoline. For ethanol fermented from corn, this mandate rises to 13.2 billion gallons per year by 2012, the energy equivalent of about 590,000 bpd of gasoline, which is about 4 percent of the projected demand for transportation fuels in that time frame (EIA, 2008c).

Market demands for corn as food will limit use of corn grain to produce ethanol. For example, based on U.S. Department of Agriculture (USDA) projections (Inter-agency Agricultural Projections Committee, 2007), we estimate that about one-third of the U.S. corn harvest in 2012 will be devoted to producing ethanol. This would more than double the 14 percent used in the 2005–2006 harvest year and would take an area about 85 percent of the size of Illinois to grow enough corn for this quantity of ethanol. For 2030, EIA's projection for corn-based fuel alcohol production is 660,000 bpd (gasoline energy equivalent),[3] still a small fraction of the total projected demand for transportation fuels (EIA, 2008c).

In 2006, DuPont and BP announced that they were forming a partnership to produce and market butanol (see DuPont, 2007). This effort centers on a demonstration facility in the United Kingdom that produces butanol from sugar beets. The plant is scheduled to begin operations in early 2009. In late

3 This projection is for the EIA reference case and assumes continued federal subsidies.

2007, the DuPont/BP partnership plans to deliver "market development quantities" of biobutanol to the United Kingdom (DuPont, 2007).

Butanol is far superior to ethanol as an automotive fuel. It has an energy density much closer to that of gasoline. It offers superior environmental performance because its vapor pressure is lower than ethanol's. It can be blended with gasoline at refineries, thereby eliminating the need for the special handling required by ethanol. It does not require modifications to automobile engines, even when used in high-butanol blends. These factors give butanol a significant competitive edge over ethanol. If butanol fermentation technology progresses to the stage at which it is competitive with etha- nol fermentation, demand for alcohol fuels produced from food crops should increase because of butanol's superior properties. Nonetheless, the lack of suitable acreage for additional crop production to satisfy demands for both food and fuel alcohol will likely limit food-based alcohol fuel production (ethanol and butanol) in the United States to less than one million bpd, gasoline energy equivalent.

Fuels Derived from Renewable Oils

To date, vegetable-oil-based fuels have seen limited application in the United States. The dominant U.S. feedstock for the production of biodiesel fuel is soybean oil. Other feedstocks for biodiesel include sunflower oil, rapeseed (canola) oil, beef tallow and other animal fats, and waste cooking grease. In 2007, the production and consumption of biodiesel in the United States averaged 32,000 bpd (EIA, 2008c, Table 10.3), which is about 0.02 percent of U.S. petroleum demand for transportation.

The renewable-fuel standard established by the Energy Independence and Security Act of 2007 (P.L. 110–140) mandates that, by 2012, biomass-based diesel use will average about 65,000 bpd, about twice the estimated biodiesel use in 2007. The costs and market impacts of this mandate, including impacts on food costs, remain highly uncertain.

The resource base of soybeans as a feedstock for biofuel production is limited. Soybeans are one of the major crops grown in the United States. Dedicating the equivalent of the entire soybean crop to biodiesel production would yield 296,000 bpd of fuel.[4] This small amount of production would require the cultivation of a very large amount of land, an area just slightly smaller than the states of Illinois and Iowa combined. Because soybeans are often planted in rotation with corn, increased corn plantings are expected to reduce the acreage available for soybeans. For these reasons, the World Agricultural Outlook Board of USDA estimated, prior to the passage of the Energy Independence and Security Act (P.L. 110–140), that biodiesel production will level off at approximately 700 million gallons per year (46,000 bpd), using 23 percent of soybean-oil production but displacing only 1.4 percent of the projected demand for diesel fuel (Interagency Agricultural Projections Committee, 2007; EIA, 2008c). In the foreseeable future, vegetable oils will likely provide no more than a few tens of thousands of barrels of fuel per day, considering the importance of soybeans to human food supplies, production costs, environmental impacts (Hill, et al., 2006), and fuel suitability.

A small amount of additional production can be achieved from animal fats. For example, in 2007, Tyson Foods announced plans for two ventures that, if fully exploited, might eventually produce about 16,000 bpd of diesel fuel.[5] Based on these state-

4 This result is based on the following data: Total 2005 soybean production was 3.06 billion bushels grown on 71.4 million acres. Biodiesel-fuel yield from soybeans is estimated at 3.8 barrels per 100 bushels. Average soybean yield in 2005–2006 was 43.0 bushels per acre (Interagency Agricultural Projections Committee, 2007).

5 This estimate is based on 175 million gallons per year through an alliance with ConocoPhillips (Tyson Foods, 2007a) and 75 million gallons per year from the initial facility built under a joint venture with Syntroleum Corporation (Tyson Foods, 2007b).

ments, it is reasonable to assume that a few tens of thousands of barrels of fuel per day might eventually be produced by processing animal fats, depending on future prices for diesel fuel and animal fat and continuing government subsidies and support.

EMERGING UNCONVENTIONAL FUELS

Oil Shale

The largest and richest oil shale deposits in the world are located in the Green River Formation, which covers portions of Colorado, Utah, and Wyoming. Total potentially recoverable resources are estimated at roughly 800 billion barrels, which is more than triple the total petroleum reserves of Saudi Arabia. This is also significantly more than the amount of liquid fuel that could be produced from the proven recoverable coal reserves of the United States, especially considering the role of coal in generating electric power.

In a recent publication (Bartis, et al., 2005), RAND researchers reviewed the prospects and policy issues associated with oil shale development in the United States. That book concluded that the prospects are uncertain. They depend primarily on the successful development of in-situ methods that offer improved economics and reduced adverse environmental impacts, as compared to mining the oil shale and retorting it above ground. Since publication of that book in 2005, important technical progress has taken place. A number of highly qualified firms have either publicly announced or indicated to us their interest in developing oil shale. In December 2006, the Bureau of Land Management (BLM) announced that it was issuing to three firms—Shell, Chevron, and EGL Resources—small lease tracts in Colorado for the purposes of conducting research, development, and demonstration (RD&D) of in-situ methods for producing fuels from oil shale. In April 2007, BLM announced an additional RD&D lease in Utah to a fourth firm—Oil Shale Exploration Company—that is interested in developing a method involving mining and surface retorting. Other firms have expressed interest in participating in the BLM program if a second round of RD&D leases becomes available.

Based on our knowledge of firms interested in oil shale development, none—with the possible exception of Shell—will be prepared to make a financial commitment to a pioneer commercial-scale facility for at least five and, in some cases, as many as ten years. Accordingly, we do not anticipate that commercial production will exceed 100,000 to 200,000 bpd for at least the next 15 years.

It is possible that oil shale might ultimately make a significant contribution to the U.S. transportation-fuel supply. However, a production level of one million bpd is probably more than 20 years in the future, and three million bpd is probably more than 30 years in the future (Bartis, et al., 2005).

U.S. Tar Sands

The most recent information on U.S. tar sand resources is contained in USGS reports that summarize resource estimates made in the 1970s, 1980s, and 1990s (Meyer, Attanasi, and Freeman, 2007; USGS, 2006) and a report by the University of Utah prepared for the U.S. Department of Energy Office of Fossil Energy and the National Energy Technology Laboratory (Utah Heavy Oil Program, 2007). Major U.S. bitumen deposits (more than 100 million barrels) can be found in Alabama, Alaska, California, Kentucky, New Mexico, Oklahoma, Texas, Utah, and Wyoming.[6] Measured resources of natural bitumen in place are 36 billion barrels (USGS, 2006). There are an additional 18 billion to 40 billion barrels of

6 Bitumen is the very viscous organic liquid found in tar sands.

speculative resources in place.[7] The largest and best-defined tar sand deposits in the United States are in Utah, which holds about 20 billion barrels (measured and speculative) in large deposits. The estimates of tar sand resources in place do not address how much of that resource may be recoverable and at what cost.

Exploitation of U.S. tar sands would conceivably use a combination of surface mining and in-situ extraction methods. Such methods are being applied to extract bitumen from the more abundant and richer Canadian tar sands (often referred to as oil sands) located in Alberta. However, U.S. tar sands are hydrocarbon wetted and often contained in solid rock, whereas the more abundant and richer Canadian tar sands located in Alberta are water wetted and contained in loose sand (Task Force on Strategic Unconventional Fuels, 2007).[8] For these and other reasons (Utah Heavy Oil Program, 2007), extraction techniques for U.S. tar sands are likely to be different and more costly than those used in Alberta.

Most of the interest in developing U.S. tar sand resources appears to emanate from Washington, D.C. For example, Section 369 of the Energy Policy Act of 2005 (P.L. 109–58) directs the U.S. Department of Energy to make public lands available for lease to promote R&D and to prepare a programmatic environmental-impact statement regarding the production of liquid fuels from oil shale and tar sands. Because of the extensive oil-sand development occurring in Alberta, numerous firms have experience in producing liquid fuels from tar sands. Of this group, or among firms of similar capabilities, none has announced interest in U.S. tar sand development.[9]

While small-scale development of U.S. tar sand resources might occur for other purposes, such as materials for road surfacing (Foy, 2006), we conclude that, in the foreseeable future, U.S. tar sand deposits are not likely to be a viable source for the production of significant quantities of liquid fuels.

Biomass from Nonfood Crops

Two approaches are potentially available for producing liquid fuels from renewable bio- mass resources other than food crops: alcohol production via cellulosic fermentation and gasification followed by FT or MTG synthesis. With either approach, between 0.6 and 0.7 tons of biomass should yield one barrel of liquid fuels with roughly the energy equivalent of one barrel of crude oil. By expanding the biomass resource base beyond food crops, these two approaches substantially increase the potential liquid-fuel production level beyond one million bpd, the upper bound for fuels derived from food crops.

Appropriate biomass resources for cellulosic fermentation or gasification include dedicated energy crops that would be cultivated specifically for the purpose of producing liquid fuels. Examples include switchgrass and poplar bred to maximize mass yield. Recent attention has been directed at the environmental benefits of cultivating mixed prairie grasses, which are mixtures of indigenous grasses that require few agricultural inputs (Tilman, Hill, and Lehman, 2006). The second broad category of biomass resources is residues. Included in this category are agricultural residues, such as corn stover, wheat

7 Speculative resource estimates are highly uncertain because they are typically the result of extrapolating observations of surface and near-surface resources to resources that might be located deeper underground.

8 In hydrocarbon-wetted sands, the bitumen is in direct contact with the sand grains. In water-wetted sands, a thin layer of water separates the bitumen from the sand grains. Steam-based methods, such as those employed in Alberta, are generally less effective when applied to hydrocarbon-wetted sands.

9 A recent report promoting the development of oil shale and tar sands identified only six firms interested in tar sands, none of which had experience in building and operating commercial processing systems similar to those that would be associated with tar-sand development (DOE, 2007b).

straw, and manure; limbs and other tree parts left over from logging operations; forest thinnings; and municipal wastes, including municipal wood waste and yard trimmings, municipal solid wastes, and sewage sludge.

Several recent analyses have investigated the potential biomass resource that exists in the United States. A study by Oak Ridge National Laboratory (Perlack, et al., 2005) estimated that approximately 330 million dry tons per year of forest residue and thinnings, urban wood waste, and agricultural residue are available annually, assuming no changes in land use or increases in agricultural yields. Assuming modest changes in agricultural yields and cultivation of energy crops on idle land, the National Renewable Energy Laboratory (Milbrandt, 2005) performed a similar analysis and estimated that 360 million dry tons per year of such biomass resources exist. The Oak Ridge estimate increases to about 1.4 billion tons annually under the following conditions: more intensive collection efforts, significantly improved agricultural yields, land-use changes, and inclusion of perennial bioenergy crops.[10] Among estimates supported by analytic methodology, the Oak Ridge estimate forecasts the highest potential tonnage of biomass resources for energy use. This estimate is based solely on land availability and does not consider the cost of collecting and delivering biomass to sites where it would be used to produce electric power, useful chemicals, or liquid fuels. Assuming that half of this biomass can be economically collected and delivered to liquid fuel-production facilities, biomass resources would be able to support a fuel-production level of roughly three million bpd.

As compared to fossil energy resources, biomass must be collected from large areas. Annual yields of biomass crops range from two to ten dry tons per acre. An FT BTL plant producing 4,000 barrels of liquids per day would require delivery of between 800,000 and 900,000 dry tons per year. That level of demand would require dedicated cultivation of between 100,000 and 400,000 acres. Taking into account the density of land available for biomass production, along with field and storage losses, the land area over which a single plant would be supported could exceed one million acres, which is roughly the area contained in a circle with a radius of 25 miles.

Alcohol Fuels from Cellulosic Feedstocks. The sugars and starches that are often the focus of food-crop cultivation are generally a relatively small fraction of the weight of a plant. Most of the material in plants is cellulose, hemicellulose, or lignin. None of these substances is amenable to the fermentation procedures that are commonly used to produce ethanol. To overcome this problem, researchers have devised approaches that break down the cellulose and hemicellulose into sugars. These sugars would then be converted to ethanol, and possibly butanol, through fermentation.

Most of the federal effort to derive liquids from cellulosic feedstocks centers on producing ethanol via fermentation. Primary emphasis is on using enzymes to break down the cellulosic materials into the simple sugar glucose. In February 2007, the U.S. Department of Energy announced federal funding of up to $385 million for commercial demonstration of six initial cellulosic-ethanol plants (DOE, 2007a), five of which involve fermentation.[11] The six demonstration facilities are to be constructed between 2007 and 2011, and their target capacities range from 440 to 2,040 bpd of ethanol. The processes underlying the proposed facilities have been demonstrated at the pilot scale (i.e., on the order of 1/100th of the proposed capacities).

The cost of producing ethanol fuels from these initial plants will be very expensive, but this should be expected. After all, they are demonstration facilities operating at scales well beyond prior experience.

10 Under this estimate, perennial bioenergy crops would be grown on about 55 million acres of cropland, idle cropland, and cropland pasture.

11 One of the selected projects involves gasification of biomass and conversion of the resulting synthesis gas to mixed alcohols.

At present, insufficient information is available to allow us to predict the costs of producing ethanol from commercial facilities that might be built using the information obtained during the ongoing demonstrations. Several years of operating experience of these and other facilities should eventually resolve uncertainties regarding the technical, economic, and environmental viability of the various conversion processes.

Biomass Gasification and Liquids Synthesis. A second approach for converting cellulosic biomass into liquid fuels is to gasify the biomass, produce synthesis gas, and, after cleaning, convert that gas to liquids in an FT reactor (Boerrigter and van der Drift, 2004; Boerrigter and Zwart, 2005) or via the MTG process. Most of the steps in this approach are identical to those that would be encountered in a CTL plant or a natural-gas-to-liquids plant. For this reason, the FT and MTG BTL approaches are of much lower technical risk than all approaches involving alcohol production from cellulosic materials via fermentation. We have not found convincing evidence that either process—gasification and liquids synthesis or fermentation—offers economic advantages over the other.

For the FT and MTG BTL approaches, the principal development need is for improved technology for handling, preparing, and feeding biomass into a pressurized gasifier.

The use of a combination of biomass and coal may be an approach that significantly lowers the costs and investment risks involved in building plants that use biomass to produce liquid transportation fuels. First, biomass yields are very sensitive to weather, especially rainfall. Cofiring a gasifier with both biomass and coal offers a means of smoothing out this inherent fluctuation in biomass availability. Second, plants that receive only biomass are limited in size. Otherwise, the land area supporting the plant becomes very large, delivery of biomass to the plant becomes expensive, and the amount of truck traffic required to deliver the biomass to the plant may be viewed as unacceptable. Cofiring with both biomass and coal offers a way to build larger plants that can capture economies of scale while avoiding the problems of very large biomass-collection areas. Third, combining coal-biomass cofiring with capture and sequestration of carbon dioxide offers a way to reduce greenhouse-gas emissions to well below those associated with conventional-petroleum fuels, as discussed in Chapter Three.

Missing from the federal R&D portfolio are any research efforts to establish the commercial viability of a few techniques for the combined use of coal and biomass. The most pressing near-term research requirement centers on developing an integrated fuel-processing and gasification system capable of handling both biomass and coal, as discussed in Chapter Three.

SUMMARY

Looking ahead to the next 30 to 50 years, there is no single unconventional-fuel technology capable of meeting U.S. demand for liquid transportation fuels and continued dependence on imported crude oil. At present, FT and MTG are two of only three unconventional-fuel technologies that are commercially ready and capable of producing significant amounts of fuel. The other is food-crop fermentation to produce ethanol. However, the ultimate potential of ethanol derived from food crops is less than one million bpd.

Biomass and oil shale, once commercial, might together be capable of supplying up to six million bpd, but building to this level of production would likely require 30 years. Coal offers the possibility of producing almost three million bpd by 2030. Most importantly, FT and MTG CTL technologies are ready for initial commercial applications. Moreover, FT and

MTG CTL may also enable the broader exploitation of nonfood biomass resources.

What the United States does at home will have implications abroad. Successful development of an unconventional-fuel industry based on coal, biomass, and oil shale should promote similar developments in other nations that also have appreciable amounts of these resources. Considering both domestic and international opportunities, it is not unreasonable to assume that unconventional-fuel development, if pursued, could reduce world demand for crude oil by 15 million to 20 million bpd from what it would otherwise be after 2030.

Section IX

Water Resources

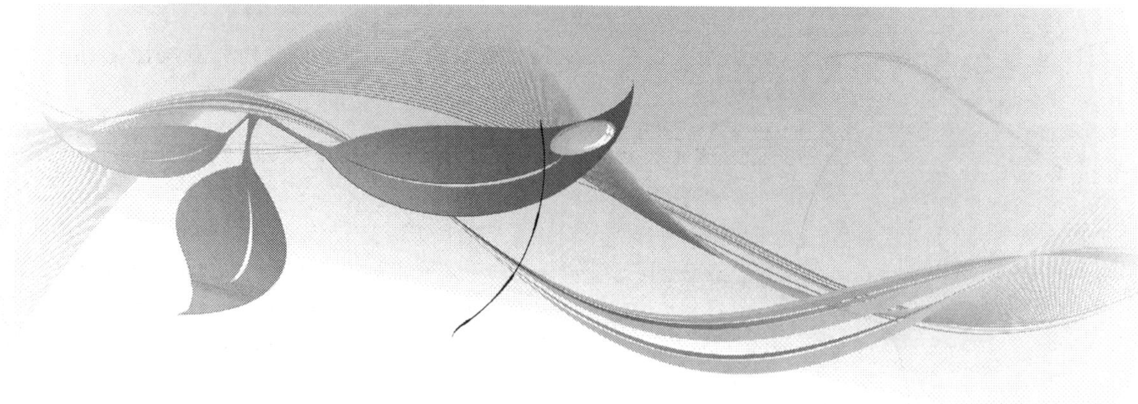

By Anne Marie Zimeri

Water is a unique compound with several properties that make it well able to be the medium for most biological reactions in our bodies. It is also so plentiful on the earth that it is influential in maintaining the earth's temperature. Water is cycled around the earth in a process called the hydrologic cycle. This cycle includes condensation, transpiration, precipitation, and runoff (Figure 9.1).

If we look closely at the quantity of water in each location in the hydrologic cycle, we see that the bulk of the earth's water resides in the oceans. Though this water is important for temperature maintenance, and certainly for marine ecosystems, salt water cannot be used for most of our day-to-day needs. Freshwater must be used for drinking, irrigating agriculture, and in industrial processes. Of the earth's water, less than 3% is freshwater, most of which is tied up in glaciers. The remainder of the freshwater is in surface water, groundwater, and the atmosphere.

Human health is closely linked to access to clean water. Water that is free of toxins and harmful pathogenic organisms is essential for maintaining public health. Unfortunately, it has been estimated that more than 1.7 billion of the 7 billion people on the planet lack access to clean drinking water. Not only is drinking water not safe to drink in many areas of the world; it is also in short supply. Global demand for water has been increasing at about 2.3% annually, which means that in just a little more than two decades, the demand for freshwater will double. Addressing water scarcity and safety issues will require worldwide cooperation as many water sources cross national, state, and county boundaries. For example, Lake Chad, in northern Africa, is surrounded and used by 30 million people in four different countries.

Like other resources discussed in this text, water use must be transitioned to sustainable water, i.e., water usage that allows for current need to be met while allowing its maintenance as a resource for future generations. Effective water management is being incorporated into many national, state, and city plans. Many cities now employ a "water conservation coordinator" tasked with directing water use in a municipality in a sustainable way. These coordinators often make decisions about how the community may use water by placing water restrictions on community members. For example, when reservoirs are low, many water conservation coordinators work with cities to put residents on a schedule for watering lawns in order to reduce the strain on their water supply.

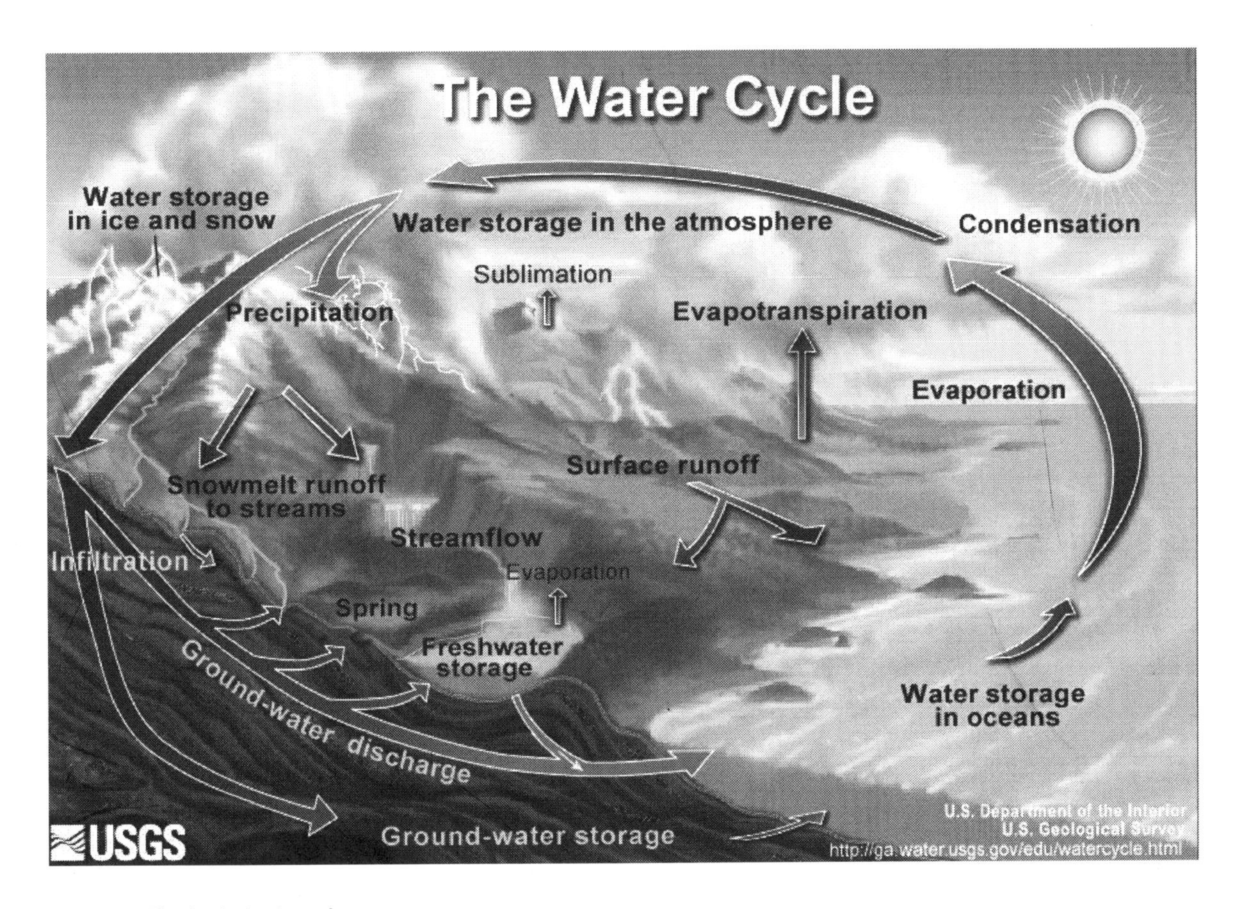

Figure 9.1. The hydrologic cycle

Water Usage

Water use can be categorized into the following major sectors: domestic (household), agricultural, and industrial. When U.S. water usage is totaled each day and divided by the number of residents (307 million), we find that it takes almost 180 gallons of water per person from public supplies to get through the day.

Domestic Water Use

Domestic water usage can vary based on the number of people in a household, whether they have efficient showerheads, and whether they have extensive landscaping, and even a swimming pool. Generally, a family of four diverts almost 30% of its water use to flushing the commode. Old commodes use about 5 gallons per flush. However, newer commodes may use as little as 1.6 gallons per flush. Replacing commodes in homes can have an enormous impact on water consumption. Another large category for domestic use is lawn watering and filling swimming pools (28%). Cutting-edge landscape architects have been replacing exotic ornamentals with native plants in their landscape designs in part to reduce the amount of water necessary to maintain plant growth with natural rainfall. Bathing, showering, and laundry are necessary to be sanitary but are often performed inefficiently with respect to water use. Old showerheads may release more than 8 gallons of water per minute, while new showerheads can reduce that to 1.5 gallons of water per minute. Filling washing machines to capacity when laundering clothes can also make more efficient use of household water. Relatively little water, less than 2%, is used for drinking and cooking; therefore, when streamlining water use, there is no reason to drink less.

Agricultural Water Use

This category is the largest consumer of freshwater. In order to feed the growing world population, farmers must irrigate their cropland. However, there is much promise and room for improvement in this category. Currently, 60% of water used to irrigate land is lost to evaporation and runoff due to flood-irrigation practices. As newer water-delivery systems are employed, such as downward facing sprinklers and drip irrigation, water use in this category can become more efficient.

Industrial Water Use

Industry uses large quantities of water throughout manufacturing and sanitation processes. Industry also uses a substantial amount of water in cooling towers. Total industrial water use in the world is about 22%, with developed countries using 59% and LDCs using 8%.

Water Pollution

Keeping the freshwater we have pristine is of extreme importance worldwide. Protecting water from contaminants is key to supporting ecosystems and to providing access to clean drinking water. Pollutants enter the water system from a point source, or non-point source. Point-source pollution sources are easy to monitor because they are from a well-defined source such as a pipes or sewer outfall. Non-point-source pollution enters the water system from a dispersed area. Agricultural fields are one of the most common non-point-source pollution contributors. These fields allow for fertilizers, herbicides, and pesticides to enter a water system such as a river along miles of the riverbank rather than from just one point. Atmospheric deposition is another example of non-point-source pollution, which is difficult to monitor and regulate.

Water Pollution and the City

By Lisa Benton-Short and John Rennie Short

Water is essential to urban life. In cities there are two main water issues, the contamination of water sources and ensuring the supply of clean water. In this chapter we will focus both issues. At the outset we will examine the case of water pollution in US because it has some of the most well-developed clean water regulations in the developed world. We will also highlight water supply and management issues in cities in the developing world.

WASTEWATER IN US CITIES

In the US, as in other rapidly industrializing-urbanizing nations, one of the major reforms to the "industrial city" of the nineteenth century was the establishment of sewer systems to collect wastewater. While many cities built wastewater treatment facilities in the nineteenth and early twentieth centuries, population growth meant that by the middle of the twentieth century, the volume of sewage and stormwater exceeded the processing ability of most treatment plants. This was particularly noticeable during heavy rains.

Combined Sewage Overflow (CSO) refers to the temporary direct discharge of untreated water. CSOs occur most frequently when a city has a combined sewer system (CSS) that collects wastewater, sanitary wastewater and stormwater runoff in various branches of pipes, which then flow into a single treatment facility (see Figure 8.1). This type of sewer system is prevalent in the urban centers in the Northeast and Great Lakes regions. CSSs serve approximately 772 municipalities with approximately 45 million people. During dry weather, CSSs transport wastewater directly to the sewage treatment plant. However, rainwater or urban storm runoff is not directed separately, but co-mingles with household wastes and industrial wastes. When it rains, few facilities can handle the sudden increase in volume of water and,

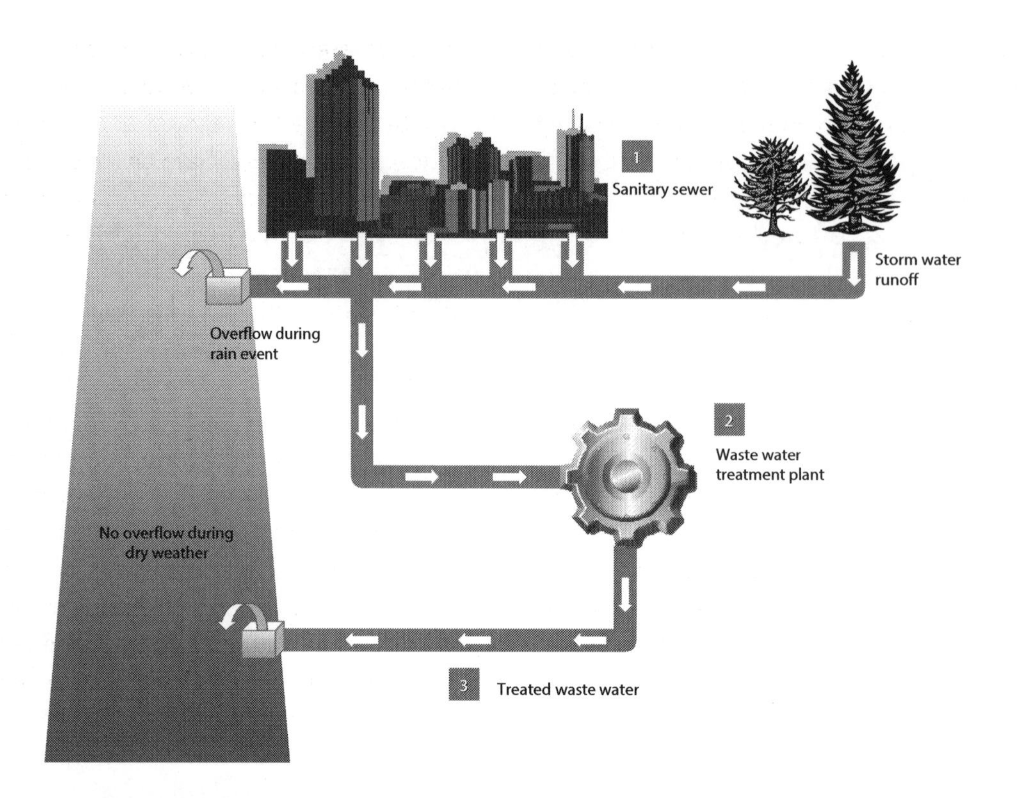

Figure 8.1. Combined Sewer Systems. In cities with combined sewer systems, stormwater runoff from rain combines with sewage in the same pipe system and is discharged directly into river, creeks or estuaries prior to any treatment. The discharge consists of many pollutants, untreated sewage and debris. Source: Lisa Benton-Short

as a result, the excess volume of sewage, clean water and stormwater is discharged untreated into rivers, lakes, tributaries and oceans. CSOs contain not only stormwater but also untreated human and industrial waste, toxic materials and debris. CSOs are among the major sources responsible for beach closings and shellfish restrictions and the contamination of drinking water. CSOs caused some 150 beach closures from 1999–2002. Table 8.1 lists the CSO releases for selected US cities. Residents of these cities are often warned to avoid contact with river water or beach water for several days after periods of heavy rainfall.

The other major type of public sewer system is the Sanitary Sewer Systems (SSSs), built in cities since the start of the twentieth century. SSS systems have separate pipes that collect sewage and stormwater

Table 8.1. Combined Sewer Overflow (CSOs) releases, 1999

CITY	BILLIONS OF GALLONS
Chicago	27.2
Atlanta	5.3
New York	84.5
Philadelphia	20.4
San Francisco	1.7
Washington DC	2.2
Richmond	4.1
Cleveland	5.9

Source: Adapted from Environmental Protection Agency (2006) Report to Congress on the Impacts and Control of CSOs and SSOs. Executive Summary

separately. However, when it rains the excess stormwater can overload the system and so stormwater is discharged untreated (see Figure 8.2). Recent studies have found that during the initial rainfall, the concentration of pollutants in urban stormwater rivals and in some cases exceeds sewage plants and large factories as a source of damaging pollutants, beach closures and shellfish decline.[167]

An EPA study conducted from 2001-3 found that each year CSO events discharge 850 billion gallons and SSOs discharge another 10 billion gallons per year.[168] Pollutants associated with CSOs and SSOs and other municipal discharges include nutrients (which can stimulate the growth of algae that deplete dissolved oxygen in surface water, thereby "asphyxiating" fish), bacteria and other pathogens (which may impair drinking water or disrupt recreational uses) and metals and toxic chemicals from industrial and commercial activities and households.

Once sewage has reached a treatment plant, physical process can separate biological solids from wastewater. Screens or filters catch raw sewage and debris, channeling them to sludge ponds where the organic materials are broken down by bacteria. Often sludge that has dried is shipped to landfills.

By the early twentieth century, some cities began to install secondary treatments in their facilities. Secondary treatment involves aeration and finer filtration. It is a more controlled way of producing bacterial growth to break down biological solids. It is estimated that secondary treatment eliminates more than 95 percent of disease-carrying bacteria from the water. Since 1972, Congress has provided $69 billion to assist cities in constructing secondary wastewater treatment plants.[169] State and local

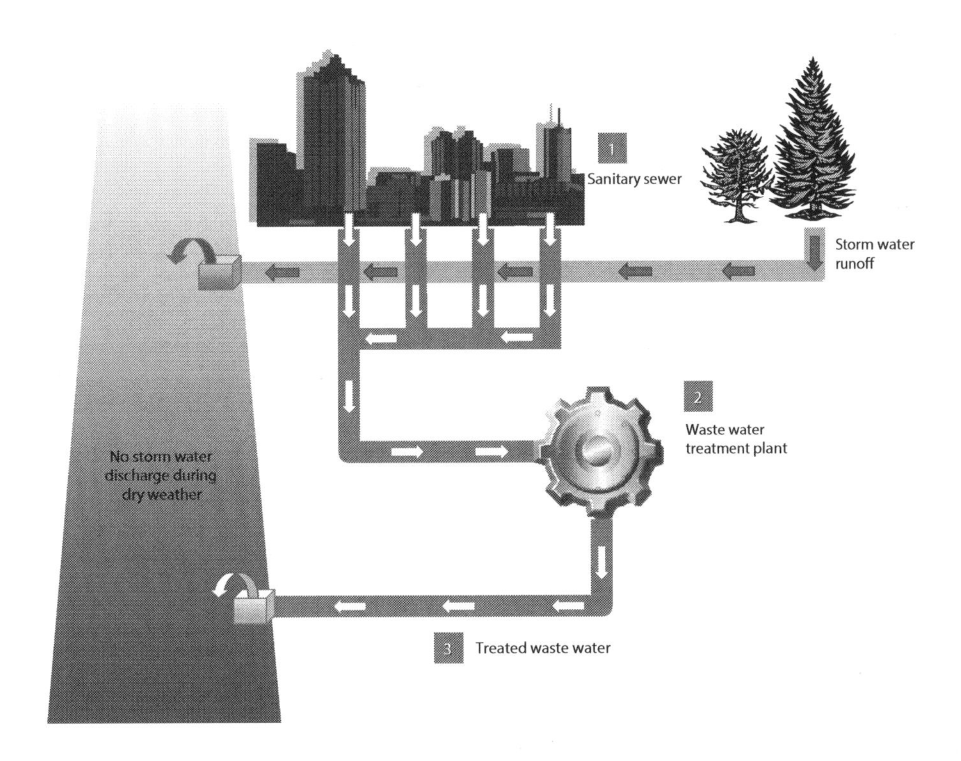

Figure 8.2. Separate Sanitary Systems. For cities that have separate sanitary and stormwater systems, stormwater runoff during rains are collected separately from sewage. The stormwater runoff, which contains pollutants and debris, is diverted from the wastewater treatment plant and discharged directly into rivers, creeks or estuaries. Source: Lisa Benton-Short

POINT AND NONPOINT POLLUTANTS

There are two main sources of water pollution, point and nonpoint. Point sources are those where there is a clear discharge mechanism such as effluent pipes or outfalls. According to the US Environmental Protection Agency, point sources are defined as any "discernible confined and discrete conveyance including but not limited to any pipe, ditch, channel, tunnel, conduit, well, discrete fissure, container, rolling stock, or concentrated animal feeding operation, or vessel or other floating craft from which pollutants are or may be discharged." These stationary devices can be measured for the amount of pollution discharged. The main point sources of pollution in cities are industrial and municipal facilities. Point sources of water pollution are relatively easy to monitor and control. The National Pollution Discharge Elimination System is a permit program of the Clean Water Act. As a result, point source pollution in the US has declined significantly since the implementation of the clean water legislation.

Nonpoint sources are any source from which pollution is discharged which is not identified as a point source. There are two main types of nonpoint sources of water pollution in cities. The first is urban runoff, a term that refers to the various pollutants that accumulate in soil and on roadways that are washed into the sewage systems during floods or rains. As rainwater makes it journey over roads, parking lots and other urban structures, it picks up a variety of pollutants. The EPA notes that urban runoff is now the largest source of pollution to estuaries and beaches. The second type of nonpoint include wastes and sewage from residential areas. These wastes are collected in the larger sewer system, and it is impossible to tell their origin. These pollutants are harder to regulate and reduce. In addition, nonpoint sources are often intermittent and diffuse, making it hard to quantify individual contributions. The Clean Water Act 1997 Amendments, Section 319, requires states and territories to develop programs to deal with nonpoint source pollution. The EPA has set aside $370 million in funds to implement nonpoint source pollution controls.

Point Sources (primarily industrial wastes)	**Nonpoint Sources**
Oil and grease	Salt
Heavy metals	Oil
Organic chemicals	Gasoline
Acids and alkalides	Antifreeze
Salts	Floatables and Debris (plastics, cans, bottles)
Solvents	Pesticides and Fertilizers
Organic matter	Organic Matter (including human and animal wastes)
Suspended solids	Microbial pathogens
Heat/thermal pollution	Solids and sediment
	Sewage
	Heavy Metals (chromium, copper, lead and zinc)

Sources: US EPA, (2007), "Polluted runoff (nonpoint source pollution: managing urban runoff)," Document EPA841-F-96-004G retrieved at www.epa.gov/owow/nps/facts/point7.htm, March 2007, and EPA, (2006), "Polluted runoff: nonpoint source pollution: the nation's largest water quality problem," retrieved at www.epa.gov/owow/nps/facts, March 2007.

governments have spent more than $25 billion. Yet it is also estimated that cities require another $140 billion to achieve the goals established.

WATER QUALITY LEGISLATION

In the United States, there were a series of legislative precedents for water quality including the 1948 Federal Water Pollution Control Act, the 1956 Water Pollution Control Act, and the Water Quality Act of 1965. These were among the first comprehensive statements of federal interest in clean water, but mounting frustration over the slow place of pollution cleanup efforts along with increased public interest in environmental protection set the stage for the 1972 amendments. Most environmental experts point to the 1972 Federal Water Pollution Control Act Amendments (commonly called the 1972 Clean Water Act) as setting the framework for the last 35 years of water pollution policy. It was remarkable for several reasons. It is the principal law governing pollution of the nation's surface waters. For the first time, a federal agency was in charge of water pollution, taking some control from the states and establishing basic federal standards, with which all states had to comply. The Clean Water Act (CWA) of 1972 required that all states set water quality standards, something that had been previously left to each state to decide. The CWA also provided technical tools and some financial assistance to address the many causes of water pollution. The CWA was highly ambitious: its major goals were to eliminate the discharge of water pollutants, restore water to "fishable and swimable" levels, and completely eliminate all toxic pollutants. Today the three ambitious goals of the CWA of 1972 have not been achieved, yet they continue to provide the framework for how the federal government (and municipalities) deal with water pollution.

Two major components of the CWA impact cities directly. Title II and Title VI authorize federal financial assistance for municipal sewage treatment plant construction and set out regulatory requirements that apply to industrial and municipal dischargers. Both of these are critical to understanding the relationship between cities and water pollution reform. With federal assistance, many cities were able to upgrade sewage treatment plants, adding secondary or tertiary treatment, or increasing the volume the treatment plants could process. Title VI, which applied mainly to industrial and municipal discharge, resulted in monitoring and management programs for point source pollution. The result has been a decrease in industrial point pollution.

ASSESSING THE CWA

We now have a more complete picture of the volume and types of municipal pollution. Thirty-five years after the initial Clean Water Act of 1972, the results are mixed. Point source pollution from industrial sources has been reduced. In many rivers and lakes, oxygen levels have recovered due to the filtering out of organic wastes. Some pollutants have declined, but others are on the increase. In 1970, more than 66 percent of rivers, lakes and estuaries were too polluted for fishing or swimming. By the late 1990s, 40 percent remained too polluted. But Lake Erie, once "dead" has recovered. And in the Cuyahoga River, once a stark symbol of the plight of America's rivers, the blue herons have returned, and the city now boasts new marinas lining a river walk and up-market sidewalk cafes. In 1970, only 85 million Americans had a wastewater treatment plant. By 2000, 173 million did (more than 70 percent of the population).[170] And yet more than 15 large cities continue to discharge some 850 billions of gallons of untreated sewage and runoff each year during heavy rains and floods (see again Table 8.1). In some cities, beach closures have increased in places where the combined sewers are older and inadequate for the population. And while the total releases of toxic pollutants have decreased (particularly with regard to mercury and DDT), more than 47 states still post fish consumption advisories for contamination due to mercury, PCB, dioxins and DDT.

The mixed results reflect several factors. First, new technologies impact on our ability to measure pollution in smaller and smaller amounts. Scientists are able to document levels of pollution in parts per million or billion, levels that we were unable to detect previously. In addition, our knowledge and understanding of the impact of pollution on human health and environmental health is improving, and this adds new "pollutants" to the list and can require a readjustment of allowable exposure standards. There are now approximately 80,000 chemicals registered with the EPA: few have been comprehensively tested, many resist breakdown and some accumulate in fat tissues. Another factor is that the EPA has tended to focus on "end of the pipe" solutions—that is regulating discharges at the source of the polluter rather than focusing on ways to encourage "front end" controls or preventive measures. The EPA has been less successful at encouraging the reduction of waste in the first place. While millions of dollars have been available to assist in the development and adoption of technological solutions, less is available for programs that encourage reduction or reusing of pollutants in the first place.

In 2000, the EPA established a set of federal policies and incentives for reducing urban storm runoff and to help smaller cities (of under 100,000 population) develop adequate wastewater treatment facilities. In addition, the 1987 Clean Water Act amendments directed states to develop and implement nonpoint pollution management programs. Under section 319 of the Act, $400 million in grants was made available for states to assess the extent of nonpoint source water quality impairments and to develop and implement plans for managing nonpoint sources.

WATER IN DEVELOPING CITIES

There are several trends common to many urban areas of the developing world. First is the water scarcity that comes with continued population growth. Second, many cities are noting record volumes of water pollution from both household and industrial sources. Water pollution problems, such as pathogenic organisms in water, continue to pose immediate threats to human health. In addition to the issues of both quantity and quality, another problem that may arise in many developing cities is the issue of equity, an issue linked closely with poverty and economic development.

Continued urban growth means tremendous stress on already inadequate resources, while also exacerbating the social and economic challenges to infrastructure investment and pollution control. Many cities in the developing world have much more serious nonpoint sources of water pollution than cities in the developed world because large sections of their population are not served by sewers, drains or solid-waste collection. The right to safe water and adequate sanitation remains a promise unfulfilled. In 2000, the World Bank estimated that about 380 million urban residents in developing cities still did not have access to sanitation and at least 170 million still lacked access to safe drinking water.[171] Unsafe drinking water continues to be responsible for more than 80 percent of diseases and 30 percent of deaths in the developing world.[172] And at any given time, close to half the people in the developing world (3 billion plus) are suffering from one or more of the main water-linked disease such as diarrhea, cholera, entric fevers, guinea worm and trachoma. Some

THE LEGACY OF THE CUYAHOGA RIVER FIRE

In June 1969, the Cuyahoga River in the city of Cleveland caught fire. The river, filled with kerosene and other flammable material was probably ignited by a passing train that provided the spark. Although it burned for only thirty minutes, the incident and the famous photograph of the river on fire became a pivotal part of an emerging environmental movement. The Cuyahoga became the symbol of urban water pollution and the need for the federal government to become involved in cleanup and regulation. Ironically, the photograph associated with the 1969 fire was actually from an earlier fire in 1952. The Cuyahoga had a history of fires, but only the fire in 1969 drew national attention. Below, Jonathan H. Adler, Associate Professor of Law at Case Western Reserve University remarks on the interesting legacy of the Cuyahoga:

> The Cuyahoga fire was a powerful symbol of a planet in disrepair and an ever-deepening environmental crisis, and it remains so to this day. That a river could become so polluted as to ignite proved the need for federal environmental regulation. Following on the heels of several best-selling books warning of ecological apocalypse and other high-profile events such as the Santa Barbara oil spill, the 1969 Cuyahoga fire spurred efforts to enact sweeping federal environmental legislation. The burning river mobilized the nation and became a rallying point for the passage of the Clean Water Act.
>
> Why didn't states act earlier? In the 1950s, let alone in 1910 or the 1930s, environmental issues did not yet rank with concerns for economic development, technological progress and other social ills. Policymakers at all levels of government knew little about the health effects of pollution and paid it little heed. While the environmental problems that plagued Cleveland and other parts of the nation are obvious in hindsight, the nature and extent of these problems were not always readily apparent at the time. Once the demand for greater pollution control emerged, action began.
>
> The 1969 fire was not evidence of nationally declining water quality either. To the contrary, early cleanup efforts on the Cuyahoga appear representative of state and local efforts nationwide. Throughout the 1950s and 1960s states began to recognize the importance of environmental quality and adopted first-generation environmental controls. As the nation became wealthier, and the knowledge base improved, attention to environmental matters increased. It is well established that wealthier societies place greater importance on environmental protection. They also have greater means to protect environmental values. ... And contrary to the common fable, in most cases state and local governments were the first to act. The 1969 fire was a catalyst for change because it was the wrong event at the right time. It was neither an impressive fire, nor one with a significant ecological impact. It may have brought greater attention to the serious environmental problems of the time, but it did not represent a continuing decline in water quality, let alone worsening environmental depredation nationwide. Contrasted with the relevant

indifference to burning rivers in decades past, the public outcry over the 1969 fire signified that increasingly wealthy Americans now wanted to devote greater resources to environmental protection—and they would likely have even in the absence of federal regulations.

Source: Adler, J.H. (2004). Smoking out the Cuyahoga fire fable: smoke and mirrors surrounding Cleveland. *National Review*, June 22.

experts argue that good-quality drinking water and proper wastewater treatment during the past 15 years has been progressively deteriorating, not improving.[173]

MEXICO CITY: SINKING

While many are familiar with Mexico City's infamous air pollution problem, an equally pressing issue centers on its water supply. Mexico City is built on top of a vast underground aquifer. Approximately 72 percent of Mexico City's water supply comes from this aquifer.[174] In general, aquifers can be an excellent source of clean water since many pollutants are filtered out as the water passes through the soil and rocks. However, aquifers are slow to replenish. Water is now being extracted twice as fast from the aquifer than it is being replaced. As a result, land subsidence has become a serious problem for Mexico City. It is estimated that over the past century, Mexico City has sunk by as much as 10 meters in some areas.[175] Since the 1970s, the rate of sinking in the central part of the city is about 6 centimeters each year. Some of the areas on the outskirts of the city are sinking the fastest; many of these are among the poorest areas in the city.

Land subsidence weakens foundations to buildings and other infrastructure including damaging the water and sewage pipes. Pipes are ruptured or develop small leaks. It is estimated that 30 percent of water is lost in the pipes from leaks before reaching users—enough to provide water to more than 4 million people.[176] Leaking sewage pipes also contaminate the groundwater with heavy metals and micro-organisms.

Land subsidence has also caused flooding in the valley particularly after heavy rains. Originally, Texcoco Lake was three meters (9 feet) lower than central Mexico City; today the lake is over two meters (6.5 feet) higher than the city.[177] Dikes had to be built to confine the stormwater flow and pumps are needed to lift the drainage water under the city to the level of the drainage canals. In addition, because of subsidence, gravity no longer takes sewage and runoffs to the Grand Canal; the city has had to install pumps in order to remove sewage to Texcoco Lake. The cost of pumping water into, within and out of the Basin of Mexico amounts to almost $900,000 per day.[178]

Because of irregular water supply, many households use water storage tanks, located on rooftops. These are often left uncovered and are not always cleaned regularly, which enables bacteria to flourish.

Mexico City also has a significant part of the population with no access to sewage facilities. Some 30 percent of Mexico City residents do not have toilets, and as much as 93 percent of waste water is discharged untreated as it flows out of the city.[179] About 75 percent of the city's residents have access to the city's current wastewater system of unlined sewer canals, sewers, rivers, reservoirs, lagoons, pumping stations and deep drainage systems. During the rainy season, when domestic wastewater and industrial wastewater are mixed with stormwater runoff, more than 1.5 million tons pass through the city's sewage system untreated.[180] A plan to build four treatments plants has floundered for lack of financing and no serious proposals for the construction of new treatment plants have been made since the 1990s. As a result, the city lacks the capacity to reuse significant volumes of water for industrial and agricultural production within the valley, a practice that would also alleviate the demand for freshwater.[181]

Mexico City, like many developing cities, is challenged by problems related to water management that may cause social conflicts. The wealthier population of Mexico City consumes up to 40 times more water than that used by the poorer sector. Compounding the issue is that many of the poorest areas of the city not only receive water of unacceptable quality, they also suffer the inconvenience of rotating schedules for water deliveries, whether through the water network or from tanker trucks.[182]

THE GANGES: A SACRED RIVER

The Ganges winds 1,500 miles across Northern India, from the Himalaya Mountains to the Indian Ocean, and through 29 cities with populations over 100,000. Known as Ganga Ma (Mother Ganges), the river is revered as a goddess whose purity cleanses the sins of the faithful and helps the dead on their path toward heaven. Hindus believe that if the ashes of their dead are deposited in the river, they will be ensured a smooth transition to the next life or freed from the cycle of death and rebirth. It is said that a single drop of Ganges water can cleanse a lifetime of sins. In cities along the Ganges, daily dips are an important ritual among the faithful. Many cities are pilgrimage sites.

Despite its spiritual importance, the physical purity of the river has deteriorated dramatically. While industrial pollutants account for some of the river's pollution, the majority of Ganges' pollution is organic waste: sewage, trash, food, human wastes and animal remains. Today nearly half a billion people live in the basin of the Ganges and more than 100 cities dump their raw sewage directly into the river. Pollutants have been measured at 340,000 times permissible levels. Not surprisingly, waterborne illnesses are common killers, accounting for the deaths of some 2 million Indian children each year.

In the city of Varanasi, one of India's oldest and holiest cities, some 40,000 cremations are performed each year; those unable to afford the cost of traditional funerals often dump the body into the river. In addition, the carcasses of thousands of dead cattle, which are considered sacred to Hindus, are tossed into the river each year. The city also pumps some 80 million gallons of sewage waste daily into the river. And yet thousands of Hindus continue to come to Varanasi and the Ganges.

The Ganges presents an interesting case study. Because the river is holy, it attracts tens of thousands of pilgrims each day for ritual bathing, exposing large numbers of people to untreated, contaminated water. As one man commented, "There is a struggle and turmoil inside my heart. I wanted to take a holy dip. I need it to live. The day does not begin for me without the holy dip. But, at the same time, I know what is B.O.D and I know what is fecal coliform."[183] It is estimated that 40 percent of the people who take a dip in the river regularly have skin or stomach ailments.

In 1985, the Indian government initiated the Ganga Action Plan, to clean up the river in selected areas by installing sewage treatment plants. Twenty years later, the plan has been relatively unsuccessful. Some have blamed the failure on the adoption of expensive multi-million dollar Western-style treatment plants. The government spent $335 million. But many of the treatment plants malfunctioned, were designed improperly or could not handle the bacterial load. In addition corruption and ineffective monitoring contributed to the problems. There is growing criticism over the adoption of Western-style technology to solve developing world issues. Western-style technology tends to be very expensive, relies on highly trained engineers and workers to maintain the technology, and requires a stable and consistent supply of electricity. In addition, Western-style waste treatment plants were engineered for use in countries where there are no monsoon rains, and where the population does not drink directly from the water source. Few considered the radically different ways that people use rivers in India. Another criticism is that many local communities along the Ganges were not included in the planning process and therefore failed to participate in the recovery of the river.

As an alternative to the high-technology treatment plants, Veer Bhadra Mishra, a Hindu priest and civil engineer, collaborated with University of California Berkeley engineer, William Oswald, to develop a non-mechanized, low-tech sewage treatment plan. This plan is more compatible with the climate of India and replaces the high-tech solution with a wastewater oxidation pond system that would store sewage in a series of ponds and use bacteria and algae to break down waste and purify the water. The ponds allow waste to decompose naturally in water. Bacteria grow on the sewage and decompose it; the algae feed on the nutrients released by the bacteria and produce oxygen for the water. This alternative treatment does not require electricity, but relies on sunshine to speed up the decomposition. The pond system is much less expensive than mechanical treatment plants.

The debate between Western-style technology and lower-cost alternatives in the case of the Ganges highlights general themes applicable to many of the solutions to pollution in developing cities. While the 1980s and 1990s saw large-scale investments by governments in high-technology solutions, the lack of results has caused many to seek alternatives that are locally sensitive and economically affordable. More recently, experts in both the developing world and developed north contend that solutions should respond to local demands and should be as simple, sturdy and inexpensive as possible. Low-cost, low-technology solutions such as the pond system or pour-flush latrines, or even improved pit latrines, have been successful. An important element in pollution control is the role of public participation. The involvement of the local community and households is now seen as a crucial component to success.

Water issues in developing cities are not all problematic. Some cities are developing and protecting vital water resources. Figure 8.3 shows the river gardens of Beijing. The river gardens attract tourists and at the same time provide an important ecosystem within the rapidly growing city.

CONCLUSIONS

Since the 1970s many countries and municipalities have enacted pollution reform measures. Some cities have seen dramatic decreases in certain types of pollutants, while other pollutants have increased. Some cities have successfully implemented pollution laws and regulations, while many cities (and countries) lack the resources to adequately enforce such measures. The twin pressures of population growth and increasing consumption of resources have, in some cases, offset new laws and regulations designed to decrease or prevent pollution.

GUIDE TO FURTHER READING

Adler, R., J. Landman, and D. Cameron. (1993). *The Clean Water Act: 20 years Later.* Washington, DC: Island Press.

RECLAIMING TORONTO'S DON RIVER

Michael Hough recounts how the city of Toronto developed by exploiting the Don River and then turned its back on this resource. Recently residents and local government have been involved in reclaiming the river for public use and ecological restoration.

> Toronto was first settled in 1787 ... and settlers harnessed the river's energy, built mills for lumber, flour, wool and paper, and mined the valley's clay and shale for brick-making, from which much of the early city was built. In less than 150 years, they cleared the lower valley of merchantable trees. The Don River was also perceived as a threat and an obstacle. Floods swept away mills and bridges, the river was an obstacle to the eastward expansion of the city and the great wetland, its mouth reviled as unhealthy swamp, lent credibility to the argument that straightening out the river and filling in the marshes would "secure the sanitary condition ... to the said river." By the end of the century, engineers had turned the last 5 kilometers of the river's meanders, where it dropped its sediments, into a canal. The railways were built in the valley, and the Ashbridges Bay marshes were filled in to create the port lands, the most massive engineering project on the continent in its time, forcing the Don into a right-angle turn into the harbour. By the mid-twentieth century, the city had turned its back on the river, a gap between places rather than a place in itself. As a sensory experience it had become a forgotten place; unloved and unused.
>
> Moves to restore the river became an act of faith by the citizens of Toronto that grew out of the concerns of many people for the natural heritage of their city. Beginning as an informal citizen's organization, the "Task Force to Bring Back the Don" was formalized and supported by Toronto City Council in 1990. Its purpose was to begin the process of renewal of the most degraded part of the river that flows through the city of Toronto, and ultimately to imitate the restoration of the entire watershed ...
>
> As an ongoing process of renewal and healing, the Don strategy involves key principles, including a fundamental understanding of process as a biological idea that is also integrated with social, economic and political agendas, economy of means where the most benefits are available for minimum input in energy and effort, and environmental education, where the understanding of nature in cities becomes part of a learning experience that begins with community empowerment and action.

Source: Hough, Michael. (1995, 2nd edn) *Cities and Natural Processes: a Basis for Sustainability.* New York: Routledge. Quotes from pp. 39, 42, 54.

Gumprecht, B. (2001). *The Los Angeles River: Its Life, Death and Possible Rebirth.* Baltimore: The Johns Hopkins University Press.

Lewin, T. (2003). *Sacred River: The Ganges of India.* Boston, MA: Houghton Mifflin/Clarion Books.

Melosi, M.V. (2001). *Effluent America: Cities, Industry, Energy, and the Environment.* Pittsburgh, PA: University of Pittsburgh Press.

Swyngedouw, E. (2004). *Social Power and the Urbanization of Water: Flows of Power. Oxford: Oxford University Press.*

Uitto, J. and A. Biswas, (eds.). (2000). *Water for Urban Areas: Challenges and Perspectives.* Tokyo and New York: United Nations University Press.

White, R. (1994). *Urban-Environmental Management: Environmental Change and Urban Design.* New York: John Wiley.

For a good account of environmental policymaking, see:

Desfor, G. and K. Roger. (2004). *Nature and the City: Making Environmental Policy in Toronto and Los Angeles.* Tucson: Arizona University Press.

The Forgotten Infrastructure:

Safeguarding Freshwater Ecosystems

By Sandra L. Postel

The water strategies of the 20th century helped to supply drinking water, food, flood control and electricity to a large portion of the human population. These strategies largely focused on engineering projects to store, extract and control water for human benefit. Indeed, it is hard to fathom today's world of 6.6 billion people and more than $65 trillion in annual economic output without the vast network of dams, reservoirs, pumps, canals and other water infrastructure now in place. These projects, however, have often failed to distribute benefits equitably and have resulted in the degradation, or outright destruction, of natural freshwater ecosystems that in their healthy state provide valuable goods and services to society.

As water stress and the risks of climate change deepen and spread around the world, policies and strategies designed to meet human needs, while protecting ecosystem health, will become increasingly critical to human well-being. Scientific understanding of the components of freshwater ecosystem health has advanced markedly over the last decade, but incorporation of this knowledge into water policy and management has lagged. A number of nations and regions—including Australia, the European Union, South Africa and the Great Lakes—are pioneering policies that establish boundaries on human degradation of freshwater with an aim of safeguarding ecosystem health. Although imperfect, and facing tough implementation obstacles, these policies offer promising ways of better harmonizing human uses of water with protection of valuable ecosystems.

THE DECLINE OF ECOLOGICAL INFRASTRUCTURE

Water infrastructure typically refers to the collection of dams, levees, canals, pipelines, treatment plants and other engineering works that help provide water

Sandra L. Postel, "The Forgotten Infrastructure: Safeguarding Freshwater Ecosystems," *Journal of International Affairs*, vol. 61, no. 2, pp. 75-90.

services to the human population. There is another class of infrastructure that also delivers valuable services to society: the aquatic ecosystems that perform nature's work. Healthy rivers, floodplains, wetlands and forested watersheds supply much more than water and fish (see Table 1). When functioning well, this "eco-infrastructure" stores seasonal floodwaters, helping to lessen flood damages. It recharges groundwater supplies, which can ensure that water is available during dry spells. It filters pollutants, purifies drinking water and delivers nutrients to coastal fisheries. Perhaps most importantly, it provides the myriad habitats that support the diversity of plants and animals that perform so much of this work.

For millennia, human societies grew and flourished by relying on this time-tested work of nature. The ancient Egyptians, for instance, thrived for several thousand years on the ecological services provided by the annual flood of the Nile River, which delivered water and nutrients to their farm fields, carried off harmful salts that had accumulated in the soil and supported a diversity of fish.[1] During the 20th century, however, such reliance on nature's services was supplanted by engineering projects that provided hydroelectric power, intensive irrigation, flood control and other benefits demanded by burgeoning populations and economies.

Since most of nature's services lie outside of commercial markets and are not priced in conventional ways, they are grossly undervalued. While the benefits of dams and other water projects are measured in familiar metrics—kilowatt-hours generated and hectares irrigated and populations served—the ecological downsides of these engineering approaches have largely been left out of the cost-benefit calculus. As a result, ecological infrastructure has been dismantled and degraded at a rapid rate. An estimated 25 to 55 percent of the world's wetlands have been drained, 35 percent of global river flows are now intercepted by large dams and reservoirs and more than 100 billion tons of nutrient-rich sediment that would otherwise have replenished deltas and coastal zones sits trapped in reservoirs.[2] River flows are turned on and off like plumbing works, eliminating the natural flow patterns and habitats upon which myriad life forms depend.[3]

It is difficult to place a dollar value on any one piece of eco-infrastructure, but in 2005, scientists participating in the Millennium Ecosystem Assessment estimated that wetlands alone provide services worth $200 to 940 billion per year.[4] Following the Great Midwest Flood of 1993, U.S. researchers estimated that restoration of 5.3 million hectares of wetlands in the upper portion of the Mississippi-Missouri watershed, at a cost of $2 to 3 billion, would have absorbed enough floodwater to have substantially reduced the $ 16 billion in flood damages that resulted from that one major flood episode.[5] And when Hurricane Katrina struck the U.S. Gulf Coast in August 2005, an important piece of nature's protective infrastructure was partially missing: coastal wetlands and barrier islands that could reduce the power of storm surges. The state of Louisiana alone has lost 492,000 hectares of coastal wetlands since the 1930s, and continues to lose them at a rate of more than 6,200 hectares per year—approximately one football field every forty-five minutes.[6] It is impossible to know how many lives and homes might have been saved had natural protections along the coast remained in place. But surely one of Katrina's lessons is to enlist nature's help in mitigating future disasters rather than simply assigning it blame when disasters occur.

Indeed with climate change impacts unfolding more rapidly than scientists had predicted even five years ago, the value of protecting and restoring ecological infrastructure is rising. Global warming and its anticipated effects on the hydrological cycle will make the robustness and resilience of nature's way of mitigating disasters all the more important, as tropical storms, spring flooding and seasonal droughts increase in frequency and/or intensity.

Table 1: Life-Support Services Provided by Rivers, Wetlands, Floodplains and Other Freshwater Ecosystems

- Provision of water supplies for irrigation, industries, cities, and homes
- Provision of fish, waterfowl, mussels, and other foods for people and wildlife
- Water purification and filtration of pollutants
- Flood mitigation
- Drought mitigation
- Groundwater recharge
- Water storage
- Provision of wildlife habitat and nursery grounds
- Soil fertility maintenance
- Delivery of nutrients to deltas and estuaries
- Delivery of freshwater flows to maintain estuarine salinity balances
- Provision of aesthetic, cultural, and spiritual values
- Provision of recreational opportunities
- Conservation of biodiversity, which preserves resilience and options for the future

Source: Postel, Sandra. *Liquid Assets: The Critical Need to Safeguard Freshwater Ecosystems* (Washington, D.C.: Worldwatch Institute, 2005).

SOUTH AFRICA AND AUSTRALIA PIONEER ENVIRONMENTAL FLOW POLICIES

To suggest that the maintenance and repair of ecological infrastructure should be a core principle of water policy and planning might sound about as necessary as suggesting a building's foundation be secure before constructing twenty stories on top of it. In reality, however, the systems of water law and policy that guide water allocation rarely give ecosystems the water they need in order to carry out their functions. However, at least two nations—South Africa and Australia—are advancing a new policy framework that places ecological health and the water required to sustain it squarely at the center of water allocation and management.

Simply framed, the old water mindset held that water acquires value only when it is extracted from the natural environment and put to use by a farm, factory or home. The evolving new mindset recognizes water's value when left in place to do its ecological work. Perhaps no nation is working harder than South Africa to shift from the old way of thinking about water to the new, more environmentally intelligent view. After coming to power in 1994, Nelson Mandela's post-apartheid government undertook a rewriting of the country's constitution and laws, and water reform was near the top of the agenda. "There was a desire to reshape water management so as to transform South African society," according to Evan Dollar, a river scientist with South Africa's Council for Scientific and Industrial Research. "We were given a unique historical opportunity to do so."[7]

South Africa's National Water Act of 1998 was the result of that process.[8] The law was grounded firmly in the doctrine of public trust—the recognition that governments hold certain rights and entitlements in trust for the people and are obligated to protect them for the common good. One of the innovative features of the law is the establishment of a water reserve consisting of two parts. The first is a non-negotiable water allocation to meet the basic drinking, cooking and sanitary needs of all South Africans.[9] The second part of the reserve is an allocation of water to support ecosystem functions so as to secure the valuable services they provide to South Africans. Specifically, the act says:

> The quantity, quality and reliability of water required to maintain the ecological functions on which humans depend shall be reserved so that the human use of water does not individually or cumulatively compromise the long-term sustainability of aquatic and associated ecosystems.[10]

The water determined to constitute this two-part reserve has priority over irrigation and other licensed uses, and only reserve water is guaranteed as a right. What the South African law says, in effect, is that both people and ecosystems must get the water they need to be healthy before other water demands are fulfilled. Not surprisingly, the pioneering law is far easier to express on paper than to implement on the ground. Because the human reserve amounts to little more than 25 liters per person per day (certainly better than no access to safe drinking water at all, but a sparse daily allotment), many poor black South Africans view the law as a perpetuation of historical inequities.

The ecological reserve has solid scientific underpinnings but is difficult to implement. Just as doctors check blood pressure, cholesterol levels and heart rate to see if these values fall within ranges essential for good human health, scientists assess certain ecosystem attributes to determine whether they fall within ranges essential for good ecological health. With sufficient information about a particular river system, scientists can develop an "environmental flow prescription"—a description of the quantity and timing of flows required to sustain an ecosystem's important functions. The approach calls for water managers to sustain or replicate a river's natural pattern of variable flows—the pattern of high and low flows, as well as periodic floods and droughts—that the river historically exhibited and to which the myriad life forms in the river have become adapted.[11] The approach does not call for or require a return to the "natural" state, but it does entail maintaining a flow regime that resembles the natural historical one to a sufficient degree to sustain the ecological functions of the aquatic system.

South African scientists have done pioneering work in the development of environmental flow methodologies, and these are informing both the policy and its implementation. However, tying flows to the provision of specific ecosystem goods and services—for example, maintaining populations of floodplain fisheries that local people rely on for protein—is complicated. Nonetheless, South Africa's scientists and citizens are tackling these issues. According to Dollar, more than 300 reserves, about 30 percent of the total needed, have been established and await implementation by water managers.[12]

Australia also began a major move toward more ecologically minded water management in 1994, when state premiers signed on to a new Council of Australian Governments (COAG) Water Reform Framework Agreement that aims to "sustain and where necessary restore ecological processes, habitats and biodiversity in water dependent ecosystems."[13] A key piece of this reform package calls for states to recognize the environment as a legitimate user of water and to allocate water specifically to freshwater ecosystems. Among the twenty guiding implementation principles—which cover issues ranging from assessing ecological flow requirements to accountability and community involvement—is one stating explicitly that environmental water provisions should be legally recognized. Another says that when environmental water allocations are not sufficient to prevent significant ecological harm, extractions of water from that river basin "should be capped."

All eight Australian states have passed new water laws to reflect the COAG goals and they are now in the process of setting environmental flow requirements for their rivers. Under the nation's constitution, the commonwealth (or federal) government has limited authority over water matters; primary responsibility rests with the states and territories. Implementation of these water reforms may, thus, vary considerably among the states. The state of Western Australia, for example, has established a water allocation policy similar to South Africa's ecological reserve: the water required to support ecosystem health gets top priority; the remainder can then be licensed for other uses.[14] Since Western Australia's rivers generally are not yet over-allocated, the setting of these environmental flows early on may avoid the contentious issues that inevitably arise in river basins that are already stressed or over-allocated.[15]

Indeed, in the grip of a multi-year drought, many Australian river basins are running squarely into a major hurdle when it comes to implementing environmental flow strategies—consumptive water-use entitlements that are expressed as specific quantities rather than as shares of the pool of water actually available. In heavily allocated river systems, the only way to ensure that ecosystems receive their sustaining flows is if private water rights or permits within the river system are reduced during droughts to an equitable proportion of the water actually available in the system once the ecological requirements are met. Even clearly defined environmental flow allocations will not be respected if property rights in water are not adjusted during periods of drought-induced water shortages—a situation likely to become much more frequent in some regions as climate change unfolds.[16]

Many of these issues are coming to a head in the Murray-Darling River Basin, Australia's largest watershed and home to 70 percent of the nation's irrigated land.[17] After a tripling of withdrawals between 1944 and 1994, river flows dropped to ecologically harmful levels. Wetlands shrank and fish populations declined, while salinity levels and the frequency of algal blooms increased. Severe low-flows now occur in the lower Murray River in nearly two years out of three (and every year during the ongoing drought), compared with 5 percent under natural conditions. The Murray's flow has dropped so low during recent drought years that its mouth has became clogged with sand.[18]

The Murray-Darling watershed spans parts of four states and all of the Australian Capital Territory. Through the Murray-Darling Basin Commission (MDBC), these political entities work cooperatively to manage the river. In 1997, in response to the rapid deterioration of the river's health, the Ministerial Council (which consists of resource ministers from each basin state or territory plus the commonwealth) placed a cap on diversions from the basin. According to the MDBC, 96 percent of the water consumed within the basin in 2003–2004 was within the cap.[19]

With a lid on extractions, new water demands in the Murray-Darling basin are met primarily through conservation, efficiency improvements and water trading. Most of the early buying and selling of water entitlements has occurred within states, but the MDBC is now piloting a program in the southern portion of the basin to allow permanent water trades across state boundaries.[20] The initial two-year review of this scheme found that it had enabled fifty-one trades collectively worth about 10 million Australian dollars, which had transferred nearly 10 million cubic meters of water between states.[21] With virtually all of the traded water going to higher-value uses, water marketing is boosting the basin's money economy. Indeed, a 1999 study projected a doubling of the basin's economic value over twenty-five years with the cap and water reforms in place.[22]

The ecological benefits of the cap, however, are far from certain. The cap was pegged to a level of withdrawals that had allowed serious degradation of the river's health. So while it may prevent further deterioration, the cap is not sufficiently stringent to revitalize the river. Moreover, the prolonged drought has exacerbated the decline in the river's health.

In early 2007, the commonwealth government responded to the dire situation of the Murray-Darling Basin—the drought in general—by passing a National Plan for Water Security under which the government will invest up to $3 billion over ten years to address the over-allocation of water in the basin. The intention appears to be to buy back between 15 and 30 percent of water entitlements in the southern part of the basin in order to return flows to the river. By any accounting, this is a big move. The volume of water that could be purchased (at current market prices) through the planned buyback of entitlements is at least fifteen times greater than the total amount of permanent water entitlements that have ever been traded in a given year.[23] The MDBC has announced that it is prepared to buy water from willing sellers and has set up a mechanism for entitlement holders to express their interest in doing so. However, it is still unclear whether the large volumes of water sought for the Murray River environment will be forthcoming voluntarily or whether mandatory measures will be needed.

ENVIRONMENTAL FLOWS AND A CAP IN THE U.S. STATE OF TEXAS

In the United States, two centuries of dam building, levee construction and straightening of river channels have left very few river segments in anything close to their natural state: Only 2 percent of U.S. rivers and streams remain free-flowing.[24] Conflicts over the allocation of water between human needs and ecosystem needs have been intensifying across the country, from west to east and north to south. Nevertheless, despite widespread degradation of its river systems, the United States has no overarching vision or goal to secure river flows that support the diversity of freshwater life and that sustain ecological functions. Historically, the federal government has deferred to the states in matters of water allocation, use and management. Consequently, water policy innovations have tended to emerge from state and regional authorities rather than from above.

One perhaps unlikely policy pioneer in the establishment of environmental flows is the state of Texas. In 2007, Texas passed one of the most comprehensive statewide environmental flow laws in the country, the Environmental Flows Allocation Process. It calls for the setting of flow standards for every major river system in the state and establishes a process for soliciting input from both scientists and stakeholders in each watershed. Texas boasts a $2 billion a year recreational and commercial fishing economy along its coast, and this valuable asset depends on adequate freshwater flows into its bays and estuaries.[25] The Texas coastline also offers premier bird habitat and supports a world-renowned diversity of bird species. Among these species is the endangered whooping crane, which spends winters near the mouth of the Guadalupe and San Antonio rivers. Bird enthusiasts flocking to the Texas coast provide another source of tourist dollars that depends on healthy rivers.

Like the Murray River in Australia, the Rio Grande in Texas runs dry before reaching the Gulf of Mexico in times of drought, and with the state's population growing rapidly, more rivers and estuaries are at risk of water deprivation. The new legislation emerged from negotiations between environmental groups and water suppliers and stands a good chance of passage. The law calls for each river basin in the state to be guided by a team of scientists (the Environmental Flows Science Committee) and a team of stakeholders (the Environmental Flows Advisory Group) representing diverse interests, and for the state environmental commission to consider the recommendations of both groups in setting formal environmental flow standards. In doing so, the commission may decide to set aside some or all of the water not yet permitted for other uses to secure future environmental flows. Where rivers are already over-allocated, the commission will examine recommendations from the stakeholder groups as to how to make up the difference, which could include, for instance, dedicating municipal wastewater return flows to environmental flow purposes or soliciting donations and voluntary sales of existing water rights.

A main goal of the flow standards is to ensure sufficient freshwater flows to protect associated estuarine ecosystems, including during times of drought.

In November 2007, the flow-setting process was to begin with the Sabine Lake and the Galveston Bay estuaries and their contributing river basins.[26] As in South Africa, the implementation of this Texas law may prove challenging in light of the need to solicit and coordinate stakeholder and scientific involvement. This participation is critical, however, to widespread support and long-term success.

Another water policy initiative in Texas mirrors the cap-and-trade approach being tried in Australia's Murray-Darling Basin. However, in the Texas case, the cap applies to groundwater pumping from an underground aquifer rather than river withdrawals. The Edwards Aquifer is a major source of irrigation water in south-central Texas and of drinking water for the city of San Antonio.[27] By the early 1990s, heavy pumping from the aquifer had substantially reduced flows in San Marcos and Comal Springs, which harbor seven species listed under the U.S. Endangered Species Act—including the Texas Blind Salamander and the Fountain Darter. The Sierra Club and others filed a lawsuit under the act to limit pumping so as to sustain flows in the springs. In response, the Texas legislature established the Edwards Aquifer Authority in 1993 and set a 555.3 million cubic meter cap on annual pumping from the aquifer through 2007 and a more stringent cap of 493.6 million cubic meters by 2008.[28] In addition, the authority is to have enforceable procedures in place by 2012 to ensure continuous minimum flows for the two springs.

As in the Murray-Darling Basin, the cap on withdrawals from the Edwards Aquifer has fostered an active water market. Most of the trades, which include both permanent sales and temporary leases of water, involve irrigators selling water to San Antonio. As of 2005, irrigators had traded some 185.1 million cubic meters of water per year to urban users.[29] The cap has also encouraged more conservation in San Antonio, where per capita domestic use is now considerably lower than in most Texas cities.[30]

The institution of the Edwards Aquifer cap represents a marked departure from Texas's long-standing "rule of capture"—sometimes called the "rule of the biggest pump"—which essentially allows landowners to withdraw as much groundwater from beneath their land as they want, as long as they put it to some beneficial use. Harm to neighbors or the environment does not constrain pumping rights under the rule of capture. This antiquated rule still governs much of the groundwater in Texas, but perhaps experience with the Edwards Aquifer will encourage broader policy reform.

INTERNATIONAL INITIATIVES: THE GREAT LAKES AND THE EUROPEAN UNION

The setting of criteria for ecological health within international watersheds is fraught with complexity. Within the last eight years, however, two important initiatives have emerged that offer the potential for large-scale protection of rivers, lakes and aquifers in a portion of North America and much of Europe. One is the Great Lakes Charter Annex and related agreements. The other is the European Union's Water Framework Directive.

The Great Lakes-St. Lawrence River Basin is a vast watershed that spans portions of eight U.S. states (Illinois, Indiana, Michigan, Minnesota, New York, Ohio, Pennsylvania and Wisconsin) and two Canadian provinces (Ontario and Quebec). About 95 percent of North America's surface freshwater supply is contained within this basin. Since 1985, the ten Great Lakes Basin (GLB) states and provinces have abided by principles set forth in the Great Lakes Charter, a bilateral agreement intended to protect the Great Lakes ecosystems. However, with increased talk about the possibility of bulk exports of water from the Great Lakes to drier parts of North America and the world, the GLB states and provinces entered into a supplementary agreement in 2001, the Great Lakes Charter Annex, which commits the ten parties to develop stronger protections for the waters and ecosystems of the Great Lakes.

Several years of public and official discussions followed, and on 13 December 2005, the eight Great Lakes governors and two premiers signed agreements to reform the rules for taking water from the aquifers, rivers or lakes in their respective states and provinces within the GLB. The first is a good-faith agreement among all ten parties; the other is a binding compact among the eight U.S. states.[31] The big gorilla of these agreements is a ban on most water diversions from the Great Lakes ecosystems, with an eye toward meeting a no-net-degradation goal for the Great Lakes ecosystem. Exceptions to the export ban are made for relatively small water withdrawals and for communities that straddle the watershed's boundaries. However, even small- and mid-size withdrawals are subject to review based on environmental criteria and sustainable water use provisions. The compact says, for example, that even withdrawals meeting the criteria for exception from the ban can be allowed only if the water withdrawn is returned to the source watershed (less some allowance for consumptive use of the water). Moreover, it largely precludes parties from satisfying this return requirement with waters from outside the basin to protect against the introduction of invasive species.[32]

Implementation of these provisions hinges on all the U.S. states ratifying the compact and passing virtually identical legislation to implement it—certainly a nontrivial hurdle. Although the compact was signed more than two years ago, so far only two states have ratified it—Minnesota (February 2007) and Illinois (August 2007). Ratification is pending in the other states. Once all the states have ratified it and passed implementing legislation, the U.S. Congress would need to approve the compact for it to become enforceable federal law. In Canada, the ban on diversions is much less controversial than it is in the United States, and so full implementation by Ontario and Quebec is expected. Ontario already meets or exceeds most of the requirements of the Charter Annex agreements. The province has already passed strict laws banning diversions out of the province's three major water basins, which include the Great Lakes-St. Lawrence River Basin.[33]

Even as progress continues on the legislative front, actions are being taken to ensure effective implementation of the agreements. For example, on 4 December 2007, the regional body overseeing implementation of the agreements adopted regional water conservation and efficiency goals and objectives. The water conservation initiative will now assist the ten Great Lakes states and provinces in working together to develop more specific conservation goals, and will serve as an ongoing forum for the involvement of tribes, first nations, regional stakeholders and others.[34] It is important to note that application of the "precautionary principle" underpinned the development of this protective water policy for the Great Lakes. As applied to ecological health, the precautionary principle essentially says that given the rapid pace of ecosystem decline, the irreversible nature of many of the resulting losses and the high value of freshwater ecosystem services to human societies, it is wise to err on the side of protecting too much rather than too little of the freshwater habitat that remains. It operates like an insurance policy—society buys extra protection in the face of uncertainty. In what is perhaps the strongest recognition of the precautionary principle by an international water institution, the International Joint Commission adopted this as a guiding principle for protecting the Great Lakes. In a 1999 report to both governments, the commission cites the precautionary approach as one of five principles, noting:

> Because there is uncertainty about the availability of Great Lakes water in the future—[a]nd uncertainty about the extent to which removals and consumptive use harm, perhaps irreparably, the integrity of the Basin ecosystem—caution should

> be used in managing water to protect the resource for the future. There should be a bias in favor of retaining water in the system and using it more efficiently and effectively.[35]

With passage of its Water Framework Directive (WFD) in 2000, the European Union has also taken an important step toward the protection of freshwater ecosystems. Up until that time, the EU had primarily focused on water quality concerns, but with the new WFD a more comprehensive approach to freshwater ecosystem health is squarely on the agenda. A key feature of the directive is the establishment of criteria for classifying the ecological status of rivers (and other water bodies) as high, good, moderate, poor or bad, depending upon how much the water body's ecological characteristics deviate from a natural or undisturbed condition.[36] Member countries are then to take measures to ensure that at least a "good status of surface water and groundwater is achieved ... and that deterioration in the status of waters is prevented."[37] Each member country has responsibility for translating the directive into legislation and for adopting implementation measures, which are likely to include controls on water withdrawals and flow alterations. Importantly, the directive establishes criteria for classifying the ecological status of rivers, including river flow and channel characteristics.

Of the 110 river basins identified within EU member states, forty cross national borders. With these forty international basins constituting 60 percent of the EU's territory, cooperation among countries is critical for successful implementation of the WFD.[38] Considerable progress in international coordination has already been achieved in the Elbe, Meuse, Odra, Rhine and Scheldt basins—and especially within the Danube basin, where initiatives were well under way long before the directive's passage. Rising in Germany's Black Forest, the Danube joins thirteen countries and over 80 million people within its watershed as it runs approximately 2,800 kilometers eastward to the Black Sea. Over the past two centuries, the river has been badly degraded by channelization, draining of wetlands, rampant pollution and the construction of numerous dams—including fifty-nine along the river's first 1,000 kilometers. A good portion of the Danube's riparian forests are gone and former floodplains have dried out. In Bulgaria, for instance, 90 percent of the river's flood-plain wetlands have disappeared.[39] Moreover, the Danube Delta—Europe's largest wetland ecosystem and home to 320 species of birds—has been seriously degraded by the heavy pollution loads carried downstream and by the reduction of cleansing flood flows.[40]

After the breakup of the Soviet bloc and the fall of the Iron Curtain, new opportunities for cooperation and collaboration arose for the nations of the Danube watershed. They did not waste much time. In September 1991, the environment ministers from a number of Danube countries met in Sofia, Bulgaria to begin planning the river's restoration and protection. In June 1994, a majority of Danube countries, plus the European Commission (the decisionmaking arm of the EU) signed the Convention on Cooperation for the Protection and Sustainable Use of the Danube River, a legal instrument that directly calls for sustainable and equitable water management.[41] Later that year, in Bucharest, Romania, the environment ministers and the EU's environmental commissioner endorsed the Danube Strategic Action Plan, which states that "[c]onservation, restoration and management of riverine habitat and biodiversity is important for maintaining the natural capital of the basin...and to establish its natural purification and assimilative capacity."[42]

Remarkably, the Danube collaboration brings together former communist countries with their Western counterparts, and relatively rich nations such as Germany and Austria with poor ones such as Bulgaria and Romania. A key benefit of a cooperative river-basin framework is that it allows international agencies and groups to fund and help implement projects. For instance, the governments of Bulgaria,

Romania, Moldova and Ukraine have pledged to create a network of at least 600,000 hectares of floodplain habitat along the lower Danube, the Prut River and in the Danube Delta[43] This joint effort is part of a project called Green Corridor for the Danube, initiated in June 2000 by the World Wildlife Fund, a private conservation organization. With funding from United Nations agencies and others, the project aims initially to demonstrate how healthy floodplains can provide habitat, reduce pollution loads and enhance fisheries. The hope is to extend the Green Corridor program to the entire length of the Danube.[44]

For Romania, which harbors more than 80 percent of the Danube Delta, restoration of the ecosystem must be accompanied by a reinvigoration of its weak economy. There—as in South Africa and most poor countries—citizens are only likely to support ecosystem restoration efforts if their livelihoods improve at the same time. Whether an upstream-downstream collaboration can meet this difficult test of sustainable development remains to be seen. One key to success will likely be acknowledging the value of revitalized ecosystem services. According to one estimate, a $275 million investment in wetland restoration in Romania alone would be recouped within six years from the ecosystem goods and services provided by the delta—including reduced pollution loads, flood control and regenerated fisheries.[45] In addition to encouraging greater international cooperation for ecosystem restoration, the WFD is giving new impetus to more ecologically sound flood management. A new directive on the Assessment and Management of Flood Risks came into force in November 2007 and calls for the preparation of flood hazard maps and flood risk management plans that are aligned with the WFD's environmental objectives.[46] The flood directive comes on the heels of several years of severe and costly flood damages. Between 1998 and 2002, European countries experienced a hundred major floods—including extreme events along the Danube and Elbe rivers—that collectively caused 700 deaths, the displacement of half a million people and economic losses totaling 25 billion euros.[47] More than 10 million people live in areas at risk of extreme floods along the Rhine River, a heavily channelized waterway that no longer meanders but flows artificially straight between engineered embankments.[48] In its upper reaches, the river is cut off from 90 percent of its original floodplain.[49] The Rhine now flows twice as fast as it did before, and flooding in the basin has grown more frequent and damaging. The European Commission estimates that assets worth 165 billion euros are potentially at risk from flooding of the Rhine.[50]

It is too early to judge the effectiveness of the EU's WFD. Nonetheless, most member states have made significant progress since the directive came into force. The Common Implementation Strategy has created a network of one hundred experts from over thirty countries and twenty-five pan-European stakeholder and other organizations that are providing input into the process and providing a platform for building implementation capacity. Among the shortcomings to date is the inadequate transposition of the WFD into national law. The identification of water bodies at risk of failing to achieve the WFD objectives is a critical part of the knowledge base required to develop effective river basin management plans. In general, however, there is insufficient data for member states to evaluate these risks for a large share of water bodies and, in some cases, insufficient evidence that states are even committed to gathering the needed data.[51]

On paper, however, the EU's WFD offers great promise for protecting and restoring ecological flows for Europe's rivers. It has already provided new standards against which organizations and agencies can judge proposed water projects and plans. For example, opponents of the large-scale river diversions that have been part of Spain's national hydrological plan have pointed out that the plan—for which Spain had sought European Community funding—is at odds with the principles and objectives established by the new European water policy directive.[52]

MOVING FORWARD

Rivers, floodplains, wetlands and watersheds constitute ecological infrastructure of increasing value that warrants protection. Policies that safeguard ecological services aim to maximize the full value of water to society, taking into account both extractive uses of water as well as water's functions within the natural environment. By setting boundaries on the degree to which human activities degrade ecosystems and their services, these policies can help maintain this ecological infrastructure into the future. Moreover, establishing these boundaries will unleash the potential of water conservation and efficiency measures to meet new water demands without extracting more water from natural ecosystems. As such, it will drive up water productivity—the value society derives from each liter of water extracted—while keeping ecological infrastructure intact.

With the impacts of climate change unfolding faster than scientists had predicted even five years ago, adopting policies that preserve the robustness and resilience of nature's way of mitigating floods, droughts and other disasters will be especially critical. Maintaining and expanding ecological infrastructure—including healthy rivers, wetlands and floodplains—is a key element of effective climate adaptation strategies.

Over the last decade, a number of national, regional and international governing bodies have adopted water policies specifically aimed at protecting freshwater ecosystems and their services. The implementation of these policies—within South Africa, Australia, the Great Lakes basin, U.S. states such as Texas and the European Union—warrant support and attention.

NOTES

- The author would like to acknowledge the research assistance of Priyanka Mehrotra in the preparation of this article.

1. Postel, Sandra. *Pillar of Sand: Can the Irrigation Miracle Last?* (New York: W.W. Norton & Co., 1999).
2. Rabbinge, Rudy and Prem S. Bindraban. Poverty, Agriculture, and Biodiversity in *Conserving Biodiversity*, ed. John A. Riggs (Washington, DC: The Aspen Institute, 2005), 65–77; Vorosmarty, Charles J. and Dork Sahagian. Anthropogenic Disturbance of the Terrestrial Water Cycle. *Bioscience* 50, no. 4 (September 2000), 753–65. Percentages calculated by author assuming 40,000 cubic kilometers per year of global runoff; Syvitski, James P.M., et al., Impact of Humans on the Flux of Terrestrial Sediment to the Global Coastal Ocean. *Science* 308, no. 5720 (15 April 2005), 376–80.
3. Postel, Sandra and Brian Richter. *Rivers for Life: Managing Water for People and Nature* (Washington, D.C.: Island Press, 2003), 2.
4. Millennium Ecosystem Assessment. *Ecosystems and Human Well-Being: Current State and Trends.* (Washington, D.C.: Island Press, 2005).
5. Rykiel, E. Ecosystem Science for the Twenty-First Century. *Bioscience* (October 1997), 705–08.
6. Louisiana Department of Natural Resources, Office of Coastal Restoration and Management, http://www.dnr.state.la.us/crm/coastalfacts.asp (accessed 31 August 2005).
7. Dollar, Evan. Balancing Use and Protection of Water Resources: Democratizing Water Management in South Africa. (keynote address, *Massachusetts Water Resources Research Center's 4th Annual Conference and associated Workshop on South Africa's Water Policy*. University of Massachusetts-Amherst, Amherst, Mass.: 9–10 April 2007).
8. South African National Water Act No. 36. 1998. (note 11), Pan 3: The Reserve and Appendix 1:

Fundamental Principles and Objectives for a New Water Law in South Africa.

9. When the government changed hands in 1994, some 14 million poor South Africans lacked water for these basic needs.
10. South African National Water Act.
11. Postel and Richter (2003); N. LeRoy, Poff, et al. The Natural Flow Regime. *Bioscience* 47, no. 11 (December 1997), 769–84.
12. Dollar (2007).
13. Agriculture and Resource Management Council of Australia and New Zealand, and the Australian and New Zealand Environment and Conservation Council. *National Principles for the Provision of Water for Ecosystems.* Revised Draft, November 2001.
14. Environmental Water Provisions Policy for Western Australia. *2000 Statewide Policy No. 5.*
15. The Delicate Balance of Sharing Water. *World Water and Environmental Engineering,* July–August 2001.
16. Gardner, A. Environmental Water Allocations in Australia. Paper. *International Symposium on Sustainable Water Management.* Canberra: 15–16 September 2005, as presented in Nevill, Jon. Australian Progress in Conjunctive Water Management: Comment on Recent Progress in the Development of Water Balance Planning, and the Supply of Environmental Flows to Groundwater Dependent Ecosystems. Manuscript submitted to *Ecological Management and Restoration.* December 2007.
17. Murray-Darling River Basin Commission, http://www.mdbc.gov.au/about/basin_statistics.
18. Blackmore, Don J. The Murray-Darling Basin Cap on Diversions: Policy and Practice for the New Millennium. *National Water* (15–16 June 1999), 1–12; Drying Out. *Economist* (12 July 2003), 38.
19. Murray-Darling Basin Commission (MDBC). *Annual Report 2003–04.* (Canberra: MDBC, 2004). The actual diversions allowed under the cap vary from year to year depending on climatic and hydrologic conditions, but are pegged to 1993-1994 withdrawal levels as described in *The Cap* (Canberra: MDBC, 2004).
20. MDBC, "Pilot Interstate Water Trading Project," http://www.mdbc.gov.au/naturalresources/water-trade/pilot_watertrade.htm.
21. Young, Mike, et al., Inter-State Water Trading: A Two Year Review. (Canberra, ACT: Commonwealth Scientific and Industrial Research Organization. *[CSIRO] Land and Water,* December 2000).
22. Study cited in Blackmore (1999).
23. Young, Mike and Jim McColl. The Unmentionable Option: Is there a Place for an Across-the-Board Purchase? *Droplet* 8 (July 2007).
24. Benke, Arthur C. A Perspective on America's Vanishing Streams. *Journal of the North American Benthological Society* 9 (March 1990), 77–88.
25. National Wildlife Federation, "Texas' Groundbreaking Environmental Flow Legislation," Summer 2007; see also http://www.texaswatern-natters.org/flows.htm.
26. National Wildlife Federation, Fall 2007.
27. Kelly, Mary. *A Powerful Thirst: Water Marketing in Texas* (Austin, Tex.: Environmental Defense, 2004).
28. Edwards Aquifer Authority, website, http://www.edwardsaquifer.org.
29. Kelly (2004).
30. Ramirez, Jorge A. SAWS, Extension Partner to Help Folks in San Antonio Conserve Water. *AgNews,* 23 January 2002.
31. The official names of these agreements are the Great Lakes-St. Lawrence River Basin Sustainable Water Resources Agreement and the Great Lakes-St. Lawrence River Basin Water Resources Compart. For more information, see the Council of Great Lakes Governors website http://www.cglg.org'projectVwater/index.asp.
32. "Great Lakes-St. Lawrence River Basin Water Resources Compact," 13 December 2005, 17, http://www.cglg.org/projectsAvater/Compact Implementation.asp.
33. Government of Ontario, Natural Resources Information Centre, "The Great Lakes Charter

Annex Agreements Backgrounder," 13 December 2005, http://www.mnr.gov.on.ca/Mnr/csb/news/2005/-decl3bg_05.html.

34. Council of Great Lakes Governors, "Great Lakes Water Conservation and Efficiency Initiative," http://www.cglg.org/piojects/water/ConservationEfficiencylnitiative.asp.
35. International Joint Commission. *Protection of the Waters of the Great Lakes: Interim Report to the Governments of Canada and the United States.* (Washington, D.C. and Ottawa: 1999), 28.
36. This classification approach is similar to that used in South Africa.
37. European Parliament and Council of the European Union. Directive 2000/60/EC Establishing a Framework for Community Action in the Field of Water Policy. *Official Journal of the European Communities* 22 December 2000, L 327: 1–72.
38. Commission of the European Communities, Commission Staff Working Document, Accompanying Document to the Communication from the Commission to the European Parliament and the Council, "Towards Sustainable Water Management in the European Union, First Stage in the Implementation of the Water Framework Directive 2000/60/EC (COM [2007] 12T final) (SEC [2007]) 363), http://ec.europa.eu/environment/water/water-framework/implrep2007/index_en.htm.
39. Postel and Richter (2003), 187.
40. Ibid.
41. Maigesson, Rhoda. Reducing Conflict over the Danube Waters: Equitable Utilization and Sustainable Development. *Natural Resources Forum* 21, no. 1 (February 1997), 23–38.
42. World Wide Fund for Nature, Living Waters Program-Europe. A Green Corridor for the Danube. As described in Postel and Richter (2003).
43. Schmidt, Karen F. A True-Blue Vision for the Danube. *Science* 294 (16 November 2001), 1444–1447.
44. WWF, Living Waters Program-Europe.
45. Schmidt.
46. U.K. Department for Environment, Food and Rural Affairs, "EU Floods Directive," http://www.defra.gov.uk/environ/fcd/eufldir/default.htm.
47. *European Commission Press Release.* Flood Protection: Commission Proposes Concerted EU Action (12 July 2004).
48. European Commission, "Towards a European Action Program on Flood Risk Management," http://ec.europa.eu.int/comm/environment/water/flood_risk/index.htm.
49. Abramovitz, Janet N. Unnatural Disasters. *WorldWatch Paper No. 158* (Worldwatch Institute: Washington, D.C., 2001).
50. "Toward a European Action Program on Flood Risk Management," (working paper, European Commission), http://ec.europa.eu/environment/water/floodjisk/key_docs.htm.
51. Commission of the European Communities. *Commission Staff Working Document* (2007).
52. World Wildlife Fund-Europe, "Pipe Dreams for Spain's Water," 17 July 2002; World Wildlife Fund-Europe, "Seven Reasons to Stop the Spanish National Hydrological Plan," 27 June 2002; World Wildlife Fund-Europe, "WWF Supports Protests against Spanish Hydrological Plan," 7 September 2001. See articles at httpy/www.panda.oig/about_ww(7what_we_do/freshwater/news»/news.cfm?uNewsID=2753.

Water Management in India:

An Offspin of Scarcity?

By S. Venkata Seshaiah

India is blessed with many rivers. Monsoon rules most part of the country months long. Yet studies carried out on water management have confirmed scarcity of water in many parts of India. The issue of water scarcity needs careful handling by the economic agents such as the policy-makers, producers, politicians and consumers. If the oil prices rise, the economic agents can reduce oil consumption by sticking to "stacations[1]," resorting to car pooling, purchasing more fuel efficient vehicles or shifting closer to their workplaces. But this is not the case with water-related issues as water is the very basis of food and livelihood. Agriculture, real estate boom, urbanisation, demographic features, behaviour patterns are among the various factors that account for huge pressure on water. India being a predominantly agricultural country, there is an enormous need for water to raise a variety of crops. Nearly 84% of available water is used for the agricultural sector in India and the remaining 16% is used for the purposes of industrial and household consumption. Farmers in India largely produce paddy which consumes a lot of water. If they are unable to cultivate paddy, they assume that they have no resources even if they have huge bank balances. This is due to the fact that in rural areas, paddy is treated as an embodiment of the goddess of wealth. The people in rural India are overly dependent on agriculture for their living. In this scenario, it is very difficult to transform the economy from one based on agriculture sector to one based on manufacture and service-sectors. Most of the tanks and lakes have been converted into construction sites for housing which has further intensified the water problem. The frequent droughts, floods and disguised unemployment across rural areas—reflective of policy failures at various levels in tackling the rural issues in India—are the major causes of migration from rural to urban areas in India. Migration of this sort has been continuously building pressure on demand for water across India. Due to rise in population, demand for drinking water in India has stood at 20 billion Cubic Metre (BCM) per year. With 450 millon Indians going to cross the age of 19 by the end of this year the future drinking water demand of young India is projected to be 51.33 BCM per year. Indians now-a-days tend to instal and use european style toilets which consume a large amount of water. In thickly populated countries like India, water is becoming a scarce resource year after year. Hence efficient water management has become the need of the hour.

PRESENT USE AND FUTURE REQUIREMENT OF WATER

According to the statistics of the Government of Maharashtra, 95% of the urban population depend on tap water as their source of drinking water. Of the rural population, 50% depend on tap water, 27% on wells and 20% on hand pumps (Exhibit I).

The usage of water increased across all the sectors and was projected to keep this upward trend. Irrigation continues to exert a major demand on water(Exhibit II).

The demand for drinking water is continuously increasing and will compound in the future. It is projected to rise from 49,935 BCM to 140,650 BCM as per the UN report and from 43,065 BCM to 100,755 BCM as per the Census report within a span of 50 years between 2001 and 2050 (Exhibit III).This amounts to roughly threefold increase in the demand for drinking water in India.

Exhibit I. Distribution of Households by Major Source of Drinking Water(%)

MAJOR SOURCE	RURAL	URBAN
Tap	50.40	95.00
Tube well/Hand pump	20.60	3.10
Well	27.20	1.60
Other	1.80	0.30
Total	100.00	100.00

Source: Economic Survey of Maharashtra 2002–03. Directorate of Economics and Statistics, Planning Department, Government of Maharashtra

While the dam-wise water level in India has been declining for all the water reservoirs except Matatila reservoir (Exhibit IV), waste water generation in India keeps on increasing decade after decade (Exhibit V).

In less than two decades (1977–1995) waste water generation in India has more than doubled from 7,007 mld to 16,662 mld.

SITUATION ANALYSIS AND STRATEGY AGENDA

As per the projected statistics given in Exhibit II, the population of India could be between 1.7 billion by the year 2050. With the rise in population, the demand for water is going to intensify for domestic use, farm irrigation and industrial purposes. It is, therefore, imperative that everybody involved in the process of consumption of water should coordinate and take measures to make optimal use of scarce water resource. The municipal corporations should focus on providing water services and other government agencies should handle the development and monitoring aspects of water management. Through proper distribution of functions and responsibilities, efficient water management can be carried out. The usage characteristic of water has both the necessity and luxury dimensions. Hence, there is a need for introducing user fee. While imposing the user fee, the government should pay attention to charging lower fee for necessary consumption and higher fee for luxury consumption.

Proper awareness programmes need to be designed and conducted for the public by appointing water saving and usage committees in villages, towns, cities and metros. These committees should explain the repercussions of water wastage and water pollution on health, environment and overall

Exhibit II. Sector-Wise Present Water-Use and Future Requirement (1990-2050)

YEAR		SECTOR WISE WATER-USE AND FUTURE REQUIREMENTS (MILLION HECTARE-METRES)				
	POPULATION (MILLION)	IRRIGATION	DOMESTIC AND LIVESTOCK	INDUSTRY	THERMAL POWER	TOTAL
1900	800	46.0	2.5	1.5	3.0	53
2000	1,000	63.0	3.4	3.6	5.0	75
2025	1,400	77.0	5.0	12.0	16.0	110
2050	1,700	70.0	6.0	20.0	16.0	112

Note: Figures on population, past water-use and future water requirements are approximate.
Source: Urban Statistics, Hand Book 2000. National Institute of Urban Affairs Statistics, Planning Department, Government of Maharashtra

Exhibit III. Future Drinking Water Demand in India

YEAR	TOTAL WATER DEMAND*		BCM/YEAR	
	BASED ON PAST CENSUS	BASED ON UN PROJECTION	BASED ON PAST CENSUS	BASED ON UN PROJECTION
2001	43,065	49,935	15.72	16.03
2011	54,810	63,555	20	23.2
2021	66,555	83,375	24.29	30.43
2025	71,340	91,350	26.04	33.34
2050	100,755	140,650	36.77	51.33

* The water demand has been worked out @ 170 lpcd for 65% of the urban population presumed to be living in Class 1 cities and @ 100 litre per capita daily (lpcd) for balance 35% of the urban population living in Class 2 cities. (*Report of the National Commission for Integrated Water Resource Development Plan.)*
Source : National Commission for Women

Exhibit IV. Dam-wise Water Level in India (2007–2008)

DAM	RESERVOIR LEVEL (FT.)		RESERVOIR LEVEL (FT.)		DEVIATION (-)/(+) OF LEVEL W.R.T. LAST YEAR
	DATE	LEVEL	DATE	LEVEL	
Bhakra	6.3.08	1,553.53	6.3.07	1,596.10	(-)42.57
Pong	6.3.08	1,297.98	6.3.07	1,344.69	(-)46.71
Thein Dam (Ranjit Sagar)	6.3.08	1,632.02	6.3.07	1,648.98	(-)16.96
Rihand	6.3.08	842	6.3.07	843.70	(-)1.7
Matatila	5.2.08	997.11	5.2.07	994.19	(+)2.92
Gandhi Sagar	6.3.08	1,273.98	6.3.07	1,291.21	(-)17.23
Bansagar	28.2.08	1,079.13	28.2.07	1,084.78	(-)5.65
Indira Sagar	6.3.08	815.94	6.3.07	817.78	(-)1.84
Bargi	23.2.08	1,353.35	23.2.07	1,356.96	(-)3.61

Source: Lok Sabha unstarred question no. 2192, dated on 14.03.2008.
Data may be reproduced for research, analysis, survey, review, studies and such other academic purposes with due acknowledgement.

Exhibit V. Population and Waste Water Generation in India

YEAR	URBAN POPULATION	LITRES/CAPITA/DAY (LPCD)	GROSS WASTE WATER GENERATION (MLD)
1977–1978	60	116	7007
1989–1990	102	119	12,145
1994–1995	128	130	16,662
2001	285	–	–
2011	373	–	–
2021	488	121 (A)	59,048 (P)
2031	638	121 (A)	77,198 (P)
2041	835	121 (A)	101,035 (P)
2051	1,093	121 (A)	132,253 (P)

Source : Ministry of Environment and Forests, Government of India
P: Indicates Projected A: Indicates Assumed

economic activity. These committees should also raise awareness among the farmers on cash crops and encourage them to cultivate the crops that consume less water. It is very difficult to manage water problem in India until and unless water usage is guided by the principles of market driven economy. But it is debatable whether market price of water—reflective of its scarcity—holds the promise for its efficient allocation.

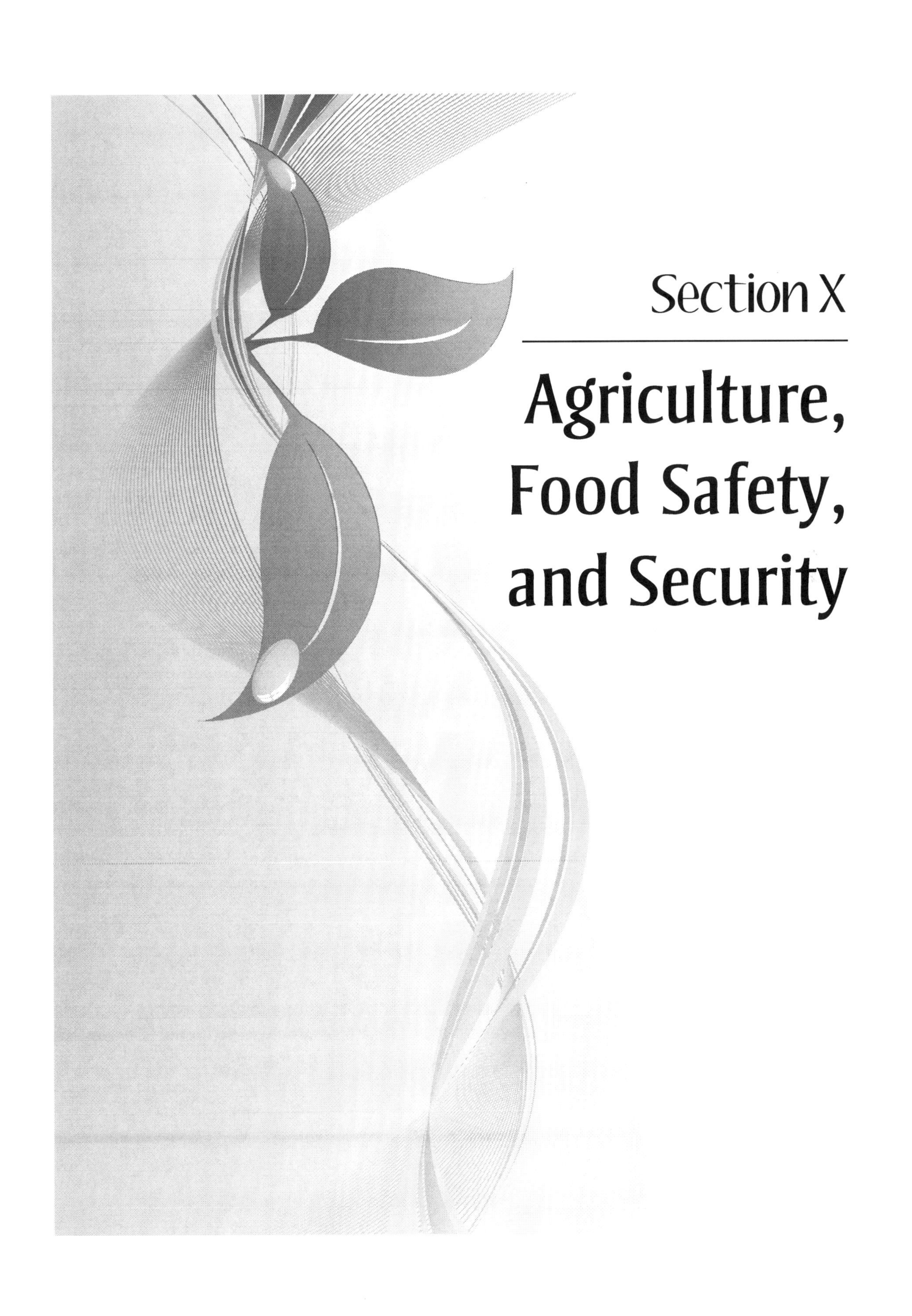

Section X

Agriculture, Food Safety, and Security

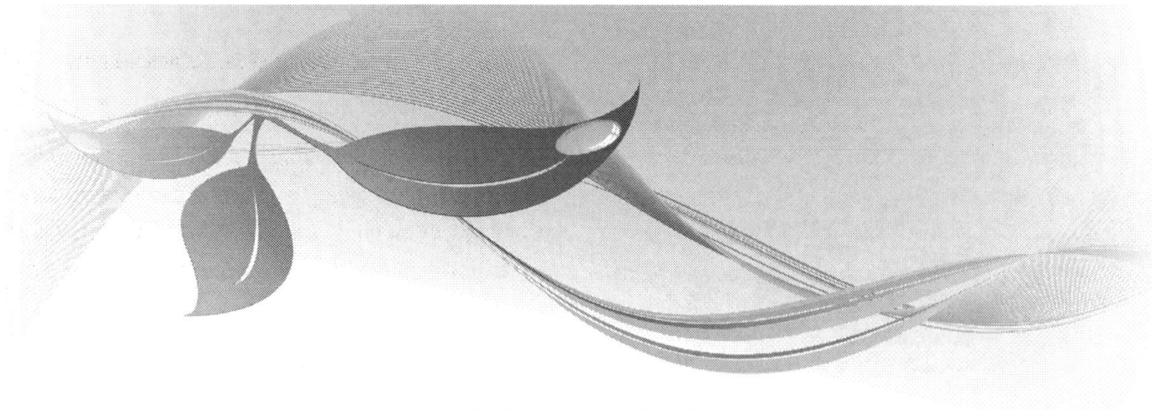

By Anne Marie Zimeri

Food security for all would exist when all people at all times had access to sufficient, safe, nutritious food to maintain a healthy and active life, according to the World Health Organization. However, our world is currently far from being food secure. Even developed nations face food-security issues. Three main aspects to food security affect both developed nations or countries and lesser developed nations or countries (LDCs):

1. **Food Availability:** sufficient quantities of food must be available on a consistent basis. At issue are producing enough food on the land available to feed the number of people of the planet *and* distributing the food. Currently, if we take calorie production and divide it by the number of people on the planet, there are enough calories to sustain us. Those calories are not distributed equally, however, and many populations do not have enough to eat, whereas others consume far too many calories. The future will be even more challenging because the 0.7 acres of farmland we currently have per person are expected to shrink to 0.4 acres per person in the next few decades due to population growth. We know that our future needs cannot be met by current production, so we will need to solve this problem by addressing population control and applying new technology to our agriculture.
2. **Food Access:** the population must have sufficient income to afford food for a nutritious diet on a regular basis and in the face of a crisis (economic or climactic).
3. **Food Use:** we must use, prepare, and grow food in an appropriate way. This includes diversifying agriculture.

Challenges to Food Production

Below, two of the main challenges to food production are briefly discussed. Many of these problems will be addressed by solutions that the current younger generation proposes in the global food market.

Soil Loss

Soil is one of our most valuable resources because it is where we cultivate our crop plants. Ongoing soil erosion and expanding urbanization are contributing to the loss of cropland in the United States and around the world. Of the millions of acres of cropland removed from agricultural use each year, more than two million are lost to erosion, while another million are lost to urbanization (building transportation networks, urban sprawl, etc.).

How we use our cropland is extremely important in the United States where more than 99% of our food originates (only 1% comes from aquatic systems). Of our 400 million acres of arable land now under cultivation in the United States, 215 million are planted with grains. These grains, or commodity crops, can be stored and transported quite easily. A large portion of these grains (68 million acres' worth) is farmed for livestock feed. About 80 million acres' worth of grain is exported. Can we expand this system? In the United States, only 20% of our 2.3 billion acres is sufficiently fertile enough for agriculture. The remainder is too cold, too wet, too dry, or too steep for crop production. The solution is to use the land we have in an efficient, sustainable way. Soil erosion removes about $20 billion worth of plant nutrients from U.S. agricultural soils each year. These nutrients then must be replaced by applications of fertilizers, which require the use of fossil fuels to mine and to apply. Sustainable farming techniques, including crop rotation, no tilling, and increased human labor rather than machines and pesticides, may hold the key to producing more without taking away from future production.

Water Resources

Agriculture is the major consumer of water in the United States. It uses about 70%–85% of all U.S. freshwater resources. It is necessary because plants use water to transport nutrients from the roots to the shoots and to the edible portions through the plant's vasculature. Plants must also respire as part of their growth and metabolic processes. This respiration can be massive. For example, a medium-sized corn crop that produces 120 bushels per acre will consume more than 500,000 gallons of water during the growing season. To calculate that on a smaller scale, one pound of corn grain requires about 175 gallons of water for its production.

Experience-based Measures of Food and Water Security

By Craig Hadley and Amber Wutich

INTRODUCTION

Food and water security are recognized as basic human rights in the Universal Declaration of Human Rights (UN 1948) and the International Covenant on Economic, Social, and Cultural Rights (UN 1999, 2002). The international community affirmed the importance of reducing food and water insecurity in initiatives sponsored by the United Nations, the World Health Organization, the World Water Forum, and other major agencies. Despite these efforts, there are currently 791 million people estimated to be food insecure and 1.1 billion people estimated to be water insecure around the world (FAO 2005; WHO and UNICEF 2006). Food insecurity is associated with reduced growth, immune function, cognition, and mental health (Struble and Aomari 2003). Water insecurity is associated with 80 percent of illnesses and 30 percent of deaths in developing countries (Elhance 1999). While food insecurity is expected to decline globally, it is projected to increase in sub-Saharan Africa—affecting every third person by 2010 (Wiebe, Ballenger, Pinstrop-Andersen 2001). By 2020, it is projected that between 34 and 76 million people will die of preventable water-related illnesses (Gleick 2004).

Given the threats that food and water insecurity pose to human health and well-being, it is vital that these trends be monitored accurately. The tools available to measure food and water security, however, have conceptual and practical limitations (Kennedy 2002; Satterthwaite 2003; Webb et al. 2006). There is increasing support for a shift toward experience-based approaches to measuring food insecurity (e.g., Frongillo 1999; Webb et al. 2006; Wolfe and Frongillo 2001), although measurement of water insecurity has not yet reached this level of sophistication. We suggest that anthropologists, while not highly engaged in this literature, are uniquely positioned to contribute to the growing need to develop valid and locally-appropriate tools to measure the experiential dimensions of food

Craig Hadley and Amber Wutich, "Experience-based Measures of Food and Water Security: Biocultural Approaches to Grounded Measures of Insecurity," *Human Organization*, vol. 68, no. 4, pp. 451-460.

and water insecurity. In this paper, we present case studies from Tanzania and Bolivia that (1) detail the steps used to adapt or create experience-based measures and (2) validate these measures using a suite of established approaches. We believe that these case studies illustrate general strategies that will be helpful to researchers working in a variety of research settings.

DEFINING FOOD AND WATER INSECURITY

Food and water insecurity occur when there is insufficient and uncertain access to adequate food and water for an active and healthy lifestyle (FAO 2005). As this definition illustrates, there are three key dimensions of insecurity: adequacy, access, and lifestyle. Measures designed to determine the adequacy of food and water consumption often assess absolute intake. Such measures include caloric consumption, measures of daily water use, and anthropometric measures. In general, daily human biological requirements for food and water include approximately 2,200 kcal for a 150 lb. person engaging in moderate levels of physical activity level and 50 liters of water for drinking, cooking, and sanitation in a low- technology environment (Gleick 1996). Food or water needs can vary widely depending on the activity level or biocultural requirements of the people under study, which means that absolute measures of intake may not provide comparable data on food or water adequacy across field sites (Sawka, Cheuvront, and Carter III 2005; WHO 1985). In recognition of this, anthropologists also employ anthropometric indicators, such as weight for age, height for age, and body mass index, to provide insight into which households are potentially experiencing difficulties with food insecurity (e.g., Crooks, Cligget, and Cole 2007). None of these existing approaches measure the experience of food or water insecurity (Frongillo 1999, Himmelgreen, et al. 2000; Webb, et al. 2006; Wolfe and Frongillo 2001).

An approach to measuring the access dimension of insecurity developed out of Sen's (1981) finding that inequitable access to food—rather than the absolute scarcity of food—produced food insecurity. Methods for assessing food access include the distance to food markets, the availability of land or livestock to produce food locally, household income, socioeconomic status, and food prices (Morris, et al. 2000). Methods for assessing water access include the distance to water sources (Sugita 2004), seasonality in water availability (Billig, Bendahmane, and Swindale 1999), water expenditures, time spent acquiring water (Pattanayak, et al. 2005), and storage capacity (Wutich 2006). These methods, while important, have been critiqued because they capture neither the adequacy of the quantity of food/water acquired nor the security of access to those resources (Webb, et al. 2006). As a result, it can be difficult to make the link between inequities in access and their health impacts.

Lifestyles, livelihoods, identities, and life ways are literally embodied in an individual's food and water needs (Krieger 2005). For example, small stature and light weight of millions of children in developing countries is the embodiment of poverty, social exclusion, limited health infrastructure, repeated bouts of illness, and, when viewed more broadly, tremendous inequality in health and well-being. Conversely, cultural adaptations allow humans to thrive under conditions that might otherwise be considered highly food or water insecure (see Moran 2000). As a result, the approaches outlined above that focus on physiological demands for food and water do not take into account culture, experience, and perceptions which may result in an incomplete picture of insecurity and its health consequences. Recognizing this, there has been a shift by nutritionists to develop experience-based measures of food insecurity (Webb, et al. 2006; Wolfe and Frongillo 2001). These measures, in particular the Radimer/ Cornell Scale (Radimer 1990) and the United States Department of Agriculture (USDA) food security module (Kennedy 2002), seek to assess the experience of food insecurity, rather than the absolute intake of food or specific nutrients. Thus, by their very nature, they recognize that food/water needs may differ in quality or quantity from basic physiological needs (McElroy and Townsend 2004; Pike and Williams 2006; Wiley 1992; Wolfe and Frongillo 2001). While these types of scales are increasingly used in public health nutrition research on food insecurity, they have not been widely used in the anthropological literature.

Here, we present two cases that adapt or build experience-based scales to show how human food and water needs reflect biological, social, and cultural requirements. Experienced-based scales seek to measure an individual's experience of insecurity, rather than proxy measures. In the first case, we show how an existing food security scale can be adapted for a specific cultural context. In the second case, we show how a water insecurity scale can be developed from the ground up. Because of the paucity of experiential water insecurity scales, we argue that scale creation is the most fruitful approach for developing new water insecurity measures.

CASE STUDY 1: ADAPTING AN EXISTING MEASURE TO MEET ETHNOGRAPHIC REALITIES OF FOOD INSECURITY IN TANZANIA

Tanzania has a high prevalence of food insecurity; estimates suggest that 44 percent of Tanzanian children suffer from hunger and its sequelae (FAO 2004). There is, however, considerable heterogeneity in food insecurity across regions and within communities. Here, we explore this heterogeneity in a single set of villages that were primarily comprised of two ethnic groups, Pimbwe and Sukuma. Pimbwe are primarily horticulturalists and Sukuma are agropastoralists; this difference is just one of many cultural, social, and economic differences that distinguish these two groups (Hadley 2005). Ethnographic and pilot research revealed the presence of a hunger season, during which time a substantial portion of the population appeared to experience food insecurity. During this period, respondents were observed seeking food, selling their labor for food, altering the amount and types of food they ate, and frequently lamented that children were "going to bed hungry." More specifically, meals were consumed less frequently and corn was consumed often without previous milling, which informants claimed made them feel fuller on less corn. These expressions of food insecurity declined rapidly with the first harvest as is often found in seasonal subsistence economies. Clearly insecure access to food and hunger were key dimensions of poverty and possible predictors of health in this area, but how would food insecurity be measured in a systematic, valid, and reliable fashion?

Initially, we sought to measure the date at which food stores ran out—this made sense in a setting where households harvested a large portion of their yearly food supply over the course of a few weeks, and then drew on those stores until they were exhausted or the next harvest occurred. We quickly realized that this measure was meaningless for community members who ran small shops or did government work rather than farming. Additionally, some households had food stocks that ran out early, but did not actually *experience* food insecurity because they had large kin or social networks that provided them with food. We also learned that food-stressed households employed a limited number of coping strategies in the face of food shortfalls. These observations led to the development of a coping strategies index (Maxwell 1996). We sought to assess the severity and distribution of food stress by asking individuals whether they had engaged in a variety of locally appropriate behaviors (Hadley, Mulder, and Fitzherbert 2007). This measure too had several conceptual problems.

One major issue was that not all individuals had equal opportunity to engage in each strategy. Selling assets was a common strategy but one that was not available to those who had no assets left to sell. Similarly, people could ask relatives for money only if they had relatives nearby. The greater conceptual problem with the coping strategy index was that it did not measure the experience of food insecurity. A respondent could endorse all of the behaviors on the coping strategy inventory, but *not* experience food insecurity. If the coping strategies were successful, then food insecurity would potentially not result; thus, two households could engage in the same strategies but, at least theoretically, have two very different responses. The measure, therefore, failed to capture the experience of food insecurity.

We then translated the USDA Household Food Security Survey Module (Bickel, et al. 2000), which we selected because the experience of food insecurity in this part of the world appeared to be conceptually similar to the experience that was depicted in the USDA's instrument. Since we carried out our survey, several studies have used this measure in rural, under-developed contexts (e.g., Coates, et al. 2006a; Coates, et al. 2006b; Melgar-Quinonez, et al. 2006; Perez-Escamilla, et al. 2004; Swindale and Bilinsky 2006). However, use by biocultural anthropologists still remains rare. This scale asks respondents whether they have experienced a variety of scenarios over a predefined time period. For instance, respondents are asked whether they ran out of food in the three months prior to the survey, whether they ever felt hungry but could not eat because there was not enough food or money to buy food; or whether their children ever had to skip meals.

All items, with slight modification, resonated well with respondents, and were widely confirmed to represent the key aspects of food insecurity in the study villages. This supports one aspect of the content validity, the face validity, of the instrument and corroborates much of our earlier work that showed that reductions in number, frequency, and diversity of meals were experiences of the food insecure households, and that worry and anxiety over not having sufficient food or more commonly running out of food prior to the next harvest were common occurrences. Several modifications were also made to the instrument, which included altering the items so that they were read as questions, removal of three items on frequency, changing the wording to allow for acquisition of food through means other than purchasing, and altering the questions about meal frequency to more closely map onto local dietary patterns (e.g., two meals is the most common meal frequency). Assessments of the internal consistency reliability revealed acceptable levels in both the wet and dry seasons (alpha>80); this was true for both horticulturalists and agriculturalists (i.e., Pimbwe and Sukuma).

Because no gold standard of food insecurity exists, we were unable to assess criterion validity. To assess construct validity, we created a series of hypotheses about the relationship between the food insecurity scale and measures of (1) season, (2) ethnicity, (3) coping strategies, (4) socioeconomic status, and (5) dietary diversity. Results supported the construct validity and performance attributes of the scale. First, our previous work revealed a strong seasonal dimension to food insecurity and this was captured by the food insecurity scale. In a longitudinal study of 173 households, the average food insecurity score was 2.2 (3.1) in the dry season but increased to 3.8 (3.9) in the wet season survey; higher scores reflect greater endorsement of food insecurity items and this difference was statistically significant ($p<0.0001$). Second, our previous work also suggested a group-level difference in vulnerability to food insecurity and this difference was captured by the measure. There was no difference in the dry season food insecurity score between Sukuma and Pimbwe. During the generally food secure dry season, Sukuma respondents had an average food insecurity score of 1.97 (2.9), whereas Pimbwe respondents had a score of 2.38 (3.2; $p=0.39$). In contrast, during the wet season survey period significant differences in mean food insecurity values emerged. In this season, Sukuma households reported a mean food insecurity value of 2.4 (2.8) and Pimbwe households reported a mean value of 5.1 (4.3); this ethnic difference was statistically significant ($p<0.0001$; Figure 1). Paired tests show that Sukuma households experienced no overall change in their food insecurity status ($p=0.21$) while Pimbwe respondents showed a significant increase indicating higher levels of food insecurity ($p<0.0001$). Third, we compared scores on the food insecurity measure among those who had and had not engaged in several locally-appropriate coping strategies during the food insecure wet season. Respondents who reported having to borrow money, having to borrow food, or having to sell their labor for food in the last three months because they had run out of food scored higher on the food insecurity measures ($p<0.0001$). Fourth, the food insecurity measure was strongly associated with more traditional measures of wealth and material ownership: households that produced more corn, more rice, and planted more types of crops received significantly lower scores on the food insecurity measure, indicating greater food security (all $p<0.05$). This is consistent with our expectations because wealthier households and those that produce more should be less likely to be food insecure. Households that reported owning more material items, or assets, also scored lower on the food insecurity measure (all $p<0.0001$), and households that reported owning more animals had scores indicative of greater food security ($p<0.0001$). Fifth, we assessed whether the food insecurity measure was associated with one dimension of nutritional intake—low dietary diversity. Using the wet season data which had concurrent dietary information, the food insecurity measure was associated with mothers' ($r=-0.24$, $p=0.0008$) and children's ($r=-0.32$, $p<0.0001$) dietary diversity scores: as food insecurity scores increased, dietary diversity decreased, again as expected given that we would expect food insecure individuals to have poorer quality diets. All associations held between both ethnic groups. Subsequent analysis using multivariate regression techniques revealed that food insecurity remained significantly associated with dietary diversity even after controlling for other potentially confounding factors. This suggested that there is added value of the food insecurity measure.

Finally, we assessed the extent to which the measure of food insecurity was associated with an indicator of stress. This provided an opportunity to examine how well our measure assesses the broad health impacts of food insecurity. The analysis revealed a strong association between the experience of food insecurity and symptoms of anxiety and depression. Even after controlling for individual-level covariates, we found that women living in food-insecure households reported greater psychosocial stress (Hadley and Patil 2006).

Consistent with the overall theme of our argument, the experience of food insecurity—even moderate food insecurity that did not result in outright reductions in food intake—were embodied by these women and expressed as symptoms of common mental disorders. While indicators of nutritional status showed limited signs of undernutrition, the uncertainty and worry that surrounds the food quest for these mothers in these Tanzania villages impacts on well-being as measured as indicators of mental health. In this setting where women produce and prepare most food and where they are the primary child caregivers, insecure access to food and the threat of not being able to provide for their young children appears to bring with it a considerable psychosocial burden, and this is evident in women's comments that they feel hungry, eat the same food day after day, never consume any meat, and that they must put their children to be bed hungry. Future studies with men are obviously needed.

In this case, we used our previous ethnographic knowledge to select an existing scale to measure food insecurity. Given the impressively large number of existing scales available to researchers, we believe this approach of scale adaptation to be a practical, important, and feasible one. In many cases, no adequate measure will be available and new measures must be constructed and validated. Below, we review a study of water insecurity that demonstrates this process.

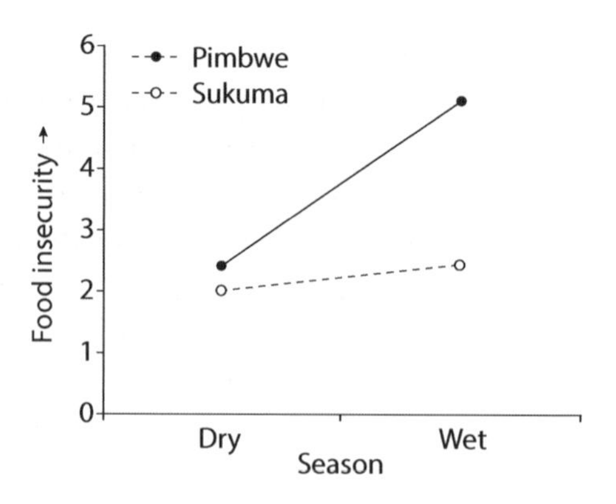

Figure 1. Seasonal Changes in Food Insecurity Scores among Mothers in Two Ethnic Groups Living in the Same Communities in Tanzania. Higher scores on the y-axis indicate greater food insecurity. The wet season reflects a period of the year when food availability is generally low and market prices high.

CASE STUDY 2: CREATING AN ETHNOGRAPHICALLY- GROUNDED MEASURE OF WATER INSECURITY IN BOLIVIA

The city of Cochabamba is located in a semi-arid region of the Bolivian Andes. Like many fast-growing cities, Cochabamba lacks sufficient water resources and water delivery infrastructure to meet its population's water needs. In 2003, at least 38 percent of the city's population lacked municipal water service (Terhorst 2003). Such households must find alternative water sources, such as rainwater, surface water, groundwater wells, and private vendors to acquire water for drinking, cooking, bathing, and cleaning. Because many of these alternative sources are vulnerable to climactic variability, year-round water insecurity intensifies during the dry season.

This case study was conducted in Villa Israel, a water-scarce squatter settlement located in the impoverished south side of Cochabamba. Participant-observation conducted during the dry season of2003 and the wet season of2004 indicated (1) that Villa Israel residents experienced water insecurity year-round and (2) that there was a clear seasonal dimension to the water insecurity. At the height of the wet season, for example, people used a small, highly-contaminated creek for activities as varied as car washing, clothes washing, bathing, and excretion. During the dry season, the creek disappeared entirely and people felt water insecurity even more acutely. We conducted interviews about water insecurity as part of a panel study of 72 randomly-selected Villa Israel households over five two-month cycles in 2004–2005. A detailed discussion of the research design, including the sampling frame and survey protocols, is available online (Wutich 2006). Because of the dearth of established methods for measuring water insecurity and related concepts (Gleick 2003; Satterthwaite 2003), we tested a number of different approaches.

During the first round of survey interviews, we measured per capita water consumption using a diary method. Although Villa Israel households clearly did not meet international water provision standards, water use did not appear to be a good indicator of variation in water insecurity within the community. This is because people can develop cultural adaptations that attenuate the effects of water scarcity. For instance, elderly women from the Altiplano used the least water for bathing (10 liters/person/week). According to them, distant water sources and cold weather led people in the rural highlands to bathe relatively infrequently and with less water. In contrast, migrants from the tropical lowlands had very high levels of water use for bathing (70 liters/person/

week) despite having relatively few resources to acquire water. In Villa Israel, then, it appeared that migrants from the water-scarce highlands were able to perform routine water use tasks with far less water than people who had come from cities or the water-rich lowlands. To capture the extent to which households had the water they needed to complete essential water use tasks, we developed an experience-based measure of water insecurity using Guttman's (1944) scalogram analysis method.

The Guttman method has long been used by anthropologists to create ethnographically-grounded scales that capture variation in a progressive, unidimensional concept across a set of indicators (Bernard 2006; Peregrine, Ember, and Ember 2004). For the data to scale adequately, scores on the indicators must be cumulative. This means that, if respondents score "yes" on an item of high difficulty, then they must also score "yes" on all the preceding items of lower difficulty. The scale is constructed by continually testing the fit of the data, eliminating indicators to produce a unidimensional scale in a theoretically-informed way, and refitting the data until a unidimensional scale is produced. The scale's fit is evaluated using the following criteria: (1) a coefficient of reproducibility (CR, the measure of scaling errors) greater than .90, (2) a coefficient of scalability greater than .60, and (3) a minimal marginal reproducibility of less than .90; Guttman scales should also contain at least nine indicators to avoid having a high CR by chance (Bernard 2006; Peregrine, Ember, and Ember 2004).

At the onset of the study, the research team spent five months living in the community, learning to use water in culturally-appropriate ways, and observing how water insecurity affected people's lives. Drawing from observations made during this research phase, we identified six domains in which water insecurity affected Villa Israel residents: water quantity, water quality, water acquisition, water conflicts, economic issues, and health outcomes. We designed a 33-item (yes/no) questionnaire to explore these content domains for the week preceding the interview (see Wutich 2006). Table 1 shows example questions from the six content domains. To assess the content validity, we pre-tested the questionnaire with key informants in Spanish and Quechua using the cognitive interviewing method (DeMaio and Rothgeb 1996). After pretesting was complete, we incorporated closed-ended questions into rounds 2–5 of the panel study.

Before we began the data analysis, we designed a theory-driven procedure for Guttman scale construction. The general plan was to eliminate sets of indicators in a way that continuously narrowed the definition of water insecurity, following the analysis plan in Table 1, until we reached a unidimensional set of indicators. If, in the last round of analysis, the data still did not scale, we would conclude that our data fail the Guttman test for unidimensionality. Critics of Guttman scales have observed that, if items are dropped simply because they are error-prone or to achieve the highest CR possible, the concept under study has no theoretical value. For this study, we were careful to design a theory-driven procedure that anyone could reproduce in the field. The analyses were conducted using the Guttman scaling routine in ANTHRO-PAC 4.0 (Borgatti 1996). In the first iteration of analysis, the data did not scale (CR=.809, CS=.317, MMR=.720). In the second iteration of analysis, the coefficient of reproducibility actually decreased (CR=.805, CS=.299, MMR=.721). In the third iteration of analysis, the coefficient of reproducibility continued to decrease (CR=.794, CS=.299, MMR=.707). In the fourth and final round of analysis, the coefficient of reproducibility was .881 (CS=.578, MMR=.717), which is smaller than is conventionally accepted (>.90). However, in a recent discussion of the anthropological literature on Guttman scaling, Bernard (2006), suggested that a CR>.85 was acceptable. We concluded that internal consistency reliability was acceptable, although we believe that further refinement of the indicators would be desirable in future studies.

Because there are no preexisting scales of water insecurity, much less an existing gold standard, we

Table 1. Content Domains, Order of Elimination in Iterative Guttman Scale Creation, and Example Questions

CONTENT DOMAIN	EXAMPLE QUESTION	ELIMINATION ORDER
Water conflicts	Did you argue with a member of your household in the last week due to water scarcity?	Iteration 1
Economic issues	Did you lack the money you needed to buy water in the last week?	Iteration 1
Health outcomes	Has anyone in your household gotten sick in the last week as a result of water scarcity?	Iteration 1
Water acquisition	Did you borrow water from anyone in Villa Israel in the last week?	Iteration 2
Water quality	Did you reuse greywater in the last week?	Iteration 3
Water quantity	Were you unable to cook because you didn't have enough water in the last week?	Iteration 4

were unable to test criterion validity. To assess construct validity, we tested a series of hypotheses regarding the association between the water insecurity scale and measures of (1) season, (2) water use, (3) income level, (4) water storage capacity, and (5) coping strategies. First, we found that there was a significant difference in water insecurity for four time periods across the dry and wet seasons ($F=4.24, p=.009$). At the onset of the dry season the average water insecurity score was high (3.83); it increased slightly at the height of the dry season (3.90). At the onset of the wet season the average water insecurity score decreased (3.47) and it was lowest during the height of the wet season (2.81). Second, we tested the association between water insecurity scores and per capita water use. As we noted during the first cycle of panel interviews, there was no significant association between diary reports of water use and water insecurity scores ($p=.11$). However, we found that there was a significant negative association between water insecurity scores and *self-reported* water use estimates, which we collected using free recall ($r=-.32, p=.006$) and prompted recall ($r=-.36, p=.002$) methods. This finding underscores the experiential nature of water insecurity. As we noted in the example of highland and lowland migrants' bathing water needs, people's perception of the adequacy water provision is shaped by past experiences and culturally-defined patterns of water use. Third, we found a significant negative association between income level and water insecurity scores ($r=-.55, p<.0001$). This is consistent with the observation that those with higher income can purchase water from private vendors. The water insecurity scale seems to capture the access dimension of water insecurity relatively well. Fourth, we found a significant negative relationship between household water storage capacity and water insecurity scores ($r=-.21, p=.02$), which indicates that people with more ability to store water are less likely to experience water insecurity. Fifth, we found that water insecurity scores had a significant positive association with two local coping strategies—borrowing water from a neighbor ($r=.43, p=.009$) and begging to be sold water from private vendors ($r=.49, p<0001$).

Data from 11 husband-wife pairs also revealed gender differences in the experience of water insecurity (Figure 2). Women reported significantly more water insecurity than men (Eta-sq.=.583, $p=.02$) with significant seasonal differences (Lambda=.654, Eta-sq.=.346, $p=.05$), particularly during the dry season and wet-to-dry season transition. Examining specific cases helps illustrate why this is the case. In one Villa Israel household, the husband worked downtown while the wife cared for their children at home. The wife was solely responsible for household water acquisition and was, therefore, much more sensitive to changes in water availability. In another household, both the wife and husband worked outside of the home. However, the wife acquired the household's water through a reciprocal arrangement with her parents. The wife was more aware when water was scarce or access was insecure because she experienced tensions or conflicts with her parents over water. In a third household, the wife was the primary wage earner while the husband stayed home with the children. Even so, the wife administered all household expenses, including water purchases. For this reason, she was very aware during the dry season, when rainwater collection was impossible, of how much of her income went toward water purchases. These examples demonstrate that, in households with very different economic and task structures, women generally bear the burden of water acquisition and, therefore, are more likely than men to perceive water insecurity during dry periods.

Finally, we assessed the extent to which the water insecurity scale was associated with emotional distress. Again, this analysis indicates how well our measure assesses the broader health impacts of water insecurity. The indicator of emotional distress is a four-point scale, developed from respondent narratives, that measures people's experiences of bother, fear, worry, and anger over water insecurity (alpha=.74). We found a strong positive association between the experience of water insecurity and emotional distress (r=.77,p<.0005). In Villa Israel, people's struggles to acquire secure access to adequate quantities of water were closely linked to their feelings of stress (Wutich and Ragsdale 2008). Clearly, the Guttman scale was effective in tapping the relationship between perceptions of water insecurity and the experience of emotional distress. The results of this and the previous hypothesis tests suggest that the Guttman scale for water insecurity has an acceptable degree of criterion validity.

To create a measure of water insecurity, we used our ethnographic knowledge to develop a unique scale from the ground up. We used participant-observation, in-depth interviews, cognitive interviewing, and Guttman analysis to ensure that the scale reflected respondents' perceptions of water insecurity. We tested criterion validity using a number of hypothesis tests. As with the food insecurity scale above, we concluded that the items tapped the construct of interest and reflected local responses to water insecurity.

Table 2. Idealized Guttman Scale for Water Insecurity (with 9 indicators)

Unable to wash dishes	No	No	No	No	No	No	No	No	No	**Yes**
Unable to cook a meal	No	No	No	No	No	No	No	No	Yes	**Yes**
Unable to wash laundry	No	No	No	No	No	No	No	**Yes**	**Yes**	**Yes**
Unable to clean the house	No	No	No	No	No	No	**Yes**	**Yes**	**Yes**	**Yes**
Unable to bathe	No	No	No	No	No	**Yes**	**Yes**	**Yes**	**Yes**	**Yes**
Conserved water to cook	No	No	No	No	**Yes**	**Yes**	**Yes**	**Yes**	**Yes**	**Yes**
Conserved water to wash laundry	No	No	No	**Yes**	**Yes**	**Yes**	**Yes**	**Yes**	**Yes**	**Yes**
Conserved water to clean house	No	No	**Yes**	**Yes**	**Yes**	**Yes**	**Yes**	**Yes**	**Yes**	**Yes**
Conserved water to bathe	No	**Yes**	**Yes**	**Yes**	**Yes**	**Yes**	**Yes**	**Yes**	**Yes**	**Yes**
Score	0	1	2	3	4	5	6	7	8	9

Key

Score	**Definition**
0	Not water insecure
1 to 3	Somewhat water insecure
5	Moderately water insecure
6 to 8	Very water insecure
9	Extremely water insecure

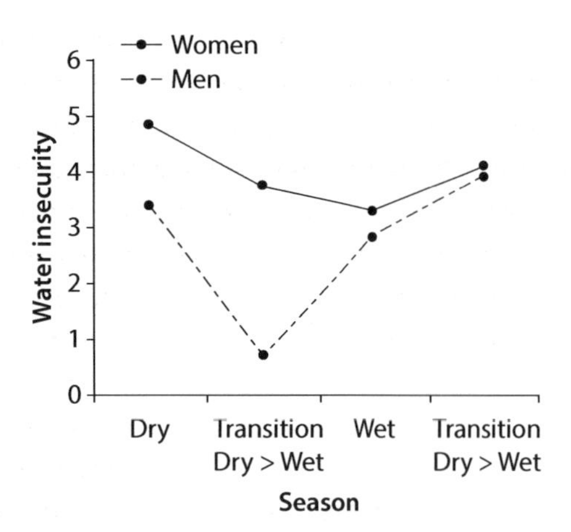

Figure 2. Seasonal Changes in Water Insecurity Scores among Women and Men in Bolivia. Higher scores on the y-axis indicate greater water insecurity.

DISCUSSION

We have argued that insecure access to food and water has important consequences for humans and is increasingly important to understand uncertainties in the lives of millions of people. We have further argued that measures of physical well-being or access alone are likely to underestimate the impact of insecurity on health and well-being. We have illustrated how insecurity measures can contribute by reporting broad associations between measures of insecurity and measures of psychosocial stress. Although the argument that experiences of insecurity impact on multiple measures of health may seem obvious and some might think that this is already well recognized by anthropologists, this is not well supported empirically in the literature. Existing measures that capture food insecurity are rare and rarely used in anthropology (although not in allied fields such as public health nutrition), and there is even less literature addressing water insecurity measures. Widely used scales that tap aspects of socioeconomic status certainly draw attention to the association between poverty and poor health, but they do not capture the biocultural and experiential dimensions of food and water insecurity or illuminate specific pathways through which poverty impacts health.

Even when nuanced scales that do capture insecurity are used, they may not reveal the range of health impacts associated with insecurity if measures of stress are not included. For example, in a well-designed study of food insecurity and indicators of nutritional status among 85 women in Columbia, Dufour, et al. (1999) reported high levels of food insecurity. Yet, they also showed that, among this sample, nutritional status was normal and, thus, concluded that coping strategies were usually adequate. It would have been interesting to compare these women on a measure of anxiety, depression, or distress to assess the extent to which their coping strategies were effective in other health domains. We would hypothesize that measures in other health domains would reveal that the experience of food insecurity exacts a tremendous mental heath burden for these women. Studies of water insecurity face an even more difficult task because there is no commonly accepted and rapid measure of adequate water intake like there is for nutritional status (e.g., anthropometrics). For water insecurity, then, measures of psychosocial stress are critical. In the Bolivia case study, our observations suggested considerable distress associated with even moderate experiences of water insecurity. These examples illustrate how the approach we espouse differs from common practice by recognizing that biological food and water demands likely differ from biocultural demands, and that food and water demands cannot meaningfully be divorced from biocultural needs; these are intimately linked—indeed embedded—in cultural systems.

Although we have argued that there are sound reasons, both theoretical and logistical, for using an experientially-based measure of insecurity, the approach is not without limitations. First, it is not yet clear the extent to which a single individual's response is correlated with other responses in the household and how this might vary across salient domains such as age, gender, and class. This is problematic because many studies rely on the information from a household head or other representative. Second, it is not yet clear how differential appraisal of "stressful" events might influence responses or outcomes. Some studies suggest a gender bias in appraisal of stressful events (Jose and Ratcliffe 2004; Tolin and Foa 2006), which presumably defines insecure access to food and water. Data from Bolivia presented here indicate that female household heads experience more water insecurity than male household heads within the same household, although differences were large only during the transition from the dry to wet season. Differential appraisal is only a problem if we are interested in food or water intake rather than the experience of insecurity. If we are interested in the psychosocial impact of perceived insecurity, differential appraisal may actually be informative and may reflect biocultural stress; that is, it may simply reflect the embodied nature of stress and health (Krieger 2005). The work and the cross-cultural literature reviewed above predict that because women are often responsible for food and water that they may differentially experience insecurity; this further suggests that elevated rates of depression and anxiety may be one psychosocial manifestation of this. A third and more general limitation is that vulnerable individuals or depressed individuals may be differentially appraising their food or water insecurity situation. This would be particularly problematic when attempting to measure the health consequences of insecurity. Several ways to get around this methodological dilemma exist. These include using the food or water insecurity assessment of another household member to predict the mood or well-being of the participant, longitudinal studies, and "anchoring scales."

As illustrated in the case studies, particularly the water insecurity case study, there are often tradeoffs that must be made when assessing scale validity. These tradeoffs reflect the possible tension between peoples' models of the world and statistical models of the world. Rasch modeling and Guttman scaling are statistical tools to assess the difficulty of items in a scale, but they assume unidimensionality. Scales built from the ground up and located in individual experience may not conform to this assumption; the investigator is forced to make a decision as to which information should be privileged. This tradeoff is also illustrated in a study by Coates, et al. (2006b) who report that items suggested by qualitative work were ultimately left off of the USDA food security scale because they failed to meet the assumptions of the Rasch model. In their own studies in Bangladesh, Coates, et al. (2006b) report similar conflicting information: real people report experiencing the world in a multidimensional fashion but Rasch model assumptions disallow this. In the end, they privileged the ethnographic data in their scale. In the Bolivia example, we set a high value on creating a simplistic unidimensional scale because the concept we explored, water insecurity, has not yet been studied rigorously. However, we lost a great deal of information about the experience of water insecurity as a result. This underscores the importance of understanding the local cultural context when unpacking socioeconomic and testing the impacts of uncertainty and insecurity.

Despite these challenges, we believe that anthropological approaches have a great deal to contribute to the conceptualization and measurement of food and water security, as they have in other fields. The nature of the scales reflects the intimate connections between human health and biocultural requirements for food and water. Future research could develop in several directions. First, there is a need for scales to be adapted or developed in local contexts that represent a range of cultures, economic systems, and ecologies. Second, once such scales have been developed and validated, we can assess the magnitude and distribution of food and water insecurity using an experience-based biocultural approach—one that takes into account interrelated physiological and cultural needs for food and water; this would demand an accompanying measure of psychosocial stress. Third, we can develop an understanding of how adaptations and coping strategies affect the incidence and experience of food and water security across cultures.

Measures that simply compare individuals against a standard biological requirement may not be tuned to recognize the social and cultural requirements that are so critical to human health and well-being. The approaches outlined above explicitly recognize this distinction, in part because they allow respondents to define when insecurity is experienced. Part of the human experience that we must understand and document is what it means for health and well-being when people are placed under severe constraints. A step towards this lies in recognizing that biological and biocultural requirements may not be isomorphic, and may, therefore, require different measurement tools. Anthropologists concerned with health and well-being can provide an ethnographically grounded, cross-cultural understanding of the experience, and the causes and consequences of food and water security.

REFERENCES

Bernard, H. Russell. 2006. *Research Methods in Anthropology: Qualitative and Quantitative Approaches.* Walnut Creek, Calif.: Altamira Press.

Bickel, Gary, Mark Nord, Cristofer Price, William Hamilton, and John Cook. *2000 Guide to Measuring Food Insecurity, Revised 2000.* Alexandria, Va.: United States Department of Agriculture, Food, and Nutrition Service.

Billig, Patricia, Diane Bendahmane, and Anne Swindale. *1999 Water and Sanitation Indicators Measurement Guide.* Washington, D.C.: Food and Nutrition Technical Assistance Project of the United States Agency for International Development.

Borgatti, Stephen P. 1996. *ANTHROPAC 4.0*. Natick, Mass.: Analytic Technologies.

Coates, Jennifer, Edward A. Frongillo, Beatrice Lorge Rogers, Patrick Webb, Parke E. Wilde, and Robert Houser. 2006a. Commonalities in the Experience of Household Food Insecurity Across Cultures: What Are Measures Missing? *Journal of Nutrition* 136(5):1438S–1448S.

Coates, Jennifer, Parke E. Wilde, Patrick Webb, Beatrice Lorge Rogers, and Robert F. Houser. 2006b. Comparison of a Qualitative and a Quantitative Approach to Developing a Household Food Insecurity Scale for Bangladesh. *Journal of Nutrition* 136(5):1420S–1430S.

Crooks, Deborah L., Lisa Cligget, and Steven M. Cole. 2007. Child Growth as a Measure of Livelihood Security: The Case of the Gwembe Tonga. *American Journal of Human Biology* 19(5):669–675.

DeMaio, Theresa J., and Jennifer M. Rothgeb. 1996. Cognitive Interviewing Techniques: In the Lab and in the Field. In *Answering Questions: Methodology for Determining Cognitive and Communicative Processes in Survey Research*. Schwartz, Norbert and Seymour S. Sudman, (eds.). Pp. 177–196. San Francisco, Calif.: Jossey-Bass.

Dufour, Darna L., Lisa K. Staten, Carol I. Waslien, Julio C. Reina, and G. B. Spurr. 1999. Estimating Energy Intake of Urban Women in Columbia: Comparison of Diet Records and Recalls. *American Journal of Physical Anthropology* 108(1):53–63.

Elhance, Arun P. 1999. *Hydropolitics in the Third World*. Washington, D.C.: United States Institute of Peace Press.

Food and Agriculture Organization (FAO). 2004. *The State of Food Insecurity in the World*. Rome: Food and Agriculture Organization.

2005. *The State of Food Insecurity in the World*. Rome: Food and Agriculture Organization.

Frongillo, Edward A. 1999. Validation of Measures of Food Insecurity and Hunger. *Journal of Nutrition* 129(2):506–509.

Gleick, Peter. 1996. Basic Water Requirements for Human Activities: Meeting Basic Needs. *Water International* 21(2):83–92.

2003. Water Use. *Annual Review of Environment and Resources* 28(1):275–314.

2004. *The World's Water 2004–2005*. Washington, D.C.: Island Press.

Guttman, Louis. 1944. A Basis for Scaling Qualitative Data. *American Sociological Review* 9(2):139–150.

Hadley, Craig. Ethnic Expansions and Between-group Differences in Children's Health: A Case Study from the Rukwa Valley, Tanzania. *American Journal of Physical Anthropology* 128(3):682–692.

Hadley, Craig, Monique Borgerhoff Mulder, and Emily Fitzherbert. 2007. Seasonal Food Insecurity and Perceived Social Support in Rural Tanzania. *Public Health Nutrition* 10(6):544-551.

Hadley, Craig and Crystal Patil. Food Insecurity in Rural Tanzania is Associated with Maternal Anxiety and Depression. *American Journal of Human Biology* 18(3):359–368.

Himmelgreen, David A., Rafael Perez-Escamilla, Sofia Segura-Millan, Yu-Kuei Peng, Anir Gonzalez, Merrill Singer, and Ann Ferris. Food Insecurity among Low-income Hispanics in Hartford, Connecticut: Implications for Public Health Policy. *Human Organization* 59:334–342.

Jose, Paul E. and Verity Ratcliffe. 2004. Stressor Frequency and Perceived Intensity as Predictors of Internalizing Symptoms: Gender and Age Differences in Adolescence. *New Zealand Journal of Psychology* 33(3): 145–154.

Kennedy, Eileen. Qualitative Measures of Food Insecurity and Hunger. In *Measurement and Assessment of Food Deprivation and Undernutrition*. Washington, D.C.: International Food Policy Research Institute.

Krieger, Nancy. Embodiment: A Conceptual Glossary for Epidemiology. *Journal of Epidemiology and Community Health* 59(5):350–355.

Maxwell, Daniel G. 1996. Measuring Food Insecurity: The Frequency of Coping Strategies. *Food Policy* 21(3):291–303.

McElroy, Ann, and Patricia K. Townsend. *Medical Anthropology in Ecological Perspective*. Boulder, Colo.: Westview Press.

Melgar-Quinonez, Hugo R., Ana C. Zubieta, Barbara MkNelly, Anastase Nteziyaremye, Maria Filipinas D. Gerardo, and Christopher Dunford. Household Food Insecurity and Food Expenditure in Bolivia, Burkina Faso, and the Philippines. *Journal of Nutrition* 136(5): 1431S–1437S.

Moran, Emilio F. 2000. *Human Adaptability. Boulder, Colo.: Westview Press.*

Morris, Saul Sutkover, Carletto Calogero, John Hoddinott, and Luc J.M. Christiaensen. 2000. Validity of Rapid Estimates of Household Wealth and Income for Health Surveys in Rural Africa. *Journal of Epidemiology and Community Health* 54(5):381–387.

Pattanayak, Subhrendu K., Jui-Chen Yang, Dale Whittington, and K.C. Bal Kumar. Coping with Unreliable Public Water Supplies: Averting Expenditures by Households in Kathmandu, Nepal. *Water Resources Research* 41(2): 1–12.

Peregrine, Peter N., Carol R. Ember, and Melvin Ember. 2004. Universal Patterns in Cultural Evolution: An Empirical Analysis Using Guttman Scaling. *American Anthropologist* 106(1):145–149.

Perez-Escamilla, Rafael, Ana Maria Segall-Correa, Lucia Kurdian Maranha, Maria de Fatima Archanjo Sampaio, Leticia Marin-Leon, and Giseli Panigassi. 2004. An Adapted Version of the United States Department of Agriculture Food Insecurity Module is a Valid Tool for Assessing Household Food Insecurity in Campinas, Brazil. *Journal of Nutrition* 134(8): 1923–1928.

Pike, Ivy L. and Sharon R. Williams. Incorporating Psychosocial Health into Biocultural Models: Preliminary Findings from Turkana Women of Kenya. *American Journal of Human Biology* 18(6):729–740.

Radimer, Kathy L. 1990. *Understanding Hunger and Developing Indicators to Assess It*. Ph.D. dissertation, Cornell University.

Satterthwaite, David. The Millennium Development Goals and Urban Poverty Reduction: Great Expectations and Nonsense Statistics. *Environment and Urbanization* 15(2):181–190.

Sawka Michael N., Samuel N. Cheuvront, and Robert Carter III. Human Water Needs. *Nutrition Review*. 63(6):S30–39.

Sen, Amartya. 1981. *Poverty and Famines: An Essay on Entitlement and Deprivation*. Oxford, UK: Clarendon Press.

Struble, Marie B. and Laurie L. Aomari. Position of the American Dietetic Association: Addressing World Hunger, Malnutrition, and Food Insecurity. *Journal of the American Dietetic Association* 103(8): 1046–1057.

Sugita, Eri Woods. *Domestic Water Use, Hygiene Behavior, and Children's Diarrhea in Rural Uganda*. Ph.D. dissertation, University of Florida.

Swindale, Anne and Paula Bilinsky. Development of a Universally Applicable Household Food Insecurity Measurement Tool: Process, Current Status, and Outstanding Issues. *Journal of Nutrition* 136(5):1449S–1452S.

Terhorst, Philipp. 2003. *Public-Popular Organizations: The Case of Cochabamba, Bolivia*. Master of Science thesis, Loughborough University.

Tolin, David F. and Edna B. Foa. 2006. Sex Differences in Trauma and Posttraumatic Stress Disorder: A Quantitative Review of 25 Years of Research. *Psychological Bulletin* 132(6):959–992.

United Nations (UN). *1948 Universal Declaration of Human Rights*. G.A. res. 217A (III), U.N. Doc A/810 at 71. Geneva: United Nations.

1999 UN Doc. E/C. 12/1999/5, 12 May 1999, *CESCR GeneralCommentNo.12ontherighttofood*(ICESCR, Article 11), paragraph 20. Geneva: United Nations.

2002 UN Doc. E/C.12/2002/11,20 January 2003, *CESCR General Comment No. 15 on the right to water* (ICESCR, Articles 11 and 12), paragraph 49. Geneva: United Nations.

Webb, Patrick, Jennifer Coates, Edward A. Frongillo, Beatrice Lorge Rogers, Anne Swindale, and Paula Bilinsky. 2006. Measuring Household Food Insecurity: Why It's So Important and Yet So Difficult To Do. *Journal of Nutrition* 136(5): 1404S–1408S.

Wiebe, Keith, Nicole Ballenger, and Per Pinstrup-Andersen, (eds.). 2001. *Who Will Be Fed in the 21 st Century? Challenges for Science and Policy.* Baltimore, Md.: John Hopkins University Press.

Wiley, Andrea S. 1992. Adaptation and the Biocultural Paradigm in Medical Anthropology: A Critical Review. *Medical Anthropology Quarterly* 6(3):216–236.

Wolfe, Wendy S. and Edward A. Frongillo. 2001. Food Security: Building Household Food-security Measurement Tools From the Ground Up. *Food and Nutrition Bulletin* 22(1):5–12.

World Health Organization (WHO). 1985. Energy and Protein Requirements. *Report of a Joint FAO/WHO/UNU Expert Consultation.* World Health Organization Technical Report Series Number 724. Geneva, Switzerland: World Health Organization.

World Health Organization and United Nations Children's Fund (WHO and UNICEF). 2006. *Meeting the MDG Drinking Water and Sanitation Target: The Urban and Rural Challenge of the Decade*. Geneva, Switzerland: World Health Organization.

Wutich, Amber. 2006. *The Effects of Urban Water Scarcity on Reciprocity and Sociability in Cochabamba, Bolivia*. Ph.D. dissertation, University of Florida.

Wutich, Amber and Kathleen Ragsdale. 2008. Water Insecurity and Emotional Distress: Coping With Supply, Access, and Seasonal Variability of Water in a Bolivian Squatter Settlement. *Social Science and Medicine* 67(12):2116–2125.

ERRATUM

Christopher McCarty's name was inadvertently left off the final version of the paper, "Homeless Women's Personal Networks: Implications for Understanding Risk Behavior," in Volume 68, No. 2. The full list of authors should read, Joan S. Tucker, David Kennedy, Gery Ryan, Suzanne L. Wenzel, Daniela Golinelli, James Zazzali and Christopher McCarty.

Food Safety

By Robert H. Friis

LEARNING OBJECTIVES

By the end of this chapter the reader will be able to:

- State three measures for preventing foodborne illness
- Discuss 10 microbiological agents that are implicated in foodborne illness
- Explain the significance of foodborne illness for the world's population
- List five categories of contaminants in the food supply
- Describe one major regulation for protecting the safety of food from carcinogens

INTRODUCTION

The focus of this chapter is on food safety, including foodborne diseases, foodborne infections, and foodborne outbreaks. The term foodborne diseases (which includes foodborne intoxications and foodborne infections) refers to "illnesses acquired by consumption of contaminated food; they are frequently and inaccurately referred to as food poisoning." A foodborne outbreak indicates "the occurrence of a similar illness among two or more people which an investigation linked to consumption of a common meal or food items, except for botulism (one case is an outbreak)."

The problem of foodborne illness is well known to the general public, due to the media's frequent coverage of outbreaks. For example, your local newspaper will occasionally print stories about foodborne illness outbreaks that happen in restaurants in your community; even the major restaurant chains are not immune to such incidents. Another example comes from media reports of dramatic foodborne illness outbreaks on cruise ships; often these outbreaks devastate the vacations of passengers who must terminate their cruises prematurely. Other foodborne outbreaks are associated with foreign travel; at one time or another, many travelers to foreign countries have experienced illnesses caused by unsanitary or improperly prepared meals. Travelers to exotic locales may acquire foodborne pathogens—uncommon in their usual place of residence—that challenge health care providers who are unfamiliar with the resulting diseases.

Foodborne illness can be both acute and long term. In this chapter, we will learn that some of the causes of acute foodborne illness are microbiologic agents and toxic chemicals such as pesticides and heavy metals. In addition, other contaminants that may be present in food are suspected of affecting human health adversely; these contaminants include food additives, antibiotics used to promote growth in animals used for food, and low levels of potential carcinogens. Sources of food contamination are pollutants in water used to process foods, chemicals used by the agricultural industry, and even pollutants found in the air. The role that food contaminants play in causing ill-defined gastrointestinal and other diseases is suspect, yet unclear. Many cases of such illness probably go unreported, so their scope is unknown. A pressing question is whether, and to what extent, human exposure to low levels of toxic contaminants in foods increases the risk of adverse health effects or is entirely benign.

In addition to providing information about causes of foodborne illness, the chapter also will cover foodborne disease prevention, methods for food

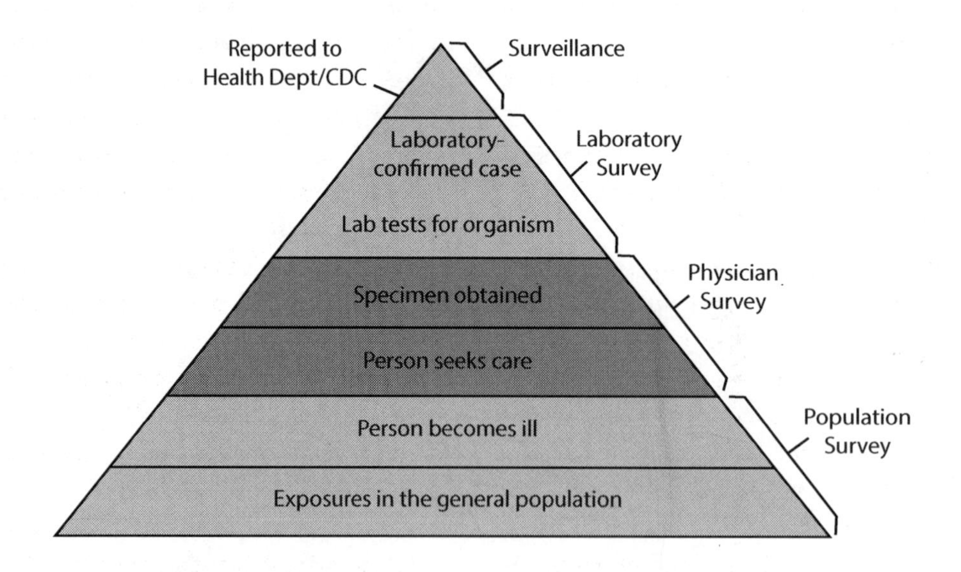

Source: Reprinted from the Centers for Disease Control and Prevention. FoodNet. Available at: http://www. cdc.gov/foodnet/surveillance_pages/burden_pyramid.htm. Accessed March 22, 2010

Figure 11.1. FoodNet surveillance—burden of illness pyramid.

safety inspection, and enforcement of government regulations. One of the important responsibilities of local and federal government agencies is to ensure the quality of foodstuffs. This goal is supported by cadres of government-employed environmental health workers who inspect the food supply and investigate foodborne illness outbreaks. The chapter will conclude with a review of new trends in production of foods and possible implications for human health.

THE GLOBAL BURDEN OF FOODBORNE ILLNESS

From the global perspective, foodborne illness is a major cause of morbidity (and occasionally mortality). Illnesses transmitted by foods can cause adverse birth outcomes, chronic illnesses, and disabilities (e.g., miscarriages, neurologic sequelae from meningitis, and kidney failure). Within the past few years, the incidence of foodborne illnesses has increased in industrialized nations. This increase has been attributed to changes in agricultural and food processing methods, globalization in food distribution, and other social and behavioral changes among the human population.

According to the US Centers for Disease Control and Prevention (CDC), foodborne illness continues to be a public health problem of great importance for the United States. Some estimates suggest that foodborne illnesses (reported and unreported) affect almost one-quarter of the population each year in the United States. Other estimates indicate that foodborne illnesses cause 9,000 deaths annually with an economic cost of $5 billion. In recent years, the incidence of foodborne illness in the United States has tended to remain stable over time, with recent declines in some conditions (e.g., *E. coli* 0157 infection) and the need to sustain these declines.

Surveillance of Foodborne Illness

In the United States, the CDC maintains responsibility at the federal level for surveillance of foodborne illness. For many types of foodborne diseases, only a small proportion may be reported by so-called passive surveillance systems, which rely on the reporting of cases of foodborne illness by clinical laboratories to state health departments and ultimately to the CDC.[7] In contrast, the CDC Foodborne Diseases Active Surveillance Network (FoodNet) is an active system whereby public health officials maintain frequent direct contact with clinical laboratory directors to identify new cases of foodborne illness. The focus of the FoodNet program is diarrheal illnesses caused by emerging infectious disease agents; foodborne diseases are monitored in several US sites, which are part of the CDC's Emerging Infections Program. Examples of the foodborne diseases that are monitored are those caused by parasites such as *Cyclospora* and *Cryptosporidium* (see Chapter 9) and bacterial agents such as *Campylobacter, E. coli* 0157:H7, *Listeria monocytogenes, Salmonella, Shigella,* and *Vibrio.* Infections with *Vibrio parahaemolyticus* and *Vibrio vulnificus* often are associated with consumption of undercooked oysters taken from waters that have been contaminated by these bacteria. The other conditions are described later in the chapter.

The CDC has developed a model, called the burden of illness pyramid, for describing how reporting of foodborne disease takes place. Figure 11.1 illustrates the burden of illness pyramid.

> This illustration (of the burden of illness pyramid) shows the chain of events that must occur for an episode of illness in the population to *be* registered in surveillance. At the bottom of the pyramid, 1) some of the general population is exposed to an organism; 2) some exposed persons become ill; 3) the illness is sufficiently

troubling that some persons seek care; 4) a specimen is obtained from some persons and submitted to a clinical laboratory; 5) a laboratory appropriately tests the specimen; 6) the laboratory identifies the causative organism and thereby confirms the case; 7) the laboratory-confirmed case is reported to a local or state health department. FoodNet conducts laboratory surveys, physician surveys, and population surveys to collect information about each of these steps.

Table 11.1 tallies the number of outbreaks that occurred in 2006 by cause; during 2006 a total of 1,270 outbreaks and 27,634 illnesses were reported.

Figure 11.2 shows data reported from 1990 to 2002 on combined foodborne disease outbreaks for the causes shown in Table 11.1. Although some major bacterial foodborne illnesses caused by agents such as *Campylobacter* and *Listeria* have shown a sustained decline over time, others caused by *Vibrio* have not (2009 data). These trends suggest that continued efforts will be required to reduce further the incidence of some types of foodborne illness.

CATEGORIES OF FOOD HAZARDS

The "big three" categories of food hazards are biological, physical, and chemical.[9] In addition, nutritional hazards are those associated with the presence of nutrients and other food constituents in excessive or deficient amounts that lead to disease. Table 11.2 provides a list of contaminants that may be present in food. Physical hazards encompass foreign objects in food such as stones, glass, metal (e.g., bullets), and pieces of wood. Food hazards may be introduced when foods are harvested, processed, shipped, or stored.

COMMON MICROBIAL AGENTS OF FOODBORNE ILLNESS

The media and the CDC publish frequent reports of disease outbreaks caused by microbial agents. Some microbial pathogens (e.g., *Salmonella*) are more common agents of foodborne illness than are others (e.g.,

Table 11.1. Number and Percentage of Reported Foodborne Disease Outbreaks and Outbreak-Associated Illnesses, by Etiology—United States, 2006

ETIOLOGY	NO. OF OUTBREAKS	NO. OF ILLNESSES
Bacterial	295	7,241
Chemical	66	267
Parasitic	12	147
Viral	511	14,855
Single etiology	884	22,510
Multiple etiologies	23	794
Unknown etiology	363	4,330
Total 2006	1,270	27,634

Totals include confirmed and suspected etiology

Source: Adapted from the Centers for Disease Control and Prevention. Surveillance for foodborne disease outbreaks—United States, 2006. *MMWR.* 2009;58:611.

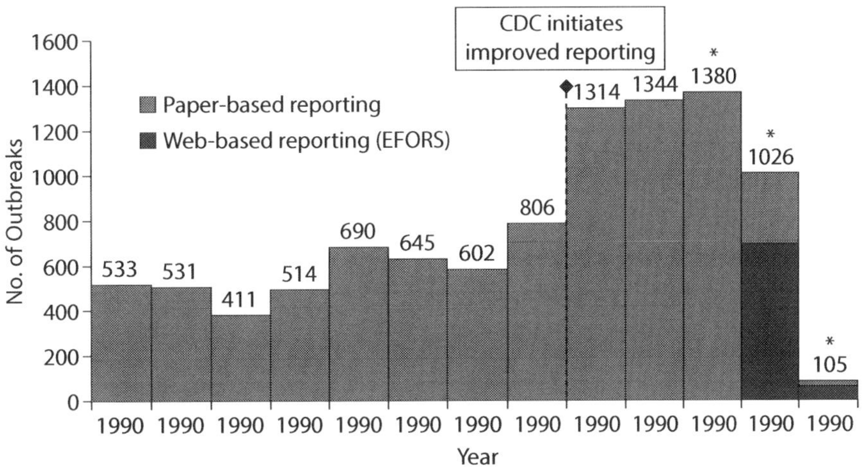

*Preliminary data; not all states had completed reporting.

Source: Adapted and reprinted from the Centers for Disease Control and Prevention. Surveillance for Foodborne-Disease Outbreaks—United States, 1998-2002. Surveillance Summaries, November 10, 2006. *MMWR.* 2006;55(No. SS-10):8.

Figure 11.2. Foodborne-disease outbreaks reported to the CDC January 1, 1990, through March 15, 2002.

Table 11.2. Contaminants That May Be Present in Food

- Pathogenic microbial agents
  - Examples: bacteria, worms, protozoa, fungi, viruses, prions, and (in certain cases) toxins from such organisms
- Chemically related foodborne hazards
  - Marine toxins, mushroom toxins
  - Heavy metals
  - Pesticides, herbicides, and fungicides
  - Preservatives and additives
- Residues of medicines administered to food animals
  - Antibiotics, growth-promoting hormones
- Foreign objects and other physical contaminants
  - Natural components: bones, shells, seeds
  - Other: glass and metal fragments, stones Radioactive materials
- Materials used in packaging
  - Residues of plastics
  - Waxes
- Miscellaneous contaminants
  - Debris from insects (insect parts, ova) and from rodents (fecal material, fur)
  - Cleaning agents used in the food processing environment

Clostridium botulinum). In addition, a type of microbial agent referred to as an emerging foodborne disease pathogen increasingly is causing foodborne infections. Emerging foodborne disease pathogens are a subset of agents from the general category of emerging infectious disease agents, which cause emerging infections. These infections, which include cholera, Rift Valley fever, and Lyme disease, "have newly appeared in the population, or have existed but are rapidly increasing in incidence or geographic range. ... Emerging foodborne disease pathogens include *E. coli* O157:H7, *Campylobacter jejuni,* and *Listeria monocytogenes.* As they increase in frequency and scope, emerging foodborne diseases are a major concern of environmental health workers across the globe.

Chapter 5 focused mainly on zoonotic emerging infectious diseases. Some of these same emerging infectious disease pathogens represent hazards to the food supply. The rise in newly recognized pathogens has altered the epidemiology of foodborne illness by increasing the occurrence and sequaelae of emerging foodborne diseases. Some of the microbial agents that cause foodborne illness have adapted to changes in food production, causing their reemergence and the development of new pathogens. Another factor that has contributed to the rise of these pathogens is globalization of the food supply, as explained in an editorial in the *New England Journal of Medicine:*

> The surge in livestock production around the world, including most parts of Asia, has resulted in unprecedentedly large populations of closely confined animals, particularly pigs and chickens. ... Today, large hog-raising operations, with tens of thousands of animals on a single farm, are common throughout parts of Asia, North and South America, and Europe. ... A number of infectious agents tend to be associated with an increasing intensity of production and concentration of animals in limited spaces.

The editorial also indicates that these infectious agents "pose a potentially serious threat to human health." Other potential hazards arise from fruits and vegetables that originate in developing countries and are consumed in developed countries. In addition, the processes (e.g., cooking, treatment, and pasteurization) designed to deactivate pathogens "can and do fail."

Table 11.3 provides an abbreviated list of microbial agents responsible for foodborne illness. The agents form three general classes (bacteria, parasites, and viruses) plus a fourth class called "other." Some also are classified as agents of emerging foodborne disease. Excluded from this section are parasitic protozoal organisms, which were discussed in Chapter 9. Microbes vary greatly in the dose that is required to produce human illness.[11] In some instances the ingestion of only a single virus particle can cause disease; in others only a few organisms (viruses, bacteria, or parasites) need be consumed.

BACTERIAL AGENTS

The pathogens *Salmonella, Clostridium botulinum, Staphylococcus aureus,* and *Clostridium perfringens* are bacteria—unicellular organisms that have characteristic shapes, such as rod shaped (in some instances a slender, curved rod), spherical (cocci), or spiral. Bacteria can be moving (motile)—propelled by flagella or projections from cells—or nonmoving (nonmotile).

Bacteria may be classified as *gram-positive* or *gram-negative,* terms that refer to Gram's method, which is a staining technique for classifying bacteria. Using this methodology, technicians stain bacteria

Table 11.3. Abbreviated Listing of Foodborne Pathogens and Diseases

TYPE OF ORGANISM	NAME OF DISEASE
Bacteria: Pathogenic Bacteria	
Salmonella	Salmonellosis
Clostridium botulinum	Botulism
Clostridium perfringens	*Clostridium perfringens* food intoxication
Staphylococcus aureus	Staphylococcal food intoxication
Campylobacter jejuni	Vibrionic enteritis
Shigella	Shigellosis
Listeria monocytogenes	Listeriosis
Bacteria: Enterovirulent *Eschericia Coli* Group	
Escherichia coli 0157:H7 enterohemorrhagic (EHEC) serotype	*E. coli* 0157:117 infection
Parasites: Worms	
Trichinella spiralis	*Trichinellosis, trichinosis*
Taenia solium, Taenia saginata	Taeniasis, cysticercosis *(T. solium)*
Viruses	
Norovirus (formerly called Norwalk virus group)	Epidemic viral gastroenteropathy
Other Pathogenic Agents	
Prions	Bovine spongiform encephalopathy (BSE), commonly referred to as mad cow disease

with a dye known as gentian violet; then they expose the bacteria to Gram's solution (a mixture of iodine, potassium iodide, and water) and other chemicals. After washing, bacteria that retain the gentian violet are gram-positive and those that do not retain it are gram-negative.

Foodborne infections should be differentiated from foodborne intoxications. Foodborne infections are induced by infectious agents such as some bacteria (e.g., *Salmonella*) that cause foodborne illness directly; symptoms of such infection are variable and may include nausea, diarrhea, vomiting, headaches, and abdominal pain. Foodborne intoxications result from other agents that do not cause infections directly but produce spores or toxins as they multiply in the food; these toxins then can affect the nervous system (neurotoxins) or the digestive system. An example of such a bacterial agent is *Staphylococcus aureus*.

Salmonella

Known as the genus *Salmonella*, this organism embodies approximately 2,000 serotypes that can cause human illness. Used to identify subspecies of bacteria, the process of serotyping involves determining whether bacterial isolates react with the blood serum from animals that have formed antibodies against specific types of bacteria. Bacteria that react with specific sera are called members of that serotype.

The causative agents for the foodborne infection salmonellosis, *Salmonella* are described as rod-shaped, motile, gram-negative, and nonsporeforming bacteria. Figure 11.3 shows an image of *Salmonella* bacteria. (Note that some serotypes of *Salmonella* are nonmotile.)

The two most common serotypes in the United States are *Salmonella* serotype Enter- itidis and *Salmonella* serotype Typhimurium, although the numbers of isolates of these serotypes have declined since 1996. In 2006, a total of 34% of laboratory-confirmed cases of salmonellosis were caused by these two serotypes. Four serotypes (Enteritidis, Typhimurium, Newport, and Heidelberg) accounted for 45% of isolates. A variety of *Salmonella* called *S.* Typhi causes typhoid fever. (This disease, which can be transmitted by contaminated food and water, was discussed in Chapter 9.)

Salmonellosis is ranked among the most frequent types of foodborne illness in the United States. Approximately 2 to 4 million cases of salmonellosis (reported and nonreported) are estimated to occur each year in the United States; about 500 deaths each year are caused by *Salmonella* infections. The number of cases of salmonellosis (excluding typhoid fever) reported each year ranges between 40,000 and 50,000. Until the late 1980s, the United States and other developed countries experienced a rising incidence of salmonellosis. Since then, the rates of this condition have tended to stabilize in the United States. The number of S. Enteritidis isolates taken from human specimens has tended to increase relative to the other three common isolates. Refer to Figure 11.4 for numbers of reported cases of salmonellosis in comparison with cases of shigellosis (covered later in the chapter). Figure 11.4 also shows the rates of four common serotypes of *Salmonella*.

Salmonella are known to occur widely in wild and domestic animal reservoirs. Animals that are used for human consumption—poultry, swine, cattle—may harbor *Salmonella*. Pet animals such as cats, dogs, and turtles also can be reservoirs for *Salmonella*. Some animals and birds are chronic carriers of the bacteria. *Salmonella* bacteria may be transferred to environmental surfaces at work and at home (e.g., the kitchen) from raw meats, poultry, and seafood; from animal feces; and from contaminated water and soil. Symptoms and characteristics of *Salmonella* infections are shown in the text box.

Salmonella infections have been associated with diverse scenarios. Here are five examples of the many documented foodborne disease outbreaks caused by *Salmonella:*

1. Consumption of raw, unpasteurized milk was initially implicated in the cases of two children hospitalized with *Salmonella* infection. Subsequent investigations identified a total of 62 affected persons. The outbreak was linked to a dairy farm that served food and had a petting zoo. The farm legally sold raw milk and products made with raw milk.
2. Eating cantaloupe imported from Mexico was associated with multistate outbreaks of *Salmonella* infections. The outbreaks affected more than 100 persons in 12 US states and Canada. Traceback investigations revealed that sanitary conditions at Mexican farms where the cantaloupes originated could have been responsible (e.g., possible sewage-contaminated irrigation water and poor hygienic practices of workers).
3. During winter of 1995–1996, ingestion of contaminated alfalfa sprouts was linked to 133 cases of salmonellosis in Oregon and British Columbia. Investigations suggested that the source of the *Salmonella* bacteria was contaminated alfalfa seeds.
4. A multistate outbreak of *Salmonella* serotype Agona infections affected more than 200 persons in 11 states during April through May 1998. The outbreak was linked to Toasted Oats cereal that had been contaminated with this serotype of *Salmonella* bacteria.
5. During 1985, the CDC became aware of the largest number of culture-confirmed cases of *Salmonella* ever reported for a single outbreak—5,770 by April 16, 1985. This large number of cases was responsible for the increase shown in Figure 11.4. The cause of the incident was pasteurized milk, which ordinarily would be free from *Salmonella*. Apparently, the pasteurization process was inadequate, or the milk became contaminated after pasteurization.

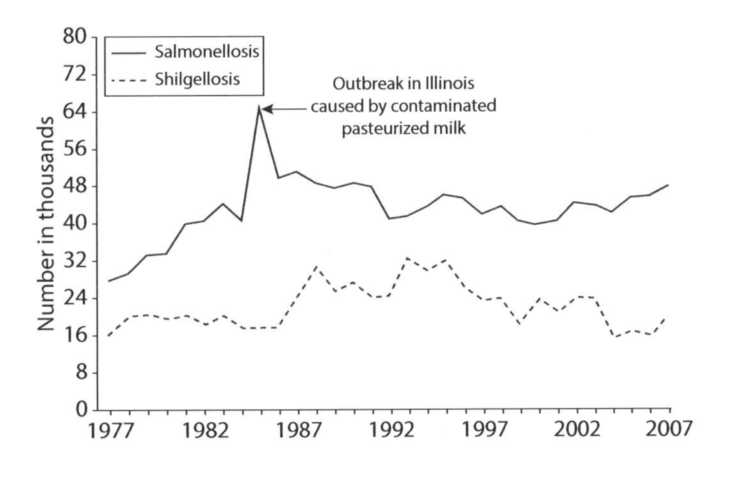

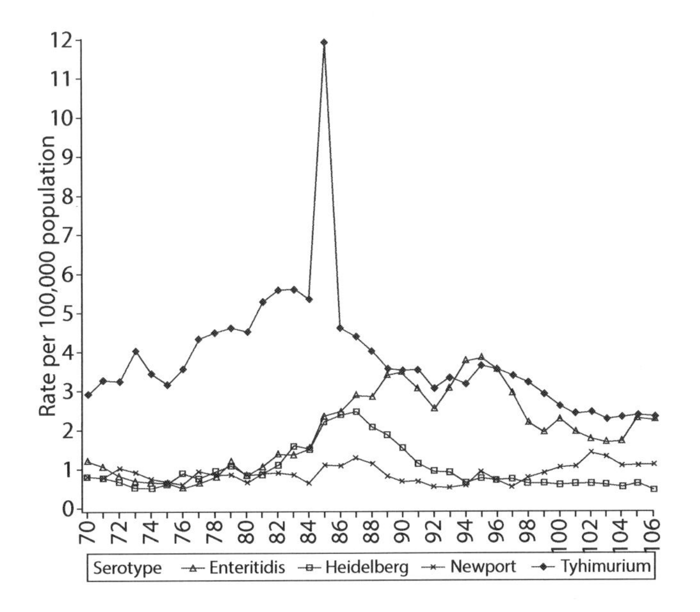

Source: Reprinted from the Centers for Disease Control and Prevention. Summary of Notifiable Diseases—United States, 2007. Published July 9, 2009 for *MMWR* 2007;56(No. 53):68; and the Centers for Disease Control and Prevention. Salmonella Surveillance: Annual Summary, 2006. Atlanta, GA: US Department of Health and Human Services, CDC, 2008:85.

Figure 11.4. Salmonellosis and shigellosis. Number of reported cases, by year—United States, 1977-2007, shown on the top. *Salmonella*, serotype isolates rates in the United States per 100,000 population, 1970-2006, shown on the bottom.

Several of the strains of *Salmonella* have developed antibiotic resistance, making their treatment more difficult. The development of antimicrobial-resistant *Salmonella* has been tied to the administration of antibiotics to animals consumed for food; these resistant forms of *Salmonella* are capable of being transmitted to the human population. For example, during the first four months of the year 2002, the CDC reported the characteristics of 47 cases of antibiotic-resistant *Salmonella* infections that

Salmonellosis (*Salmonella* spp.)

Nature of disease: The acute symptoms of salmonellosis are nausea, vomiting, abdominal cramps, diarrhea, fever, and headache. Chronic consequences include arthritic symptoms that may follow 3–4 weeks after onset of acute symptoms. The onset time is 6–48 hours.

Infective dose: As few as 15–20 cells can cause infection; occurrence of infection depends upon age and health of host, and strain differences among the members of the genus.

Duration of symptoms: Acute symptoms may last for 1 to 2 days or may be prolonged, again depending on host factors, ingested dose, and strain characteristics.

Diagnosis of human illness: Serological identification of culture isolated from stool.

Associated foods: These include raw meats, poultry, eggs and egg shells, milk and dairy products, fish, shrimp, frog legs, yeast, coconut, sauces and salad dressing, cake mixes, cream-filled desserts and toppings, dried gelatin, peanut butter, cocoa, and chocolate. Various Salmonella species have long been isolated from the outside of egg shells.

Target populations: All age groups are susceptible, but symptoms are most severe in the elderly, infants, and the infirm. AIDS patients suffer salmonellosis frequently (estimated 20-fold more than general population) and suffer from recurrent episodes.

Source: Adapted and reprinted from US Food and Drug Administration. Center for Food Safety and Applied Nutrition, Bad Bug Book—*Salmonella* spp. Available at: http://www.fda.gpv/ Food/FoodSafety/FoodborneIllness/FoodborneIllnessFoodbornePathogensNaturalToxins/BadBugBook/ucm069966.htm. Accessed April 30, 2010.

occurred in five US states. These cases were caused by a serotype *of Salmonella* called *Salmonella* serotype Newport. The source of many of these cases was raw or undercooked ground beef.

Clostridium botulinum

Clostridium botulinum causes the foodborne disease botulism, a form of foodborne intoxication. C. *botulinum* grows in an anaerobic (oxygen-free) environment and produces a potent toxin (a neurotoxin) that affects the nervous system. The organism is rod shaped, as shown in Figure 11.5.

The organism forms spores that are resistant to heat; the spores are able to survive in foods that have been incorrectly or minimally processed. However, the organism can be eradicated by heating at a temperature of 80°C (176°F) for 10 or more minutes. When present in foods, the toxin produces a very severe disease that has a high mortality rate, particularly if not treated promptly. The disease has a low incidence in the United States; about 10 to 30 outbreaks occur each year. The form of the disease called infant botulism has been linked to consumption of honey by infants.

Other facts regarding botulism are listed in the following text box.

Exhibit 11.1 presents two case studies of botulism that were reported by the CDC.

Clostridium perfringens

Clostridium perfringens causes perfringens food poisoning, a common source of foodborne illness in the United States. The CDC estimates that about 10,000 cases occur each year. C. *perfringens* is anaerobic, rod shaped, and forms spores. The organism occurs commonly in the environment (e.g., in soil and sediments), especially in those areas contaminated with feces; C. *perfringens* is a frequent resident of the intestines of

Clostridium botulinum

- Symptoms: A very small amount (a few nanograms) (or billionths of a gram) of toxin can cause illness. Onset of symptoms in foodborne botulism is usually 18 to 36 hours after ingestion of the food containing the toxin, although cases have varied from 4 hours to 8 days. Early signs of intoxication consist of marked lassitude, weakness and vertigo, usually followed by double vision and progressive difficulty in speaking and swallowing. Difficulty in breathing, weakness of other muscles, abdominal distention, and constipation may also be common symptoms.
- Associated Foods: The types of foods involved in botulism vary according to food preservation and eating habits in different regions. Any food that is conducive to outgrowth and toxin production, that when processed allows spore survival, and is not subsequently heated before consumption can be associated with botulism. Almost any type of food that is not very acidic (pH above 4.6) can support growth and toxin production by *C. botulinum*. Botulinal toxin has been demonstrated in a considerable variety of foods, such as canned corn, peppers, green beans, soups, beets, asparagus, mushrooms, ripe olives, spinach, tuna fish, chicken and chicken livers and liver pate, and luncheon meats, ham, sausage, stuffed eggplant, lobster, and smoked and salted fish.

Source: Adapted and reprinted from US Food and Drug Administration. Center for Food Safety and Applied Nutrition, Bad Bug Book—*Clostridium botulinum*. Available at: http://www.fda.gov/Food/FoodSafety/FoodborneIllness/FoodborneIllnessFoodbornePathogens NaturalToxins/BadBugBook/ ucm070000.htm. Accessed April 26, 2010.

humans and animals. The following text box contains some facts regarding C. *perfringens*.

Here is an example of an outbreak: In 1993, C. *perfringens* caused illness outbreaks after diners consumed St. Patrick's Day meals of corned beef in Ohio and Virginia. The Ohio incident sickened at least 156 persons. The Virginia outbreak caused 115 persons to become ill. In both situations, the causative factor was C. *perfringens* intoxication, apparently from consuming meat that had been kept at improper holding temperatures, which allowed the spores and bacteria to proliferate.'

Staphylococcus aureus

Staphylococcus aureus is a spherically shaped bacterium that causes a foodborne intoxication with rapid onset. Figure 11.6 illustrates a grape-like cluster of *S. aureus* bacteria. During growth in foods, *S. aureus* elaborates a toxin that is extremely resistant to cooking at high temperatures. The organism can thrive in an environment that has heavy concentrations of salt or sugar (ham is a frequent vehicle for foodborne *S. aureus* poisoning); most other bacteria are unable to tolerate such an environment. Frequently, *S. aureus-associaled* foodborne outbreaks are linked to foods that have not been stored in a safe temperature range (below 45°F [7.2°C] and above 140°F [60°C]).

EXHIBIT 11.1

CASE STUDY: Botulism

1. On June 30,1994, a 47-year-old resident of Oklahoma was admitted to an Arkansas hospital with subacute onset of progressive dizziness, blurred vision, slurred speech, difficulty swallowing, and nausea. ... He developed respiratory compromise and required mechanical ventilation. The patient was hospitalized for 49 days, including 42 days on mechanical ventilation, before being discharged. The patient had reported that, during the 24 hours before onset of symptoms, he had eaten home-canned green beans and a stew containing roast beef and potatoes. Although analysis of the leftover green beans was negative for botulism toxin, type A toxin was detected in the stew. The stew had been cooked, covered with a heavy lid, and left on the stove for 3 days before being eaten without reheating.24(pp200–201)
2. On November 23, 1997, a previously healthy 68-year-old man became nauseated, vomited, and complained of abdominal pain. During the next 2 days, he developed diplopia, dysarthria, and respiratory impairment, necessitating hospitalization and mechanical ventilation. Possible botulism was diagnosed, and ... antibotulinum toxin was administered. ... A food history revealed no exposures to home-canned products; however, the patient had eaten pickled eggs that he had prepared 7 days before onset of illness; gastrointestinal symptoms began 12 hours after ingestion. The patient recovered after prolonged supportive care. The pickled eggs were prepared using a recipe that consisted of hard-boiled eggs, commercially prepared beets and hot peppers, and vinegar. The intact hard-boiled eggs were peeled and punctured with toothpicks then combined with the other ingredients in a glass jar that closed with a metal screw-on lid. The mixture was stored at room temperature and occasionally was exposed to sunlight. Cultures revealed Clostridium botulinum type B, and type B toxin was detected in samples of the pickled egg mixture. 25(pp778–779)

Source: The Centers for Disease Control and Prevention. Foodborne botulism from eating home-pickled eggs—Illinois, 1997. *MMWR.* 2000;49:778–780; the Centers for Disease Control and Prevention. Foodborne botulism—Oklahoma, 1994. *MMWR.* 1995;44:200–202.

Some of the additional characteristics of staphylococcal food poisoning are listed in the following text box.

The CDC reported the occurrence of an *S. aureus*-associated foodborne outbreak in 1997. Refer to the case study in Exhibit 11.2.

Escherichia coli 0157:H7 (£. *coli* 0157:H7)

E. coli 0157:H7 is one of four classes of *E. coli* that are designated enterovirulent, meaning that they cause inflammation of the intestines and the stomach. The disease caused by *E. coli* 0157:H7 is called hemorrhagic colitis, because one of the dramatic symptoms of the illness produced by the toxin from *E. coli* 0157:H7 is bloody diarrhea. For the same reason, *E. coli* 0157:H7 is referred to as an enterohemorrhagic

(EHEC) strain of *E. coli*. Figure 11.7 shows an electron micrograph of *E. coli* bacteria.

Nonpathogenic strains of *E. coli* reside harmlessly in the intestinal tracts of animals and humans and are part of the normal bacterial flora. Apparently, they retard the growth of pathogenic bacteria and synthesize vitamins. However, a few strains of *E. coli* are pathogenic, an example being the rare strain *E. coli* 0157:H7.

The agent *E. coli* 0157:H7 gained notoriety when it was associated with consumption of undercooked hamburger meat at a fast-food restaurant. The organism, found in the intestines of healthy cattle, may invade meat during the slaughter process. Outbreaks have occurred among visitors to petting farms. In the United States, this organism—which causes both bloody and nonbloody diarrhea—has been linked to approximately 73,000 cases and 61 deaths annually. Among children, infection with *E. coli* 0157-.H7 is the leading cause of hemolytic uremic syndrome (HUS), which produces such adverse consequences as acute kidney failure, end-stage renal disease, and hypertension. Young children are particularly vulnerable to HUS caused by *E. coli* 0157:H7. Refer to Table 11.4 for a chronology of the *E. coli* 0157:H7 pathogen, and refer to the following text box for the characteristics of *E. coli* Ol57:H7-associated foodborne illness.

Among the foodborne illness outbreaks associated with *E. coli* 0157:H7 is an incident that affected the patrons of a hamburger chain and impacted four western states. Refer to Exhibit 11.3.

Shigella

Shigella bacteria, which are highly infectious, are the cause of shigellosis (bacillary dysentery); an example of this agent is *Shigella sonnei*. The bacteria can be transmitted via the fecal-oral route through contaminated water and foods that are handled in an unsanitary manner. In the United States, approximately 300,000 cases of shigellosis occur each year; the percentage of these cases that is caused by food is unknown. However, fewer than 10% of foodborne disease outbreaks in the United States are believed to be caused by *Shigella*. Other facts regarding *Shigella* are presented in the following text box.

Clostridium perfringens (C. perfringens)

- Onset and Symptoms: The common form of perfringens poisoning is characterized by intense abdominal cramps and diarrhea which begin 8-22 hours after consumption of foods containing large numbers of those C. perfringens bacteria capable of producing the food poisoning toxin. The illness is usually over within 24 hours but less severe symptoms may persist in some individuals for 1 or 2 weeks.
- Associated Foods: In most instances, the actual cause of poisoning by C. perfringens is temperature abuse of prepared foods. Small numbers of the organisms are often present after cooking and multiply to food poisoning levels during cool down and storage of prepared foods. Meats, meat products, and gravy are the foods most frequently implicated.

Source: Adapted and reprinted from US Food and Drug Administration. Center for Food Safety and Applied Nutrition. Bad Bug Book—*Clostridium perfringens*. Available at: http://ww.fda.gov/Food/FoodSafety/Foodbornelllness/FoodbomelllnessFoodbornPathogensNatural Toxins/BadBugBook/ ucm070000.htm. Accessed April 26, 2010.

Staphylococcal Food Poisoning

- Symptoms: The onset of symptoms in staphylococcal food poisoning is usually rapid and in many cases acute, depending on individual susceptibility to the toxin, the amount of contaminated food eaten, the amount of toxin in the food ingested, and the general health of the victim. The most common symptoms are nausea, vomiting, retching, abdominal cramping, and prostration.
- Foods Incriminated: Meat and meat products; poultry and egg products; salads such as egg, tuna, chicken, potato, and macaroni; bakery products such as cream-filled pastries, cream pies, and chocolate eclairs; sandwich fillings; and milk and dairy products. Foods that require considerable handling during preparation and that are kept at slightly elevated temperatures after preparation are frequently involved in staphylococcal food poisoning.
- Reservoir: Staphylococci exist in air, dust, sewage, water, milk, and food or on food equipment, environmental surfaces, humans, and animals. Humans and animals are the primary reservoirs. Staphylococci are present in the nasal passages and throats and on the hair and skin of 50 percent or more of healthy individuals. (They may exist in great numbers in skin lesions and lacerations.)

Source: Adapted and reprinted from US Food and Drug Administration. Center for Food Safety and Applied Nutrition, Bad Bug Book—*Staphylococcus aureus*. Available at: http://www. fda.gov/Food/FoodSafety/FoodborneIllness/FoodborneIllnessFoodbornePathogensNaturalToxins/ BadBugBook/ucm070015.htm. Accessed April 26, 2010.

Campylobacter jejuni

Campylobacter jejuni is responsible for an illness known as campylobacteriosis. Among the bacterial causes of foodborne infections, C. *jejuni* is among the most commonly reported agents. About 2.1 to 2.4 million cases of campylobacteriosis occur each year in the United States. Shown in Figure 11.8 are data on the relative rates of laboratory-confirmed infections with *Campylobacter*, Shiga toxin-producing *Escherichia* coli (STEC) 0157, *Listeria, Salmonella,* and *Vibrio* reported by the CDC's FoodNet program.

Some of the characteristics of foodborne campylobacteriosis are listed in the text box titled "*Campylobacter jejuni.*"

Among the *C. jejuni*-associated outbreaks reported by the CDC was the occurrence of 75 cases of illness in December 2002 related to consumption of unpasteurized milk in Wisconsin. The unpasteurized milk was distributed by a retail dairy farm store. Unpasteurized milk can serve as a vehicle for C. *jejuni* and several other pathogens. The CDC noted that

> The facility that supplied milk to patients was a Grade A organic dairy farm with 36 dairy cows. The farm also had a retail store in which milk and other food products were available. In addition, farm operators provided unpasteurized milk samples at community events and to persons who toured the farm, including children from childcare facilities. Because unpasteurized milk cannot be sold legally to consumers in Wisconsin, the dairy distributed

EXHIBIT 11.2

CASE STUDY: S. *aureus* Food Outbreak Associated with Ham

On September 27, 1997, a community hospital in northeastern Florida notified the St. Johns County Health Department about several persons who were treated in the emergency department because of gastrointestinal illnesses suspected of being associated with a common meal. Self-administered questionnaires were distributed to the 125 attendees to document food histories, illnesses, and symptoms. A case was defined as nausea and/or vomiting in a person who attended the party or consumed food served at the party and who became ill within 8 hours after eating. Leftover food was collected and submitted for laboratory analysis. Food preparers were interviewed about the purchase and preparation of food served at the party.

Of the approximately 125 persons who attended the party, 98 completed and returned questionnaires. Of these, 31 persons attended the event but ate nothing, and none of them became ill; they were excluded from further analysis. A total of 18 (19%) persons had illnesses meeting the case definition, including 17 party attendees and one person who ate food brought home from the party. Ill persons reported nausea (94%), vomiting (89%), diarrhea (72%), weakness (67%), sweating (61%), chills (44%), fatigue (39%), myalgia (28%), headache (11%), and fever (11%). Onset of illness occurred at a mean of 3.4 hours after eating (range: 1–7 hours); symptoms lasted a median of 24 hours (range: 2–72 hours). Seven persons sought medical treatment, and two of those were hospitalized overnight.

Illness was strongly associated with eating ham. Of the 18 ill persons, 17 (94%) had eaten ham. The ill person who had not attended the party had eaten only leftover ham. None of the other foods served at the party were significantly associated with illness. One sample of leftover cooked ham and one sample of leftover rice pilaf were analyzed and were positive for staphylococcal enterotoxin type A. Samples of stool or vomitus were not obtained from any ill persons, and cultures from nares or skin were not obtained from the food preparers.

On September 25, a food preparer had purchased a 16-pound precooked packaged ham, baked it at home at 400°F (204°C) for 1.5 hours, and transported it to her workplace, a large institutional kitchen, where she sliced the ham while it was hot on a commercial slicer. The food preparer reported having no cuts, sores, or infected wounds on her hands. She reported that she routinely cleaned the slicer in place rather than dismantling it and cleaning it according to recommended procedures and that she did not use an approved sanitizer. All 16 pounds of sliced ham had been placed in a 14-inch by 12-inch by 3-inch plastic container that was covered with foil and stored in a walk-in cooler for 6 hours, then transported back to the preparer's home and refrigerated overnight. The ham was served cold at the party the next day. The rice pilaf was prepared the day of the party by a different person.

Source: The Centers for Disease Control and Prevention. Outbreak of Staphylococcal food poisoning associated with precooked ham—Florida, 1997. *MMWR.* 1997;46:1189–1191.

Table 11.4. *E. coli* 0157:H7 (chronology)

1982	First recognized as a pathogen
1985	Associated with hemolytic uremic syndrome
1990	Outbreak from drinking water
1991	Outbreak from apple cider
1993	Multistate outbreak from fast-food hamburgers
1995	Outbreak from fresh produce
1996	Outbreak in Japan
	Multistate outbreak from unpasteurized apple juice

Source: Adapted and reprinted from the Centers for Disease Control and Prevention. Public Health Image Library, ID #107. Available at: http:// phil.cdc.gov/Phil/ details.asp. Accessed May 1, 2010.

unpasteurized milk through a cow-leasing program. Customers paid an initial fee to lease part of a cow. Farm operators milked the cows and stored the milk from all leased cows together in a bulk tank. Either customers picked up milk at the farm or farm operators had it delivered. On December 8, investigators obtained a milk sample from the farm's bulk milk tank, and cultures of the milk samples grew *C. jejuni* with a PFGE pattern that matched the outbreak strain. Farm operators were ordered to divert all milk to a processor for pasteurization.

Characteristics of *E. coli* 0157:H7-Associated Foodborne Illness

- Symptoms: The illness is characterized by severe cramping (abdominal pain) and diarrhea which is initially watery but becomes grossly bloody. Occasionally vomiting occurs. Fever is either low-grade or absent. The illness is usually self-limited and lasts for an average of 8 days. Some individuals exhibit watery diarrhea only. ...
- Associated Foods: Undercooked or raw hamburger (ground beef) has been implicated in many of the documented outbreaks; however, E. coli 0157: H7 outbreaks have implicated alfalfa sprouts, unpasteurized fruit juices, dry-cured salami, lettuce, game meat, and cheese curds. Raw milk was the vehicle in a school outbreak in Canada.

Source: Adapted and reprinted from US Food and Drug Administration. Center for Food Safety and Applied Nutrition, Bad Bug Book—*Escherichia coli* O157:H7. Available at: http://www.fda.gov/Food/FoodSafety/ FoodborneIllness/ FoodborneIllnessFoodbornePathogensNaturalToxins/ BadBugBook/ucm071284.htm. Accessed April 26, 2010.

Listeria monocytogenes

The illness listeriosis is caused by the bacterium *Listeria monocytogenes,* which has been found in some domestic mammals and several species of birds. The bacterium occurs widely in plant-related materials such as vegetables that may have been contaminated with sewage used as a fertilizer. These vegetables may be consumed raw or after minimal processing. In the United States, *L. monocytogenes* is associated with about 2,500 cases of illness and 500 fatalities per year. The characteristics of listeriosis are listed in the above text box.

A decline in the incidence of listeriosis has occurred in recent years, although sporadic outbreaks have been reported. A 2002 outbreak in eight northeastern U.S. states was responsible for 46 confirmed cases, 7 fatalities, and 3 stillbirths or miscarriages. The multistate outbreak was associated with eating sliceable turkey deli meat.

EXHIBIT 11.3

CASE STUDY: Multistate *E. coli* 0157:H7 Outbreaks

From November 15, 1992, through February 28, 1993, more than 500 laboratory-confirmed infections with *E. coli* 0157:H7 and four associated deaths occurred in four states—Washington, Idaho, California, and Nevada.

On January 13, 1993, a physician reported to the Washington Department of Health a cluster of children with hemolytic uremic syndrome (HUS) and an increase in emergency room visits for bloody diarrhea. During January 16–17, a case-control study comparing 16 of the first cases of bloody diarrhea or postdiarrheal HUS identified with age- and neighborhood-matched controls implicated eating at chain A restaurants during the week before symptom onset. On January 18, a multistate recall of unused hamburger patties from chain A restaurants was initiated.

As a result of publicity and case-finding efforts, during January–February 1993, 602 patients with bloody diarrhea or HUS were reported to the state health department. A total of 477 persons had illnesses meeting the case definition of culture-confirmed *E. coli* 0157: H7 infection or postdiarrheal HUS. Of the 477 persons, 52 (11%) had close contact with a person with confirmed *E. coli* 0157:H7 infection during the week preceding onset of symptoms. Of the remaining 425 persons, 372 (88%) reported eating in a chain A restaurant during the 9 days preceding onset of symptoms. Of the 338 patients who recalled what they ate in a chain A restaurant, 312 (92%) reported eating a regular-sized hamburger patty. Onsets of illness peaked from January 17 through January 20. Of the 477 patients, 144 (30%) were hospitalized; 30 developed HUS, and three died. The median age of patients was 7.5 years (range: 0–74 years). (During approximately the same time period, additional cases occurred in Idaho, California, and Nevada. These cases also were linked to the chain A restaurants.)

During the outbreak, chain A restaurants in Washington linked with cases primarily were serving regular-sized hamburger patties produced on November 19, 1992; some of the same meat was used in "jumbo" patties produced on November 20, 1992. The outbreak strain of *E. coli* 0157: H7 was isolated from 11 lots of patties produced on those two dates; these lots had been distributed to restaurants in all states where illness occurred. Approximately 272,672 (20%) of the implicated patties were recovered by the recall.

A meat traceback by a CDC team identified five slaughter plants in the United States and one in Canada as the likely sources of carcasses used in the contaminated lots of meat and identified potential control points for reducing the likelihood of contamination. The animals slaughtered in domestic slaughter plants were traced to farms and auctions in six western states. No one slaughter plant or farm was identified as the source.

Source: Adapted and reprinted from the Centers for Disease Control and Prevention. Update: multistate outbreak *of Eschcricliia coli* 0157:H7 infections from hamburgers—western United States, 1992-1993. *MMWR.* 1993;42:258–259, 261.

Shigella

- Symptoms: Abdominal pain; cramps; diarrhea; fever; vomiting; blood, pus, or mucus in stools. The onset time for shigellosis ranges from 12 to 50 hours. Infections are associated with mucosal ulceration, rectal bleeding, drastic dehydration; fatality may be as high as 10% to 15% with some strains.
- Associated Foods: Salads (potato, tuna, shrimp, macaroni, and chicken), raw vegetables, milk and dairy products, and poultry.
- Risk Groups: Infants, the elderly, and the infirm are susceptible to the severest symptoms of disease, but all humans are susceptible to some degree. Shigellosis is a very common malady suffered by individuals with acquired immune deficiency syndrome (AIDS).

Source: Adapted and reprinted from US Food and Drug Administration. Center for Food Safety and Applied Nutrition, Bad Bug Book—*Shigella* spp. Available at: http://wvmfda;gov/Food/FoodSafety/htm. Accessed April 26, 2010.

Campylobacter jejuni

- Symptoms: *C. jejuni* infection causes diarrhea, which may be watery or sticky and can contain blood. Other symptoms often present are fever, abdominal pain, nausea, headache and muscle pain. The illness usually occurs 2–5 days after ingestion of the contaminated food or water. Illness generally lasts 7–10 days, but relapses are not uncommon (about 25% of cases). Most infections are self-limiting and are not treated with antibiotics.
- Associated Foods: *C. jejuni* frequently contaminates raw chicken. Surveys show that 20 to 100% of retail chickens are contaminated. This is not overly surprising since many healthy chickens carry these bacteria in their intestinal tracts. Raw milk is also a source of infections. The bacteria are often carried by healthy cattle and by flies on farms. Non-chlorinated water may also be a source of infections. However, properly cooking chicken, pasteurizing milk, and chlorinating drinking water will kill the bacteria.

Source: Adapted and reprinted from US Food and Drug Administration. Center for Food Safety and Applied Nutrition, Bad Bug Book—*Campylobacter jejuni*. Available at: http://www. fda.gov/Food/FoodSafety/FoodborneIllness/FoodborneIllnessFoodbornePathogensNaturalToxins/BadBugBook/ ucm070024.htm. Accessed April 30, 2010.

WORMS

Trichinella

Trichinosis is a foodborne disease associated with eating meat that contains a nematode (also called a roundworm) from the genus *Trichinella*. The classic agent of trichinosis is *Trichinella spiralis*, which can be found in many carnivorous and omnivorous animals (e.g., pigs, bears, walruses, rodents, and cougars). The disease may be transmitted from animals such as these to humans when meat that contains the cysts of *T. spiralis* has not been cooked adequately. For example, outbreaks have been associated with eating pork from local farms and from consumption of grizzly bear meat.42

Formerly trichinosis was common in the United States, because garbage that contained raw meat was used routinely as feed for swine. Trichinosis outbreaks have been prevented following the

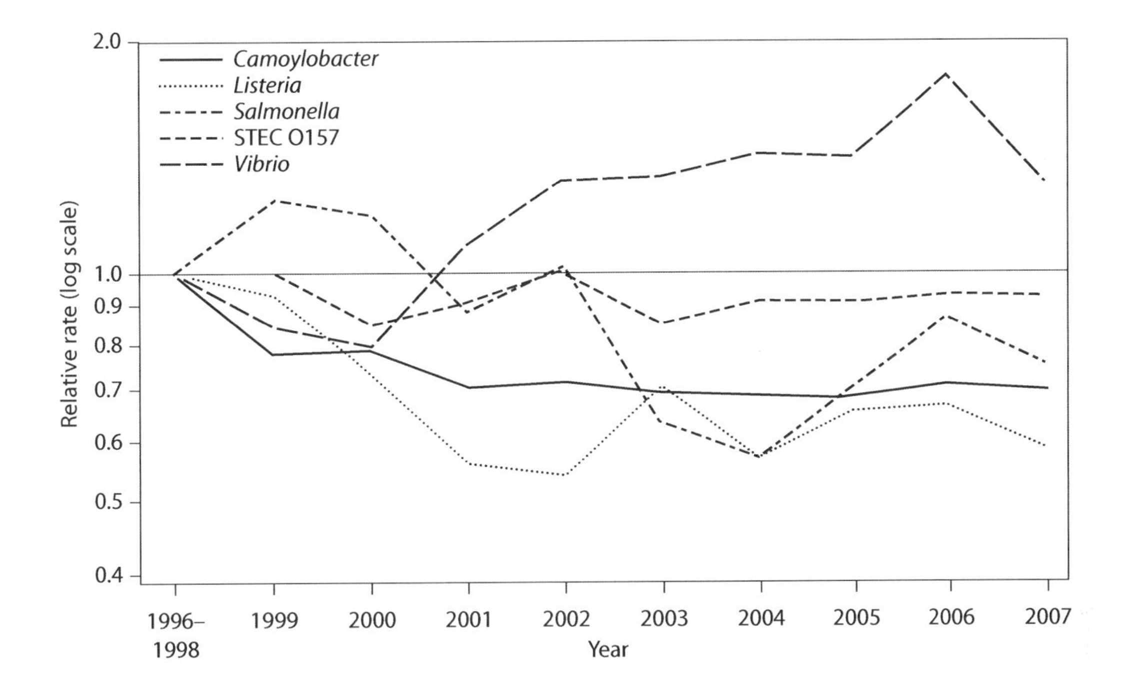

* Shiga toxin-producing *Escherichia coil.*

† The position of each line indicates the relative change in the incidence of that pathogen compared with 1996–1998. The actual incidences of these infections can differ.

Source: Reprinted from the Centers for Disease Control and Prevention. Preliminary FoodNet data on the incidence of infection with pathogens transmitted commonly through food—10 states, 2007. *MMWR.* 2008;57:368.

FIgure 11.8. Relative rates of laboratory-confirmed infections with *Campylobacter,* STEC* *0157, Listeria, Salmonella,* and *Vibrio* compared with *1996–1998* rates, by year—Foodborne Diseases Active Surveillance Network, United States, *1996-2007.t*

elimination of this practice. Consumers can protect themselves further by cooking pork and wild game products adequately; also, freezing kills the worms of *T. spiralis.*

Infection with *T. spiralis* can be asymptomatic, as in the instance of light infections, or can include gastrointestinal effects in more severe infections. After about one week, when the organism migrates into muscle tissues, symptoms can range from fever and facial edema to life-threatening effects upon the heart and central nervous system.

The life cycle of *Trichinella* is shown in Figure 11.9. Following ingestion of meat that contains cysts, the larvae are released due to the action of stomach acids on the cysts. The larvae subsequently may invade the small intestine and develop into adult worms. The female adult worms release larvae that can migrate into muscle tissue and become encysted. The human host is a dead-end host, meaning that the disease cannot be spread to other humans.

Listeria monocytogenes

- Symptoms: Septicemia, meningitis (or meningoencephalitis), encephalitis, and intrauterine or cervical infections in pregnant women, which may result in spontaneous abortion (2nd/3rd trimester) or stillbirth. The onset of the aforementioned disorders is usually preceded by influenza-like symptoms including persistent fever. Gastrointestinal symptoms such as nausea, vomiting, and diarrhea may precede more serious forms of listeriosis or may be the only symptoms expressed.
- Associated Foods: *L. monocytogenes* has been associated with such foods as raw milk, supposedly pasteurized fluid milk, cheeses (particularly soft-ripened varieties), ice cream, raw vegetables, fermented raw- meat sausages, raw and cooked poultry, raw meats (all types), and raw and smoked fish. Its ability to grow at temperatures as low as 3°C permits multiplication in refrigerated foods.

Source: Adapted and reprinted from US Food and Drug Administration. Center for Food Safety and Applied Nutrition, Bad Bug Book—*Listeria monocytogenes.* Available at: http://www.fda.gov/Food/FoodSafety/FoodborneIllness/FoodborneIllness FoodbornePathogensNaturalToxins/BadBugBook/ucm070064.htm. Accessed April 30, 2010.

Tapeworms

Taeniasis is a parasitic disease caused by tapeworms. One form is caused by the beef tapeworm (*Taenia saginata*) and the other by the pork tapeworm *(Taenia solium).* These organisms may induce human illness following the consumption of raw or undercooked infected beef or pork. Usually the symptoms of taeniasis are limited to mild abdominal distress; one of the main symptoms is the passage of the proglottids (the section of the worm that contains eggs) of *T. saginata* and *T. solium* in stools. *T. solium* can cause a serious condition termed cysticercosis, in which the organism migrates to muscle or brain tissue and forms cysts. The worms of *T. saginata,* which reside in human intestines, can be quite long. (See Figure 11.10.)

VIRAL AGENTS

Hepatitis A Virus (HAV)

Chapter 9 indicated that HAV can be transmitted by both contaminated food and water. Foods that are associated commonly with HAV outbreaks include fruits, sandwiches made with cold cuts, dairy products, vegetables, and shellfish. A common mode for contamination of foods is by HAV-infected workers in food processing plants and restaurants. Of the 23,000 cases of hepatitis A reported in the United States annually, about 7% are believed to be food- or waterborne.

One instance of a large hepatitis A outbreak occurred from February to March 1997; this incident caused 213 cases in Michigan and 29 cases in Maine, and seemed to be linked to sporadic cases in several other states. The source of the outbreak was frozen strawberries fed to schoolchildren in schools that had purchased the contaminated product from the same commercial processor. Other examples of hepatitis A outbreaks include those linked to sandwich shops and to national hamburger chains.

Hepatitis A outbreaks that involved almost 1,000 cases occurred in Tennessee, North Carolina, Georgia, and Pennsylvania from September to November 2003. The outbreak in Pennsylvania was associated with a single restaurant and was responsible for more

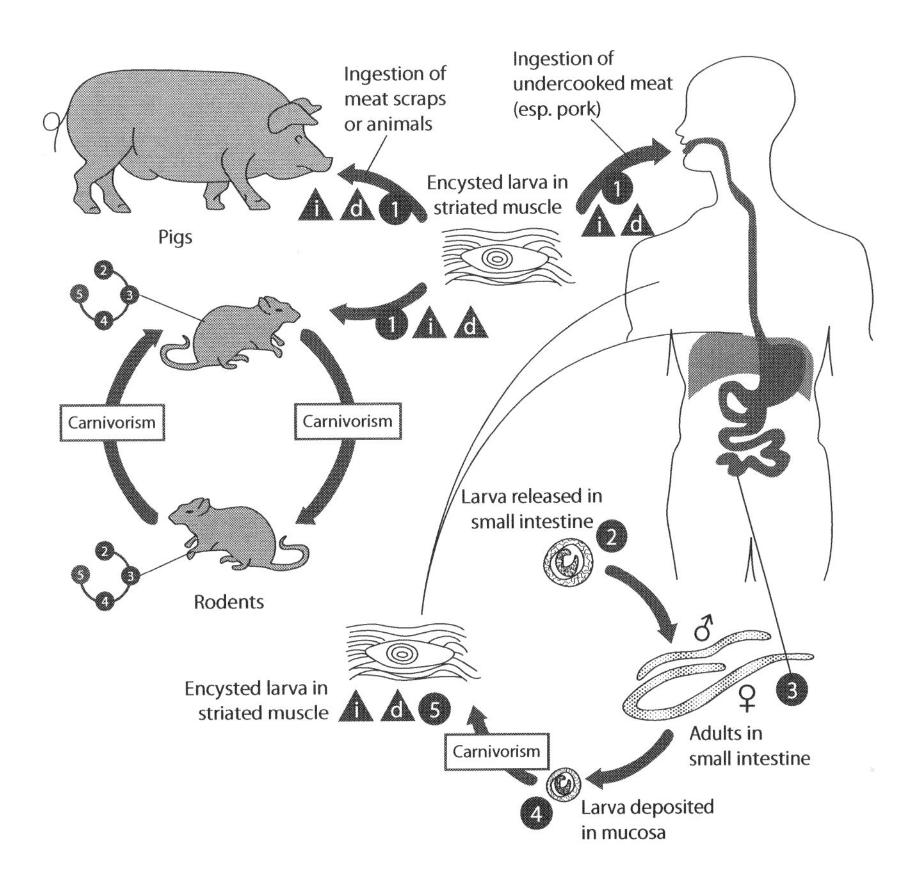

Source: Reprinted from the Centers for Disease Control and Prevention. Parasites and Health: Trichinellosis. Available at: http://vvwvv.dpd.cdc.gov/dpdx/HTML/Trichinellosis.htm. Accessed April 30, 2010.

Figure 11.9. Life cycle of Trichinella.

than 600 infected persons and 3 deaths. This outbreak was the largest episode of foodborne hepatitis A that, to date, ever has been reported in the United States. The Pennsylvania outbreak, as well as those in the other three states, was associated with green onions that had been imported from Mexico. This widespread occurrence of hepatitis A demonstrates the need to cook green onions thoroughly instead of serving them raw.

Norovirus

Chapter 9 presented the topic of noroviruses, particularly with respect to waterborne norovirus infections. This virus is transmitted easily within closed environments such as cruise ships. (See Figure 11.11.) Some cases may be brought on board by passengers who fall ill just before embarking on a cruise. It is also possible that crew members and shipboard environmental contamination can act as disease reservoirs.

Outbreaks of gastrointestinal illness have been responsible for sickening hundreds of passengers on cruise ships. The culprit for many outbreaks is believed to be a virus similar to the Norwalk virus, which is also called the winter vomiting virus. During one month in fall 2002, the toll was estimated to be 1,400 infected passengers and crew members on 10 ships. One outbreak that occurred on the *Amsterdam* (operated by Holland America) afflicted 163 passengers and 18 crew members out of 1,905 primarily elderly persons who went on the cruise. Illness symptoms consisted of vomiting and diarrhea as well as associated dehydration and abdominal cramps. These symptoms lasted from one to two days.

Just as the occurrence of gastrointestinal illness has continued to be reported throughout the United States, the CDC expects to continue to see increases in the number of reported gastrointestinal illness cases on some cruises. Annually, about 6.8 million passengers in North America took cruises during the late 1990s. During the six-year period from 1993 to 1998, the number of passengers who embarked on cruises from the United States increased by 50%. The number of North American cruise passengers increased by 11% from 2003 to 2004 (a record year with nearly 9 million passengers) and by 9% between 2004 and 2005.

U.S. government regulations state:

> Cruise vessels sailing to U.S. ports are required to notify the CDC of all reported gastrointestinal illnesses that have been reported to the ships' medical staff. This report must be filed 24 hours prior to arrival at a U.S. port, from a foreign port. If the number of ill passengers or crewmembers reaches 2%, the vessel is required to file a special report. The CDC continues to closely monitor illness reports on a daily basis. An "outbreak" of gastrointestinal illness is defined as having 3% or more of either passengers or crew reported with a gastrointestinal illness.

The CDC operates the Vessel Sanitation Program (VSP), which aims to keep the risks of diarrheal disease outbreaks among passengers at a minimum. A component of the VSP program is twice-yearly inspections of cruise ships that enter U.S. waters; these inspections are not announced in advance. Refer to the text box for a description of outbreaks that occurred on four cruise ships: *Sundream, Olympia Voyager, Carnival Spirit,* and *Sun Princess*.

Outbreaks on Cruise Ships

Sundream (Sun Cruises, UK): January 20 to February 3, 2003. This vessel reported 95 of 1,085 passengers (8.8%) and 12 of 403 crew (3.0%) with gastrointestinal illness. The vessel made one US port of call in St. Thomas, US Virgin Islands on January 31, and then proceeded onto [sic] Venezuela. A UK public health laboratory confirmed gastrointestinal illness from the ill passengers and crews' stool samples. The Sundream crew conducted extensive cleaning and disinfection of the vessel.

Olympia Voyager (Royal Olympic Cruises, Greece): January 15 to February 3, 2003. This vessel reported 35 of 756 passengers (4.6%) and 5 of 356 crew (1.4%) ill with gastrointestinal illness. The cruise began in Port Everglades Florida and concluded in Houston, Texas. Extensive cleaning and disinfection was initiated [sic] during the first week of the cruise as the initial cases were identified. No new cases were identified the second week.

Carnival Spirit (Carnival Cruise Lines): January 27 to February 4, 2003. Carnival Spirit reported 102 of 2143 passengers (4.8%) and 10 of 902 crew (1.0%) were ill with a gastrointestinal illness. This vessel sails a 7-day cruise from Miami. Stool specimens were collected and submitted to [the] CDC for analysis. Clinical presentation of the illness [was] consistent with a noroviral outbreak. Carnival Spirit crew [conducted] extensive cleaning and disinfection aboard the vessel and additional staff will board in Miami on February 4th to assist.

Sun Princess (Princess Cruise Lines): January 25 to February 9, 2003. On January 28, the VSP was notified that 267 of 2029 passengers (13.1%) and 29 of 877 crew (3.3%) reported ill with a gastrointestinal illness. The cruise sails from Los Angeles to Hawaii, and returns to Los Angeles. On February 4th, Princess Cruises made the decision to terminate the 15 day cruise in Hawaii and not continue with the remaining five days transit from Hawaii to Los Angeles. This measure [was] taken to protect the health and safety of both passengers and crew by adopting every precaution possible to avoid the risk of any further transfer of this illness. All passengers [were] flown home.

Cruise ship travelers are reminded that simple hygienic practices, such as frequent and thorough hand washing and avoiding contact with other passengers when ill, are important measures to prevent the spread of disease.

Source: Reprinted from the Centers for Disease Control and Prevention, Vessel Sanitation Program. Investigation update on the *Carnival Legend*— February 4, 2003. Available at: http://mvw.cdc.gov/nceh/vsp/surv/outbreak/2003/update_feb4.htm. Accessed March 22, 2010.

We Have No Bananas

By Mike Peed

Darwin, the capital of Australia's Northern Territory, is more than a thousand miles northwest of the country's largest banana plantations, which are centered around Innisfail, on the eastern seaboard. A ramshackle place, Darwin is known for its many impoverished indigenous residents, entertainment attractions like Crocosaurus Cove (where visitors are lowered, via "the Cage of Death," into a crocodile-filled tank), and, as one local puts it, "not partying, exactly, but certainly drinking." To Robert Borsato, a fruit farmer, the area looked like an ideal place to grow bananas. In 1996, he began farming a thousand acres in Humpty Doo, which is on the road between Darwin and Kakadu National Park

To bear fruit, banana plants need at least fourteen consecutive months of frost-free weather, which is why they are not grown commercially in the continental United States. Darwin offered this, and more. As one of Borsato's workers told me recently, "You came up here and saw the consistency that you've got between the blue sky, the sunshine, the water, the fucking soil. You *knew* you were going to beat everybody else, hands down." There were a few nuisances: crocodiles wandered onto the property, Asian buffalo trampled young plants, and dingoes chewed the sprinklers. Before long, though, the Darwin Banana Farming Company was growing lush ten-foot plants with as many as a hundred and seventy bananas on each stalk. In 2006, Cyclone Larry decimated ninety percent of the Innisfail plantations; banana prices soared from ten dollars a carton to a hundred and thirty, and Borsato became a multimillionaire.

More than a thousand kinds of banana can be found worldwide, but Borsato specialized in a variety called Cavendish, which a nineteenth-century British explorer happened upon in a household garden in southern China. Today, the Cavendish represents ninety-nine percent of the banana export market. The vast majority of banana varieties are not viable for international trade: their bunches are too small, or their skin is too thin, or their pulp is too bland. Although Cavendishes need pampering, they are the only variety that provides farmers with a high yield of palatable fruit that can endure overseas trips without ripening too quickly or bruising too easily. The Cavendish, which is rich in Vitamins B_6 and C, has high levels of potassium, magnesium, and fibre; it is also cheap—about sixty cents a pound. In 2008, Americans ate 7.6 billion pounds of Cavendish bananas, virtually all of them imported from Latin America. Each year, we eat as many Cavendish bananas as we do apples and oranges combined. Your supermarket likely sells many varieties of apples, but when you shop for bananas you usually have one option. The world's banana plantations are a monoculture of Cavendishes.

Several years ago, Borsato noticed a couple of sick-looking plants on a neighbor's property. The leaves turned a soiled yellow, starting at the edges and rapidly moving inward; necrotic patches appeared and, a few weeks later, the leaves buckled. What had once formed a canopy now dangled around the base of the plant, like a cast-off grass skirt. Inside the plant, the effects were even worse. Something was blocking the plants' vascular system, causing rot, and tissue that should have been as ivory as the inside of a celery stalk was a putrefying mixture of brown, black, and blood-red. When the plants were cut open, they smelled like garbage, and their roots were so anemic that the plants could barely stay upright.

Borsato feared that he was seeing the symptoms of a pestilence that had wiped out the Cavendish across Asia: Tropical Race Four. A soil-borne fungus that is known to be harmful only to bananas, it can survive for decades in the dirt, spreading through the transportation of tainted plants, or in infected mud stuck to a tractor's tire or a rancher's boot. It cannot be controlled with chemicals. Tropical Race Four appeared in Taiwan in the late eighties, and destroyed roughly seventy percent of the island's Cavendish plantations. In Indonesia, more than twelve thousand acres of export bananas were abandoned; in Malaysia, a local newspaper branded the disease "the H.I.V. of banana plantations." When the fungus reached China and the Philippines, the effect was equally ruinous.

Australia was next. Over the following three years, Borsato watched as the other banana farmers in Darwin succumbed to the disease. "A lot of people were in denial," he recalled. "Most growers tried to hide the fact that they had it. It would've devalued their property immensely. But today nearly everybody is out. The guy across the street now grows melons." He went on, "The government tried to put in a quarantine. You couldn't move equipment around. There were footbaths to wash your shoes in. Stuff like that. We put a new car park in, paved it all up for the sake of the quarantine, went to that level of expense, and you know what happened? The first idiot to drive up went screaming past all our new signs. Drove right up on the muddy road, up to the shed! It was a bloody idiot government official."

Scientists believe that Tropical Race Four, which has caused tens of millions of dollars' worth of damage, will ultimately find its way to Latin America—and to the fruit that Americans buy. "I don't have a crystal ball," Randy Ploetz, a plant pathologist at the University of Florida, who was the first researcher to identify Tropical Racc Four, said. "People are bringing stuff in their luggage, moving stuff around the world that they shouldn't be. I hope it doesn't happen, but history has shown that this kind of stuff does happen." Borsato was more blunt: "Shit's gonna *move.* Americans are snookered. They'd better wake up and realize it, or they're not going to have any bananas to eat."

We were prepared to give up," Borsato told me one afternoon, as he and his farm manager, Mark Smith, showed me around their plantation. "But you just can't get excited about melons."

Borsato, who is fifty, has a fringe of white hair and the shoulders of a rugby fullback. (He used to play in high school.) He told me that he had taken much of the money he had earned from the Cyclone

Larry shortage and invested it back into his farm. For a while, he and Smith tried to grow bananas only in soil that Tropical Racc Four had apparendy not yet reached. They soon ran out of virgin land. Then they planted cassava and pinto peanuts, hoping to rejuvenate the soil, and applied quicklime to lower the soil's acidity, these efforts failed to counteract the blight. A few years ago, Borsato leased ninety acres of a seemingly unsullied property twenty miles away. To insure that infected machinery wasn't used, Smith and a small crew planted forty-three thousand banana stalks by hand. Within eight months, Tropical Race Four had appeared. Today, Borsato farms only a quarter of his land, and every week he and Smith chop down two hundred infected plants. "In another month, that'll be three hundred," Smith said.

As we walked through the fields, Tropical Race Four seemed as abundant as the mosquitoes circling our heads. "There's one," Smith said, pointing. "That's two. You can see that one there? He's coming out. There's another one." Some plants were just turning yellow; others were a desiccated mass of raw umber. At one point, Smith unsheathed a cane knife, which is similar to a machete, but with a shorter, wider blade. An axe is not needed to cut down a banana plant, which is not a tree but, rather, the world's largest herb. The part that is usually called the trunk is the pseudostem—a barkless staff composed only of leaves waiting to unfurl. In one stroke, Smith sliced through a diseased plant. The inside resembled a crushed-out cigar, and the fetid odor was overwhelming. Smith said, "You smell that, and you think, Ah, fuck."

Borsato shook his head. "Cruel," he said. "Just cruel." Lately, he had been obtaining fresh plants from a laboratory that cultivated the seeds in antiseptic petri dishes. But, because the fungus is in his soil, he could get only one or two bunches before the plants died.

Smith was wearing a blue baseball cap that depicted a banana above the slogan "Get Bent Into Shape." He removed it and wiped his brow. "You see one plant, and you know pretty soon you'll be up shit creek," he said. "All this work we're doing—it's not viable. The only way to keep going is to breed a disease-resistant variety, one with commercial potential. That's the only way."

Borsato knew that attempts to replace the Cavendish through traditional breeding—crossing two bananas to create a third, disease-resistant fruit—had failed. After a series of phone calls, Internet searches, and chance encounters, he found James Dale, a professor at Queensland University of Technology, in Brisbane, who experiments with genetically modified crops. In 1994, Dale produced one of the first genetically transformed Cavendishes. He and his team members inserted what is known as a marker gene; the resulting banana, when placed under ultraviolet light, glowed fluorescent green. More recently, in research supported by the Gates Foundation, Dale has been trying to increase the provitamin-A content of locally grown bananas in Uganda, where villagers eat several bananas daily.

This spring, Dale expects to plant on Borsato's land four acres of banana plants that have been genetically modified to resist the blight. The Australian Research Council will pay for much of the field trial, but Borsato is also investing a quarter of a million dollars. "And we'll go to the millions," he told me. "Someone has to do this work. Otherwise, there'll be grief."

The Queensland University of Technology borders Brisbane's City Botanic Gardens, where students read under weeping figs and cabbage-tree palms. Dale's office, at the university's Centre for Tropical Crops and Biocom- modities, is less picturesque. "Our lab was once voted the ugliest building in Brisbane," Dale told me. "It was painted pink and gray—appalling, just dreadful. The university's solution was to give each floor a different color scheme. Now we've got a lime-green-and-red interior."

Dale leads a team of a dozen scientists, and on the day I visited he informed them of the Australian

Research Council's decision to fund the Darwin field trial. "It's all go!" he said. The news aroused only muffled enthusiasm.

"They know that the plants have to perform," Dale explained later. "Otherwise, we don't keep getting money." Dale, an affable, diminutive man with a wispy white beard, noted another challenge that banana scientists faced: bad jokes. He said, "I hate when people ask, 'Hey, James. How are you going to straighten the banana?' Yeah, we get it: bananas are phallic. I always say, 'Well, you dickhead ...'"

Despite the fruit's priapic associations—the refrain from an old blues song contains the line "Let me put my banana in your fruit basket"—the bananas we eat are sterile. Unlike wild bananas, the Cavendish doesn't have seeds, because it has three sets of chromosomes; it's what biologists call a triploid—in this case, the haphazard product of two wild, seeded diploids that mated thousands of years ago. (Wild bananas, which have flinty seeds the size of peppercorns, can be found across Asia.) Cavendish and other domesticated banana plants produce fruit without fertilization. In a healthy nine-month-old Cavendish plant, a secondary stalk rises from the center of the pseudostem and, a few months later, droops with a single eighty-pound bunch. A bunch consists of a dozen "hands," and each hand has some twenty "fingers," or individual bananas. (Fingers do not hang down but, instead, curl toward the sky.) Meanwhile, small suckers poke out from the plant's roots. When growers harvest the bunch, they cut down the "mother" plant and all but the heartiest sucker. In another year, that sucker sprouts a new bunch. In this way, commercial banana plants can produce genetically identical fruit for decades. Tropical Race Four has upended this efficient method of cultivation; one of the main ways that farmers spread the disease is by uprooting contaminated suckers that appear to be clean and replanting them elsewhere.

On a wall in Dale's office, there was a painting of a gyrating Josephine Baker, naked except for a pendulous skirt of bananas. "It's a real leap of faith," he said of the Cavendish project. "We trust the gene's in there, but until the plant grows up we don't know if it will be blight- resistant. That's the mystery and the magic."

Despite the danger posed by Tropical Race Four, only a handful of scientists are working to modify bananas. Whereas other biotechnology researchers have focused on trying to insert an antifungal gene into the Cavendish, Dale wants to insert a gene that will starve the fungus to death. For years, scientists believed that the fungus injected toxins into the plant, killing cells and gorging on the waste. "Now there's good evidence that these toxins don't actually kill," Dale told me. "Instead, they switch on a certain mechanism in the plant and the plant actually kills itself." That mechanism is known as programmed cell death. In stressful situations, plants fortify themselves by, say, dropping leaves; they kill weaker cells so that stronger ones may live to fight. "Our thinking," Dale said, "is that we can insert a gene that inhibits this process, that tells the plant *not* to kill its own cclls."

Once a year, Dale explained, his lab assistants extract cells from sterilized banana plants and attempt to multiply them in liquid media. Their success rate is five percent. The cells that survive are selected for gene transformation, which Dale accomplishes by making clever use of a common soil bacterium, *Agrobacterium tumefaciens*. When the pathogen invades the nucleus of a cell, it installs a few of its own genes, hoodwinking the cell into making food for it. To sneak his desired genes into a banana plant, Dale swaps a few of the bacterium's genes for the ones he wants. The bacterium then installs those genes into the cells. Later, the bacterium is killed off with an antibiotic. "It's a natural genetic engineer," Dale said. "It still amazes me that we can do it."

Two years ago, Dale and his team inserted into banana cells one of nine genes, which were taken from life-forms as diverse as rice, thale cress, and an armyworm. Each gene was known to impede programmed cell death. "The toxin will still get in, but the cells don't die," Dale theorized. "And if the cells don't die the fungus hasn't got anything to live on."

That afternoon, Dale and I drove east of Brisbane to a state-owned greenhouse. A yellow biohazard sign was posted on the front door. Inside, on waist-high tables, young banana plants grew in six-inch plastic pots. Several staff members, wearing white lab coats and white surgical gloves, were slicing open two hundred and fifty genetically altered banana plants and checking for disease resistance. After digging a plant out of its pot, they used a chef's knife to split the plant's stem lengthwise. The two halves were photographed, side by side, on a plastic tarp.

"The sap from the stems gets everywhere, and it never comes out," a research fellow said. "We didn't used to have the tarp. Management got a bit pissed off."

Four months earlier, the plants in the greenhouse had been infected with two hundred times as much fungus as they would face in the field. The fungus, however, was not Tropical Race Four but an earlier, more widespread strain of the disease called Race One. "We're not allowed to bring the Race Four fungus down here," Dale said. "That's how paranoid people are." This problem had led to another: Cavendish is resistant to Race One. So the banana plants here were actually a variety known as Lady Finger, which is susceptible to Race One. When I asked Dale how his science could be sound when he was experimenting with a different fungus on a different banana, he replied, "Race One and Race Four are very, very closely related, and we believe that they use the same mechanism of killing. If they both use programmed cell death, then the same mechanism should stop both. That's our hypothesis, anyway."

We were talking in a side room of the greenhouse when a young researcher suddenly called to us. "Oh, they've got something!" Dale said. In the main room, the group members were bent over the tarp, examining a pair of bifurcated plants; with their coats and gloves and camera, they looked like forensics experts at a crime scene. The two leading gene candidates, I was told, were taken from the armyworm and the thale cress. A plant harboring the armyworm gene was soot-colored, its roots a mush of corrosion. "Ewww," Dale said. A plant with the thale-cress gene was as spotessly white as a wedding gown. "Well done, guys!" Dale said. "We'll definitely be taking this one up to Darwin."

As we drove back to Brisbane, he said, "No one's ever taken any of these genes and put them in and seen that they've given resistance. What you saw today was a world first." He went on, "It looks like we've got the potential to produce a product that uses a plant gene rather than an animal gene. That makes me happy. The general public will be much more willing to accept a plant gene."

Bananas, which Alexander the Great introduced to the West in 327 B.C., initially came from jungles in India, China, and Southeast Asia. There are fuzzy bananas whose skins are bubble-gum pink; green-and-white striped bananas with pulp the color of orange sherbet; bananas that, when cooked, taste like strawberries. The Double Mahoi plant can produce two bunches at once. The Chinese name of the aromatic Go San Heong banana means "You can smell it from the next mountain." The fingers on one banana plant grow fused; another produces bunches of a thousand fingers, each only an inch long.

Many of these varieties are known as plantains; they are starchy and inedible until cooked. In 1870, when a Cape Cod fishing-boat captain named Lorenzo Dow Baker imported a hundred and sixty bunches of bananas into Jersey City—the first bananas in the U.S.—he chose a kind of banana that is sugary and eaten raw, and that he'd seen growing in Jamaica. Baker's variety, called the Gros Michel, offered a sweet and complex flavor, and its skin was resilient. Baker could throw the bunches directly into the hold of his ship without worrying that he'd bruise the fruit or hasten its ripening. When the bunches arrived in stores, shopkeepers hung them up and, at a customer's request, cut off the desired number of bananas. As Dan Koeppel notes in "Banana: The Fate of the Fruit That Changed the World," by 1900 Americans were eating fifteen million bunches of Gros Michcls every year; by 1910, the number was forty million. Two decades later, Baker's company, renamed United Fruit—and today called Chiquita—was worth more than two hundred million dollars.

In converting a tropical fruit into a global commodity, United Fruit amassed land across Latin America, from Guatemala to Colombia, replacing virgin jungle with vast tracts of Gros Michels. Poorly compensated workers, battling malaria, dengue fever, tarantulas, pythons, and jaguars, constructed miles of railroad track, telecommunications lines, and irrigation canals. By the nineteen-sixties, United Fruit controlled nearly seven hundred million acres of land. "Tropical nature left to herself creates foodless jungles and miasmic swamps," a historian wrote at the time. "The banana of commerce is one of Man's proud triumphs over Nature."

United Fruit eventually commanded ninety percent of the American banana market, and in Latin America it became known as El Pulpo—the Octopus. When a head of state tried to thwart its progress, the company often responded with militaristic force. It clandestinely aided the 1911 coup in Honduras and the 1954 coup in Guatemala. At the company's urging, Colombia's Army launched a campaign against striking workers, which culminated in a massacre. In 1975, the company's chairman, Eli Black, jumped to his death from the forty-fourth floor of the Pan Am Building, in New York; the Securities and Exchange Commission soon discovered that he had given a high-ranking Hon duran official a one-and-a-quarter-million-dollar bribe.

Early in its ascendancy, United Fruit began contending with crop disease. "FRUIT BLIGHT COSTS MILLIONS IN COSTA RICA AND PART OF PANAMA: NO REMEDY IS AVAILABLE," the *Times* reported in 1927. "As much mystery surrounds the banana disease as the plague in medieval times and so far it has not been possible for modern science to cope with it." This was Race One. Over the next thirty years, a hundred thousand acres of Gros Michels were wiped out across Latin America, and the industry lost $2.3 billion. As the historian John Soluri has pointed out, before United Fruit arrived Latin Americans farmed small, diffuse tracts of land. But as jungle diversity was replaced with monolithic fields of Gros Michels, funguses like Race One were provided with many more hosts. In the words of one reporter, "Acts of God have not been wholly unsolicited."

In the nineteen-forties, a distant rival, Standard Fruit, reacted to the blight by shifting to Cavendish bananas, the Chinese variety, which were growing in the private greenhouse of the Duke of Devonshire, in Chatsworth, England. The bananas proved naturally resistant to Race One. In every other way, however, the Cavendish was less desirable than the Gros Michel. The Cavendish was susceptible to other diseases, which were controllable only with costly pesticides. It had a tendency to bruise, which meant that, rather than shipping bananas directly on the stalk, Standard Fruit had to box them in elaborate new packing houses. The Cavendish also needed special ripening rooms, where the green bananas, after arriving in the U.S., were helped along with doses of ethylene gas. When the bananas made it to stores, they lasted only a week before spoiling. And to those who knew the Gros Michel the flavor of the Cavendish was lamentably bland.

United Fruit, fearing that consumers would reject the taste of the Cavendish, was slow to switch, and saw its profits drop from $66 million in 1950 to $2.1 million in 1960. Standard Fruit (now known as Dole) became the country's largest seller of bananas, and remains so today. In 1960, United Fruit opened a research center in La Lima, Honduras, and hired Phil Rowe, a former rice breeder from Arkansas, to breed the perfect export banana: flavorful, hardy, and disease-resistant. The company also wanted the new plant to sprout big bunches, and to be sturdy enough to withstand the high winds that occasionally blew through Latin America. Because domesticated bananas are sterile,

Rowe was forced to cross wild diploids that offered a grab bag of good and bad traits. In four decades of work, he grew twenty thousand hybrids, but he never found a replacement for the Cavendish. His leading candidate, called Goldfinger, withstood Race One, but consumers rejected it as acidic and starchy. In the end, the unrelenting capriciousness of his work proved too much. One morning in 2001, Rowe walked into his experimental-banana fields and hanged himself from a tree. United Fruit was stuck growing the Cavendish.

"Cavendish *is* fairly bland," James Dale told me. "But, if you have no option but one, then that option looks pretty good." Today, there are millions more acres planted with the Cavendish than ever were planted with the Gros Michel.

This past fall, I flew to Honduras to visit the La Lima research station, which is now a nonprofit traditional breeding facility known as the Fundacion Hondurena de Investigacion Agricola, or FHIA. The head banana breeder is a man named Juan Fernando Aguilar, and outside his office hangs a sun-blanched picture of Phil Rowe. Aguilar, a thickset, gregarious Guatemalan, with a salt-and-pepper mustache and glasses with light-sensitive lenses, picked me up one morning at my hotel—the Banana Inn—and gave me a tour of a gated residential community called the Zona Americana. "This is where all of United Fruit's workers lived," he said. "La Lima is ugly, but the American Zone is beautiful!" Aguilar pointed out the old company swimming pool and two golf courses, which are still in use, and stopped the car in front of a white mansion. "This was the house of the general manager," he said. "Look at the size! And two tennis courts!" When we arrived at FHIA's headquarters, down the road, a dead mule lay near the entrance, its distended belly rising above the weeds. I remembered that a United Fruit executive had once joked, "A mule costs more than a Honduran deputy."

Rowe's death, ten years ago, coincided with advances in biotechnology, and many researchers now dismiss the traditional breeding of bananas as too reliant on happenstance. In 2003, the director of research at Chiquita told a British journalist, "We supported a breeding program for forty years, but it wasn't able to develop an alternative to Cavendish. It was very expensive and we got nothing back." Aguilar, however, has crucial leverage: nearly fifty percent of Americans, and sixty percent of Europeans, oppose genetically modified food. For this reason, Chiquita reversed its position in 2004 and signed a confidential agreement with FHIA, hiring the center that it once owned to naturally engineer a better banana. (The contract is said to be worth two million dollars.) Dole and FHIA are negotiating a similar deal. "We never left traditional breeding," a spokesman for Chiquita told me. "In our core markets, in America and Europe, a genetically modified banana would never be marketable. At the end of the day, we're interested in continuing to sell bananas." Jorge Gonzales, Dole's senior vice-president of agricultural research, said, "Traditional breeding is getting closer. This may be a shot in the dark, but if you don't take the shot you've got absolutely zero chance of hitting the target."

Aguilar's operation has the rusticity of a summer camp; the buildings have rough-hewn wooden exteriors, tin roofs, and chicken-wire windows. "I am glad you have come," Aguilar told me. "The hybrid plants are like women. To look at a woman from afar is not to know the woman. To know her, you must be with her. And to know the hybrids you must be with the hybrids." Aguilar begins his work by planting twenty thousand plants of a single variety. Once the plants have flowered, at nine months, workers manually dust them with the pollen of another banana plant that has a desirable trait, such as disease resistance. Three months later, Aguilar harvests the bunches, in the hope that the forced fertilization has impelled the plants to produce seeds. Every Monday, local women peel a hundred thousand bananas. Two days later, after the bananas have fermented and softened, the women smash them on a sieve, let the pulp ooze through, and retrieve any seeds.

On average, Aguilar recovers one seed from every ten thousand bananas—about ten seeds a week If any of those seeds provide a working embryo—the odds aren't great—Aguilar might be able to grow a new hybrid. But even if that plant acquires the trait of blight resistance, it will likely pick up several other, less desirable attributes, such as a low yield of fruit. One round of this exercise lasts three years. "The people in Australia don't like this—it's too time- consuming for them," Aguilar said. "Many people call me crazy, but I'm very confident that I can develop a Cavendish replacement."

Like Phil Rowe, Aguilar focuses on diploid bananas, which have an unusual capacity to accept pollen and produce seeds. Recently, however, Aguilar forced a seed out of a Cavendish—the first banana breeder in history to do so— after hand-pollinating tens of thousands of plants. He now finds one viable seed out of every million Cavendishes. In November, at an international banana conference in Medellin, Colombia, Aguilar presented a paper arguing that the Cavendish is not actually sterile. Still, because Aguilar finds seeds only through achingly artificial means and, even then, finds so few of them, any fertility seems to be a human-induced aberration.

Aguilar led me to a small trial plot. He was unable to share many details, he explained, because of confidentiality agreements, but he said that these few plants represented his best shot at success. We approached a tall, thick plant, with a dozen suckers growing around it. A robust bunch hung from the mother, but the bananas were only four inches long. "This one is strong, vigorous, full of power," Aguilar said. "It has a high yield, but the texture of the pulp inside is very poor, just mushy." He bounded off. "Look at this one!" he said. "This one has the same mother as that one before, but a different father. You can see it's much thinner. The plant is healthy, but not as vigorous. I don't know why." He walked to a far corner of the field, near a barbed-wire fence, and paused next to the tallest banana plant I'd ever seen. "Nine metres tall!" he called out. "Here you get very good resistance to winds, and a very good taste, but the bunch is small."

Finally, Aguilar inspected a stubby plant in the center of the field. It was sectioned off by a shin-high fence made of broken sticks and lolling twine. A tag read "06-04-333." He crushed under his foot the cigarette he'd been smoking and breathed deeply. Dragonflies buzzed, and a stray cat pawed at the dirt. Aguilar said sombrely, "This is *mi esperanza.* My hope and my wish. Give me six years. I will have to taste it and test it, but my dream is for this to be my Cavendish replacement." The plant was by far the smallest in die field—six feet tall. Several leaves were dying. But, Aguilar said, "the mother of this plant is Cavendish." Its "blood," as he put it, was promising.

Aguilar tapped a fist against his chest. "The field is spiritual for me," he said. "Plant breeding, it must be a part of you, part of your emotions. Biotechnology came along, and they suddenly labelled us 'traditional breeding.' It's a way to diminish our work But we are not in competition with anyone. We are doing what we do, and they are doing what they do. The difference is that my tools are in the field, not in the laboratory."

Near the end of my stay in Brisbane, James Dale invited me to his house for dinner. He lives with his wife, Ged, and their son, Jordan, on ten acres west of the city, in a bucolic suburb called Moggill. The Brisbane River meanders past their backyard, and on the property he and Ged grow oranges, avocados, pumpkins, pomelos, blueberries, papayas, and finger limes. There's not a banana to be seen," Dale said. "It's very embarrassing. But they'd just get blown over if we planted them."

"That's our excuse, anyway," Ged said.

"It's not an excuse," Dale said. "It's a reason."

We were sitting on their front porch, cracking macadamia nuts pulled from the trees that line the driveway, drinking beer, and watching the sun drop. Flying foxes, the largest bats in the world, were swooping, warming up for their nightly hunt. Since obtaining the results from the greenhouse test, Dale and his colleagues had been waiting for the Office of the Gene Technology Regulator, which oversees the safety of the country's genetically modified crops, to approve the field trial. A verdict was expected soon. Dale's team had begun preparations to transport the modified plants to Darwin—airplane seats would be bought for them—where they would toughen up for a few months in "hu- midicribs." In April, the plants would be released into the fungus-infested soil, and Dale hoped to have initial results sometime next fall.

Robert Borsato, meanwhile, had been looking to buy new land, confident that he would soon be able to expand his operation. He had also been travelling to Papua New Guinea, searching for new banana varieties that might thrive in the marketplace. "I found a little banana about three inches long, the color of tomato juice," he told me. "I think I could sell it to hotels, for a breakfast banana. It's acid-sweet, and there are lots of people who only want half a banana for breakfast."

As word spread of the planned field trial in Darwin, local news outlets had begun asking Dale what, exactly, he intended to put in the ground. Critics of genetically modified foods, such as Greenpeace and Earth First!, have claimed that such foods not only violate the sanctity of nature but also could spark antibiotic resistance and allergic reactions. (In 1996, researchers found that soybeans modified with a Brazil-nut gene triggered allergies in test subjects; the product, which had been designed to have a higher amino-acid content, was never commercialized.) Opponents also fear that foreign genes might be transferred into related species, through the uncontrolled exchange of pollen. Ronnie Cummins, the co-author of "Genetically Engineered Food: A Self-Defense Guide for Consumers," told me, "Scientists often think technologies are safe that turn out not to be. It's not that you're going to keel over and die as soon as you eat genetically modified papaya. It's the unpredictable long-term health and environmental effects that concern us."

As we sat down to dinner, Dale, who helped write the federal guidelines for genetically modified food in Australia, said, "The public isn't necessarily wrong to be wary of them, and everything should be regulated—that's extremely important. But you can't name a single time when G.M.O.s did something really bad, either to humans or to the environment. And don't forget that bananas are sterile. They don't have seeds, and they can't cross-fertilize. So even if we were putting something dangerous into banana plants—which we are not—there's absolutely no way for those genes to exit the banana and enter the wider world." Dale added that he thought it would take more than a hundred years for traditional breeding to solve the Cavendish problem. "Cavendish is a very, very well-accepted cultivar," he said. "The taste of Cavendish, the method of growing Cavendish, the method of harvesting and transporting Cavendish—it's all extremely well worked out. A Cavendish plant with an extra gene for Tropical Race Four—well, I think it's a more elegant method." Dale also seemed pleased that neither Chiquita nor Dole would own his creation. Borsato would no doubt want a return on his investment, but eventually he and Dale would offer a blight-resistant banana to the world.

For now, the limits of science force Dale to express foreign genes throughout the entire banana plant, but he hopes eventually to confine them to the roots, where the fungus attacks, and away from the fruit. Furthermore, a French institute called Genoscope is sequencing the banana genome, and once that is complete Dale might be able to insert a disease-resistance gene from a wild banana. "Then we'll be putting a banana gene into a banana," he said gleefully. "The public will have to love that."

Dignified as Dale made this sound, a new Cavendish banana still didn't seem like a panacea. The cultivar may dominate the world's banana export market, but, it turns out, eighty-seven percent of bananas are eaten locally. In Africa and Asia, villagers grow such heterogeneous mixes in their back yards that no one disease can imperil them. Tropical Race

Four, scientists now theorize, has existed in the soil for thousands of years. Banana companies needed only to enter Asia, as they did twenty years ago, and plant uniform fields of Cavendish in order to unleash the blight. A disease-resistant Cavendish would still mean a commercial monoculture, and who's to say that one day Tropical Race Five won't show up?

Dale once remarked to me that his favorite bananas are "the little sweet ones from Uganda called *sukali ndizi,* or sugar banana. They're absolutely fabulous." Now he added, "What we really want to do is make the Gros Michel resistant to both Race One and Race Four, and then someday, maybe, we'll put these disease-resistance genes in any variety we want." It was tempting to envision a reen- gineered supermarket that afforded Americans a broader replication of tropical bounty: next to the McIntoshes, Granny Smiths, and Honey-crisps would sit the Cavendishes, Gros Michels, and *sukali tidizis.* Of course, before this could happen the more exotic bananas would also have to be genetically modified for seaworthiness. "Those kinds of genes—we're years away," Dale said.

At Dale's house that evening, we'd seen peacocks strutting about and a hare bouncing through the field. The chickens that Ged raises clucked occasionally, and, even though the sun had set, the richness of the couple's gardens remained visible through a kitchen window. "From a biological perspective, I love genetic diversity," Dale said. "That's what makes the world safe, what makes it thrive. It means that everything is terribly healthy. And when you see the narrowing of genetic culture, that's when you know things are going to die."

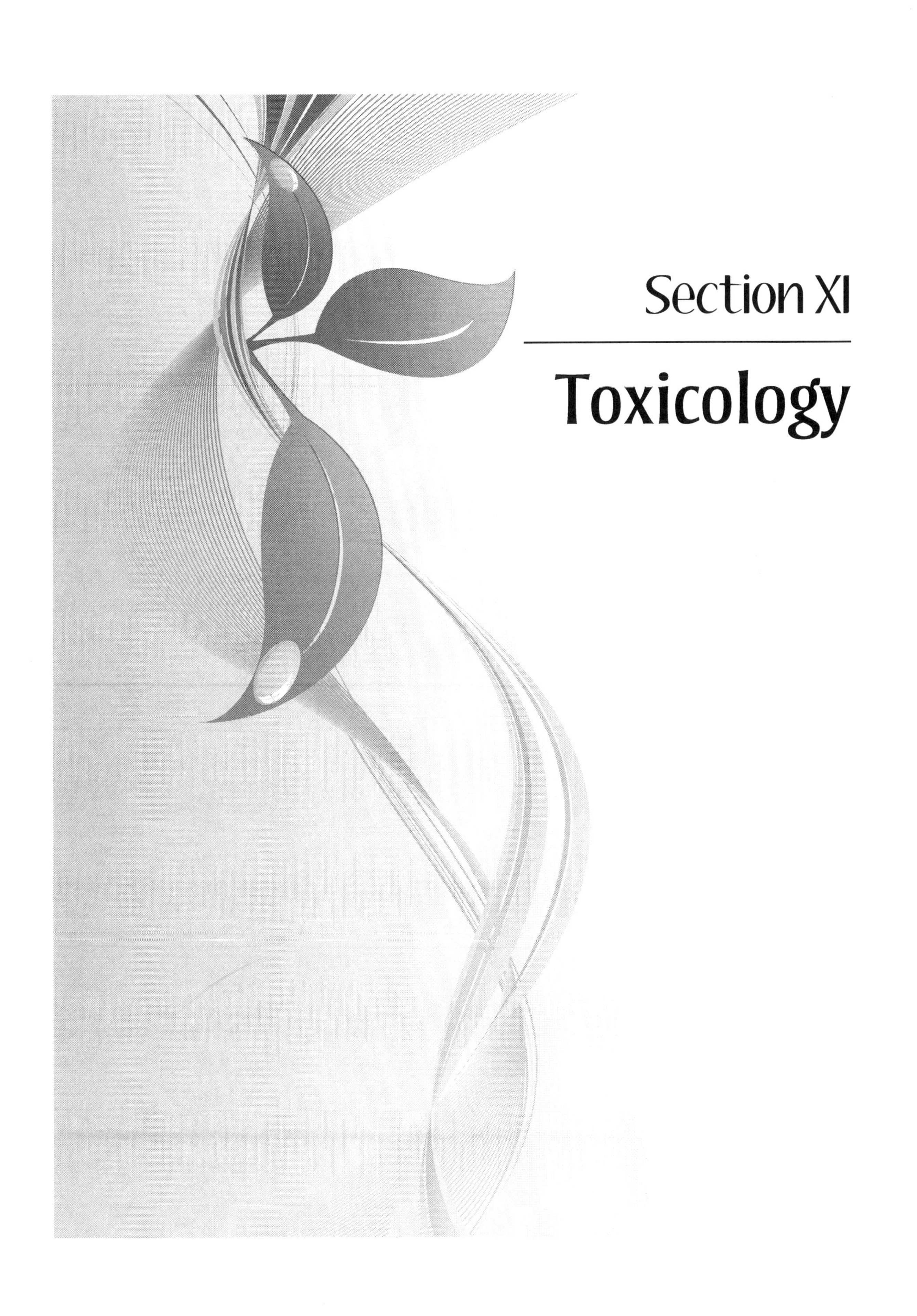

Section XI

Toxicology

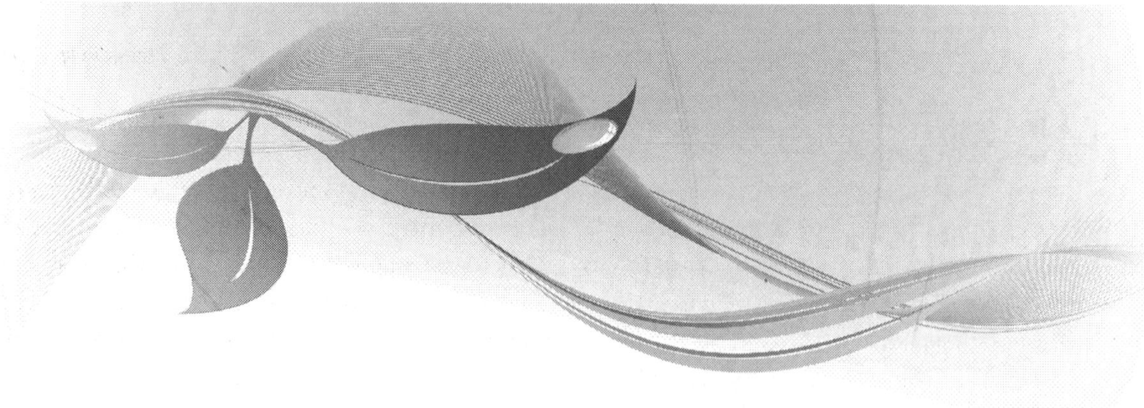

By Anne Marie Zimeri

Toxicology is the study of the adverse effect of chemicals on living organisms. It is an integral part of environmental health because humans and other living organisms are affected by exposure to toxins. There are ranges of effects that can occur from exposure to these toxins, often depending on the dose. Chemicals can incite an allergic response. Allergic responses can be mild and involve histamines and/or they can be severe enough to cause anaphylactic shock. Other chemicals/toxins/drugs can suppress the immune system, which may make a person more susceptible to an infectious agent.

Toxicants can even be neurotoxins, i.e., they affect the central nervous system. Mercury is a classic example of a neurotoxin. It can cause "the shakes" and even insanity. The Mad Hatter in *Alice in Wonderland* was deemed mad because of mercury exposure. In the past, the felting process used to produce the materials for hats involved exposure to mercury vapor that would poison craftsmen who made hats, or hatters.

Other toxins can actually alter DNA such that symptoms can arise from cellular processes that cannot occur properly. Oftentimes, when DNA that codes for tumor-suppressor genes is altered, tumors arise. These chemicals are considered "**mutagenic**." If a mutagenic compound can pass the placental barriers and alter the DNA of a developing embryo or fetus, it is considered "**teratogenic**."

What are these damaging compounds and where do humans encounter them? Many of the top 50 most hazardous substances are heavy metals that occur naturally, such as mercury as previously mentioned. Arsenic is another such example. Arsenic-rich rocks that line underground water sources such as aquifers can leach arsenic into the water supply upon exposure to oxygen. Humans can be exposed when they drink this water directly or eat food that was grown in fields irrigated with arsenic-laden water. Arsenic is also quite toxic across the six major taxa and can inhibit growth of microorganisms such as molds. Because of this toxicity, wood is treated with arsenic to prevent microorganism growth on exterior wood. To minimize arsenic exposure, many wood companies are switching to wood embedded with copper. Other toxins are man-made. Many are pesticides such as DDT.

The most common route to human exposure to toxins is through ingestion. Humans eat and drink foods that have pesticides, preservatives, and heavy metals present. Other forms of exposure can also cause illness when a threshold dose is exceeded. Inhalation of toxins in the air outside, in the home, or in the workplace can affect our health. Dermal exposure can cause an effect as well.

BIOACCUMULATION

Many toxins are persistent in the environment for a long period of time. The half-life of DDT, for example, is 15 years. Therefore it has the ability to bioaccumulate in the food web. Bioaccumulation occurs when low concentrations of toxins in the environment reach a dangerous level inside organisms over the course of their lifetime. For example, when DDT is present at a low, nontoxic level in freshwater, it can be taken in and remain in the tissues of organisms at a low trophic level. This is the case with plankton. The plankton are then eaten by the next trophic level and continue to accumulate DDT because it is recalcitrant and does not degrade during the course of this time period. Therefore, by the time a top carnivore (or human) eats something that has been consuming organisms at lower trophic levels, the level of DDT can have increased by several orders of magnitude (Figure 11.1).

Not all toxins are persistent; therefore, they can be broken down, mineralized, or detoxified before they can cause harm. In humans, the primary site of detoxification is the liver. Once toxins make their way to the liver, they are exposed to a series of enzymes that can metabolize organic molecules sometimes completely. Certainly some toxins can make their way to other organs and tissues where they can injure cells, but more often than not, the liver can detoxify low levels of toxins so they can be harmlessly excreted.

Ideally, before humans are exposed to a chemical, we have some idea of that chemical's toxicity. Toxicity tests have historically been performed on animals. Typically, model animals, frequently rats or mice, are selected, and are then given increasing doses of a potential toxin in order to illicit a response. This system is not without its flaws because the researchers are making the large assumption that humans will respond to the toxins in the same way as the model animals. That is not always the case, likely because mice and humans have not had a common ancestor for more than 60 million years and have therefore had the opportunity to evolve ways to manage exposure to toxins that may not be represented in both species. These tests are also time-consuming as well as often cruel and inhumane. To address these issues, an inexpensive and more ethical test was developed by Bruce Ames in the 1970s. This test looks specifically at whether a compound is mutagenic. It works well because it determines the ability of a compound to affect DNA, which is common across all the taxa. Because of this commonality, it is assumed that a chemical that alters the DNA found in a bacterial cell will also alter the DNA found in a human cell. Bacterium used in the Ames test is a strain of *Salmonella typhimurium* that carries a defective (mutant) gene making it unable to synthesize the amino acid histidine (His) from the ingredients in its culture medium. However, some types of mutations, including this one, can undergo a reversal, termed a back mutation, resulting in the gene regaining its function. These **revertants** are able to grow on a medium lacking histidine.

The Ames test exposes these bacteria to a test compound and then looks for mutagenicity by visualizing whether revertants appear. If revertants are numerous, then the compound in the test was mutagenic (Figure 11.2).

DOSE RESPONSE CURVES

Toxicologists work to establish dose response curves to help inform policy makers of the levels of a toxin that will illicit a response in the majority of the population. Because of genetic variation among humans, there will almost always be a small percentage of the population that will be extremely sensitive to a toxin and will show a response at a low level. Other members of the population will be quite tolerant

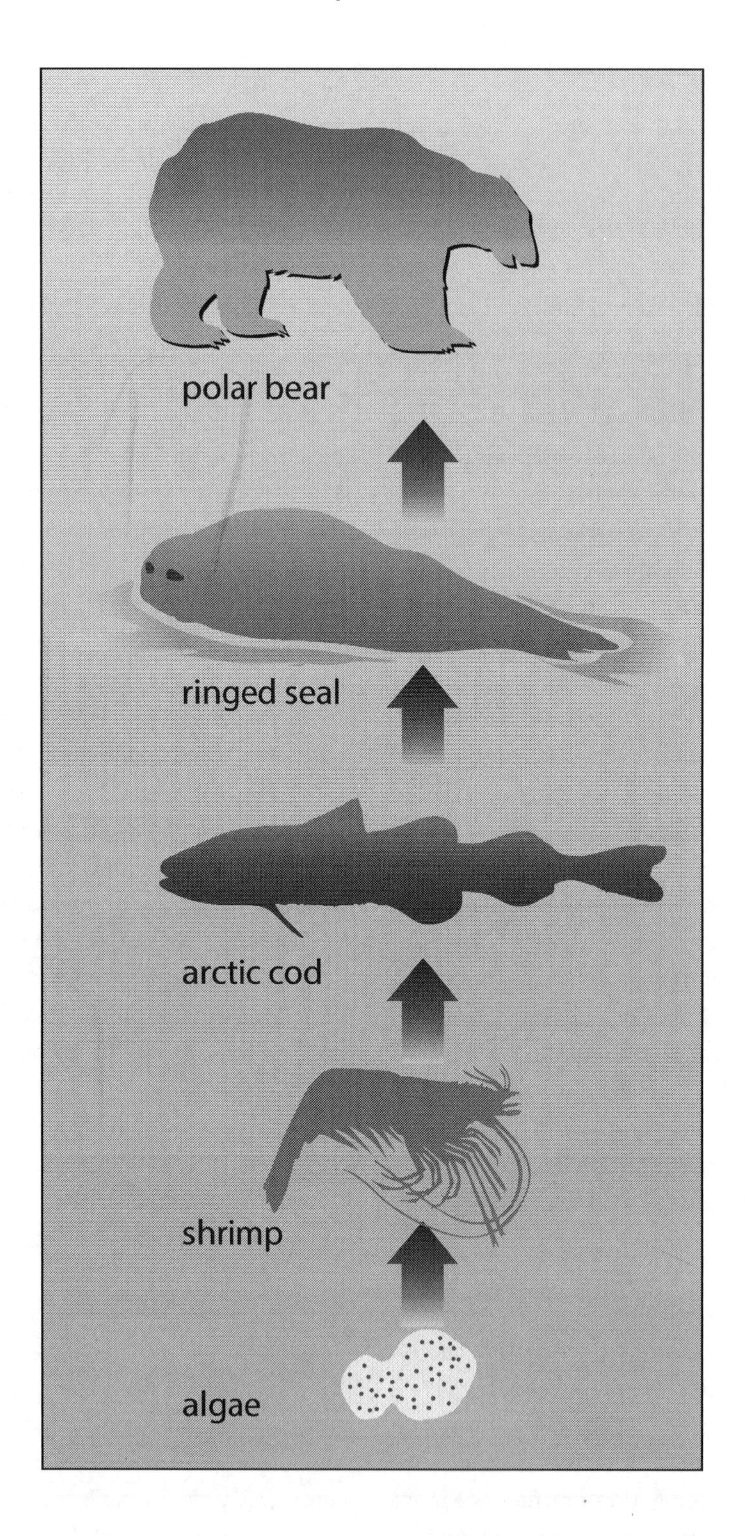

Figure 11.1. As higher trophic-level animals eat lower trophic-level animals, the level of contamination in the food continues to add to the level of contamination already in their body.

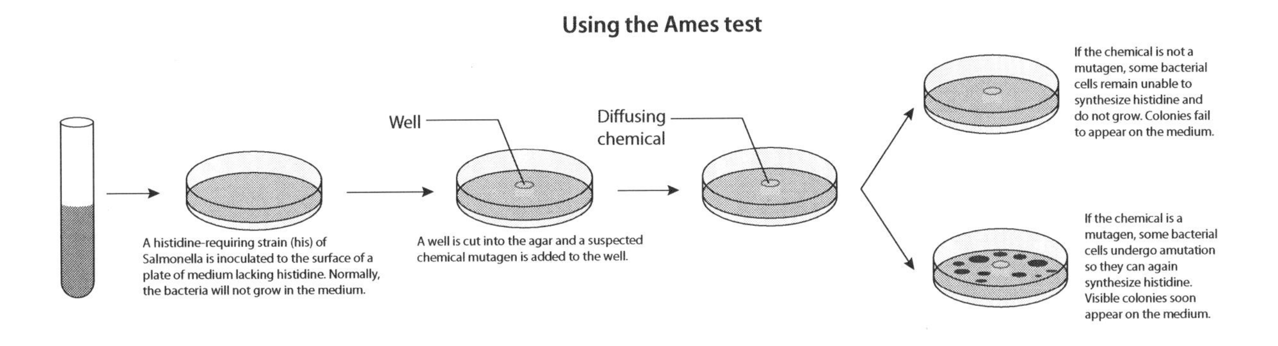

Figure 11.2. The Ames test is a biological assay used to determine the mutagenicity of a compound.

to the toxin and will require a higher exposure to a toxin to illicit a response. These dose response curves can be shown in humans when there has been an accidental release of a chemical, but they are most often established in a model system. The data is then extrapolated to humans based on a comparable dose by body weight. Model systems are also used to establish the LD_{50}, or lethal dose at which 50% of the test population is killed by exposure to the toxin. The lower the LD_{50}, the more toxic a compound. LD_{50} is usually expressed in mg of toxin per kg. of body weight (Figure 11.3). If the dosage of a toxin is not lethal, it can have either an acute effect or a long-term chronic effect. Acute effects are immediate health effects caused by a single dose of a toxin. Chronic effects are long lasting and more likely than acute effects to be permanent. Chronic effects can be caused by a high single dose or by continuous lower doses of a toxin.

RISK

Given the data and information on many of these toxins, humans are generally put at some level of risk by activities that would lead to exposure. Risk in this case can be defined as the probability of harm times the probability of exposure. Humans take risks every day and are willing to accept these risks, especially if they are voluntary. The activity that poses the most risk to college students is usually riding in or driving a car where the lifetime chances of dying can be 1 in 100. A far greater voluntary risk would be smoking cigarettes, which poses a lifetime chance of 1 in 4 of dying a death related to smoking. Typically, we are more likely to tolerate a risk if it is voluntary. If a risk is involuntary, meaning we are exposed to a chemical whether we want to be or not, we are less likely to accept it. Risk assessors are professionals who analyze data to determine the statistical risks associated with exposure to a chemical so that policy makers can be better informed when setting threshold exposure limits or release values. Establishing public policy can be complex because considerations must be made for the combined effects of exposures from many different sources as well as different sensitivities of members of the population.

BIOLOGICAL RISK/EMERGING INFECTIOUS DISEASE

At the top of the list of causes of death worldwide is infectious disease. Infectious diseases are caused

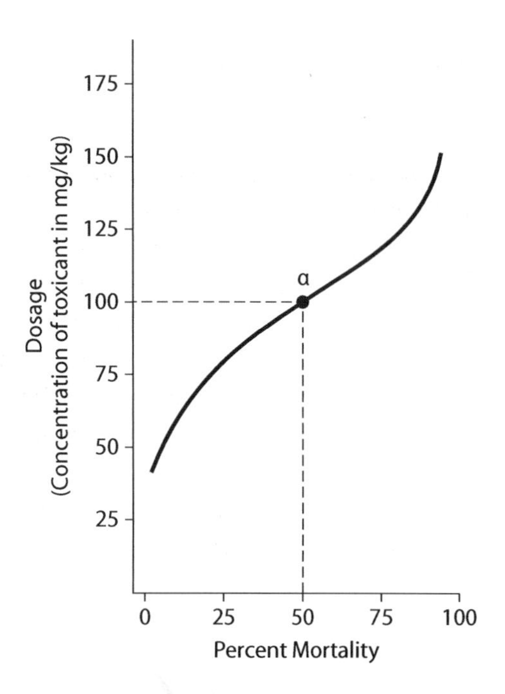

Figure 11.3. Dose response curve showing an LD_{50} at the 100 dosage, i.e., the dosage where 50% of the population is killed.

by infectious agents, which include bacteria, viruses, and parasites. Infectious disease is on the rise, despite major medical advances, due to several factors. First, there are more humans on the planet than ever before, and as cities are becoming more crowded, the average person comes into contact with more potentially infectious people than in years past. Populations are more well-traveled, and along with travel and commerce comes the potential for the spread of disease to areas that would otherwise not be exposed. Ecological changes mostly associated with global climate changes have broadened the geographical areas of many vector-borne diseases such as malaria. Some cities in the past have been protected from malaria, which is spread by mosquitoes, because the cities were constructed above the mosquito line. Harare, in Zimbabwe, for example, was built at an altitude where the temperatures were not conducive to mosquito breeding. Due to climate change, the mosquito line has risen in altitude and now encompasses part of the city. Evidence for microbial resistance is also on the rise as more cases of antibiotic-resistant strains of pathogenic bacteria have been reported.

ECOLOGICAL CHANGE

In addition to weather patterns that change naturally, such as el Niño, or are man-made, such as global climate change, ecological changes can occur from changes to ecological systems. The bringing of agriculture-based economic development into previously undeveloped areas can expose humans to diseases that would otherwise not be an issue. Towns and suburbs can expand into areas where a host thrives. For example in the four-corners region of the United States, hanta virus, shed in the feces of rodents, is on the rise because human activities, such as threshing, have aerosolized the virus-laden fecal matter in an area that would otherwise be undisturbed.

HUMAN POPULATION GROWTH

Increased population density certainly increases the risk of coming in contact with an infectious person, but it also can cause other risk factors for disease. A population that is growing rapidly can outgrow its sewage infrastructure. This can lead to the release of effluent from water-treatment plants that contain infectious agents, and/or can expose the population to fecal contaminants and infectious agents when sewers overflow on the way to a treatment facility. Hospital systems can also be overwhelmed and unable to effectively treat an influx of patients as a result of population growth.

MICROBIAL RESISTANCE TO ANTIBIOTICS

Antimicrobial agents have been widely distributed since their discovery in the last century. Wide use and prophylactic use of these agents can contribute to the generation of bacterial strains that are resistant to antibiotics. Under normal circumstances there is some genetic variability in any population of bacteria due to random and spontaneous mutations. Some of these mutations may lead to the ability of a cell to resist the challenge of an antibiotic. Without challenge, the cells will proliferate along with the bulk of the nonresistant cells until the human host's immune stem can clear the body of these cells. When antibiotics are given, the bulk of the nonresistant cells are killed or forced to slow their growth enough that the host can clear them. The resistant cells, however, will continue to grow and thrive. They have less competition for resources once the sensitive cells have been killed, so now we have created a scenario where there is a large population of antibiotic-resistant cells.

Antibiotic resistance can be transferred from cell to cell in many cases where bacteria mate, or conjugate. In conjugation, two bacteria cells form a conjugation tube and exchange pieces of DNA. It is in this manner that a non-harmful, antibiotic-resistant strain of bacteria can pass its antibiotic-resistant genes to a pathogenic strain. Due to both of the above-mentioned mechanisms, medical doctors have become more judicious about when to prescribe antibiotics. In fact, there are several antibiotics that are not prescribed unless many others have been shown to be ineffective so that resistance development can be minimized. Vancomycin and Rocephin are two examples of antibiotics used only in dire circumstances.

EMERGING DISEASES

Diseases that are considered "emerging" are those that have increased dramatically in the past two decades or that threaten to increase in the near future. The suite of emerging diseases differs among areas of the world. According to the Centers for Disease Control and Prevention (CDC) in Atlanta, GA, diseases currently emerging in the United States include cryptosporidium, HIV, *E. coli* infections, Lyme disease and group A strep. On the other side of the globe, in Africa, emerging diseases include HIV, malaria, Chagas, and tuberculosis.

TWO EXAMPLES OF EMERGING BACTERIAL DISEASES

Escherichia coli (*E. coli*) is a rod-shaped gram-negative bacterium that is facultative. When bacteria are facultative, they can live in the presence of oxygen or in its absence, as occurs in the human gut. Gut bacteria, or enterics, can cause gastroenteritis, urinary tract infections, and more severe diseases such as septicemia and meningitis. There is a specific strain of *E. coli* that has become more prevalent in recent years due to our current livestock-related agricultural and slaughterhouse practices. Cattle grown to market weight in concentrated animal-feeding operations (CAFOs) stand in close proximity to one another and are often in up to 6 inches of feces. *E. coli* 0157:H7 is shed in fecal waste and quickly spreads from animal to animal. Human infections have been traced by epidemiologists to the consumption of undercooked beef products and vegetables irrigated with waters contaminated with feedlot wastewater. This *E. coli* 0157:H7 strain can cause hemorrhagic colitis and severe kidney damage.

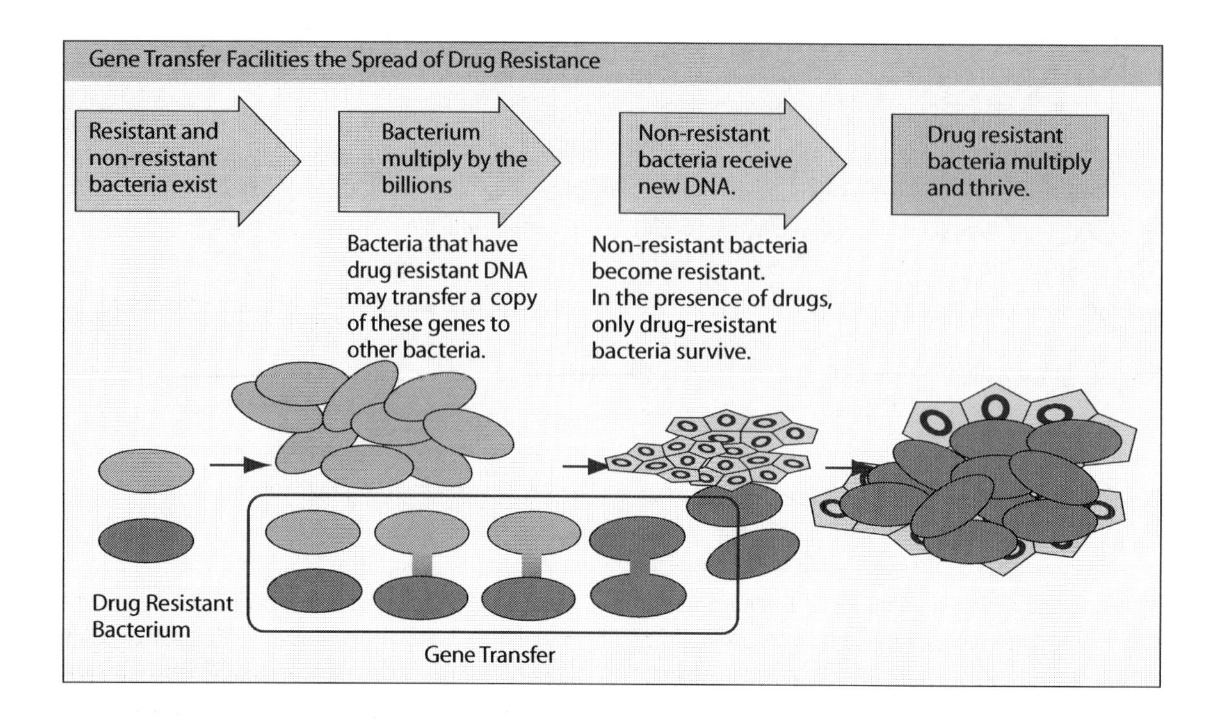

Figure 11.4. Antibiotic-resistance genes can be acquired through conjugation.

The infectious agent for tuberculosis (TB) is a strain of bacteria called Mycobacterium tuberculosis. It is a relatively slow-growing strain with cell walls that contain high levels of lipids. Mycobacterium have a dormant form resistant to antibiotics. Therefore, in order to treat a patient with TB, antibiotics must be administered regularly over a period of months to catch those cells when they break dormancy. This regime is often met with resistance by patients and can be cost prohibitive. Infection rates are on the rise in developing countries, and the WHO estimates that recently there have been close to 9 million cases of TB annually.

Global Status of DDT and Its Alternatives for Use in Vector Control to Prevent Disease

By Henk van den Berg

The Stockholm Convention seeks the elimination of 12 chemicals or classes of chemicals, one of which is dichlorodiphenyltrichloro- ethane (DDT) (United Nations Environment Programme [UNEP] 2002). DDT is used in indoor spraying for control of vectors of malaria and visceral leishmaniasis. In negotiations that led to the treaty, there was concern that a sudden ban on DDT use could adversely affect the malaria burden. Thus, DDT was permitted to be produced and used for the purpose of controlling disease vectors in accordance with recommendations and guidelines of the "World Health Organization" (WHO) and when locally safe, effective, and affordable alternatives are not available (WHO 2007a). Ironically, DDT use in Africa has increased since the Stockholm Convention came into effect (Manga, L., personal communication).

Malaria is a complex parasitic disease confined mostly to tropical areas and transmitted by mosquitoes of the genus *Anopheles*. There are an estimated 250 million clinical cases of malaria, causing nearly a million deaths, mostly of children <5 years of age and mostly in sub-Saharan Africa (WHO 2008b). Malaria-endemic countries are faced with a high cost of prevention and treatment of the disease.

Vector control is an essential component of malaria control programs. The WHO has reaffirmed the importance of vector control through indoor residual spraying (IRS) as one of the primary interventions for reducing or interrupting malaria transmission in countries in both stable and unstable transmission zones. Twelve insecticides have been recommended for IRS, including DDT. The course of action promoted by the WHO has been to retain DDT as part of the arsenal of insecticides available for IRS globally, to be able to manage insecticide resistance until suitable alternatives are available (WHO 2007a). The use of DDT for IRS is recommended only where the intervention is appropriate and effective in the local epidemiologic situation. Nonetheless, DDT has not been subjected to the WHO's Pesticide Evaluation Scheme for many years.

Henk van den Berg, "Global Status of DDT and Its Alternatives for Use in Vector Control to Prevent Disease," *Environmental Health Perspectives*, vol. 117, no. 11, pp. 1656-1663.

In this review, I present the current situation regarding the use of DDT for vector control, covering aspects of production, use, legislation, cost-effectiveness, health effects, environmental effects, insecticide resistance, monitoring, and evaluation. I provide an outline of alternative methods, strategies, and new developments; discuss cost-effectiveness, current implementation, barriers, and gaps in implementing the alternatives; and present possible solutions to reduce reliance on DDT.

This review is based largely on a document commissioned by the Stockholm Convention Secretariat, which served as background paper for a global stakeholders' meeting to review the establishment of a global partnership to develop alternatives to DDT, held 3–5 November 2008 in Geneva, Switzerland.

METHODS

Contemporary information on the production and use of DDT was obtained from a) formal questionnaires by the Stockholm Convention Secretariat, completed by national authorities; b) documents published by the Stockholm Convention; c) direct communications with national authorities; and d) information available from project proposals submitted to the Global Environment Facility (2009). Information has been supplemented with data presented by country delegates at workshops in the context of the Stockholm Convention.

I obtained information on side effects, insecticide resistance, cost-effectiveness, and alternatives from literature searches. I used the search engine Scopus (2008) to retrieve studies related to DDT and malaria, with vector control as additional search term. Because of the breadth of the subject matter, only the most relevant studies were selected, and reviews were prioritized. Old literature was accessed electronically, or hard copies were obtained from libraries. Additional information on insecticide resistance was obtained from web-based reports from the African Network on Vector Resistance (ANVR) (Vector Biology and Control 2008). Information on human exposure and health effects was based on reviews published over the past 5 years and supplemented with recent studies on exposure due to indoor spraying.

STATUS OF DDT

Production, use, and management. DDT is currently being produced in three countries: India, China, and the Democratic People's Republic of Korea (DPRK; North Korea) (Table 1). By far the largest amounts are produced in India for the purpose of disease vector control. In China, the average annual production during the period 2000–2004 was 4,500 metric tons of DDT, but 80–90% was used in the production of Dicofol, an acaricide, and around 4% was used as additive in antifouling paints. The remainder was meant for malaria control and was exported. Recent information from the DPRK (United Nations Institute for Training and Research [UNITAR[, unpublished data) indicates that 160 metric tons of DDT is produced per year, for use mainly in agriculture (which is not acceptable under the Stockholm Convention) and a small portion for use in public health. India and China both export DDT to countries in Africa, either as technical product or as a formulation, for the purpose of vector control. DDT is being formulated in Ethiopia and South Africa with ingredients imported from China. South Africa exports some of its formulated product to other countries in Africa.

An estimated 5,000 metric tons of DDT (active ingredient) was used for disease vector control in 2005 (Table 1). The primary use is for malaria control, but approximately 1,000 metric tons/year (20% of global consumption) is used for control of visceral leishmaniasis restricted to India. India is by far the largest consumer of DDT, but in 2007 use was down one-fourth from the 2005 level. Mozambique, Zambia, and Zimbabwe have recently reintroduced the use of DDT. With the possible exception of the Dominican Republic, there is no reported use of DDT for disease vector control from the Americas. Use in Ecuador, Mexico, and Venezuela was phased out in 2000. China has reported that no DDT has been used for disease vector control since 2003, and future use is reserved only for malaria outbreaks.

IRS programs are currently expanding in Africa, the main driver being the U.S. President's Malaria Initiative (PMI 2009). Pilot programs on IRS have been initiated in some African countries, and several other countries are considering reintroducing the intervention. In some of these countries, a decision has not been made on whether to use DDT in their IRS program. Hence, the use of DDT may be increasing—especially in African countries—because new countries are initiating IRS programs, including the use of DDT, and countries that are using DDT are expanding their IRS programs to stable transmission areas.

There is a paucity of data on DDT supplies. The available information indicates that large amounts of DDT are stored in many countries, but most of the stock is outdated or of unknown quality. Moreover, the transfer of DDT stock between countries is not always documented or reported, and this poses a problem in tracking quantities of the chemical and establishing the quality of DDT being used. A major multistakeholder effort is needed for the cleanup of outdated DDT stock, for example, through the Africa Stockpiles Programme (Curtis and Olsen 2004).

Many countries that use DDT have inadequate legislation or lack capacity to implement or enforce regulations on pesticide management. Unpublished information suggests that DDT is being traded on local markets for use in agriculture and termite control (UNEP 2008). Funding agencies aiding in the purchase of DDT should be obligated to provide financial assistance to ensure that regulations and monitoring capacity are in place to support proper management of DDT from the cradle to the grave, for example, by involving the environmental sector.

Cost-effectiveness of DDT. No published data exist on cost-effectiveness in terms of cost per disability-adjusted life-year averted by IRS using DDT. Statements of high cost-effectiveness of DDT have been based on the positive experience from the malaria eradication eta (Mabaso, et al. 2004) supplemented with more recent results on reductions in malaria morbidity and incidence associated with the use of DDT (Curtis 2002; Gunasekaran et al. 2005; Sharp, et al. 2007).

Both the effectiveness and costs of DDT are dependent on local settings and merit careful consideration in relation to alternative products or methods. DDT has been known as the only insecticide that can be used as single application in areas where the transmission season is >6 months. However, information is lacking on the potential variability in residual action of insecticides, including DDT (e.g., due to sprayable surface, climatic conditions, social factors).

Direct costs of IRS are the procurement and transport of insecticide, training of staff, operations, awareness-raising of communities, safety measures, monitoring of efficacy and insecticide resistance, monitoring of adverse effects on health and the environment, and storage and disposal. In 1990, the insecticide costs per house per 6 months of control were substantially lower for DDT (US$1.60) than for other insecticides (>US$3.40), but in 1998 the cost range for DDT (US$1.50–3.00) overlapped with that of alternative insecticides (> US$2.20), pyrethroids in particular (Walker 2000). This comparison will further change with the availability of new formulations of pyrethroids that have increased residual activity. Moreover, incorporating the cost of safety measures in the application of DDT will significandy change its comparative cost advantage.

Apart from the direct costs, it is essential that the unintended costs of DDT (or alternative insecticides) to human health and the environment are included in the cost assessment. In addition, contamination of food crops with DDT could negatively affect food export (Anonymous 2007). A comprehensive cost assessment of DDT versus its alternatives should include the potential costs of atmospheric transport and chronic health effects.

Proposed and ongoing projects by the WHO, United Nations Environment Programme, and United Nations Development Programme are expected to establish a more solid evidence base for the effectiveness of DDT in relation to its alternatives (WHO 2007b). The results will be crucial in future decision making on vector management strategies for prevention of malaria.

Health effects of DDT. High levels of human exposure to DDT among those living in sprayed houses, most of whom are living under conditions of poverty and often with high levels of immune impairment, have been found in recent studies in South Africa and Mexico (Aneck-Hahn, et al. 2007; Bouwman, et al. 1991; De Jager, et al. 2006; Yanez, et al. 2002), but contemporary peer-reviewed data from India, the largest consumer of DDT, are lacking. The simultaneous presence of, and possible interaction between, DDT, dichloro-diphenyldichloroethylene (DDE), and pyrethroids in human tissue is another area of concern (Bouwman, et al. 2006; Longnecker 2005). In North America, rather high levels of exposure have been recorded in biological samples collected near the time of peak use during the 1960s (Eskenazi, et al. 2009). Exposure of the fetus and young child occurs through the placenta and through lactation (Bouwman, et al. 2006); exposure of children and adults occurs through direct contact with DDT in the environment, through indoor dust (Herrera-Portugal, et al. 2005), and through the food chain. DDT accumulates in fatty tissue and is slowly released. A monitoring system is needed for the assessment of trends in exposure to DDT, allowing for the attribution of effects to IRS locally; in this regard, human milk is considered an important media to be monitored (Malisch and van Leeuwen 2003).

Studies on health effects of DDT have focused mostly on subjects in North America and Europe, who have generally been exposed to levels lower than those reported from areas with IRS. No global assessment has been made on the evidence of health risks of DDT in relation to IRS because data are scarce. As an indication, however, initial work suggests that nonoccupational exposure through IRS is associated with impaired semen quality in men (Aneck-Hahn, et al. 2007; De Jager, et al. 2006).

Health effects of DDT and DDE most commonly suggested by studies in North America and Europe are early pregnancy loss, fertility loss, leukemia, pancreatic cancer, neurodevelopmental deficits, diabetes, and breast cancer (Beard 2006; Chen and Rogan 2003; Cox, et al. 2007; Eriksson and Talts 2000; Garabrant, et al. 1992; Ribas-Fito, et al. 2006; Snedeker 2001; Venners, et al. 2005). In many cases the results have not been consistent between studies, but nevertheless these accumulating reports bear much concern, particularly in relation to chronic effects. Breast cancer has been most rigorously studied; even though the majority of results showed no causative association with DDT exposure (Brody, et al. 2007), the latest evidence indicates an increased risk in women who were exposed at a young age (Cohn, et al. 2007). In addition, experimental studies on animals have demonstrated neurotoxic, carcinogenic, immunotoxic, and reproductive effects attributable to DDT and DDE (Turusov. et al. 2002).

The adverse health effects of DDT versus the health gains in terms of malaria prevention require more attention. For example, a gain in infant survival resulting from malaria control could be partly offset by an increase in preterm birth and decreased lactation, both of which are high risk factors for infant mortality in developing countries. The WHO is conducting a reevaluation of health risks of DDT, but progress has been slow.

Environmental effects of DDT. As a persistent molecule, DDT has low to very low rates of metabolism and disposition, depending on ambient temperatures. It is degraded slowly into its main metabolic products, DDE and dichlorodiphenyldichloroethane (DDD), which have similar physicochemical properties but differ in biological activity. DDT is emitted through volatilization and runoff. It is more volatile in warmer than in colder parts of the world, which through long-range atmospheric transport results in a net deposition and thus gradual accumulation at high latitudes and altitudes (Harrad 2001).

Loss through runoff is low because DDT has a strong affinity for organic matter in soils and aquatic sediment but is virtually insoluble in water. Half-lives of DDT have been reported in the range of 3–7 months in tropical soils (Varca and Magallona 1994; Wandiga 2001) and up to 15 years in temperate soils (Ritter. et al. 1995). The half-life of each of its metabolic products is similar or longer. DDT readily binds with fatty tissue in any living organism, and because of its stability, bioconcentrates and biomagnifies with increasing trophic level in food chains (Kelly, et al. *2004).* The half-life of DDT in humans *is* >4 years; the half-life for DDE is probably longer (Longnecker 2005). Studies have shown that DDT is highly toxic to insects, shrimp, and fish (Fisk, et al. 2005; Galindo, et al. 1996; Metcalf 1973) and adversely affects the reproduction of wild birds through thinning of egg shells (Ratcliffe 1967).

DDT and its metabolic products present in the global environment have originated mostly from its previous large-scale use in agriculture and domestic hygiene. Because DDT is currently allowed only for indoor spraying for disease vector control, its use is much smaller than in the past. Nevertheless, DDT sprayed indoors may end up in the environment (e.g., when mud blocks of abandoned houses are dissolved in the rain). Data from Brazil, India, Mexico, and South Africa suggested that higher levels of DDT are found in water or soil samples in areas with DDT residual spraying than in areas without spraying (Bouwman et al. 1990; Dua, et al. 1996; Sereda and Meinhardt 2005; Vieira, et al. 2001; Yanez, et al. 2002), but these results need further verification.

Insecticide resistance. As the number and size of programs that use DDT for indoor spraying increase, insecticide resistance is a matter of growing concern. Since the introduction of DDT for mosquito control in 1946, DDT resistance at various levels has been repotted from >50 species of anopheline mosquitoes, including many vectors of malaria (Hemingway and Ranson 2000). Unless due attention is paid to the role of insecticide resistance in the breakdown of the malaria eradication campaign of the 1960s, resistance may once again undermine malaria control (Busvine 1978).

In the past, the use of DDT in agriculture was considered a major cause of DDT resistance in malaria vectors, as many vectors breed in agricultural environments (Mouchet 1988). At present, DDT resistance is thought to be triggered further by the use of synthetic pyrethroids (Diabate, et al. 2002). This is due to a mechanism of cross-resistance between pyrethroids and DDT, the so-called sodium channel mutation affecting neuronal signal transmission, which is governed by the *kdr* (knock-down resistance) gene (Martinez- Torres, et al. 1998). Vectors with the *kdr* gene are resistant to both groups of insecticides, and this has serious consequences for malaria vector control, because pyrethroids and DDT are the two main groups of chemicals used. The *kdr* gene is being reported from an increasing number of countries; thus, even in countries without a history of DDT use, resistance to DDT is emerging in populations of malaria vectors (WHO 2006).

Contemporary data from sentinel sites in Africa indicate that the occurrence of resistance to DDT is widespread, especially in West and Central Africa (ANVR 2005; Coleman, et al. 2007). The main African vector, *Anopheles gambiae s.s.*, showed resistance to DDT in the majority of tests. Further, there is recent evidence of resistance in *A. gambiae s.l.* in Ethiopia (ANVR 2005), and there are signs of DDT resistance

Table 1. Annual global production and use of DDT tin 10^3 kg active ingredient in 2003,2005, and 2007.

COUNTRY	2003	2005	2007	COMMENT	SOURCE
Produce DDT for vector control					
China[a]	450	490	NA	For export	Pd
India[b]	4,100	4.250	4,495	For malaria and leishmaniasis	Pd. Ws, Dc
DPRK	NA	NA	5	>155 metric tons for use in agriculture	UNITAR
Global production	>4.550	>4,740	>4,500		
Use DDT for vector control					
Cameroon	0	0	0	Plan to pilot in 2009	WHO
China	0	0	0	Discontinued use in 2003	SC
Eritrea	13	15	15	Epidemic-prone areas	Qu, WHO
Ethiopia	272	398	371	Epidemic-prone areas	WHO, Ws
Gambia	0	0	NA	Reintroduction in 2008	Dc
India	4,444	4,253	3,413	For malaria and leishmaniasis	WHO. Dc
DPRK	NA	NA	5	>155 metric tons used in agriculture	UNITAR
Madagascar	45	0	0	Plan to resume use in 2009	Qu
Malawi	0	0	0	Plan to pilot in 2009	WHO
Mauritius	1	1	<1	To prevent malaria introduction	Qu
Morocco	1	1	0	For occasional outbreaks	Qu
Mozambique	0	308	NA	Reintroduction in 2005	WHO
Myanmar	1	1	NA	Phasing out	Ws
Namibia	40	40	40	Long-term use	WHO
Papua New Guinea	NA	NA	0	No recent use reported	SC
South Africa	54	62	66	Reintroduction in 2000	Qu., WHO
Sudan	75	NA	0	No recent use reported	Qu., WHO
Swaziland	NA	8	8	Long-term use	WHO
Uganda	0	0	NA	High Court prohibited use, 2008	SC., Dc
Zambia	7	26	22	Reintroduction in 2000	Ws.,Qu. WHO
Zimbabwe	0	108	12	Reintroduction in 2004	WHO
Global use	>4.953	>5.219	>3.950		

Abbreviations: Dc: Direct communication with national authorities; NA, not available; Pd: project proposals submitted to the Global Environment Facility; Qu: questionnaire on DDT by the Secretariat of the Stockholm Convention completed by national autorities; SC: documents published by the Secretariat; Ws: workshop presentations by country delegates in the context of the Stockholm Convention. Further information was obtained from the WHO and UNITAR reports, as indicated. [a]The figure for 2005 was extrapolated from the total production; in addition to production for vector control, DDT is produced for Dicofol manufacture (-3,800 metric tons per year) and for antifoulant paints (200 metric tons per year), [b]DDT is also produced for dicofol manufacture 260 metric tons per year).

in *Anopheles arabiensis,* another key vector, from Uganda, Cameroon, Sudan, Zimbabwe, and South Africa. In Asia, the resistance to DDT is particularly widespread in India. Multiple resistance to DDT and other insecticides in the major vector *Anopheles culicifacies* is present in many parts of the country (Dash, et al. 2009) and has reportedly caused a major loss in effectiveness of intervention (Sharma 2003). Resistance has also been reported in *Anopheles sinensis* from China (Cui, et al. 2006) and in *Anopheles epiroticus* (formerly named *Anopheles sundaicus]* in Vietnam (Dusfour, et al. 2004).

Resistance does not necessarily result in failure to control disease. Standard testing of DDT resistance focuses on the insecticide's toxic action. However, the repellent and irritant properties of DDT also have the potential to reduce transmission of disease and relieve the selective pressure for toxic resistance (Grieco, et al. 2007; Roberts and Andre 1994). This is an area requiring more research.

An important lesson learned from the experience with oncocerciasis (river blindness), another vector-borne disease, is that the development and spread of insecticide resistance is much slower when vector populations are under effective control (Guillet, P., personal communication), suggesting that suppressing vector proliferation helps prevent or delay the development of resistance.

Effective monitoring and decision support systems can enable insecticide resistance to be detected at an early stage, which should lead to the implementation of changes in insecticide policy (Sharp, et al. 2007). However, the choice of unrelated insecticides remains limited (Nauen 2007). Even an intelligent insecticide resistance management strategy using rotations, mosaics, or mixtures may not prevent resistance development (Hemingway, et al. 1997; Penilla, et al. 2006). In a recent report from India, the Joint Monitoring Mission (JMM 2007) pointed out that the insecticide choice for IRS is rarely based on contemporary insecticide susceptibility testing.

ALTERNATIVES TO DDT

A number of vector control methods are available as alternatives to DDT. Two of these, the use of alternative insecticides in IRS and the use of insecticide-treated bed nets (ITNs), are mainstreamed because of their proven impact on the malaria burden. Other available alternatives are receiving limited attention in contemporary malaria control efforts, but have an important role to play. Table 2 summarizes alternative methods. Alternatives to DDT should pose less risk to human health and the environment and be supported with monitoring data.

Chemical methods. IRS with insecticides is an effective method of malaria control. Its strength lies in its effect on shortening the life span of adult mosquitoes near their human targets, which has a critical impact on malaria transmission (MacDonald 1957). However, there is limited information on effectiveness and operational feasibility of IRS in African countries with highly endemic malaria, some of which recendy reintroduced IRS or plan to do so. Twelve insecticides belonging to four chemical classes are recommended for IRS in vector control, which collectively address only three modes of toxic action (Nauen 2007). Pyrethroids are the most cost-effective alternatives to DDT in malaria control except where pyrethroid resistance occurs (Walker 2000).

There are two new developments with regard to IRS. First, some existing insecticides not currendy available for public health; chlorfenapyr and indoxacarb, for example, showed potential in areas with pyrethroid resistance (N'Guessan, et al. 2007a, 2007c). Second, new formulations of existing insecticides with prolonged residual activity are being developed as alternatives to DDT (Hemingway et al. 2006). Two slow-release formulations of pyrethroids are already available on the market.

The main current alternative to IRS is the use of ITNs. The insecticide enhances the protective effect for the person under the net, but also has a beneficial effect on the community at large (Hawley, et al. 2003). ITNs have been shown convincingly to cause substantial reductions in all-cause child mortality, under both experimental (Lengeler 2004) and operational conditions (Armstrong Schellenberg, et al. 2001; Fegan, et al. 2007). They are effective in highly endemic settings by reducing the risk of severe disease, particularly in infants and young children before they have acquired a certain level of natural immunity (Smith, et al. 2001). Two categories of ITNs are available: conventionally treated nets and long-lasting ITNs. The former needs regular retreatment, a follow-up action that has proven difficult to achieve at field level. The latter is a relatively new technology that retains the efficacy for at least 3 years. Pyrethroids are the only chemical group recommended for use in ITNs.

There have been several new developments in ITN technology. Research on treatment with non-pyrethroids has been conducted to cope with the problem of resistance, but safety issues are a concern. At least one insecticide with novel chemistry is being developed for ITNs (Hemingway, et al. 2006). It is critical that this unique product, once it enters the market, is reserved solely for public health purposes, thus reducing the risk of insecticide resistance in the future. New ITN products are not expected to come to market in the short term.

The relative cost-effectiveness of IRS and ITNs has been studied on several occasions. Both have been considered attractive interventions in terms of cost per disability-adjusted life-years averted (Goodman, et al. 2000), but their relative effectiveness depends on vector behavior and human sleeping habits in a given setting. ITNs are generally more cost-effective in highly endemic settings (Yukich, et al. 2008), whereas IRS operations can respond faster to epidemic situations (Curtis and Mnzava 2000).

The use of chemical insecticides as larvicides to control mosquito breeding can play targeted vector stage, the potential an important role in malaria control where this is appropriate and feasible, particularly in urban settings, but the broad-spectrum effects of most chemicals are a concern to the integrity of aquatic ecosystems. Moreover, chemical repellents could have a useful supplementary role in vector control (Rowland, et al. 2004). Innovative work is in progress on the attractiveness of human odors to malaria vectors, with potential applications as mosquito attractants and repellents for use in trapping and personal protection (Zwiebel and Takken 2004).

Nonchemical methods. "Environmental management for vector control" is the collective term for manipulating or modifying environmental factors or their interaction with humans to reduce vector breeding and vector—human contact. Before the advent of synthetic insecticides, vector control depended primarily on environmental management; a meta-analysis of data mostly from that period indicated that it substantially reduced malaria risk (Keiser, et al. 2005). Eliminating vector- breeding habitats and managing water bodies has the potential to suppress vector populations, particularly in human-made habitats or urban settings (Walker and Lynch 2007). In irrigated agriculture, vector breeding can be controlled, for example, through land leveling and intermittent irrigation (Keiser, et al. 2002). New irrigation systems or dams cause drastic changes in vector-human contact, and planning to avoid health risks is essential at the design stage.

Improvement of housing, for example, through plastering of walls or closing of eaves, contributes significantly to transmission control (Gunawardena, et al. 1998). Moreover, screening to keep mosquitoes out at night is a protective option for houses with solid walls (Lindsay, et al. 2002). However, information on the cost and feasibility of housing improvement in various settings is largely missing.

The role of aquatic predators as control agents of malaria vectors is potentially enhanced through conservation or through the introduction of agents from outside. Larvivorous fish have frequently been reared and released for controlling vector breeding in small water tanks and wells, but successes have generally been limited to more or less permanent water bodies (Walker and Lynch 2007).

The bacteria *Bacillus thuringiensis israelensis* and *Bacillus sphaericus* are used in formulations as microbial larvicides. They produce toxins that are specific to mosquitoes and that have a low risk of resistance development (Lacey 2007). Recent field trials and pilot projects have shown good potential of both bacteria to manage mosquito breeding and to reduce biting rates in certain settings (Fillinger, et al. 2008). Insect pathogenic fungi have shown promising results for controlling adult *Anopheles* mosquitoes when sprayed on indoor surfaces and have potential to substantially reduce malaria transmission (Scholte, et al. 2005). Other alternative vector control methods include the use of locally available plants or plant materials as mosquito repellents Or as larvicides

(Okumu, et al. 2007; Seyoum, et al. 2003), and the use of expanded polystyrene beads in specific breeding sites (Yapabandara and Curtis 2002). Novel methods under development are genetically engineered mosquitoes and the sterile insect technique (Catteruccia 2007).

Data on the cost-effectiveness of non-chemical methods are scarce. In a retrospective analysis of data from Zambia, Utzinger, et al. (2001) indicated that environmental management was as cost-effective as ITNs. Moreover, environmental management can benefit from local resources, reducing the need for external funds.

Current implementation of DDT alternatives. The past decade has seen a steady increase in commitment to malaria control by the international community (Snow, et al. 2008). This has caused a boost in financial and human resources available for implementation of vector control interventions, due to the support of the Global Fund, the World Bank, the U.S. President's Malaria Initiative, and many nongovernmental organizations.

China, the Solomon Islands, and Vietnam have largely replaced their IRS programs with ITNs during the past decades (Najera and Zaim 2001). Conversely, the use of IRS is on the increase in Africa, where it has been more difficult to come to grips with malaria because of aspects of vector biology and disease epidemiology. In South Asia, indoor spraying using DDT and alternative insecticides continues on a large scale, but the quality of the intervention is a critical issue (JMM 2007).

National campaigns of free or highly subsidized ITNs, often in combination with other malaria control interventions, have reportedly approached coverage levels of >50% among households in a number of African countries, resulting in dramatic reductions in the malaria incidence (Bhattarai, et al. 2007; Nyarango, et al. 2006; Otten, et al. 2009; WHO 2008b).

Nonchemical methods, such as environmental management and biological control have been promoted or tested in pilot projects. However, contemporary cases of sustained implementation are not common. Case examples include the use of intermittent irrigation in China (Liu, et al. 2004), integrated and participatory strategies in Mexico (Chanon, et al. 2003) and India (Sharma 1987), river flow management in Sri Lanka (Konradsen, et al. 1998), and the use of farmer field schools on vector management in agriculture in Sri Lanka (van den Berg, et al. 2007).

Barriers and gaps. Several barriers exist in the implementation of alternatives to DDT.

Vector resistance to insecticides is a direct threat to the sustainability of ITNs and IRS. Resistance to pyrethroids has been reported in malaria vectors from West, East, and southern Africa (ANVR 2005; Coleman, et al. 2007). Particularly, *kdr*-type cross-resistance between pyrethroids and DDT severely limits the choice of insecticide. South Africa was forced to reintroduce DDT after failure of pyrethroids, due to one of the locally extinct vectors returning and having acquired pyre- throid resistance (not *kdr-type)* elsewhere (Hargreaves et al. 2000).

There is growing concern about sustained effectiveness of ITNs because the intervention currently depends solely on pyrerhroid insecticides (Greenwood, etal. 2008). Multivillage studies in an area with highly resistant *A. gambiae* in Cote d'lvoire indicated that ITNs retained most of their effect (Chandre, et al. 2000; Henry, et al. 2005). The explanation for this finding was that resistant mosquitoes were less irritated, which resulted in a higher uptake of insecticide. More worrisome are the results of a semi-field study from an area with highly resistant vectors in Benin (N'Guessan, et al. 2007b), which showed a major loss in efficacy of ITNs locally. Without the insecticidal action, bed nets provide a much lower level of personal protection (Lengeler 2004).

Resistance is caused by the use of insecticides in agriculture (Diabate, at al. 2002) and in public health. There is evidence of increased frequencies of

Table 2. Alternative methods for malaria vector control, indicating the risk, and required resources and delivery mechanisms.

VECTOR MANAGEMENT METHOD	VECTOR STAGE	RISK[a]	RESOURCES/DELIVERY
Chemical methods			
Insecticide-treated bed nets	Adult	Resistance, toxicity	Free distribution, social marketing, private sector
Indoor residual spraying	Adult	Resistance, toxicity	Spray teams
Chemical larviciding	Larva	Resistance, effect on ecosystems	Spray teams
Repellents and attractants6	Adult	Toxicity	Local, private sector
Nonchemicals methods			
Elimination of breeding sites	Larva	—	Local
Habitat manipulation	Larva	—	Local, agriculture sector
Irrigation management	Larva	—	Local, irrigation sector
Design of irrigation structures	Larva	—	Irrigation sector
House improvement	Adult	—	Local, development programs
Predation	Larva	—	Local, programs, agriculture sector
Microbial larvicides	Larva	Resistance	Programs, private sector
Botanicals	Larva/adult	Toxicity	Local
Polystyrene beads	Larva	—	Local
Fungi[b]	Adult	—	Not applicable
Genetic methods[b]	Adult	To be studied	Not applicable

—, Negligible risk.
[a]Theoretically, (behavioral) resistance could also develop against repellents, attractants, and house improvement.
[b](Partly) under development.

resistance genes attributable to IRS or ITN programs (Karunaratne and Hemingway 2001; Stump, et al. 2004). Moreover, there are records of a change in vector behavior from indoor resting to outdoor resting in response to indoor spraying, as well as a change in daily pattern of biting and host choice in response to ITN interventions (Molineaux and Gramiccia 1980; Pates and Curtis 2005; Phillips 2001; Takken 2002). A system of sentinel sites to monitor vector density, quantify insecticide resistance, and guide informed decision making on insecticide choice still needs to be established in most disease-endemic countries (Coleman and Hemingway 2007).

Another barrier is operational capacity. The effective coverage of programs depends critically on the access and targeting of populations and vulnerable groups most at risk of malaria, the degree of compliance of the provider, and adherence by the consumer. In most countries with endemic malaria, health systems lack capacity to plan and implement programs effectively. Reforms in the health sector have led to the decentralization of planning and budgeting. Consequently, the responsibility for service provision has shifted from national to subnational or district-level health departments, requiring new skills for malaria control at each level. An analysis of case studies from four countries suggested that decentralization can potentially benefit malaria control (Barat 2006). In general, however, there is a lack of guidance on how malaria control might be implemented in a decentralized environment (World Bank 2005).

Traditionally, IRS has been managed as vertical programs, which is still the case in various countries. In some countries the transition process after health reforms has caused an erosion of the specialist skills needed for IRS (Shiff 2002). It will be a challenge for many countries to conduct and sustain effective IRS programs (Kolaczinski, et al. 2007). The delivery of ITNs has used a variety of models, including vertical programs, integrated health sector programs, and involvement of the private sector and nongovernmental organizations (Webster, et al. 2007). As the global thrust is to promote coverage with ITNs and IRS, vector control capacity is needed at the appropriate levels.

Interventions involving environmental management and other larval control methods depend on the participation of other sectors and communities. Even though decisions affecting the risk of vector-borne disease are taken in other public sectors, there is insufficient awareness of the effects. Moreover, the health sector lacks capacity to facilitate community participation and education. A possible solution is the integration of health activities with community programs that generate income (e.g., from agriculture). Rich experience with participatory approaches exists within the agriculture sector (Pretty 1995); the health sector potentially can benefit from these resources. One relevant model is the Farmer Field School on Integrated Pest Management, developed and promoted by the Food and Agriculture Organization of the United Nations (van den Berg and Knols 2006).

Integration of methods. An integrated approach to vector control has frequently been advocated (McKenzie, et al. 2002; Shiff 2002; Utzinger, et al. 2002). The need for a reduced reliance on insecticides for vector- borne disease control, as pointed out in World Health Assembly Resolution 50.13 (International Programme on Chemical Safety 1997), has been stressed further by the Intergovernmental Forum on Chemical Safety, Forum VI (Intergovernmental Forum on Chemical Safety 2008).

Various studies have demonstrated that integration of vector control methods resulted in significant reductions in transmission and morbidity rates of malaria (Chanon, et al. 2003; Dua, et al. 1997; Sharma, et al. 1991; Singh, et al. 2006; Takken, et al. 1990; Utzinger, et al. 2001). Moreover, modeling studies predicted that combinations of interventions can be much more effective in reducing malaria transmission than individual interventions and that the effect of IRS and ITNs is amplified by environmental management, even in areas of intense transmission (Killeen, et al. 2000, 2004).

Besides its direct effect on transmission intensity, the integration of methods may also contribute to resistance management. For example, larval control is expected to prevent or delay the onset of vector resistance to insecticides (Walker and Lynch 2007), whereas measures that reduce human contact with vectors, through their proximity, housing conditions, or presence of repellents, for example, will reduce the selection pressure.

Integrated vector management. Modeled on the positive experience from integrated pest management in agriculture, integrated vector management (IVM) has been defined by the WHO (2008a) as "a rational decision-making process for the optimal use of resources for vector control." The aim of IVM is to improve cost-effectiveness, ecologic soundness, and sus- tainability of disease vector control (Townson, et al. 2005; WHO 2004). In contrast to conventional vector control programs with a top-down decision-making structure, IVM emphasizes decision making at the lowest possible level in accordance with local data collection and situational analysis, and requires collaboration within the health sector and with other sectors, as well as community participation. Hence, decentralization in the health sector can potentially work in favor of IVM by facilitating tailored action at the local level (van den Berg and Takken 2007).

The Global Malaria Action Plan advocates the scaling-up of ITNs and IRS for an immediate impact on the malaria burden of populations at risk (Roll Back Malaria Partnership 2008). However, to address sustainability issues, interventions must be implemented in accordance with an IVM approach by being evidence-based and by integrating available resources and supplementary methods in an effective and ecologically sound manner. To enable the graduation from a conventional vector control program to IVM, the evidence base and human capacity needs strengthening at all relevant levels of administration. Recently, targets have been set for the elimination of malaria (Feachem and Sabot 2008). An IVM approach is important to sustain achievements and reduce transmission to critical low levels needed to eliminate malaria (Beier, et al. 2009).

CONCLUSIONS

The reported global use of DDT for disease vector control is 4–5,000 metric tons per year, with India by far the largest consumer and several countries reintroducing DDT. The insecticide is known for its long residual effect and low operational cost. However, the effectiveness of DDT depends on local settings and merits closer consideration vis-a-vis chemical and nonchemical alternatives. Legislation and capacity to enforce regulations and management practice is inadequate in most countries.

Recent evidence indicates that indoor spraying causes high levels of human exposure to DDT (e.g. Aneck-Hahn, et al. 2007). This could adversely affect human health, because the evidence base on some of the more serious and chronic health effects of DDT is growing. Moreover, the occurrence of resistance to the toxic action of DDT is common in malaria vectors and appears to be spreading. A comprehensive cost assessment of DDT versus its alternatives is needed and should include the monitoring of side effects and unintended costs to human health, the environment, and international trade.

Effective chemical alternatives to DDT for vector control are available, but the choice of insecticides is limited. Insecticides with novel chemistry will not come to market in the short term. Alternative insecticides should pose less risk to human health and the environment. The coverage of populations with ITNs and IRS has increased in recent years, particularly in Africa. However, insecticide resistance is reducing the efficacy of these methods in certain areas. To be prepared for future emergencies, the continued effectiveness of insecticides needs to be safeguarded.

A number of nonchemical methods have proven their value in malaria control in certain settings, but more work is needed on the incremental impact of methods such as environmental management or the use of microbial larvicides when used in conjunction with IRS and ITNs. Several new technologies are under development but require increased investment. To continue this development, we must foster new researchers in the field of vector control.

To reduce reliance on DDT, support is needed for integrated and multipartner strategies of vector control. IVM provides a framework for improving cost-effectiveness, ecologic soundness, and sustainability of vector control through integration with other arms of public health and other sectors. Now that malaria transmission is decreasing in a number of African countries, there is a greater prospective role for environmental management and other nonchemical methods within IVM strategies. This will increase the sustainability of control efforts and assist in achieving malaria elimination objectives.

REFERENCES

Aneck-Hahn, N.H., G.W. Schulenburg, M.S. Bornman, P. Farias, and C. De Jager. 2007. Impaired semen quality associated with environmental DDT exposure in young men living in a malaria area in the Limpopo Province, South Africa. *J Androl* 28:423–434. [Anonymous.] 2007. DDT for malaria control: the issue of trade

[Editorial]. Lancet 369:248. Armstrong Schellenberg, J.R.M., S. Abdulla, R. Nathan, O. Mukass, T.J. Merchant, and N. Kikumbih, et al. 2001. Effect of large-scale social marketing of insecticide-treated nets on child survival in rural Tanzania. *Lancet* 340:1241–1247.

Barat, L.M. 2006. Four malaria success stories: how malaria burden was successfully reduced in Brazil, Eritrea, India, and Vietnam. *Am J Trop Med Hyg* 74:12–16.

Beard, J. 2006. DDT and human health. *Sci Total Environ* 355:78–69.

Beier, J.C., J. Keating, J.L. Githure, M.B. Macdonald, D.E. Impoinvil, and R.J.Novak. 2009. Integrated vector management for malaria control. Malar, J., 7(suppl 1):S4; doi:10.1186/1475–2875-7-Sl-S4 [11 December 2008],

Bhattarai, A., A.S. Ali, S.P. Kachur, A. Martensson, A.K. Abbas, and R. Khatib, et al. 2007. Impact of artemisinin-based combination therapy and insecticide-treated nets on malaria burden in Zanzibar. *PLoS Med* 4:e30S; doi:10.1371/journal. p med.0040309 (Online 6 November 2007).

Bouwman, H., A. Coetzee, and C.H.J. Schutte. 1990. Environmental and health implications of DDT-contaminated fish from the Pongolo Flood Plain. *Afr J Zool* 104:275–286.

Bouwman, H., R.M. Cooppan, P.J. Becke, and S. Ngxongo. 1991. Malaria control and levels of DDT in serum of two populations in Kwazulu. *J Toxicol Environ Health* 33:141–155.

Bouwman, H., B. Sereda, and H.M. Meinhardt. 2006. Simultaneous presence of DDT and pyrethroid residues in human breast milk from a malaria endemic area in South Africa. *Environ Pollut* 144:902–917.

Brody, J.G., K.B. Moysich, O. Humblet, K.R. Attfield, G.P. Beehler, and R.A. Rudel. 2007. Environmental pollutants and breast cancer: epidemiologic studies. *Cancer* 109(12 suppl):2667–2711.

Busvine, J.R. 1978. Current problems in the control of mosquitoes. *Nature* 273:604–607.

Catteruccia, F. 2007. Malaria vector control in the third millennium: progress and perspectives of molecular approaches. *Pest Manag Sci* 63:634–640.

Chandre,F., F. Darriet, S. Duchon, L. Finot, S. Manguin, and P. Carnevale, et al. 2000. Modifications of pyrethroid effects associated with kdr mutation in Anopheles gambiae. *Med Vet Entomol* 14:81–88.

Chanon, K.E., J.F. Mendez-Galvan, J.M. Galindo-Jaramilloc, H. Olguin-Bernalb, and V.H. Borja-Aburto. 2003. Cooperative actions to achieve malaria control without the use of DDT. *Int J Hyg Environ Health* 206:387–394.

Chen, A.and W.J. Rogan. 2003. Nonmalarial infant deaths and DDT use for malaria control. *Emerg Infect Dis* 9:360–964.

Cohn, B.A., M.S. Wolff, P.M. Cirillo, and R.I. Sholtz. 2007. DDT and breast cancer in young women: new date on the significance of age at exposure. *Environ Health Perspect* 115:1406–1414.

Coleman, M. and J. Hemingway. 2007. Insecticide resistance monitoring and evaluation in disease transmitting mosquitoes. *J Pestic Sci* 32:69–76.

Coleman, M, B. Sharp, and J. Seocharan. Developing an evidence-based decision support system for rational insecticide choice in the control of African malarial vectors. *J Med Entomol* 43:663–668.

Cox, S., A.S. Niskar, K.M. Narayan, and M. Marcus 2007. Prevalence of self-reported diabetes and exposure to organochlorine pesticides among Mexican Americans: Hispanic Health and Nutrition Examination Survey, 1982–1984. *Environ Health Perspect* 115:1747–1752.

Cui, F., M. Raymond, and C.L.Qiao. 2006. Insecticide resistance in vector mosquitoes in China. *Pest Manag Sci* 62:1013–1022.

Curtis, C. and C.P. Olsen. 2004. The Africa Stockpiles Programme: cleaning up obsolete pesticides; contributing to a healthier future. *Ind Environ* 27(2–3)1:37–38.

Curtis, C.F. 2002. Restoration of malaria control in the Madagascar highlands by DDT spraying. *Am J Trop Med Hyg* 66:1.

Curtis, C.F. and A.E.P. Mnzava. 2000. Comparison of house spraying and insecticide-treated nets for malaria control. *Bull WHO* 78:1389–1400.

Dash, A.P., N. Valecha, A.R. Anvika, and A. Kumar. 2009. Malaria in India: challenges and opportunities. *J Biosci* 33:583–592.

De Jager, C., P. Faria, A. Barraza-Villarreal, M.H. Avila, P. Ayotte, and E. Dewailly, et al. 2006. Reduced seminal parameters associated with environmental DOT exposure and *p.p'*-DDE concentrations in men in Chiapas, Mexico: a cross-sectional study. *J Androl* 27:16–27.

Diabate, A., T. Baldet, F. Chandre, M. Akoobeto,. T.R. Guiguemde, and F. Darrie, et al. 2002. The role of agricultural use of insecticides in resistance to pyrethroids in Anopheles gambiae s.l. in Burkina Faso. *Am J Trop Med Hyg* 67:617–622.

Dua, V.K., C.S. Pant, and V.P. Sharma. 1996. Determination of levels of HCH and DDT in soil, water and whole blood from bio-environmental and insecticide-sprayed areas of malaria control. *Indian J Malariol* 33:7–15.

Dua.V.K., S.K. Sharma, A. Srivastava, and V.P. Sharma. 1997. Bioenvironmental control of industrial malaria at Bharat Heavy Electricals Ltd., Hardwar, India—results of a nine-year study (1987–95). *J Am Mosq Control Assoc* 13:278–285.

Dusfour. I., R.E. Harbach, and S. Manguin. 2004. Bionomics and systematica of the oriental Anopheles sundaicus complex in relation to malaria transmission and vector control. *Am J Trop Med Hyg* 71:518–524.

Eriksson, P. and U. Talts. 2000. Neonatal exposure to neurotoxic pesticides increases adult susceptibility: a review of current findings. *Neurotoxicology* 21:37–47.

Eskenazi, B., J. Chevrier, L.G. Rosas, H.A. Anderson, M.S, Bornman, and H. Bouwman, et al. 2009. The Pine River statement human health consequences of DOT use. *Environ Health Perspect* 117:1359–1367; doi;10.1289/ehp.11748 (Online 4 May 2009).

Feachem R. and O. Sabot. 2008. A new global malaria eradication strategy. *Lancet* 371:1633–1635.

Fegan, G.W., A.M. Noor, W.S. Akhwale, S. Cousens, and R.W. Snow. 2007. Effect of expanded insecticide-treated bednet coverage on child survival in rural Kenya: a longitudinal study. *Lancet* 370:1035–1039.

Fillinger, U., K. Kannady, G. William, M.J. Vanek, S. Dongus, and D. Nyika, et al. 2008. A tool box for operational mosquito larval control: preliminary results and early lessons from the Urban Malaria Control Programme in Dar es Salaam. Tanzania. *Malar J* 7:20; doi:10.1186/1475–2875–7–20 (Online 25 January 2008)

Fisk, A.T., C.A. de Wit, M. Wayland, Z.Z. Kuzyk, N. Burgess, R. Letcher, et al. 2005. An assessment of the toxicological significance of anthropogenic contaminants in Canadian arctic wildlife. *Sci Total Environ* 351–352:57–93.

Galindo, R.J.G., J.A. Medina, and L.C. Villagrana. 1996. Physiological and biochemical changes in shrimp larvae (Penaeus vannamei) intoxicated with organochlorine pesticides. *Mar Pollut Bull* 32:872–875.

Garabrant, O.H., J. Held, B. Langholz, J.M. Peters, and T.M. Mack. 1992. DDT and related compounds and risk of pancreatic cancer. *J Natl Cancer Inst* 84:764–771.

Global Environment Facility, 2009, "The GEF Project Database." Available: http://gefonline.org/home.cfm (accessed 22 September 2009).

Goodman, C., P. Coleman, and A. Mills, 2000, "Economic Analysis of Malaria Control in Sub-Saharan Africa," Geneva: Global Forum for Health Research, Available: http://www.doh. gov.za/issues/malarie/red_reference/cross_cuttingJ Economics/eco5.pdf (accessed 23 September 2009).

Greenwood, B.M., D.A. Fidock, D.E. Kyle, S.H.I. Kappe, P.L. Alonso, and F.H. Collins, et al. 2008. Malaria: progress, perils, and prospects for eradication. *J Clin Invest* 118:1266–1276.

Grieco, J.P., N.L. AcheeL, T. Chareonviriyaphap, W. Suwonkerd, K. Chauhan, and M.R. Sardelis, et al. 2007. A new classification system for the actions of IRS chemicals traditionally used for malaria control. *PLoS ONE* 2(8):e716; doi:10.1371/journal.pone.0000716 (Online 8 August 2007).

Gunasekaran, K., S.S. Sahu, P. Jambulingam, and P.K. Das. 2005. DDT indoor residual spray, still an effective tool to control Anopheles fluvatilis-transmitted Plasmodium falciparum malaria in India. *Trop Med Int Health* 10:160–168.

Gunawardena, D.M., A.R. Wickremasinghe, L. Muthuwatta, S. Weerasingha, J. Rajakaruna, and T. Senanayaka, et al. 1998. Malaria risk factors in an endemic region of Sri Lanka, and the impact and cost implications of risk factor-based interventions. *Am J Trop Med Hyg* 58:533–542.

Hargreaves, K., L.L. Koekemoer, B. Brooke, R.H. Hunt, J. Mthembu, and M. Coetzee. 2000. Anopheles funestus resistant to pyrethroid insecticides in South Africa. *Med Vet Entomol* 14:181–189.

Harrad, S. 2001. *Persistent Organic Pollutants: Environmental Behaviour and Pathways of Human Exposure.* Dordrecht, the Netherlands:Kluwer Academic Publishers.

Hawley, W.A., P.A. Phillips-Howard, F.O. ter Kuile, D.J. Terlouw, J.M. Vulule, and M. Ombok, et al. 2003. Community-wide effects of permethrin-treated bed nets on child mortality and malaria morbidity in western Kenya. *Am J Trop Med Hyg* 68:121–127.

Hemingway J., B.J. Beaty, M. Rowland, T.W. Scott, and B.L. Sharp. 2006. The Innovative Vector Control Consortium: improved control of mosquito-borne diseases. *Trends Parasitol* 22:308–312.

Hemingway, J, R.P. Penilla, A.D. Rodriquez, B.B. James, W. Edge, and H. Rogers, et al. 1997. Resistance management strategies in malaria vector mosquito control. A large-scale field trial in southern Mexico. *Pestic Sci* 51:375–382.

Hemingway, J. and H. Hanson. 2000. Insecticide resistance in insect vectors of human disease. *Annu Rev Entomol* 45:371–391.

Henry, M.C., S.B. Assi, C. Rogier, J. Dossou-Yovo, F. Chandre, and P. Guillet, et al. 2005. Protective efficacy of lambda-cyhalothrin treated nets in Anopheles gambiae pyrethroid resistance areas of Cote d'Ivoire. *Am J Trop Med Hyg* 73:859–864.

Herrera-Portugal, C., H. Ochoa, G. Franco-Sanchez, L. Yanez, and F. Diaz-Barriga. 2005. Environmental pathways of exposure to DDT for children living in a malarious area of Chiapas. Mexico. *Environ Res* 99:158-–163.

Intergovernmental Forum on Chemical Safety, 2008, "Forum VI, Sixth Session of the Intergovernmental Forum on Chemical Safety, Final Report," Available: http://www.who.int/ifcs/ documents/ forums/forum6/f6_finalreport_en.pdf (accessed 23 September 2009).

International Programme on Chemical Safety, 1997, "World Health Assembly Resolution 50.13," Available: http://www. who.int/ipcs/pub!ications/ wha/whares_53_13/en/index. html (accessed 22 September 2009],

JMM (Joint Monitoring Mission). 2007. *National Vector Borne Diseese Control Programme: Joint Monitoring Mission Report, February 2007*. New Delhi:World Health Organization.

Karunaratne, S.H.P.P. and J. Hemingway. 2001. Malathion resistance and prevalence of the malathion carboxylesterase mechanism in populations of mosquito vectors of disease in Sri Lanka. *Bull WHO* 79:1060–1064.

Keiser, J., B.H. Singer, and J. Utzinger. 2005. Reducing the burden of malaria in different eco-epidemiological settings with environmental management a systematic review. *Lancet Infect Dis* 5:695–708.

Keiser, J., J. Utzinger, and B.H. Singer. 2002. The potential of intermittent irrigation for increasing rice yields, lowering water consumption, reducing methane emissions, and controlling malaria in African rice fields. *J Am Mosq Control Assoc* 18:329-–340.

Kelly, B.C., F.A.P.C. Gobas, and M.S. McLachlan. 2004. Intestinal absorption and biomagnification of organic contaminants in fish, wildlife, and humans. *Environ Toxicol Chem* 23:2324–2336.

Killeen, G.F., F.E. McKenzie, B.F. Foy, C. Schieffelin, P.F. Billingsley, and J.C. Beier. 2000. The potential impact of integrated malaria transmission control on entomologic inoculation rate in highly endemic areas. *Am J Trop MBd Hyg* 62:545–551.

Killeen, G,F., A. Seyoum, and B.G. Knols. 2004. Rationalizing historical successes of malaria control in Africa in terms of mosquito resource availability management. *Am J Trop Med Hyg* 71(suppl 2):87–93.

Kolaczinski, K., J. Kolaczinski, A. Kilian, and S. Meek. 2007. Extension of indoor residual spraying for malaria control into high transmission settings in Africa. *Trans R Soc Trop Med Hyg* 101:852–853.

Konradsen, F., Y. Matsuno, F.P. Amerasinghe, P.H. Amerasinghe, and W. van der Hoek. 1998. Anopheles culicifacies breeding in Sri Lanka and for control through water management. *Acta Trop* 71:131–138.

Lacey, L.A. 2007. Bacillus thuringiensis serovariety israelensis and Bacillus sphaericus for mosquito control, *J Am Mosq Control Assoc* 23 (suppl 2):133–163.

Lengeler, C. 2004. Insecticide-treated bed-nets and curtains for preventing malaria. *Cochrane Database Syst Rev* 2:CD000363; doi:10.1002/14651858.CD000363.pub2 (Online 19 April 2004).

Lindsay. S.W., P.M. Emerson, and J.D. Charlwood. 2002. Reducing malaria by mosquito-proofing houses. *Trends Parasitol* 18:510–514.

Liu, W.H., K. Xin, C.Z. Chao, S.Z. Feng, L. Yan, and R.Z. He, et al. 2004. New irrigation methods sustain malaria control in Sichuan Province, China. *Acta Trap* 89:241–247.

Longnecker, M.P. 2005. Invited commentary: why DDT matters now. *Am J Epidemiol* 162:726–728.

Mabaso, M.L.H., B. Sharp, and C. Lengeler. 2004. Historical review of malarial control in southern African with emphasis on the use of indoor residual house-spraying. *Trop Med Int Health* 9:846–856.

MacDonald, G. 1957. *The Epidemiology and Control of Malaria*. London:oxford University Press.

Malisch, R. and F.X.R. van Leeuwen. 2003. Results of the WHO-coordinated exposure study on the levels of PCBs, PCDDs and PCDFs in human milk. *Organohalogen Compounds* 64:140–143.

Martinez-Torres, O., F. Chandre, M.S. Williamson, F. Darriet, J.B. Berge, and A.L. Devonshire, et al. 1998. Molecular characterization of pyrethroid knockdown resistance (kdri in the major malaria vector Anopheles gambiae s.s. *Insect Mol Biol* 7:179–184.

McKenzie, F.E., J.K. Baird, J.C. Beier, A.A. Lai, and W.H. Bossert. 2002. A biological basis for integrated malaria control. *Am J Trop Med Hyg* 67:571–577.

Metcaff, R.L. 1973. A century of DDT. *J Agric Food Chem* 21:511–519.

Molineaux, L. and G. Gramiccia. 1980. *The Garki project. Research on the Epidemiology and Control of Malaria In the Sudan Savanna of West Africa.* Geneva:World Health Organization.

Mouchet, J. 1988. Mini review: agriculture and vector resistance. *Insect Sci Appl* 9:297–302.

Najera, J.A.and M. Zaim. 2001. *Malaria Vector Control—Insecticides for Indoor Residual Spraying.* WHO/CDS/WHOPES/2001.3. Geneva:World Health Organization.

Nauen, R. 2007. Insecticide resistance in disease vectors of public health importance. *Pest Manag Sci* 63:628–633.

N'Guessan, R., P. Boko, A. Odjo, M. Akogbeto, A. Yates, and M. Rowland. 2007a. Chlorfenapyr, a pyrrole insecticide for the control of pyrethroid or DDT resistant Anopheles gambiae (Diptera: Culicidae) mosquitoes. *Acta Trop* 102:69–78.

N'Guessan. R., V. Corbel, M. Akogbeto, and M. Rowland. 2007b, Reduced efficacy of insecticide-treated nets and indoor residual spraying for malaria control in pyrethroid resistance area, Benin. *Emerg Infect Dis* 13:199–206.

N'Guessan, R., V. Corbel, J. Bonnet, A. Yates. A. Asidi, and P. Boko, et al, 2007c. Evaluation of indoxacarb, an oxadiazine insecticide for the control of pyrethroid-resistant Anopheles gambiae (Diptera: Culicidae). *J Med Entomol* 44:270–276.

Nyarango, P.M., T. Gebremeskel, G. Mebrahtu, J. Mufunda, and U. Abdulmumini, et al. 2006, A steep decline of malaria morbidity and mortality trends in Eritrea between 2000 and 2004: the effect of combination of control methods. *Malar J* 5:33; doi:10.1186/1475–2875–5–33 (Online 24 April 2006).

Okumu, F.O., B.G.J. Knols, and U. Fillinger. 2007. Larvicidal effects of a neem (Azadirachra indica) oil formulation on the malaria vector Anopheles gambiae. *Malar J* 6:63; doi:10.1186/1475–2875–6–63 (Online 22 May 2007).

Otten, M.,M. Aregawi, W. Were, C. Karema, A. Medin, and D. Jima, et al. 2009. Initial evidence of reduction of malaria cases and deaths in Rwanda and Ethiopia due to rapid scale-up of malaria prevention and treatment. *Malar* J 8:14; doi:10.1186/1475–2875–8–14 (Online 14 January 2009).

Pates, H. and C. Curtis. 2005. Mosquito behavior and vector control. *Annu Rev Entomol* 50:53–70.

Penilla, R.P., A.D. Rodriguez, J. Hemingway, J.L. Torres, F. Solis, and M.H. Rodriguez. 2006. Changes in glutathione S-transferase activity in DDT resistant natural Mexican populations of Anopheles albimanus under different insecticide resistance management strategies. *Pestic Biochem Physiol* 86:63–71.

Phillips, R.S. 2001. Current status of malaria and potential for control. *Clin Microbiol Rev* 14:208–226.

"President's Malaria Initiative," 2009, PMI Homepage. Available: http://www.fightingmalaria.gov/ (accessed 22 September 2009).

Pretty, J.N. 1995. Participatory learning for sustainable agriculture. *World Dev* 23:1247–1263.

Ratcliffe D.A. 1967. Decrease in eggshell weight in certain birds of prey. *Nature* 215:208–210.

Ribas-Fito, N., M. Torrent, D. Carrizo, L. Munoz-Ortiz, J. Juivez, J.O. Grimalt, et al. 2006. In utero exposure to background concentrations of DDT and cognitive functioning among preschoolers. *Am J Epidemiol* 164:955–962.

Ritter, L., K.R. Solomon, J. Forget, M. Stemeroff, and C. O'Leary. 1995. *Persistent Organic Pollutants.* Geneva:International Programme on Chemical Safety.

Roberts, D.R. and R.G. Andre. 1994. Insecticide resistance issues in vector-borne disease control. *Am J Trop Med Hyg* 50(Suppl) 61:21–34.

Roll Back Malaria Partnership, 2008, "The Global Malaria Action Plan for a Malaria-Free World," Available: http://rbm.who. int/gmap/ (accessed 22 September 2009).

Rowland, M., T. Freeman, G. Downey, A. Hadi , and M. Saeed. 2004. DEET mosquito repellent sold through social marketing provides personal protection against malaria in an area of all-night mosquito biting and partial coverage of insecticide-treated nets: a case-control study of effectiveness. *Trop Med Int Health* 9:343–350.

Scholte, E.J., K. Ng'habi, J. Kihonda, W. Takken, K. Paaijmans K, and S. Abdulla, et al. 2005. An entomopathogenic fungus for control of adult African malaria mosquitoes. *Science* 308:1641–1642.

Scopus, 2008, "Everything You Need to Know about Scopus," Available: http://info.scopus.com/ (accessed 22 September 2009).,

Sereda, B.L.and H.R. Meinhardt. 2005. Contamination of the water environment in malaria endemic areas of KwaZulu-Natal, South Africa by DDT and its metabolites. *Bull Environ Contam Toxicol* 75:538–545.

Seyoum, A., G.F. Killeen, E.W. Kabiru, B.G.J. Knols, and A. Hassanali. 2003. Field efficacy of thermally expelled or live potted repellent plants against African malaria vectors in western Kenya. *Trop Med Int Health* 8:1005–1011.

Sharma, R.C., A.S. Gautam, R.M. Bhatt, D.K. Gupta, and V.P. Sharma. 1991. The Kheda malaria project: the case for environmental control, *Health Policy Plan* 6:262–270.

Sharma, V.P. 1987. Community-based malaria control in India. *Parasitol Today* 3:222–226.

Sharma, V.P. 2003. DDT: the fallen angel. *Curr Sci (Bangalore)* 85:1532–1537.

Sharp, B.L., I. Kleinschmidt, E. Streat, R. Maharaj, K.I. Barnes, and D.N. Durrheim, et al. 2007. Seven years of regional malaria control collaboration—Mozambique, South Africa, and Swaziland. *Am J Trop Med Hyg* 76:42–47.

Shiff, C. 2002. Integrated approach to malaria control. *Clin Microbiol Rev* 15:278_293.

Singh, N., M.M. Shukla, A.K. Mishr, M.P. Singh, J.C. Paliwaland A.P. Dash. 2006. Malaria control using indoor residual spraying and larvivorousfish: a case study in Betul, central India. *Trop Med Int Health* 11:1512–1520.

Smith, T.A., R. Leuenberger, and C. Lengeler. 2001. Child mortality and malaria transmission intensity in Africa. *Trends Parasitol* 17:145–149.

Snedeker, S.M. 2001. Pesticides and breast cancer risk: a review of DDT, DDE, and dieldrin. *Environ Health Perspect* 109:35–47.

Snow, R.W., C.A. Guerra, J.J. Mutheu, and S.I. Hay. 2008. International funding for malaria control in relation to populations at risk of stable Plasmodium falciparum transmission. *PLoS Med* 5:e142; doi:10.1371/joumal.pmed.0050142 (Online 22 July 2008).

Stump, A.D., F.K. Atieli, J.M. Vulule, and N.J. Besansky. 2004. Dynamics of the pyrethroid knockdown resistance allele in Western Kenyan populations of Anopheles gambiae in response to insecticide-treated bed net trials. *Am J Trop Med Hyg* 70:591–596.

Takken, W. 2002. Do insecticide-treated bednets have an effect on malaria vectors? *Trop Med Int Health* 7:1022–1030.

Takken, W., W.B. Snellen, J.P. Verhave, B.G.J. Knols, and S. Atmosoedjono. 1990. Environmental measures for malaria control in Indonesia—an historical review on species sanitation. *Wageningen Agricultural Research Papers* 90.7. Wageningen. the Netherlands:Wageningen University.

Townson, H., M.B. Nathan, M. Zaim, P. Guillet, L. Manga, and R. Bos, et al. 2005. Exploiting the potential of vector control for disease prevention. *Bull WHO* 83:942–947.

Turusov. V., V. Rakitsky, and L. Tomatis. 2002. Dichlorodiphenyltrichloro-ethane (DDT): ubiquity, persistence, and risks. *Environ Health Perspect* 110:125–128.

UNEP. 2002. *Stockholm convention on persistent organic pollutants (POPs).* UNEP/Chemicals/2002/9. Geneva:United Nations Environment Programme.

UNEP, 2008, "Persistent Organic Pollutants. Workshop Proceedings," Geneva:United Nations Environment Programme, Available: http://www.chem.unep.ch/PQps/PQPs_Inc/ptoceedings/coverpgs/procovers.htm (accessed 22 September 2009).

Utzinger J, M. Tanner, D.M. Kammen, G.F. Killeen, and B.H. Singer. 2002. Integrated programme is key to malaria control (Letter). *Nature* 419:431.

Utzinger, J., Y. Tozan , and B.H. Singer. 2001. Efficacy and cost-effectiveness of environmental management for malaria control. *Trop Med Int Health* 6:677–687.

van den Berg, H. and B.G.J. Knols. 2006. The Farmer Field School: a method for enhancing the role of rural communities in malaria control? *Malar J* 5:3; doi: 10.1186/1475-2875-5-3 (Online 19 January 2006).

van den Berg, H. and W. Takken. 2007. A framework for decisionmaking in integrated vector management to prevent disease. *Trop Med Int Health* 12:1230–1238.

van den Berg, H., A. von Hildebrand, V. Ragunathan, and P.K. Das. 2007. Reducing vector-borne disease by empowering farmers in integrated vector management. *Bull WHO* 85:561–566.

Varca, L.M.and E.D. Magallona. 1994. Dissipation and degradation of DDT and DDE in Philippine soil under field conditions. *J Environ Sci Health* 29:25–35.

Vector Biology and Control, World Health Organization Regional Office for Africa, 2008, Vector Biology and Control Reports, Available: http://www.afro.who.int/vbc/reports/ (accessed 23 September 2009).

Venners, S., S. Korrick, X. Xu, C. Chen, W. Guang, and A, Huang, et al. 2005. Preconception serum DDT and pregnancy loss: a prospective study using a biomarker of pregnancy. *Am J Epidemiol* 162:709–716.

Viaira, E.D.R., J.P.M. Torres, and O. Malm. 2001. DOT environmental persistence from its use in a vector control program: a case study. *Environ Res* 86:174–182

Walker, K. 2000. Cost-comparison of DDT and alternative insecticides for malaria control. *Med Vet Entomol* 14:345–354.

Walker, K. and M. Lynch. 2007. Contributions of Anopheles larval control to malaria suppression in tropical Africa: review of achievements and potential. *Med Vet Entomol* 21:2–21.

Wandiga, S.O. 2001. Use and distribution of organochlorine pesticides. The future in Africa. *Pure Appl Chem* 73:1147–1155.

Webster J., J. Hill, J. Lines, and K. Hanson. 2007. Delivery systems for insecticide treated and untreated mosquito nets in Africa: categorization and outcomes achieved. *Health Policy Plan* 22:277–293.

WHO. 2004. *Global Strategic Framework for Integrated Vector Management*. WHO document WHO/CDS/CPE/PVC/2004.10. Geneva:World Health Organization.

WHO. 2006. *Malaria Vector Control and Personal Protection. Report of a WHO Study Group.* WHO Technical Report Series No. 936. Geneva:World Health Organization. Available: www.who.int/malaria/docs/WHO-TRS-936s.pdf (accessed 22 September 2009).

WHO. 2007a. *The Use of DDT in Malaria Vector Control. WHO Position Statement.* Geneva:World Health Organization.

WHO. 2007b. *Update for CoP3 on WHO Activities Relevant to Country Implementation of the Stockholm Convention on Persistent Organic Pollutants*. Geneva.World Health Organization. Available: http://www.who.int/ipcs/capacity_ building/COP3%20update.pdf (accessed 22 September 2009).

WHO. 2008a. *Statement on Integrated Vector Management WHO/HTM/NTD/VEM/2008.2.* Geneva:World Health Organization

WHO. 2008b. *World Malaria Report 2008.* Geneva:World Health Organization.

World Bank. 2005. Rolling Back Malaria: the World Bank Global Strategy and Booster Program. Washington, DC:World Bank.

Yanez L, Ortiz-Perez D, Batres LE. Borja-Aburto VH. Diaz- Barriga F. 2002. Levels of dichlorodiphenyltrichloroethane and deltamethrin in humans and environmental samples in malarious areas of Mexico. Environ Res 88:174-181.

Yapabandara AMGM, Curtis CF. 2002. Laboratory and field comparisons of pyriproxyfen, polystyrene beads and other larvicidal methods against malaria vectors in Sri Lanka. Acta Trop 81:211-223.

Yukich JO, Lengeler C, Tediosi F, Brown N, Mulfigan JA, Chavasse D. et al. 2008. Costs and consequences of large-scale vector control for malaria. Malar J 7:258; doi:10.1186/1475-2875-7-258 lOnline 17 December 2008],

Zwiebel LJ, Takken W. 2004. Olfactory regulation of mosquito- host interactions. Insect Biochem Mol Biol 34:645-652.

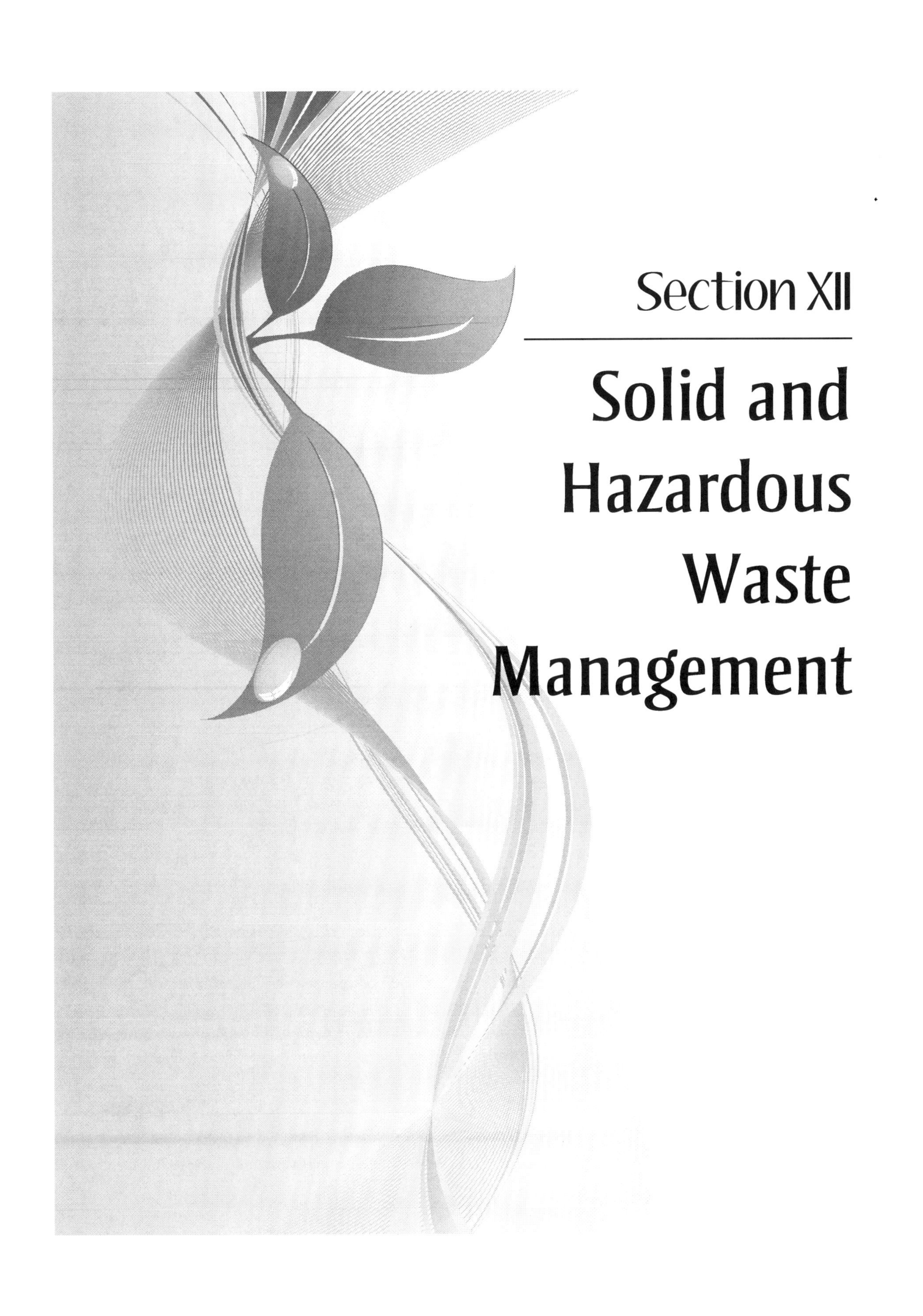

Section XII

Solid and Hazardous Waste Management

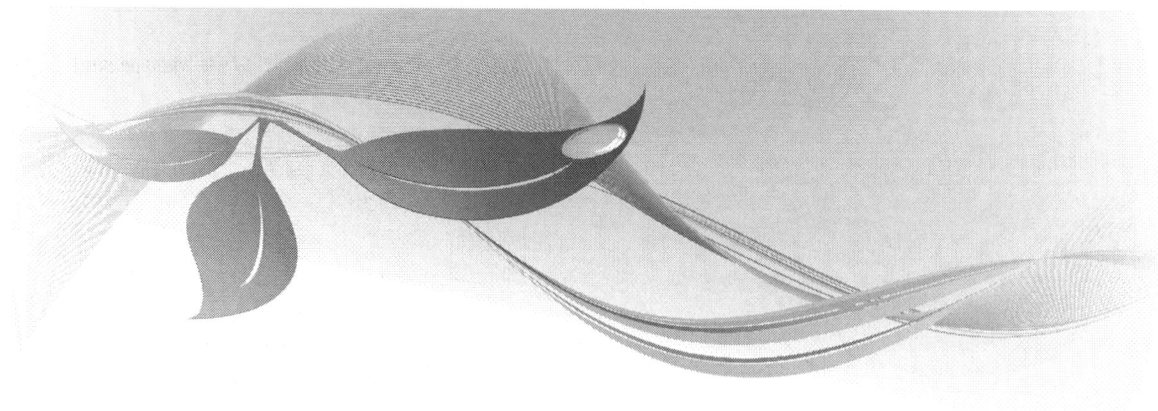

By Anne Marie Zimeri

MUNICIPAL SOLID WASTE (MSW)

As human populations expand, so must their infrastructure, including their mechanisms for handling waste. Mishandling of waste can pose a great threat to the environment and public health. Inappropriately handled waste can allow for the proliferation of disease-carrying insects and rodents, and microbes that are pathogenic. Past practices that simply dumped waste in piles or crevasses allowed for it to accumulate and smolder to the point where it would pollute ambient air. From an economic standpoint, inappropriate disposal of waste can contribute to the lowering of property values because of malodor, the spread of debris around the area, and increased traffic from companies or individuals who need to drop off trash. Most important from a public health standpoint is the risk that toxins from the waste pile will leach into groundwater and contaminate the water used for drinking and/or agriculture.

How much waste must be handled in the United States each year? Currently, almost 1000 pounds of MSW are generated per capita in the United States. This makes a total of close to 300 million tons of waste. Once our MSW began to pose a threat to public health, the U.S. government began to regulate its disposal. In 1976, the Resource Conservation and Recovery Act (RCRA) was passed. This act forbade the open dumping practices of the past and introduced the concept of landfills that are sanitary and protect public health. The U.S. Environmental Protection Agency (EPA) is responsible for enforcing RCRA and all of its statutes. The EPA also officially endorses several practices aimed at the reduction of the volume and toxicity of MSW:

- source reduction—product package reuse or redesign
- recycling and composting
- waste combustion with energy recovery

Included in MSW are paperboard, paper, wood, glass, plastic, metal, food wastes and yard trimmings. MSW can be residential, industrial, or commercial. RCRA stipulates that these wastes be taken to a Subtitle D landfill. Subtitle D landfills have several technical requirements aimed to protect public health, especially by protecting groundwater supplies. First, when a landfill is sited, it must be located in a site well above groundwater sources. Test wells must be installed to test the groundwater in the area. Once construction begins, a leachate collection system must be installed to collect liquids that make their way through the landfill that could otherwise

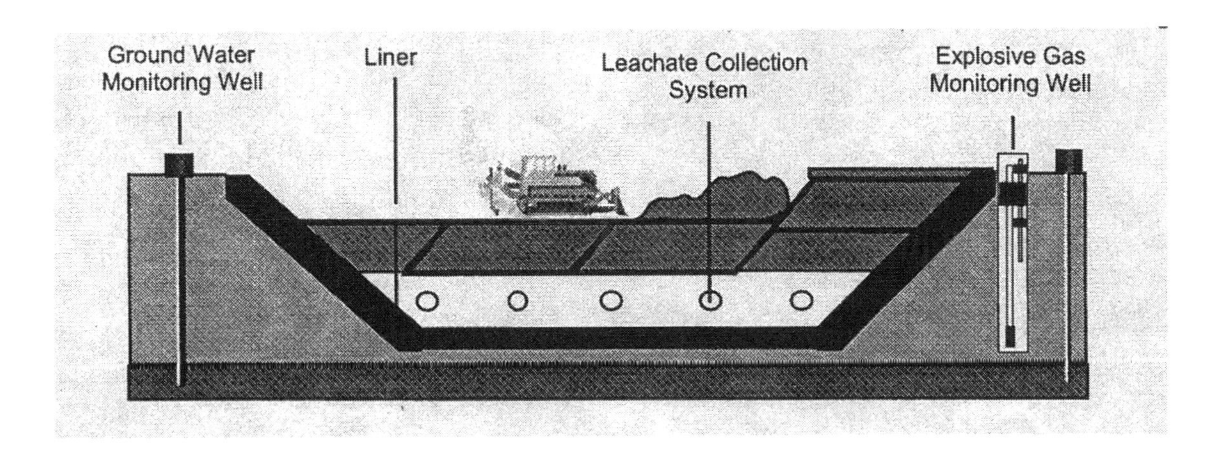

Figure 12.1. Cross section of a municipal solid waste landfill

make it into groundwater. This collection system is installed above a thick layer of compacted clay or bentonite. Above the collection system are several measures to prevent leachate from leaving the landfill. A plastic liner, geotextile mat, and washed rock must all be in place. In addition, each day, workers at the landfill must cover that day's refuse with a layer of soil, thus creating a "landfill cell."

Leachate from the landfill is typically pumped into a holding pond or large storage tank. The leachate can then be resprayed over the waste in dry weather.

Getting solid waste to the landfill is the responsibility of the government, though the government contracts out to waste disposal companies. These disposal companies use collection vehicles to remove waste from bins set out by their customers. Many modern vehicles compact the waste prior to taking it to the landfill. Compaction vehicles reduce the volume of waste in the landfill and reduce the number of trips a collection vehicle must make to the landfill. There are many rural areas that do not have collection companies because the population density does not make it possible for them to be profitable. In these cases, the community members can either take their waste directly to the landfill or use a transfer station. Transfer stations are sites where waste is collected, and sometimes compacted, before being taken to a Subtitle D landfill. Some transfer stations offer recycling bins where items can be sorted.

On a national level, about 55% of MSW is landfilled, 30% is recycled, and the remainder is combusted. Regional percentages vary quite a bit. For example, in the Southeast, almost no waste is combusted whereas, in the Northeast, a large percentage of waste is combusted. Combustion can be more expensive, but it is employed in areas of the country where land is less available for landfill construction. In the Northeast, there is less land available for landfill, so there is more combustion. In fact, of the approximately 2,400 landfills in the nation, only 8.7% are in the Northeast. To address the expense of combustion of waste, communities in the Northeast have more curbside recycling operations, thus reducing the volume of MSW.

RECYCLING

The Northeast serves 80% of its population with curbside recycling programs. California and Oregon have curbside recycling available to 85% of their residents. The Midwest and the South serve fewer than 45%

of their residents with curbside recycling. However, these numbers promise to increase. Recycling facilities collect recyclables and batch, meaning prepare, them to go back into the market place. Materials come into the facility on trucks and then are tipped onto the floor where a magnet can remove steel objects. The rest of the materials are pushed onto a conveyor belt where they are sorted: paper is sorted from aluminum and plastic (further sorted by number), and glass is sorted by color.

The profile of materials that make their way to recycling facilities is similar to the percentages of materials seen in a typical landfill. Most of the recovery, greater than 50%, is paper and paperboard, followed by metals (11%), glass (6%), and plastics (2%) by weight.

COMPOSTING

Composting is the process of degradation of organic matter by microorganisms into a humus material. The resulting material is an excellent soil conditioner because it greatly increases water retention. Compost can be useful in wetland mitigation, mulches, and even as a low-grade fertilizer.

COMBUSTION

Combustion coupled to energy production is called WTE. WTE facilities occasionally sort out recyclables, but frequently they mass burn. Facilities are well designed to minimize emissions. Materials are burned in multiple hearth furnaces and are then passed through bag houses and economizers to reduce particulate matter. Before venting to the atmosphere, the air is passed through smoke stacks with chemical scrubbers to further reduce the toxicity of emissions.

HAZARDOUS WASTES

Hazardous wastes are also regulated under RCRA. These wastes are discarded materials that are solids or liquids with substances fatal in low concentration, toxic, carcinogenic, mutagenic or teratogenic. Also included are wastes that are corrosive, explosive, reactive, and/or flammable. U.S. industries generate almost 300 million tons of hazardous waste each year, a similar amount to MSW quantities. The largest source of these wastes comes from chemical and petroleum companies. These hazardous wastes may pose a substantial threat or potential danger to human health or the environment when improperly handled. To be sure hazardous wastes are handled properly, RCRA instituted a "cradle to grave" system. This system identifies hazardous wastes, tracks them from the point of generation to the point of disposal, and assures that treatment, storage, and disposal facilities (TSDs) meet minimum national standards. In addition, RCRA requires that TSDs are properly maintained post closure.

Acts of Congress must be reauthorized periodically, and these reauthorizations are excellent opportunities to modify and improve the original act. In 1984, amendments were made to RCRA when it was reauthorized and its powers expanded. This set of amendments was called the Hazardous and Solid Waste Amendments (HSWA). The HSWA focused on protecting groundwater by 1) mandating stricter requirements for landfills accepting hazardous waste, 2) requiring that waste be treated prior to disposal 3) increasing the number of companies that fall under hazardous waste regulations. The HSWA included additional requirements, but the three noted here were some of the most important.

TREATMENT OF HAZARDOUS WASTES PRIOR TO DISPOSAL

Ideally, wastes can be reduced or minimized at the source or recycled back into manufacturing processes. However, when hazardous waste remains, it must first be treated prior to permanent disposal. Several treatment options exist, including biological, chemical, physical, and thermal. There are myriad species of living organisms, mostly microbes, with metabolic pathways that can detoxify and in some cases mineralize hazardous organic chemicals. Exposing hazardous wastes to this biological treatment is quite common, especially when waste is in the form of sludges that can easily be aerated, leading to the growth of aerobic microbes being promoted. Chemical treatment is also a valid way to lessen the toxicity of hazardous wastes. Reagents can be used to alter a characteristic of the waste, such as ignitability or corrosivity, or to form a less-toxic substance. Physical treatment of waste can be used to reduce the volume of toxic waste by separating a toxic organic layer from a nontoxic aqueous layer, for example. Finally, thermal treatment (combustion) may be employed to treat hazardous wastes.

SUPERFUND

In December 1980, Congress passed the Comprehensive Environmental Response, Compensation, and Liability Act (CERCLA). This act, also known as "Superfund," authorized the federal government to spend $1.6 billion over a five-year period for emergency cleanup activities of sites that were in existence with no clear financially responsible party at the ready. The use of the term "Superfund" is attributed to the fact that the bill created a trust fund financed primarily by excise taxes on chemicals and oil, and an environment tax on corporations.

After identification of a site that was posing an immediate threat to public health, money from the Superfund could be used for remediation. The money would be replenished by lawsuits that the EPA would file against whomever it identified as principle responsible parties. Once Superfund was enacted, however, it became apparent that there were too many sites (more than 30,000) to address at once, so the National Priority List (NPL) was established. The NPL is a list of Superfund sites that are deemed the most hazardous based on a hazard-ranking system (HRS). The factors used to rank the sites include the waste characteristics, the distance to the local population, surface water, groundwater, and drinking water supplies.

Superfund sites anywhere in the United States may be accessed using the following website: http://www.epa.gov/superfund/sites/

Excerpts from the *RCRA Orientation Manual*

By the U.S. Environmental Protection Agency

OVERVIEW

Congress enacted the Solid Waste Disposal Act of 1965 to address the growing quantity of solid waste generated in the United States and to ensure its proper management. Subsequent amendments to the Solid Waste Disposal Act, such as RCRA, have substantially increased the federal government's involvement in solid waste management.

During the 1980s, solid waste management issues rose to new heights of public concern in many areas of the United States because of increasing solid waste generation, shrinking disposal capacity, rising disposal costs, and public opposition to the siting of new disposal facilities. These solid waste management challenges continue today, as many communities are struggling to develop cost-effective, environmentally protective solutions. The growing amount of waste generated has made it increasingly important for solid waste management officials to develop strategies to manage wastes safely and cost-effectively.

WHAT IS SOLID WASTE?

- Garbage
- Refuse
- Sludges from waste treatment plants, water supply treatment plants, or pollution control facilities
- Nonhazardous industrial wastes
- Other discarded materials, including solid, semisolid, liquid, or contained gaseous materials resulting from industrial, commercial, mining, agricultural, and community activities.

RCRA encourages environmentally sound solid waste management practices that maximize the reuse of recoverable material and foster resource recovery. Solid waste is predominately regulated by state and local governments. EPA has, however, promulgated some regulations pertaining to solid waste, largely addressing how disposal facilities should be designed and operated. EPA's primary role in solid waste management includes setting national goals, providing leadership and technical assistance, and developing guidance and educational materials. The Agency has played a major role in this program by developing tools and information through policy and guidance to empower local governments, business, industry, federal agencies, and individuals to make better decisions in dealing with solid waste issues. The Agency strives to motivate behavioral change in solid waste management through nonregulatory approaches.

This section presents an outline of the RCRA solid waste program. In doing so, it defines the terms solid waste and municipal solid waste, and it describes the role EPA plays in assisting waste officials in dealing with solid waste management problems. The section will provide an overview of the criteria that EPA has developed for solid waste landfills, and will introduce some Agency initiatives designed to promote proper and efficient solid waste management.

DEFINITION OF SOLID WASTE

RCRA defines the term **solid waste** as:

- Garbage (e.g., milk cartons and coffee grounds)
- Refuse (e.g., metal scrap, wall board, and empty containers)
- Sludges from waste treatment plants, water supply treatment plants, or pollution control facilities (e.g., scrubber slags)
- Industrial wastes (e.g., manufacturing process wastewaters and nonwastewater sludges and solids)
- Other discarded materials, including solid, semisolid, liquid, or contained gaseous materials resulting from industrial, commercial, mining, agricultural, and community activities (e.g., boiler slags).

The definition of solid waste is not limited to wastes that are physically solid. Many solid wastes are liquid, while others are semisolid or gaseous.

The term solid waste, as defined by the Statute, is very broad, including not only the traditional nonhazardous solid wastes, such as municipal garbage and industrial wastes, but also hazardous wastes. Hazardous waste, a subset of solid waste, is regulated under RCRA Subtitle C. RCRA addresses solid wastes, including those hazardous wastes that are excluded from the Subtitle C regulations (e.g., household hazardous waste), and hazardous waste generated by **conditionally exempt small quantity generators (CESQGs).** [...] For purposes of regulating hazardous wastes, EPA established by regulation a separate definition of solid waste. [...]

MUNICIPAL SOLID WASTE

Municipal solid waste is a subset of solid waste and is defined as durable goods (e.g., appliances, tires, batteries), nondurable goods (e.g., newspapers, books, magazines), containers and packaging, food wastes, yard trimmings, and miscellaneous organic wastes from residential, commercial, and industrial nonprocess sources (see Figure 1).

Municipal solid waste generation has grown steadily over the past 35 years from 88 million tons per year (2.7 pounds per person per day) in 1960, to 236 million tons per year (4.4 pounds per person per day) in 2003. While generation of waste has grown steadily, recycling has also greatly increased. In 1960, only about 7 percent of municipal solid waste was recycled. By 2003, this figure had increased to 30 percent.

To address the increasing quantities of municipal solid waste, EPA recommends that communities adopt "integrated waste management" systems tailored to meet their needs. The term "integrated waste management" refers to the complementary use of a variety of waste management practices to safely and effectively handle the municipal solid waste stream. An integrated waste management system will contain some or all of the following elements: source reduction, recycling (including composting), waste combustion, and/or landfilling. In designing systems, EPA encourages communities to consider these components in an hierarchical sequence. The hierarchy favors source reduction to reduce both the volume and toxicity of waste and to increase the useful life of manufactured products. The next preferred tier in the hierarchy is recycling, which includes composting of yard and food wastes. Source reduction and recycling are preferred over the third tier of the hierarchy, which consists of combustion and/or landfilling, because they divert waste from the third tier and they have positive impacts on both the environment and economy. The goal of EPA's approach is to use a combination of all these methods to safely and effectively manage municipal solid waste. EPA recommends that communities tailor their systems from the four components in the three tiers to meet their individual needs, looking first to source reduction, and second to recycling as preferences to combustion and/or landfilling (see Figure 2).

Source Reduction

Rather than managing waste after it is generated, **source reduction** changes the way products are made and used in order to decrease waste generation. Source reduction, also called waste prevention, is defined as the design, manufacture, and use of products in a way that reduces the quantity and toxicity of waste produced when the products reach the end of their useful lives. The ultimate goal of source reduction is to decrease the amount and the toxicity of waste generated. Businesses, households, and state and local governments can all play an active role in source reduction. Businesses can manufacture products with packaging that is reduced in both volume and toxicity. They also can reduce waste by altering their business practices (e.g., reusing packaging for shipping, making double-sided copies, maintaining equipment to extend its useful life,

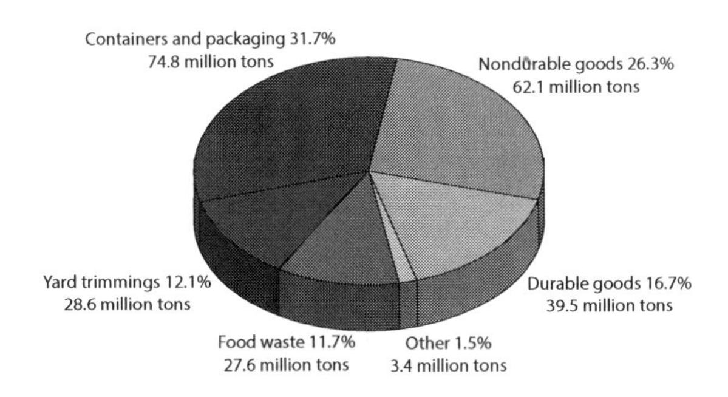

Figure 1. Products Generated in MSW by Weight, 2003 (total weight—236 million tons)

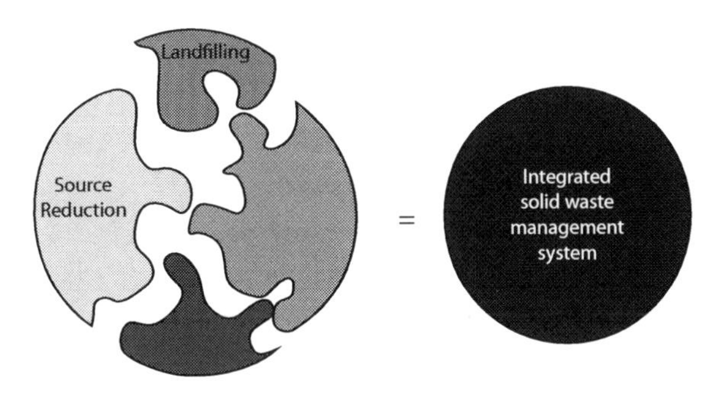

Source reduction, landfilling, recycling, and combustion are all pieces of the solid waste management puzzle. Source reduction and recycling are preferred elements of the system.

Figure 2. The Solid Waste Management Hierarchy

using reusable envelopes). Community residents can help reduce waste by leaving grass clippings on the lawn or composting them with other yard waste in their backyards, instead of bagging such materials for eventual disposal. Consumers play a crucial role in an effective source reduction program by purchasing products having reduced packaging or that contain reduced amounts of toxic constituents. This purchasing subsequently increases the demand for products with these attributes.

Recycling

Municipal solid waste **recycling** refers to the separation and collection of wastes, their subsequent transformation or remanufacture into usable or marketable products or materials, and the purchase of products made from recyclable materials. In 2003, 30.6 percent (72.3 million tons) of the municipal solid waste generated in the United States was recycled (see Figure 3). Solid waste recycling:

- Preserves raw materials and natural resources
- Reduces the amount of waste that requires disposal
- Reduces energy use and associated pollution
- Provides business and job opportunities
- Reduces greenhouse gas emissions
- Reduces pollution associated with use of virgin materials.

Communities can offer a wide range of recycling programs to their residents, such as drop-off centers, curbside collection, and centralized composting of yard and food wastes.

Composting processes are designed to optimize the natural decomposition or decay of organic matter, such as leaves and food. Compost is a humus-like material that can be added to soils to increase soil fertility, aeration, and nutrient retention. Composting can serve as a key component of municipal solid waste recycling activities, considering that food and yard wastes accounted for 23.8 percent of the total amount of municipal solid waste generated in 2003. Some communities are implementing large-scale composting programs in an effort to conserve landfill capacity.

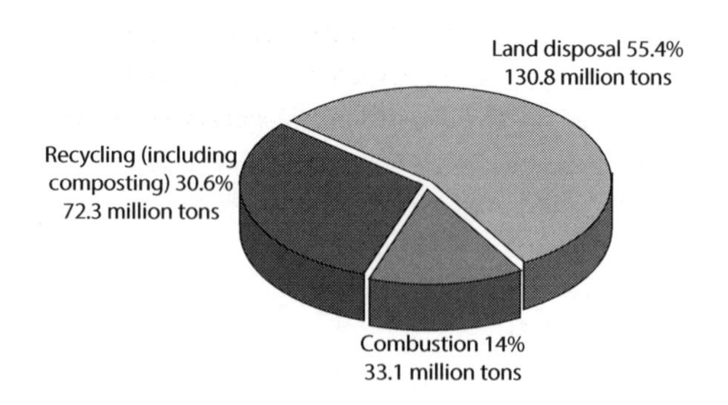

Figure 3. Management of MSW in the U.S., 2003 (total weight = 236 million tons)

For recycling to be successful, the recovered material must be reprocessed or remanufactured and the resulting products bought and used by consumers. Recycling programs will become more effective as markets increase for products made from recycled material. The federal government has developed several initiatives in order to bolster the use of recycled products. EPA's federal procurement guidelines, authorized by RCRA Subtitle F, are designed to bolster the market for products manufactured from recycled materials. The procurement program uses government purchasing to spur recycling and markets for recovered materials. [...]

Combustion

Confined and controlled burning, known as **combustion,** can not only decrease the volume of solid waste destined for landfills, but can also recover energy from the waste-burning process. Modern waste-to-energy facilities use energy recovered from combustion of solid waste to produce steam and electricity.

In 2003, combustion facilities handled 14 percent (33.1 million tons) of the municipal solid waste generated (see Figure II.3). Used in conjunction with source reduction and recycling, combustion can recover resources and materials and greatly reduce the volume of wastes entering landfills.

Landfilling

Landfilling of solid waste still remains the most widely used waste management method. Americans landfilled approximately 55.4 percent (130.8 million tons) of municipal solid waste in 2003 (see Figure 3). Many communities are having difficulties siting new landfills, largely as a result of increased citizen concerns about the potential risks and aesthetics associated with having a landfill in their neighborhoods. To reduce risks to health and the environment, EPA developed minimum criteria that solid waste landfills must meet.

INDUSTRIAL WASTE

Industrial waste is also a subset of solid waste and is defined as solid waste generated by manufacturing or industrial processes that is not a hazardous waste regulated under Subtitle C of RCRA. Such waste may include, but is not limited to, waste resulting from the following manufacturing processes: electric power generation; fertilizer or agricultural chemicals; food and related products or by-products; inorganic chemicals; iron and steel manufacturing; leather and leather products; nonferrous metals manufacturing or foundries; organic chemicals; plastics and resins manufacturing; pulp and paper industry; rubber and miscellaneous plastic products; stone, glass, clay, and concrete products; textile manufacturing; transportation equipment; and water treatment. Industrial waste does not include mining waste or oil and gas waste.

Each year in the United States, approximately 60,000 industrial facilities generate and dispose of approximately 7.6 billion tons of industrial solid waste. Most of these wastes are in the form of wastewaters (97%). EPA has, in partnership with state and tribal representatives and a focus group of industry and public interest stakeholders, developed a set of recommendations and tools to assist facility managers, state and tribal regulators, and the interested public in better addressing the management of land-disposed, nonhazardous industrial wastes.

Similarly to municipal solid waste, EPA recommends considering pollution prevention options when designing an industrial waste management system. Pollution prevention will reduce waste disposal needs and can minimize impacts across all environmental media. Pollution prevention can also reduce the volume and toxicity of waste. Lastly, pollution prevention can ease some of the burdens, risks, and liabilities of waste management. As with municipal solid waste, EPA recommends a hierarchical approach to industrial waste management: first, prevent or reduce waste at the point of generation (source reduction); second, recycle or reuse waste materials; third, treat waste; and finally, dispose of remaining waste in an environmentally protective manner (see Figure 4). There are many benefits of pollution prevention activities, including protecting human health and the environment, cost savings, simpler design and operating conditions, improved worker safety, lower liability, higher product quality, and improved community relations.

When implementing pollution prevention, industrial facilities should consider a combination of options that best fits the facility and its products. There are a number of steps common to implementing any facility-wide pollution prevention effort. An essential starting point is to make a clear commitment to identifying and taking advantage of pollution prevention opportunities. Facilities should seek the participation of interested partners, develop a policy statement committing the industrial operation to pollution prevention, and organize a team to take responsibility for it. As a next step, facilities should conduct a thorough pollution prevention opportunity assessment. Such an assessment will help set priorities according to which options are the most promising. Another feature common to many pollution prevention programs is measuring the program's progress. The actual pollution prevention practices implemented are the core of a program. The following sections give a brief overview of these core activities: source reduction, recycling, and treatment.

Source Reduction

Source reduction is the design, manufacture, and use of products in a way that reduces the quantity and toxicity of waste produced when the products reach the end of their useful lives. Source reduction activities for industrial waste include equipment or technology modifications; process or procedure modifications; reformulations or redesign of products; substitution of less-noxious product materials; and improvements in housekeeping, maintenance, training, or inventory control.

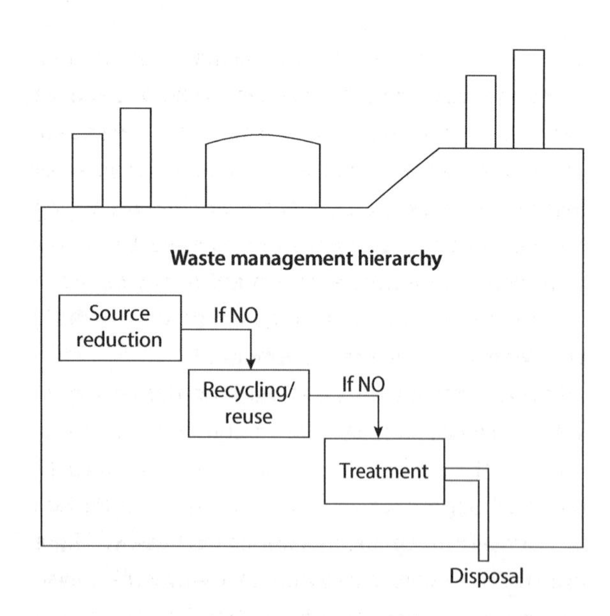

Figure 4. Waste Management Hierarchy

One source reduction option is to reformulate or redesign industrial products and processes to incorporate materials more likely to produce lower-risk wastes. Some of the most common practices include eliminating metals from inks, dyes, and paints; reformulating paints, inks, and adhesives to eliminate synthetic organic solvents; and replacing chemical-based cleaning solvents with water-based or citrus-based products.

Newer process technologies often include better waste reduction features than older ones. For industrial processes that predate consideration of waste and risk reduction, adopting new procedures or upgrading equipment can reduce waste volume, toxicity, and management costs. Some examples include redesigning equipment to cut losses during batch changes or during cleaning and maintenance, changing to mechanical cleaning devices to avoid solvent use, and installing more energy and material-efficient equipment.

In-process recycling involves the reuse of materials, such as cutting scraps, as inputs to the same process from which they came, or uses them in other processes or for other uses in the facility. This furthers waste reduction goals by reducing the need for treatment or disposal and by conserving energy and resources. A common example of in-process recycling is the reuse of wastewater.

Some of the easiest, most cost-effective, and most widely used waste reduction techniques are simple improvements in housekeeping. Accidents and spills generate avoidable disposal hazards and expenses. They are less likely to occur in clean, neatly organized facilities. Good housekeeping techniques that reduce the likelihood of accidents and spills include training employees to manage waste and materials properly; keeping aisles wide and free of obstructions; clearly labeling containers with content, handling, storage, expiration, and health and safety information; spacing stored materials to allow easy access; surrounding storage areas with containment berms to control leaks or spills; and segregating stored materials to avoid cross-contamination, mixing of incompatible materials, and unwanted reactions.

Recycling

Industry can benefit from recycling: the separation and collection of wastes, their subsequent transformation or remanufacture into usable or marketable products or materials, and the purchase of products made from recyclable materials.

Many local governments and states have established materials exchange programs to facilitate transactions between waste generators and industries that can recycle wastes as raw materials. Materials exchanges are an effective and inexpensive way to find new users and uses for a waste.

Recycling can involve substituting industrial byproducts for another material with similar properties. For example, coal combustion ash has value as a construction material, road base, or soil stabilizer. The ash replaces other, non-recycled materials, such as fill or Portland cement, not only avoiding disposal costs but also yielding a quality product and generating revenue. Other examples of industrial materials recycling include using wastewaters and sludges as soil amendments and using foundry sand in asphalt, concrete, and roadbed construction. Many regulatory agencies require approval of planned recycling activities and may require testing of the materials to be reused. Others may allow certain by-products to be designated for recycling, as long as the required analyses are completed. Generally, regulatory agencies want to ensure that recycled materials are free from constituents that might pose a greater risk than the materials they are replacing. Industrial facilities should consult with the state agency for criteria and regulations governing recycling before implementing this option.

EPA is targeting industrial materials for recycling as part of the Resource Conservation Challenge (RCC). Through the RCC, EPA forms voluntary partnerships with industries to encourage them to generate less waste and recycle by-products through environmentally sound practices. The objective is to achieve the economic and environmental benefits of recycling industrial byproducts as inputs to new products and to extend the useful life of landfills, conserve virgin materials, and reduce energy use and associated greenhouse gas emissions. EPA is pursuing four broad strategies in increasing the beneficial reuse of industrial materials: analyzing and characterizing the target materials; identifying environmentally safe and beneficial practices; identifying incentives and barriers to beneficial reuse; and increasing outreach and education on the benefits of source reduction, recycling, and beneficially using wastes/materials. [...]

Treatment

Treatment of nonhazardous industrial waste is not a federal requirement. However, it can help to reduce the volume and toxicity of waste prior to disposal. Treatment can also make a waste amenable for reuse or recycling. Consequently, a facility managing nonhazardous industrial waste might elect to apply treatment. For example, treatment might be incorporated to address volatile organic compound (VOC) emissions from a waste management unit, or a facility might elect to treat a waste so that a less stringent waste management system design could be used. Treatment involves changing a waste's physical, chemical, or biological character or composition through designed techniques or processes. There are three primary categories of treatment—physical, chemical, and biological. Physical treatment involves changing the waste's physical properties such as its size, shape, density, or state (i.e., gas, liquid, solid). Physical treatment does not change a waste's chemical composition. One form of physical treatment, immobilization, involves encapsulating waste in other materials, such as plastic, resin, or cement, to prevent constituents from volatilizing or leaching. Listed below are a few examples of physical treatment:

- Immobilization, including encapsulation and thermoplastic binding
- Carbon absorption, including granular activated carbon and powdered activated carbon
- Distillation, including batch distillation, fractionation, thin film extraction, steam stripping, thermal drying, and filtration Evaporation/volatilization
- Grinding
- Shredding
- Compacting
- Solidification/addition of absorbent material.

Chemical treatment involves altering a waste's chemical composition, structure, and properties through chemical reactions. Chemical treatment can consist of mixing the waste with other materials (reagents), heating the waste to high temperatures, or a combination of both. Through chemical treatment, waste constituents can be recovered or destroyed. Listed below are a few examples of chemical treatment:

- Neutralization
- Oxidation
- Reduction
- Precipitation
- Acid leaching
- Ion exchange
- Incineration
- Thermal desorption
- Stabilization
- Vitrification
- Extraction, including solvent extraction and critical extraction
- High temperature metal recovery.

Biological treatment can be divided into two categories—aerobic and anaerobic. Aerobic biological treatment uses oxygen-requiring microorganisms to decompose organic and non-metallic constituents into carbon dioxide, water, nitrates, sulfates, simpler organic products, and cellular biomass (i.e., cellular growth and reproduction). Anaerobic biological treatment uses microorganisms, in the absence of oxygen, to transform organic constituents and nitrogen-containing compounds into oxygen and methane gas (CH_4). Anaerobic biological treatment typically is performed in an enclosed digestor unit.

The range of treatment methods from which to choose is as diverse as the range of wastes to be treated. More advanced treatment will generally be more expensive, but by reducing the quantity and risk level of the waste, costs might be reduced in the long run. Savings could come from not only lower disposal costs, but also lower closure and post-closure care costs. Treatment and post-treatment waste management methods can be selected to minimize both total cost and environmental impact, keeping in mind that treatment residuals, such as sludges, are wastes themselves that will need to be managed.

Landfilling

As with municipal solid waste, industrial facilities will not be able to manage all of their industrial waste by source reduction, recycling, and treatment. Landfilling is the least desirable option, and should be implemented as part of a comprehensive waste management system. Implementing a waste management system that achieves protective environmental operations requires incorporating performance monitoring and measurement of progress towards environmental goals. An effective waste management system can help ensure proper operation of the many interrelated systems on which a unit depends for waste containment, leachate management, and other important functions. If the elements of an industrial waste landfill are not regularly inspected, maintained, improved, and evaluated for efficiency, even the best designed unit might not operate efficiently. Implementing an effective waste management system can also reduce long- and short-term costs, protect workers and local communities, and maintain good community relations.

Industrial waste landfills can face opposition as a result of concerns about possible negative aesthetic impact and potential health risks. To reduce risks to health and the environment, EPA developed minimum criteria that industrial waste landfills must meet. The federal criteria for nonhazardous industrial waste facilities or practices are provided in 40 CFR Part 257, Subparts A and B. The criteria for solid waste disposal facilities are discussed in the next section.

Guide for Industrial Waste Management

EPA, in close collaboration with state and tribal representatives through the Association of State and Territorial Solid Waste Management Officials (ASTSWMO), and a focus group of industry and public interest stakeholders, developed a set of recommendations and tools to assist facility managers, state and tribal regulators, and the interested public in better addressing the management of land-disposed, nonhazardous industrial wastes. The *Guide for Industrial Waste Management* (EPA530-R-03-001) provides considerations and Internet-based tools for siting industrial waste management units; methods for characterizing waste constituents; fact sheets and Web sites with information about individual waste constituents; tools to assess possible risks posed by the wastes; principles for building stakeholder partnerships; opportunities for waste minimization; guidelines for safe unit design; procedures for monitoring surface water, air, and ground water; and recommendations for closure and post-closure care.

CRITERIA FOR SOLID WASTE DISPOSAL FACILITIES

One of the initial focuses of the Solid Waste Disposal Act (as amended by RCRA) was to require EPA to study the risks associated with solid waste disposal and to develop management standards and criteria for solid waste disposal units (including landfills) in order to protect human health and the environment. This study resulted in the development of criteria for classifying solid waste disposal facilities and practices.

Criteria for Classification of Solid Waste Disposal Facilities and Practices

On September 13, 1979, EPA promulgated criteria to designate solid waste disposal facilities and practices which would not pose adverse effects to human health and the environment (Part 257, Subpart A). Facilities failing to satisfy the criteria are considered **open dumps** requiring attention by state solid waste programs. As a result, open dumps had to either be closed or upgraded to meet the criteria for sanitary landfills. States were also required to incorporate provisions into their solid waste programs to prohibit the establishment of new open dumps.

WHAT IS AN OPEN DUMP?

An open dump is defined as a disposal facility that does not comply with one or more of the Part 257 or Part 258 criteria. Using the Part 257, Subpart A criteria as a benchmark, each state evaluated the solid waste disposal facilities within its borders to determine which facilities were open dumps that needed to be closed or upgraded. For each open dump, the state completed an Open Dump Inventory Report form that was sent to the Bureau of the Census. At the end of fiscal years 1981 through 1985, the Bureau compiled all of the report forms and sent them to EPA, where they were summarized and published annually.

Technical Criteria for Solid Waste Disposal Facilities

The Part 257, Subpart A regulatory criteria used to classify solid waste disposal facilities and practices consist of general environmental performance standards. The criteria contain provisions designed to ensure that wastes disposed of in solid waste disposal units will not threaten endangered species, surface water, ground water, or flood plains. Further, owners and operators of disposal units are required to implement public health and safety precautions such as disease vector (e.g., rodents, flies, mosquitoes) controls to prevent the spread of disease and restrictions on the open burning of solid waste. In addition, facilities are required to install safety measures to control explosive gases generated by the decomposition of waste, minimize the attraction of birds to the waste disposed in the unit, and restrict public access to the facility. The criteria also restrict the land spreading of wastes with high levels of cadmium and polychlorinated biphenyls (PCBs) in order to adequately protect ground water from these dangerous contaminants.

These criteria serve as minimum technical standards for solid waste disposal facilities. As a result, facilities must meet the Part 257 standards to ensure that ongoing waste management operations adequately protect human health and the environment. If they fail to do so, the facility is classified as an open dump and must upgrade its operations or close. States have the option of developing standards more stringent than the Part 257, Subpart A criteria.

Technical Criteria for Municipal Solid Waste Landfills (MSWLFs)

Protection of human health and the environment from the risks posed by solid waste disposal facilities was an ongoing concern of Congress after RCRA was passed in 1976. As a result, the 1984 Hazardous and Solid Waste Amendments (HSWA) required EPA to report on the adequacy of existing solid waste disposal facility criteria (Part 258) and gather detailed data on the characteristics and quantities of nonhazardous municipal solid wastes.

Report to Congress on Solid Waste Disposal

In October 1988, EPA submitted a Report to Congress indicating that the United States was generating an increasing amount of municipal solid waste. The Report revealed that approximately 160 million tons of municipal solid waste were generated each year, 131 million tons of which were landfilled in just over 6,500 MSWLFs. EPA also reported that although these landfills used a wide variety of environmental controls, they may pose significant threats to ground water and surface water resources. For instance, rain water percolating through the landfills can dissolve harmful constituents in the waste and can eventually seep into the ground, potentially contaminating ground water. In addition, improperly maintained landfills can pose other health risks due to airborne contaminants, or the threat of fire or explosion.

To address these environmental and health concerns, and to standardize the technical requirements for these landfills, EPA promulgated revised minimum federal criteria in Part 258 for MSWLFs on October 9, 1991. The criteria were designed to ensure that MSWLFs receiving municipal solid waste would be protective of human health and the environment. All other solid waste disposal facilities and practices, besides MSWLFs, remain subject to Part 257, Subpart A.

Criteria for Municipal Solid Waste Landfills

A **municipal solid waste landfill** is defined as a discrete area of land or excavation that receives household waste. A MSWLF may also receive other types of nonhazardous wastes, such as commercial solid waste, nonhazardous sludge, conditionally exempt small quantity generator (CESQG) waste, and industrial nonhazardous solid waste. In 2002, there were approximately 1,767 MSWLFs in the continental United States.

The revised criteria in 40 CFR Part 258 address seven major aspects of MSWLFs (see Figure 5):

- Location
- Operation Design
- Ground water monitoring
- Corrective action
- Closure and post-closure
- Financial assurance.

The location criteria restrict where a MSWLF may be located. New landfills must meet minimum standards for placement in or near flood plains, wetlands, fault areas, seismic impact zones, and other unstable areas. Because some bird species are attracted to landfills, the criteria also restrict the placement of landfills near airports to reduce the bird hazards (i.e., collisions between birds and aircraft that may cause damage to the aircraft or injury to the passengers).

The operating criteria establish daily operating standards for running and maintaining a landfill. The standards dictate sound management practices that ensure protection of human health and the environment. The provisions require covering the landfill daily, controlling disease vectors, and controlling explosive gases. They also prohibit the open burning of solid waste and require the owner and operator of the landfill to control unauthorized access to the unit.

The design criteria require each new landfill to have a liner consisting of a flexible membrane and a minimum of two feet of compacted soil, as well as a leachate collection system. **Leachate** is formed when rain water filters through wastes placed in a landfill. When this liquid comes in contact with buried wastes, it leaches, or draws out, chemicals or constituents from those wastes. States with approved MSWLF permit programs can allow the use of an alternative liner design that controls ground water contamination. The liner and collection system prevent the potentially harmful leachate from contaminating the soil and ground water below the landfill.

In order to check the performance of system design, MSWLF facility managers must also establish a ground water monitoring program. Through a series of monitoring wells, the facility owner and operator is alerted if the landfill is leaking and causing contamination. If contamination is detected, the owner and operator of the landfill must perform **corrective action** (i.e., clean up the contamination caused by the landfill).

When landfills reach their capacity and can no longer accept additional waste, the criteria stipulate procedures for properly closing the facility to ensure that the landfill does not present any danger to human health and the environment in the future. The **closure** activities at the end of a facility's use are often expensive, and the owner and operator must have the ability to pay for them. As a result, the criteria require each owner and operator to prove that they have the financial resources to perform these closure and **post-closure** activities, as well as any known corrective action.

Solid waste disposal is overseen by the states, and compliance is assured through state-issued permits. Each state is to obtain EPA approval for their MSWLF permitting program. This approval process assesses whether a state's program is sufficient to ensure each landfill's compliance with the criteria. In addition to the minimum federal criteria, some states may impose requirements that are more stringent than the federal requirements.

Conditionally Exempt Small Quantity Generator Waste Disposal Facilities

In July of 1996, EPA promulgated standards for non-municipal, nonhazardous waste facilities that may receive conditionally exempt small quantity generator (CESQG) waste (40 CFR Part 257, Subpart B). These revisions addressed location restrictions, requirements for monitoring for ground water contamination, and corrective action provisions to clean up any contamination. [...]

Bioreactor Landfills

EPA is investigating the feasibility of improving how waste is managed in MSWLFs. Projects are being conducted to assess bioreactor landfill technology. A bioreactor landfill operates to more rapidly transform and degrade organic waste. The increase in waste degradation and stabilization is accomplished through the addition of liquid and air to enhance microbial processes. This bioreactor concept differs from the traditional "dry tomb" municipal landfill approach. Thus, decomposition and biological stabilization of the waste in a bioreactor landfill can occur in a shorter time frame than occurs in a traditional landfill, providing a potential decrease in long-term environmental risks and landfill operating and post-closure costs.

Additional information about bioreactor landfills can be found at www.epa.gov/epaoswer/non-hw/muncpl/landfill/bioreactors.htm.

ASSISTANCE TO NATIVE AMERICAN TRIBES

EPA developed a municipal solid waste strategy to assist Native American tribes in the establishment of healthy, environmentally protective, integrated solid waste management practices on tribal lands. The initial strategy was based on input from tribal focus groups convened by the National Tribal Environmental Council and discussions with tribal organizations, EPA Regional Indian Program Coordinators, other EPA offices, and other federal agencies with trust responsibilities on Native American lands. The strategy emphasizes building tribal municipal solid waste management capacity, developing tribal organizational infrastructure, and building partnerships among tribes, states, and local governments. Direct EPA support of these goals includes technical assistance, grant funding, education, and outreach.

Solid waste managers on Native American lands face unique challenges. To address issues such as jurisdiction, funding, and staffing, EPA offers several resource guides featuring in-depth information specific to Native American lands. The Agency recognizes that every solid waste management program needs funding to survive and that, in an era of tightening budgets, it may be difficult to find necessary resources. One of EPA's ongoing priorities is to make current information available to help tribes locate the funding they need to develop and implement safe and effective solid waste programs.

One such initiative is the *Tribal Waste Journal*. The journal contains in-depth information on a variety of solid and hazardous waste topics including interviews with representatives from Native American Tribes and Alaskan Native Villages. Each issue focuses on a single topic and presents ideas, approaches, and activities that other Native American Tribes and Alaskan Native Villages have successfully employed.

Additionally, EPA has initiated the Tribal Open Dump Cleanup Project to assist tribes with closure or upgrade of open dump sites. The project is part of a Tribal Solid Waste Interagency Workgroup, which is working to coordinate federal assistance for tribal solid waste management programs. The cleanup project's specific goals include assisting tribes with 1) proposals to characterize/assess open dumps; 2) proposals to develop Integrated Solid Waste Management (ISWM) Plans and Tribal Codes and regulations; 3) proposals to develop and implement alternative solid waste management activities/ facilities; and 4) proposals to develop and implement closure and post-closure programs.

Outreach and education materials are two other tools EPA provides to tribes to support environmentally sound integrated solid waste management practices. The Agency's outreach support helps tribes connect and learn from each other's experiences. Educational resources help tribal leadership as well as the general tribal community understand the importance of good municipal solid waste management. Better understanding ensures that tribal municipal solid waste programs are assigned a high priority and facilitates the communities' adoption of new and improved waste disposal practices.

OTHER SOLID WASTE MANAGEMENT INITIATIVES

Along with the Resource Conservation Challenge, [...] EPA has developed a number of solid waste management initiatives to help facilitate and promote proper waste management, and encourage source reduction by both industry and the public. Several are described below.

Jobs Through Recycling Program

The Jobs Through Recycling (JTR) program was developed in 1994 with the intent to foster recycling market development primarily by awarding grants to state government agencies, tribal authorities, and regional nonprofit organizations. However, due to funding cutbacks, JTR now operates exclusively by facilitating information exchange and providing networking opportunities via a Web site and e-mail list server. The list server, called JTRnet, allows market development officials to share insights and seek advice on problems and issues facing recycling programs in their states and regions. The Web site is available at www.epa.gov/jtr, and includes information on commodities, financing, business assistance, and profiles of the past JTR grants. It has information on the economic benefits of recycling and market development information for all 50 states.

Between 1994 and 1999, the JTR program provided "seed" funding totaling approximately $8 million through grants to states, tribes, and territories. These grants were awarded through a national competitive process, managed by a joint EPA Headquarters and regions team. Based on reported results, JTR funding helped create more than 8,500 new jobs, $640.5 million in capital investments, and 14 million tons of recovered materials. One job was created for every $1,000 of grant money invested.

Pay-As-You-Throw (PAYT)

Some communities are using economic incentives to encourage the public to reduce solid waste sent to landfills. One of the most successful economic incentive programs used to achieve source reduction and recycling is variable rate refuse pricing, or unit pricing. Unit pricing programs, sometimes referred to as pay-as-you-throw systems, have one primary goal: customers who place more solid waste at the curb for disposal pay more for the collection and disposal service. Thus, customers who recycle more have less solid waste for disposal and pay less. There are a few different types of unit pricing systems. Most require customers to pay a per-can or per-bag fee for refuse collection and require the purchase of a special bag or tag to place on bags or cans. Other systems allow customers to choose between different size containers and charge more for collection of larger containers. EPA's role in the further development of unit pricing systems has been to study effective systems in use and to disseminate documentation to inform other communities about the environmental and economic benefits that unit pricing may have for their community. The number of communities using unit pricing grew to more than 4,033 in 1999, and the population served has more than tripled since 1990 to over 35 million today.

Additional information about unit pricing or pay-as-you-throw programs is available at www. epa. gov/payt.

Full Cost Accounting for Municipal Solid Waste

Full cost accounting is an additional financial management tool that communities can use to improve solid waste management. Full cost accounting is an accounting approach that helps local governments identify all direct and indirect costs, as well as the past and future costs, of a MSW management program. Full cost accounting helps solid waste managers account for all monetary costs of resources used or committed, thereby providing the complete picture of solid waste management costs on an ongoing basis. Full cost accounting can help managers identify high-cost activities and operations and seek ways to make them more cost-effective.

EPA is continually studying these and other programs in order to assist communities in deciding whether one of these programs is right for them. In addition to these initiatives, EPA has published numerous guidance documents designed to educate both industry and the public on the benefits of source reduction, to guide communities in developing recycling programs, and to educate students on the benefits and elements of source reduction and recycling.

Additional information about full cost accounting can be found at www.epa.gov/fiillcost.

Construction and Demolition Materials

Under its Resource Conservation Challenge, EPA's Industrial Materials Recycling Program is supporting projects to reduce, reuse, and recycle waste materials generated from building construction, renovation, deconstruction, and demolition. Construction and demolition materials commonly include concrete, asphalt, wood, glass, brick, metal, insulation, and furniture. From incorporating used or environmentally friendly materials into a building's construction or renovation to disassembling structures for the reuse and recycling of their components, each phase of a building's life cycle offers opportunities to reduce waste.

Additional information about construction and demolition is available at www.epa.gov/epaoswer/non-hw/debris-new/index.htm. [...]

Industrial Ecology

The study of material and energy flows and their transformations into products, by-products, and waste throughout industrial and ecological systems is the primary concept of industrial ecology. This initiative urges industry to seek opportunities for the continual reuse and recycling of materials through a system in which processes are designed to consume only available waste streams and to produce only usable waste. Wastes from producers and consumers become inputs for other producers and consumers, and resources are cycled through the system to sustain future generations. Individual processes and products become part of an interconnected industrial system in which new products or processes evolve out of or consume available waste streams, water, and energy; in turn, processes are developed to produce usable resources.

SUMMARY

The term "solid waste" includes garbage, refuse, sludges, nonhazardous industrial wastes, hazardous wastes, and other discarded materials. Subtitle D addresses primarily nonhazardous solid waste. Subtitle D also addresses hazardous wastes that are excluded from Subtitle C regulation (e.g., household hazardous waste).

Municipal solid waste, a subset of solid waste, is waste generated by businesses and households. EPA recommends an integrated, hierarchical approach to managing municipal solid waste that includes, in descending order of preference:

- Source reduction
- Recycling
- Disposal by combustion and/or landfilling.

As part of Subtitle D, EPA has developed detailed technical criteria for solid waste disposal facilities (40 CFR Part 257), including specific criteria for MSWLFs. These criteria include specific provisions for MSWLFs (40 CFR Part 258):

- Location
- Operation
- Design
- Ground water monitoring
- Corrective action
- Closure and post-closure
- Financial assurance (i.e., responsibility).

In addition, other solid waste management initiatives have been developed by EPA to help facilitate proper waste management. These initiatives focus on the environmental and economic benefits of source reduction and recycling. These initiatives include:

- Jobs through Recycling
- Pay-As-You-Throw
- Full cost accounting
- Construction and demolition materials
- Industrial ecology.

ADDITIONAL RESOURCES

Additional information about municipal solid waste management can be found at www.epa.gov/msw.

HAZARDOUS WASTE IDENTIFICATION: OVERVIEW

What is a hazardous waste? Simply defined, a **hazardous waste** is a waste with properties that makes it dangerous or capable of having a harmful effect on human health or the environment. Unfortunately, in order to develop a regulatory framework capable of ensuring adequate protection, this simple narrative definition is not enough. Determining what is a hazardous waste is paramount, because only those wastes that have specific attributes are subject to Subtitle C regulation.

Making this determination is a complex task which is a central component of the hazardous waste management regulations. Hazardous waste is generated from many sources, ranging from industrial manufacturing process wastes, to batteries, to fluorescent light bulbs. Hazardous waste may come in many forms, including liquids, solids, gases, and sludges. To cover this wide range, EPA has developed a system to identify specific substances known to be hazardous and provide objective criteria for including other materials in this universe. The regulations contain guidelines for determining what exactly is a waste (called a solid waste) and what is excluded from the hazardous waste regulations, even though it otherwise is a solid and hazardous waste. Finally, to promote recycling and the reduction of the amount of waste entering the RCRA system, EPA provides exemptions for certain wastes when they are recycled in certain ways.

This chapter introduces the hazardous waste identification process, describes how to determine if a waste is a solid waste, and provides the regulatory definition for hazardous waste. It also discusses those wastes specifically excluded from Subtitle C regulation, and those wastes exempted when recycled.

HAZARDOUS WASTE IDENTIFICATION PROCESS

Proper hazardous waste identification is essential to the success of the RCRA program. This identification process can be a very complex task. Therefore, it is best to approach the issue by asking a series of questions in a step-wise manner (see Figure 6). If facility owners and operators answer the following questions, they can determine if they are producing a hazardous waste:

- Is the material in question a solid waste?
- Is the material excluded from the definition of solid waste or hazardous waste?
- Is the waste a listed or characteristic hazardous waste?
- Is the waste delisted?

This chapter will examine these key questions.

IS THE MATERIAL A SOLID WASTE?

The Subtitle C program uses the term solid waste to denote something that is a waste. In order for a material to be classified as a hazardous waste, it must first be a solid waste. Therefore, the first step in the hazardous waste identification process is determining if a material is a solid waste.

The statutory definition points out that whether a material is a solid waste is not based on the physical form of the material (i.e., whether or not it is a solid as opposed to a liquid or gas), but rather that the material is a waste. The regulations further define **solid waste** as any material that is discarded by being either abandoned, inherently waste-like, a certain military munition, or recycled (see Figure 7).

- Abandoned—The term **abandoned** simply means thrown away. A material is abandoned if it is disposed of, burned, or incinerated.
- Inherently Waste-Like—Some materials pose such a threat to human health and the environment that they are always considered solid wastes; these materials are considered to be **inherently waste-like.** Examples of inherently waste-like materials include certain dioxin-containing wastes.
- Military Munition—**Military munitions** are all ammunition products and components produced for or used by the U.S. Department of Defense (DOD) or U.S. Armed Services for national defense and security. Unused or defective munitions are solid wastes when aban-

doned (i.e., disposed of, burned, incinerated) or treated prior to disposal; rendered nonrecyclable or nonuseable through deterioration; or declared a waste by an authorized military official. Used (i.e., fired or detonated) munitions may also be solid wastes if collected for storage, recycling, treatment, or disposal.

- Recycled—A material is **recycled** if it is used or reused (e.g., as an ingredient in a process), reclaimed, or used in certain ways (used in a manner constituting disposal, burned for energy recovery, or accumulated speculatively). [...]

Recycled Materials

Materials that are recycled are a special subset of the solid waste universe. When recycled, some materials are not solid wastes, and therefore, not hazardous wastes, while others are solid and hazardous waste, but are subject to less-stringent regulatory controls. The level of regulation that applies to recycled materials depends on the material and the type of recycling (see Figure 8). Because some types of recycling pose threats to human health and the environment, RCRA does not exempt all recycled materials from the definition of solid waste. As a result, the manner in which a material is recycled will determine whether or not the material is a solid waste and, therefore, potentially regulated as a hazardous waste. In order to encourage waste recycling, RCRA exempts three types of wastes from the definition of solid waste:

- Waste Used as an Ingredient—If a material is directly used as an ingredient in a production process without first being reclaimed, then that material is not a solid waste.
- Waste Used as a Product Substitute—If a material is directly used as an effective substitute for a commercial product (without first being reclaimed), it is exempt from the definition of solid waste.
- Wastes Returned to the Production Process — When a material is returned directly to the production process (without first being reclaimed) for use as a feedstock or raw material, it is not a solid waste.

Conversely, materials are solid wastes, and are not exempt, if they are recycled in certain ways. If these materials are used in a manner constituting disposal; burned for energy recovery; used to produce a fuel; or contained in fuels; accumulated speculatively; or are dioxin-containing wastes considered inherently waste-like; then they are defined as solid wastes.

- Used in a Manner Constituting Disposal—**Use constituting disposal** is the direct placement of wastes or products containing wastes (e.g., asphalt with petroleum-refining wastes as an ingredient) on the land.
- Burned for Energy Recovery, Used to Produce a Fuel, or Contained in Fuels—Burning hazardous waste for fuel (e.g., **burning for energy recovery)** and using wastes to produce fuels are regulated activities. Conversely, commercial products intended to be burned as fuels are not considered solid wastes. For example, off-specification jet fuel (e.g., a fuel with minor chemical impurities) is not a solid waste when it is burned for energy recovery, because it is itself a fuel.
- Accumulated Speculatively—In order to encourage recycling of wastes as well as ensure that materials are actually recycled, and not simply stored to avoid regulation, EPA established a provision to encourage facilities to recycle sufficient amounts in a timely manner. This provision designates as solid wastes those materials that are **accumulated speculatively.** A material is accumulated speculatively (e.g., stored in lieu of expeditious recycling) if it has

no viable market or if the person accumulating the material cannot demonstrate that at least 75 percent of the material is recycled in a calendar year, commencing on January 1 (see Figure 9).

- Dioxin-Containing Wastes Considered Inherently Waste-Like—Dioxin-containing wastes are considered inherently waste-like because they pose significant threats to human health and the environment if released or mismanaged. As a result, RCRA does not exempt such wastes from the definition of solid waste even if they are recycled through direct use or reuse without prior reclamation. This is to ensure that such wastes are subject to the most protective regulatory controls.

Secondary Materials

Not all materials can be directly used or reused without reclamation. A material is **reclaimed** if it is processed to recover a usable product (e.g., smelting a waste to recover valuable metal constituents), or if it is regenerated through processing to remove contaminants in a way that restores them to their usable condition (e.g., distilling dirty spent solvents to produce clean solvents). If **secondary materials** are reclaimed before use, their regulatory status depends on the type of material. For this solid waste determination process, EPA groups all materials into five categories. These secondary materials consist of spent materials, sludges, by-products, commercial chemical products (CCPs), and scrap metal.

Spent Materials

Spent materials are materials that have been used and can no longer serve the purpose for which they were produced without processing. For example, a solvent used to degrease metal parts will eventually become contaminated such that it cannot be used as a solvent until it is regenerated. If a spent material must be reclaimed, it is a solid waste and is subject

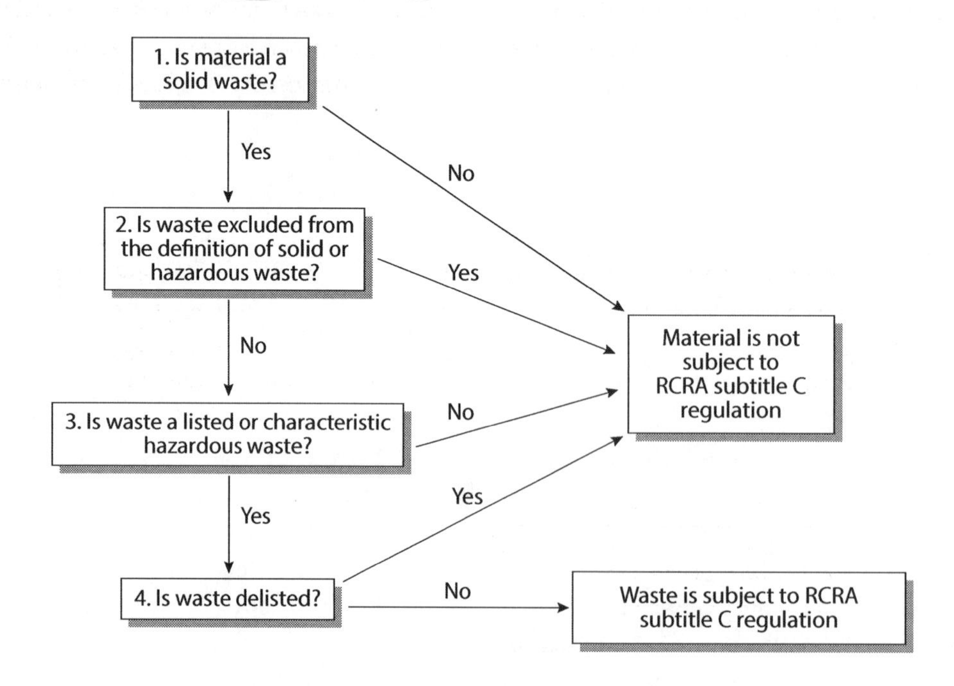

Figure 6. Hazardous Waste Identification Process

to hazardous waste regulation. Spent materials are also regulated as solid wastes when used in a manner constituting disposal; burned for energy recovery, used to produce a fuel, or contained in fuels; or accumulated speculatively (see Figure 10).

Sludges

Sludges are any solid, semisolid, or liquid wastes generated from a wastewater treatment plant, water supply treatment plant, or air pollution control device (e.g., filters or baghouse dust). Sludges from specific industrial processes or sources (known as listed sludges) are solid wastes when reclaimed; used in a manner constituting disposal; burned for energy recovery; used to produce a fuel, or contained in fuels; or accumulated speculatively. On the other hand, characteristic sludges (which are sludges that exhibit certain physical or chemical properties) are not solid wastes when reclaimed, unless they are used in a manner constituting disposal; burned for energy recovery; used to produce a fuel, or contained in fuels; or accumulated speculatively (see Figure 10). (Listings and characteristics are fully discussed later in this chapter.)

By-products

By-products are materials that are not one of the intended products of a production process. An example is the sediment remaining at the bottom of a distillation column. By-product is a catch-all term and includes most wastes that are not spent materials or sludges. Listed by-products are solid wastes when reclaimed; used in a manner constituting disposal; burned for energy recovery; used to produce a fuel, or contained in fuels; or accumulated speculatively. On the other hand, characteristic by-products are not solid wastes when reclaimed, unless they are used in a manner constituting disposal; burned for energy recovery; used to produce a fuel, or contained in fuels; or accumulated speculatively (see Figure 10).

Commercial Chemical Products

Commercial chemical products (CCPS) are unused or off-specification chemicals (e.g., chemicals that have exceeded their shelf life), spill or container residues, and other unused manufactured products that are not typically considered chemicals. CCPs are not solid wastes when reclaimed, unless they are used in a manner constituting disposal; or burned for energy

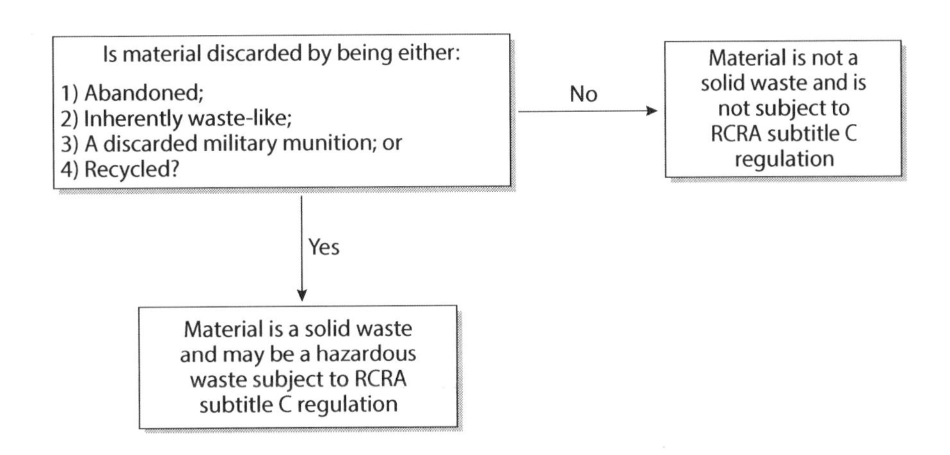

Figure 7. Is It a Solid Waste?

recovery, used to produce a fuel, or contained in fuels (see Figure 10).

Scrap Metal

Scrap metal is worn or extra bits and pieces of metal parts, such as scrap piping and wire, or worn metal items, such as scrap automobile parts and radiators. If scrap metal is reclaimed, it is a solid waste and is subject to hazardous waste regulation. [...] Scrap metal is also regulated as a solid waste when used in a manner constituting disposal; burned for energy recovery; used to produce a fuel, or contained in fuels; or accumulated speculatively. This does not apply to processed scrap metal which is excluded from hazardous waste generation entirely (as discussed later in this chapter).

SHAM RECYCLING

Sham recycling may include situations when a secondary material is:

- Ineffective or only marginally effective for the claimed use (e.g., using certain heavy metal sludges in concrete when such sludges do not contribute any significant element to the concrete's properties)
- Used in excess of the amount necessary (e.g., using materials containing chlorine as an ingredient in a process requiring chlorine, but in excess of the required chlorine levels)
- Handled in a manner inconsistent with its use as a raw material or commercial product substitute (e.g., storing materials in a leaking surface impoundment as compared to a tank in good condition that is intended for storing raw materials).

Sham Recycling

For all recycling activities, the above rules are based on the premise that legitimate reclamation or reuse is taking place. EPA rewards facilities recycling some wastes by exempting them from regulation, or by subjecting them to lesser regulation. Some facilities, however, may claim that they are recycling a material in order to avoid being subject to RCRA regulation, when in fact the activity is not legitimate recycling. EPA has established guidelines for what constitutes legitimate recycling and has described activities it considers to be illegitimate or **sham recycling.** Considerations in making this determination include whether the secondary material is effective for the claimed use, if the secondary material is used in excess of the amount necessary, and whether or not the facility has maintained records of the recycling transactions.

IS THE WASTE EXCLUDED?

Not all RCRA solid wastes qualify as hazardous wastes. Other factors must be considered before deciding whether a solid waste should be regulated as a hazardous waste. Regulation of certain wastes may be impractical or otherwise undesirable, regardless of the hazards that the waste might pose. For instance, household waste can contain dangerous chemicals, such as solvents and pesticides, but subjecting households to the strict RCRA waste management regulations would create a number of practical problems. As a result, Congress and EPA exempted or excluded certain wastes, such as household wastes, from the hazardous waste definition and regulations. Determining whether or not a waste is excluded or exempted from hazardous waste regulation is the second step in the RCRA hazardous waste identification process. There are five categories of exclusions:

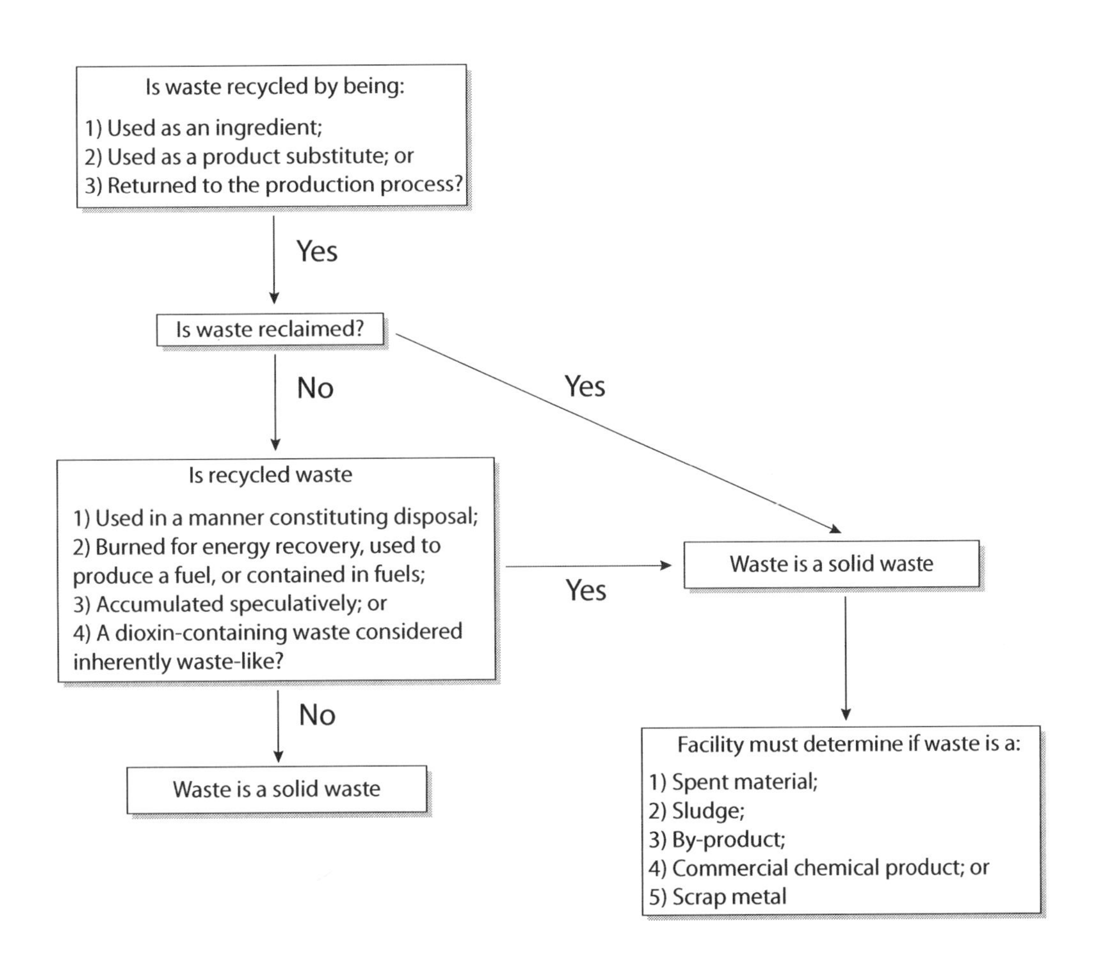

Figure 8. Are All Recycled Wastes Hazardous Wastes?

- Exclusions from the definition of solid waste
- Exclusions from the definition of hazardous waste
- Exclusions for waste generated in raw material, product storage, or manufacturing units
- Exclusions for laboratory samples and waste treatability studies
- Exclusions for dredged material regulated under the Marine Protection Research and Sanctuaries Act or the Clean Water Act.

If the waste fits one of these categories, it is not regulated as a RCRA hazardous waste, and the hazardous waste requirements do not apply.

IS THE WASTE A CHARACTERISTIC HAZARDOUS WASTE?

After a facility determines its waste is a solid waste and is not excluded from the definitions of solid or hazardous waste, it must determine if the waste is

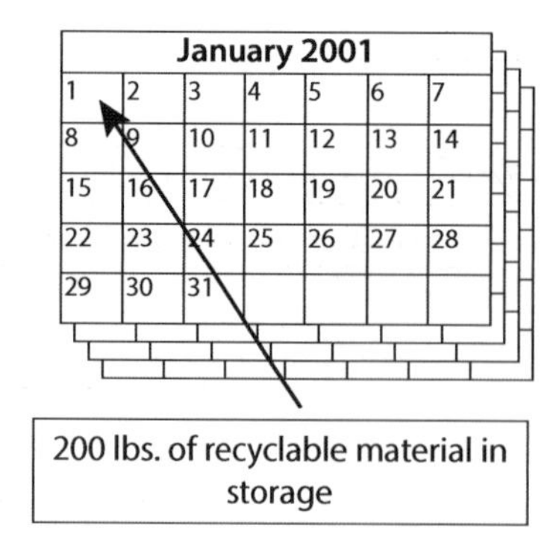

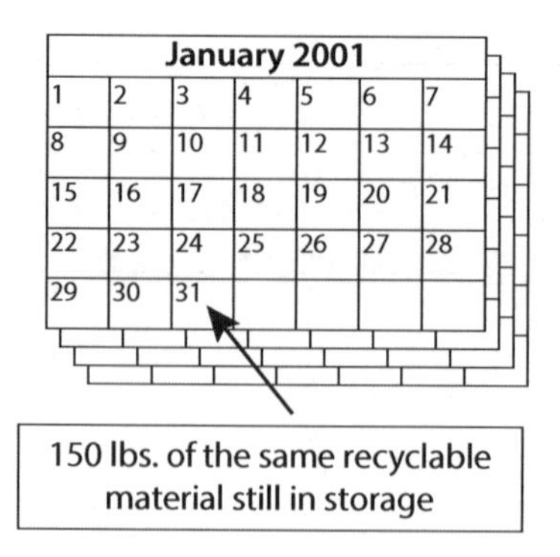

On January 1, 2001, a facility has 200 lbs. of a material that it wants to re-insert directly into its production process. Such a material is technically exempt from the definition of solid waste because it is being recycled through direct reuse without prior reclamation. However, by the end of the calendar year (December 31, 2001), less than 75 percent (i.e., less than 150 lbs.) of the material has been reclaimed or sent off site for reclamation. Therefore, the material has been speculatively accumulated and is no longer exempt from the definition of solid waste. The material may then be regulated as a hazardous waste.

Figure 9. Materials Accumulated Speculatively

a hazardous waste. This entails determining if the waste is listed, and also if the waste is characteristic. Even if a waste is a listed hazardous waste, the facility must also determine if the waste exhibits a hazardous characteristic by testing or applying knowledge of the waste.

Characteristic wastes are wastes that exhibit measurable properties which indicate that a waste poses enough of a threat to warrant regulation as hazardous waste. EPA tried to identify characteristics that, when present in a waste, can cause death or injury to humans or lead to ecological damage. The characteristics identify both acute (near-term) and chronic (long-term) hazards, and are an essential supplement to the hazardous waste listings. For example, some wastes may not meet any listing description because they do not originate from specific industrial or process sources, but the waste may still pose threats to human health and the environment. Therefore, a facility is also required to determine whether such a waste possesses a hazardous property (i.e., exhibits a hazardous waste characteristic). The characteristics are applied to any RCRA solid waste from any industry.

Even if a waste does meet a hazardous waste listing description, the facility must still determine if the waste exhibits a characteristic. If such listed wastes do exhibit a characteristic, the waste poses an additional hazard to human health and the environment, and may necessitate additional regulatory precautions. For example, wastes that are both listed and characteristic may have more extensive **land disposal restrictions (LDR)** requirements than those that are only listed (the LDR program is fully discussed in Chapter III, Land Disposal Restrictions).

EPA decided that the characteristics of hazardous waste should be detectable by using a standardized test method or by applying general knowledge of the waste's properties. Given these criteria, EPA established four hazardous waste characteristics:

- Ignitability
- Reactivity
- Corrosivity

These materials are solid wastes when...				
	Reclaimed	Used in a manner constituting disposal	Burned for energy recovery, used to produce a fuel, or contained in fuels	Accumulated speculatively
Spent materials	√	√	√	√
Listed sludges	√	√	√	√
Characteristic sludges		√	√	√
Listed by-products	√	√	√	√
Characteristic by-products		√	√	√
Commercial chemical products		√ *	√ *	
Scrap metal	√	√	√	√

* If such management is consistent with the product's normal use, then commercial chemical products used in a manner constituting disposal or burned for energy recovery, used to produce a fuel, or contained in fuels are not solid wastes.

√ Material is a solid waste

Figure 10. Regulatory Status of Secondary Materials

- Toxicity.

Ignitability

The **ignitability characteristic** identifies wastes that can readily catch fire and sustain combustion. Many paints, cleaners, and other industrial wastes pose such a hazard. Liquid and nonliquid wastes are treated differently by the ignitability characteristic.

Most ignitable wastes are liquid in physical form. EPA selected a flash point test as the method for determining whether a liquid waste is combustible enough to deserve regulation as hazardous. The flash point test determines the lowest temperature at which the fumes above a waste will ignite when exposed to flame. Liquid wastes with a flash point of less than 60°C (140°F) in closed-cup test are ignitable.

> The ignitability characteristic identifies wastes that can readily catch fire and sustain combustion.

Many wastes in solid or nonliquid physical form (e.g., wood or paper) can also readily catch fire and sustain combustion, but EPA did not intend to regulate most of these nonliquid materials as ignitable wastes. A nonliquid waste is considered ignitable only if it can spontaneously catch fire or catch fire through friction or absorption of moisture under normal handling conditions and can burn so vigorously that it creates a hazard. Certain compressed gases are also classified as ignitable. Finally, substances meeting the DOT's definition of oxidizer are classified as ignitable wastes. Ignitable wastes carry the waste code D001 and are among some of the most common hazardous wastes. The regulations describing the characteristic of ignitability are codified in 40 CFR §261.21.

Corrosivity

The **corrosivity characteristic** identifies wastes that are acidic or alkaline (basic). Such wastes can readily corrode or dissolve flesh, metal, or other materials. They are also among some of the most common hazardous wastes. An example is waste sulfuric acid from automotive batteries. EPA uses two criteria to identify liquid and aqueous corrosive hazardous wastes. The first is a pH test. Aqueous wastes with a pH greater than or equal to 12.5 or less than or equal to 2 are corrosive. A liquid waste may also be corrosive if it has the ability to corrode steel under specific conditions. Physically solid, nonaqueous wastes are not evaluated for corrosivity. Corrosive wastes carry the waste code D002. The regulations describing the corrosivity characteristic are found in 40 CFR §261.22.

> The corrosivity characteristic identifies wastes that are acidic or alkaline (basic) and can readily corrode or dissolve flesh, metal, or other materials.

Reactivity

The **reactivity characteristic** identifies wastes that readily explode or undergo violent reactions or react to release toxic gases or fumes. Common examples are discarded munitions or explosives. In many cases, there is no reliable test method to evaluate a waste's potential to explode, react violently, or release toxic gas under common waste handling conditions. Therefore, EPA uses narrative criteria to define most reactive wastes. The narrative criteria, along with knowledge or information about the waste properties, are used to classify waste as reactive.

> The reactivity characteristic identifies wastes that readily explode or undergo violent reactions.

A waste is reactive if it meets any of the following criteria:

- It can explode or violently react when exposed to water or under normal handling conditions
- It can create toxic fumes or gases at hazardous levels when exposed to water or under normal waste handling conditions meets the criteria for classification as an explosive under DOT rules
- It generates toxic levels of sulfide or cyanide gas when exposed to a pH range of 2 through 12.5.

Wastes exhibiting the characteristic of reactivity are assigned the waste code D003. The reactivity characteristic is described in the regulations in 40 CFR §261.23.

Toxicity

When hazardous waste is disposed of in a land disposal unit, toxic compounds or elements can leach into underground drinking water supplies and expose users of the water to hazardous chemicals and constituents. EPA developed the **toxicity characteristic (TC)** to identify wastes likely to leach dangerous concentrations of toxic chemicals into ground water.

In order to predict whether any particular waste is likely to leach chemicals into ground water at dangerous levels, EPA designed a lab procedure to estimate the leaching potential of waste when disposed in a municipal solid waste landfill. This lab procedure is known as the **Toxicity Characteristic Leaching Procedure (TCLP).**

DETERMINING BOTH LISTINGS AND CHARACTERISTICS

A facility must determine both listings and characteristics. Even if a waste is a listed hazardous waste, the facility must then still determine if the waste exhibits a characteristic because waste generators are required to fully characterize their listings. While some wastes may not meet any listing description because they do not originate from specific industrial or process sources, the waste may still pose threats to human health and the environment. As a result, a facility is also required to determine whether such a waste possesses a hazardous property (i.e., exhibits a hazardous waste characteristic).

The TCLP requires a generator to create a liquid leachate from its hazardous waste samples. This leachate would be similar to the leachate generated by a landfill containing a mixture of household and industrial wastes. Once this leachate is created via the TCLP, the waste generator must determine whether it contains any of 40 different toxic chemicals in amounts above the specified regulatory levels (see Figure 11). These regulatory levels are based on ground water modeling studies and toxicity data that calculate the limit above which these common toxic compounds and elements will threaten human health and the environment by contaminating drinking water. If the leachate sample contains a concentration above the regulatory limit for one of the specified chemicals, the waste exhibits the toxicity characteristic and carries the waste code associated with that compound or element. The TCLP may not be used however, for determining whether remediation waste from manufactured gas plants (MGP) is hazardous under RCRA. Therefore, MGP remediation wastes are exempt from TC regulation. The regulations describing the toxicity characteristic are codified in 40 CFR §261.24, and the TC regulatory levels appear in Table 1 of that same section.

WASTE CODE	CONTAMINANT	CONCENTRATION (MG/L)
D004	Arsenic	5.00
D005	Barium	100.00
D018	Benzene	0.50
D006	Cadmium	1.00
D019	Carbon tetracholride	0.50
D020	Chlordane	0.03
D021	Chlorobenzene	100.00
D022	Chloroform	6.00
D007	Chromium	5.00
D023	o-Cresol	200.00
D024	m-Cresol	200.00
D025	p-Cresol	200.00
D026	Total Cresole	200.00
D016	2, 4-D	10.00
D027	1, 4-Dichlorobenzene	7.50
D028	1, 2-Dichloroethane	0.50
D029	1, 1-Dichloroethylene	0.70
D030	2, 4-Dinitrotoluene	0.13
D012	Endrin	0.02
D031	Heptachlor (and its epoxide)	0.008
D032	Hexachlorobenzene	0.13
D033	Hexachlorobutadene	0.50
D034	Hexachloroethane	3.00
D008	Lead	5.00
D013	Lindane	0.40
D009	Mercury	0.20
D014	Methoxychlor	10.00
D035	Methyl ethyl ketone	200.00
D036	Nitrobenzene	2.00
D037	Pentachlorophenol	100.00
D038	Pyridine	5.00
D010	Selenium	1.00
D011	Silver	5.00
D039	Tetracholoroethylene	0.70
D015	Toxaphene	0.50
D040	Trichloroethylene	0.50
D041	2, 4, 5-Trichlorophenol	400.00
D042	2, 4, 6-Trichlorophenol	2.00
D017	2, 4, 5-TP (Silvex)	1.00
D043	Vinyl chloride	0.20

If o-, m-, and p-Cresols cannot be individually measured, the regulatory level for total cresols is used.

Figure 11. TLCP Regulatory Levels

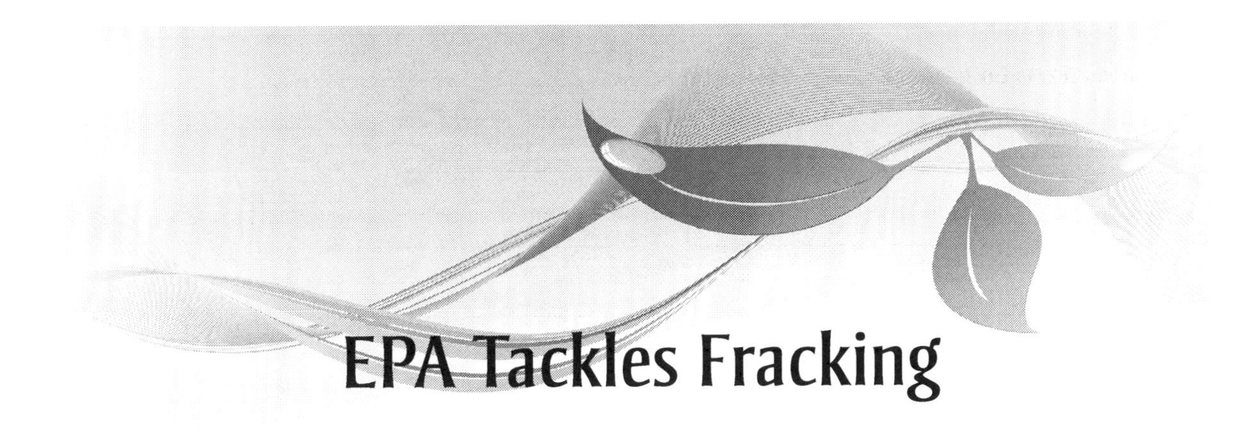

EPA Tackles Fracking

By John Manuel

With the push for energy independence and fuels that emit fewer greenhouse gases, domestically produced natural gas has been growing in popularity. But alongside this growth have come concerns that hydraulic fracturing ("fracking"), a procedure used in the extraction of natural gas and oil, may pollute ground and surface waters. Responding to increasing public pressure for federal action and a call by the U.S. House of Representatives Appropriations Conference Committee, the U.S. Environmental Protection Agency (EPA) announced 18 March 2010 it will conduct a comprehensive study to investigate the potential adverse effects of fracking on water quality and public health.

Natural gas provides almost 25% of the U.S. energy supply and could provide 50% by 2035, according to the 2010 report *Fueling North Americas Energy Future* by consultancy IHS Cambridge Energy Research Associates. In recent years, vast new deposits of natural gas have been discovered in layers of shale thousands of feet underground. Some of these deposits, such as the Marcellus Shale running under the Appalachian Basin, lie beneath watersheds supplying drinking water to millions of people. In many locations fracking—in which a mixture of water, sand, and chemicals is injected into natural gas wells under high pressure—occurs within hundreds of feet of residences that use wells for drinking water.

Recent evidence suggests fracking may have contributed to groundwater contamination with methane in some instances and that proprietary chemicals used in the procedure could theoretically pose a public health threat. However, because groundwater supplies and natural gas deposits are often separated by thousands of feet of rock and earth, and groundwater can be contaminated by many sources, it is difficult to establish a definitive connection between contaminated drinking water and fracking. Further, there has been very little in-depth research on the subject with respect to drilling in shale beds.

In 2005 Congress exempted fracking from regulation under the Safe Drinking Water Act partly on

John Manuel, "EPA Tackles Fracking," *Environmental Health Perspectives*, vol. 118, no. 5, p. Q199. Published by the National Institute of Environmental Health Sciences.

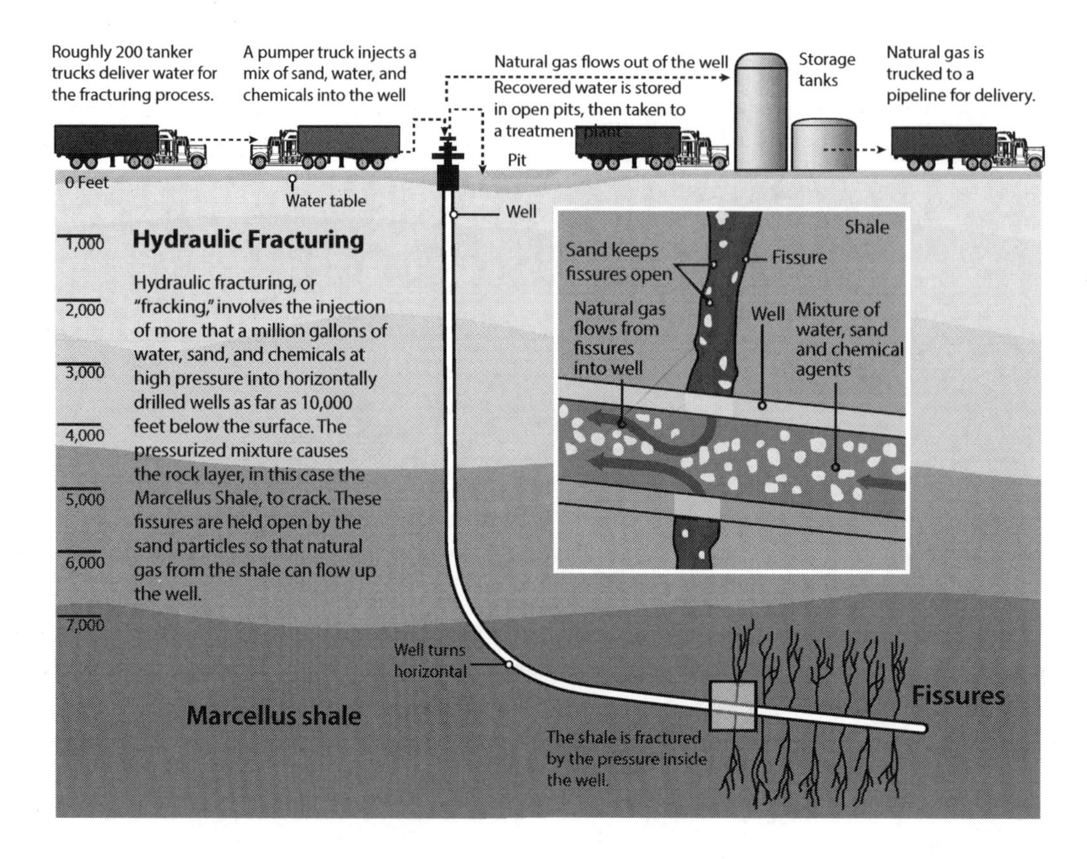

What's Getting Pumped In?

A variety of chemicals—among them methanol, formaldehyde, ethylene glycol, hydrochloric acid, and sodium hydroxide—are used for purposes such as improving fluid viscosity, inhibiting corrosion, and limiting bacterial growth.

the basis of the EPA report *Evaluation of Impacts to Underground Sources of Drinking Water by Hydraulic Fracturing of Coalbed Methane Reservoirs.* The authors of this report wrote that hydraulic fracturing poses "minimal threat" to drinking water and that "additional or further study is not warranted at this time." However, the study involved no direct monitoring of water wells but instead relied on existing peer-reviewed literature and interviews with industry and state and local government officials. It also was strictly limited to one specific type of drilling and did not address the effects in substrates other than coalbeds.

Operators surround drill holes with steel casings cemented into place to prevent groundwater contamination during fracking. In addition, a large volume of "back-flow fluids," on the order of hundreds of thousands of gallons per well, are brought to the surface during drilling and production. Back-flow fluids are typically stored in on-site pits (which, depending on state regulations, may or may not be lined) and ultimately are disposed of either by injection into EPA-approved underground wells or by delivery to municipal waste treatment facilities. In January 2010 state officials testified before the Ohio House of Representatives that standards and enforcement regarding oil and gas well construction are not always adequate to ensure proper performance.

The EPA has reallocated $1.9 million for the new study in this fiscal year and will request further funding in the President's FY2011 budget proposal. The agency's scoping document identifies three major categories for research: characterization of the fracking life cycle, potential relationships to drinking water sources, and potential health and environmental hazards. Explicit research goals, as yet undefined, will be divided into short term (1–3 years) and long term (3–5 years).

"We're very pleased that EPA is doing this study," says Amy Mall, a senior policy analyst with the Natural Resources Defense Council (NRDC), which has long pressured Congress for federal regulation of fracking. "There are communities around the country that are very concerned because their water has been contaminated or could be contaminated."

The NRDC also supports passage of the Fracturing Responsibility and Awareness of Chemicals (FRAC) Act, which was introduced in Congress in 2009. The FRAC Act would permit regulation of hydraulic fracturing under the SDWA and would require oil and gas companies to disclose the chemicals used in fracking operations.

Energy In Depth, a nonprofit organization representing the oil and gas industry, has lobbied against the FRAC Act, but supports the new EPA study in concept. "We've said from the outset that any updated study in this area should be science-based, peer-reviewed, and completely isolated from political design or interference," says Chris Tucker, spokesman for Energy In Depth. "Assuming those criteria are met, we're confident the new study will end up reaching the same conclusions that were produced by the old study."

Race, Class, and Environmental Justice

By Lisa Benton-Short and John Rennie Shortl

The city is both an environmental and a social construct. It is predicated upon ecological processes, indeed it is a complex ecological system in its own right, yet it is also a social artifact that embodies and reflects power relations and social differences. The city is at the center of a social-environment dialectic that connects the environmental and the political. In this chapter we will explore how issues of class, race and gender interconnect with environmental issues.

URBAN ENVIRONMENTS AND RACE

It is a consistent finding that toxic facilities are concentrated in minority-dominated areas of the city, and major infrastructure projects with negative environmental impacts such as urban motorways are more commonly found in minority neighborhoods. Study after study reveal a correlation between negative environmental impacts and the presence of racial/ethnic minorities. An early study in the US revealed that race was the most significant variable associated with the location of hazardous waste sites and that the greatest number of commercial hazardous facilities were located in areas with the highest composition of racial and ethnic minorities. The study also showed that three out of every five Black and Hispanic Americans lived in communities with one or more toxic waste sites. Although socioeconomic status was also an important variable in the location of these sites, race was the most significant.

Take the case of Chester, PA, a typical industrial town outside the city of Philadelphia (see Figure 11.1). It grew as a manufacturing center with steel mills, shipyards, aircraft engines plants and a Ford Motor Company plant. By the 1970s, however, deindustrialization had begun to erode much of the manufacturing base. As firms closed, workers left and the town's population became increasingly poor, older and

Figure 11.1. Chester, Pennsylvania. This former industrial city, with a predominant African American population, has been the site for the location of several waste incinerators since the 1980s. Source: Photo by John Rennie Short

black. By the time of the 2000 Census the population of 36,851 was 75 percent black compared to the state average of 10 percent, and the median household income was $13,052 compared to the state average of $20,880. A quarter of the town's population lived below the poverty line.

Eager to lure tax-generating facilities, the city government in the 1980s sought to redevelop old factory sites, attract business and generate jobs and tax revenue. Chester provided an ideal opportunity for certain industries as it was poor, desperate, and had land, while local communities had limited political power. Other cities and areas might have provided stronger resistance to polluting enterprises. A large real estate developer bought up the land rights in the old industrial area and leased space to other businesses. In 1987 the Pennsylvania State Department of Environmental Protection granted permits for three waste facilities in the city of Chester and two outside the city. By the mid-1990s the city housed the nation's largest concentration of waste facilities including a trash transfer business, an incinerator, a medical waste sterilizing facility, a contaminated soil burning facility, a rock crushing plant and wastewater plant that handled effluent from factories, and a refinery. A local citizen's group, The Chester Residents Concerned for Quality Living, claimed that the toxic emissions led to low birth rate babies and local cancer clusters. They also claimed environmental racism, taking their case against the state Department of Environmental Protection (DEP) to the courts. They argued that between 1987 and 1996 the DEP approved permits for 2.1 million tons of landfill in black areas of the city, but approved only 1,400 tons in white areas. A federal judge threw out the lawsuit in 1996 stating that there may have been a discriminatory *effect* but the residents could not prove a discriminatory *intent*. A court of appeal

Environmental Racism and Injustice?

Just because a hazardous site is situated in a minority community does not necessarily mean that environmental racism is at work. Consider the case of trash transfer in Washington, DC. The city needed sites to consolidate its garbage collection and disposal system; the narrow city streets meant that only relatively small trucks could collect garbage in the city, but larger trucks were needed to move the trash to remote incinerators and landfills. A plan emerged from an independent committee in the 1990s to build a new transfer station in Ward 8 in the city. The site was ideal; it was flat, the city owned the property, it had good accessibility for the truck traffic and was large enough to satisfy a city ordinance requiring a 500 foot buffer zone between transfer stations and residences. The new proposed site also allowed the closure of older, noisier and more polluting transfer stations in other minority neighborhoods. But Ward 8 houses a predominantly low-income minority population. In the lively debate that followed the terms "environmental racism" and "environmental injustice" were often used. A closer reading of the facts, however, show that there were few other options. Local activists could easily use the widely known terms in the polarized racial politics of Washington, DC; it was an obvious and easy rhetorical point to make. The terms environmental racism and environmental justice can be powerful words of rhetoric even when they have little explanatory purchase.

Source: Talking Trash http://www.Washep.com/archives/cover/2001/ cover0126.html (accessed June 10, 2006)

reversed the judge's decision but the Supreme Court dismissed the case in 1998. However, by even agreeing to hear the case at all the Supreme Court signaled the viability of environmental racism as a legal argument.

In Chester there was a cluster of noxious, polluting facilities in predominantly poor, black, residential areas. This is an extreme case but one that highlights the character of environmental racism. It is less a formal legal issue since the permits for the facilities had been issued correctly and formal procedures had been followed, and represents a more moral issue. Vulnerable and poor residents were dumped on both metaphorically and literally.

It is often difficult to untangle the intent and effects of environmental racism. The term "environmental racism" implies both, but in the case of Chester, while the effect was obvious the intent was more difficult to prove. In counterarguments it was argued that the city was so poor that it needed to attract tax-paying operations to fund city services. The rationale for this cost-benefit may be unbalanced; the health of residents far outweighs the contribution to the city's coffers. However, the simple causal processes implied by the cavalier use of the term racism need some caution.

Environmental negative impacts occur in racially homogeneous nations. In Japan, for example, there are 2,000 highly polluting municipal incinerators that spew out a deadly cocktail of heavy metals, nitrous oxides, carbon dioxide, toxins and furans that lead to cancer, birth defects and illness. The results are spread throughout the cities of Japan, spewing pollution on rich and poor Japanese.

In a careful study of environmental inequity, Christopher Boone and Ali Modarres examined the case of Commerce. This predominantly Latino city east of Los Angeles has a high concentration of polluting manufacturing plants. They show that the businesses located in the city because it provided vacant land and accessibility. The bulk of industry was located in the city before the demographic changes that made it a Latino city. In other words, there were toxic neighborhoods because of factors of accessibility and land availability more than because they were Mexican neighborhoods. The Boone and Modarres study suggests that immigrant, migrant and minority communities may develop around toxic areas because of the operation of the housing market, where the poorest people get the least choice and end up in the worst neighborhoods, rather than the case that toxic areas are knowingly located in minority neighborhoods.

This is not to dispute the connector between environmental quality and minority residential areas. In the US, among counties that have three or more pollutants 12 percent are majority white, 20 percent are majority African American and 31 percent are majority Hispanic. Race and ethnicity intertwine with issues of power and access to power to produce an uneven experience of environmental quality at home and in the workplace.

While the connections between intent and effect are sometime difficult to disentangle, race and ethnicity can play a role in mobilization. Robert Bullard provides case studies of community disputes in cities in the US ranging from conflict over solid waste landfill in Houston, a lead smelter in Dallas and a solid waste incinerator in Los Angeles. Race can become a site of mobilization, a shared experience on which to build resistance and to fight against environmental injustice.

URBAN ENVIRONMENTS AND SOCIOECONOMIC STATUS

Socioeconomic status plays a major role in the environmental quality of urban living. Poorer communities have less pleasant urban environments and often bear the brunt of negative externalities. It is through their neighborhoods that motorways are constructed; it is in their neighborhoods that heavy vehicular traffic caused, and in some places still causes, high lead levels in the local soil and water. There is a direct correlation between socioeconomic status and the quality of the urban environment.

The causal web is sometimes complex but often simple. Poor people get dumped on because they are poor, and they are poor because they lack the wealth to generate political power and bargaining strength. The city is a space where the best areas go to those with the most money and those with the least get what is left. On top of this historical relationship there is the current trend in the siting of noxious facilities that tends to skew their location to those with least power to resist. The poor are the least powerful and experience the worst urban environments. In more racially homogeneous cities the deciding factor is socioeconomic status.

In some cases, the correlation between race and class on the one hand and environmental quality on the other was, and in some cases still is, reinforced by political movements and economic forces that tended to discount environmental quality in favor of economic growth and employment opportunities. Many labor movements were slow to realize that environmental issues were social justice issues, not just the superstructural concerns of the affluent. And the brute economic forces of the industrial city often forced a false divide between environmental qualities versus jobs. Crenson, for example, tells the story of the lack of an environmental movement in many industrial cities in the US because of the supposed linkage between pollution and employment.

Racialized Topographies

Richmond, Virginia, whose residents in 2000 numbered 818,836, of whom 32.4 percent were African American, had a socio-spatial pattern of lower-lying, inner-city, black areas and surrounding white-dominated hills. The correlations between percent black and altitude were -0.41 in 1990 and -0.47 in 2000. Richmond has a long history of segregating races by elevation. It was one of the first southern cities to embrace racial zoning in 1910. Richmond's local elites used the state to control the expansion of black communities. African Americans were crowded into the dilapidated houses in the Jackson Ward area. East End and Church Hill absorbed a large influx of blacks after World War II. Residential discrimination extended into public policy decisions in the placement of public housing. Sixty-four percent of public housing units built prior to 1970 were placed in Church Hill and Jackson Ward. These decisions reinforced the racial topography of Richmond.

Source: Ueland, J. and B. Warf. (2006). Racialized topographies: altitude and race in southern cities. *Geographical Review* 96(1): 50–78.

Smoking chimneys signified good well-paying jobs. We have also highlighted this issue in our discussion of Onondaga Lake in Syracuse. For years the business elite promoted the issue in terms of "Does Syracuse want its people employed, or do they want the lake cleaned up?" It was only in the 1970s that a cleaned-up lake was reimagined as a vital part of a new postindustrial city. But as long as it was an industrial city, lake cleanup was a distant second to economic growth. And even today in many cities, especially in the developing world, dictates of business expansion and economic growth often outweigh the environmental concerns of, and the quality of urban life for, the majority of the people.

We can now see the poverty of this false dichotomy of jobs versus environment. Low-grade environments have their most negative health impacts on working people. It is not a case of jobs versus the quality of the urban environment but jobs *and* the quality of the urban environment. As we move further into a greener economy there are direct connections between improving the urban environment and employment opportunities. Recycling, green technologies and greening the city are all ways to create jobs in hard-pressed cities.

ENVIRONMENTAL JUSTICE

In 1994 President Clinton signed Executive Order 12898 that initiated an environmental justice program within the Environmental Protection Agency (EPA). The aim was to raise awareness of environmental justice issues, identify and assess inequitable environmental impacts and provide assistance to local areas and community groups.

Environmental justice issues arise from the obvious fact that there is a correlation between the siting of hazardous facilities and low-income communities and/or minority communities. However, as the case of Chester shows, legal redress is difficult if procedures were correctly followed. Environmental justice, like social justice, is not possible in the absence of more interventionist methods. The normal workings of a racist, classist society will produce racist, classist outcomes in the normal course of events, even without the aid of illegality or the help of corruption. In order to produce more equitable outcomes we need more positive interventions. Environmental impact statements, for example, need a more explicit assessment of equity and justice issues. Low-income communities facing environmental challenges should also receive greater resources from government, not simply equal treatment under the law. Unless more positive outcomes are engineered, the system will tend to produce inequitable results.

URBAN ENVIRONMENTS AND SOCIAL DIFFERENCE

The experience of living in the urban environment varies across the dimensions of social difference. Socioeconomic status, gender, age and level of physical ability/disability are all sources of social difference that are embodied and reflected in urban environments.

Take the case of gender. Women have played an important role in bringing environmental issues to a broader public. Lois Gibbs, for example, was a typical suburban housewife with two children living in a suburb in western New York. When her son Michael developed epilepsy, her daughter Melissa contracted a rare blood disease and her neighbors' children also got sick, she began to look into possible causes. She and others soon discovered that the area where she lived, Love Canal, was built on a toxic waste site, with more than a dozen known carcinogens including the deadly chemical dioxin. The local soil and water was poisoning the community. Lois Gibbs became one of the community organizers mobilizing public opinion and promoting state and federal involvement. The area was evacuated in 1980, the area's name joining places like Bhopal as a byword for environmental pollution. The publicity given to Love Canal strengthened the case for the Superfund legislation, in which chemical cleanup of major toxic sites was federally mandated, which was passed in 1981.

The experience of urban toxicity is often experienced by women in households and neighborhoods where women play a strong role in household care and management. They tend to children's illnesses, connect with local neighbors and may be closer to local issues. It is often women who are on the front line of local environmental issues.

The case of Love Canal in which children became very sick is also an example of the effect of age on the experience of urban environmental conditions. Embryos in the womb and very young children are especially vulnerable to adverse environmental conditions. They are the human equivalent of the canary in the mine, providing early and tragic warnings of the state of our environment. Older people are also especially vulnerable to air quality and other environmental conditions. Yet in many cities the very young and very old have the least voice but are often the most impacted. Most cities are designed by and for affluent males; the further from this profile the more marginal to most discourses of power and influence. The worst urban environments are experienced by the more marginal; the poor, young and old.

HEALTHY CITIES

In 1984, the World Health Organization sponsored a conference in Toronto entitled *Healthy Cities.* Two years later, the Ottawa Charter for Health Promotion outlined the basic requirement of improving overall health by improving the physical and social environments of cities. There are now almost 2,000 healthy city initiatives around the world. New initiatives have been promoted along with existing public health programs under the term healthy cities. In Europe, the emphasis has been on integrating health and urban planning. In Latin America, healthy cities are more associated with the provision of basic infrastructural services of water and sewerage. There has been a more explicit acceptance that healthy cities must connect to issues of environmental justice. In 2001, member counties of the EU ratified the Aarhus

Healthy City Initiatives

One obvious healthy city initiative is to encourage more vegetation growth. A careful tree-planting program, for example, can lower summertime temperatures, minimize the urban heat island and reduce air pollution. Some cities actively encourage tree planting. In Sacramento, CA, since 1990, over 375,000 shade trees have been given away to city residents with plans for 4 million more trees to be planted throughout the city. In other U.S. cities, in contrast, tree-planting programs have been cut back. A study of 24 cities in the U.S. showed there has been a 25 percent decline in tree canopy in the past 30 years. In cities such as Milwaukee the tree-planting program was reorientated away from communities to greening the downtown to make it more attractive to investors. The Sacramento experience, however, shows that tree planting not only makes for a healthier city, it also saves money. The city estimates that for every dollar spent on trees it recoups $2.80 in energy savings, pollution reduction, stormwater management and increased property values.

There has been resistance to tree planting from some utility companies who argue that it costs more to secure their power lines though vegetation. In states such as Iowa, in contrast, which mandates tree planting by utility companies, the experience has been more positive. In a time of rapid energy costs, utility companies are often grateful for the public relations bonus they receive from partnering in tree planting.

Source: Harden, B. (2006). In California city, shade crusade takes root to cut energy costs. *The Washington Post*, September 4: A and A10.

Convention that every person has the right to live in an environment adequate to his or her health and well-being, and to achieve this end citizens must be involved in decision making and have access to information. The Convention promoted early and effective participation and various evaluation criteria. Hartley and Wood examined public participation in the environmental impact assessments in four UK waste disposal case studies in the wake of the Aarthus Convention. They report that the Aarhus Convention led to a strengthening of participation procedures but that the level of improvement secured would depend upon how its ideals were interpreted and incorporated into specific legislation. In other words, it is the legislative and administrative details that will structure the implementation.

Although the pronouncements of the Aarhus Convention are easy to make, rhetoric being easier than real action, they do hint at the future connections between issues of social justice, participatory democracy and urban environmental quality. They also indicate that environmental rights may join the list of citizen rights and government obligations that mobilize communities and motivate governments.

GUIDE TO FURTHER READING

Adamson, J., M.M. Evans, and R. Stein, (eds.). (2002). *The Environmental Justice Reader*, Tucson: University of Arizona Press.

Bullard, R. D., (ed.). (2005). *The Quest for Environmental Justice: Human Rights and the Politics of Pollution*. San Francisco: Sierra Club.

Bullard, R. D. (ed.). (2007). *Growing Smarter: Achieving Livable Communities, Environmental Justice, and Regional Equity.* Cambridge, MA: MIT Press.

Pellow, D. N. and R.J. Brulle. (2005). *Power, Justice and the Environment: A Critical Appraisal of the Environmental Justice Movement*. Cambridge, MA: MIT Press.

Short, J. R. (1989). *The Humane City,* Oxford: Basil Blackwell.

Stein, R. (ed.). (2004). *New Perspectives on Environmental Justice: Gender Sexuality and Activism.* New Brunswick: Rutgers University Press.

Takanos, T. (ed.). (2003). *Healthy Cities and Urban Policy Research.* London: Spon.

Tsourous, A. (2006). *Healthy Cities in Europe.* London Spon.

Washington, S.H. (2005). *Packing Them In: An Archaeology of Environmental Racism in Chicago.* Washington, DC, and Covelo, CA: Rowman and Littlefield.